T0344877

Inorganic Chemical Biology

Inorganic Chemical Biology

Principles, Techniques and Applications

Edited by

GILLES GASSER

Department of Chemistry, University of Zurich, Switzerland

WILEY

This edition first published 2014
© 2014 John Wiley & Sons, Ltd

Registered office

John Wiley & Sons Ltd, The Atrium, Southern Gate, Chichester, West Sussex, PO19 8SQ, United Kingdom

For details of our global editorial offices, for customer services and for information about how to apply for permission to reuse the copyright material in this book please see our website at www.wiley.com.

The right of the author to be identified as the author of this work has been asserted in accordance with the Copyright, Designs and Patents Act 1988.

All rights reserved. No part of this publication may be reproduced, stored in a retrieval system, or transmitted, in any form or by any means, electronic, mechanical, photocopying, recording or otherwise, except as permitted by the UK Copyright, Designs and Patents Act 1988, without the prior permission of the publisher.

Wiley also publishes its books in a variety of electronic formats. Some content that appears in print may not be available in electronic books.

Designations used by companies to distinguish their products are often claimed as trademarks. All brand names and product names used in this book are trade names, service marks, trademarks or registered trademarks of their respective owners. The publisher is not associated with any product or vendor mentioned in this book.

Limit of Liability/Disclaimer of Warranty: While the publisher and author have used their best efforts in preparing this book, they make no representations or warranties with respect to the accuracy or completeness of the contents of this book and specifically disclaim any implied warranties of merchantability or fitness for a particular purpose. It is sold on the understanding that the publisher is not engaged in rendering professional services and neither the publisher nor the author shall be liable for damages arising herefrom. If professional advice or other expert assistance is required, the services of a competent professional should be sought

The advice and strategies contained herein may not be suitable for every situation. In view of ongoing research, equipment modifications, changes in governmental regulations, and the constant flow of information relating to the use of experimental reagents, equipment, and devices, the reader is urged to review and evaluate the information provided in the package insert or instructions for each chemical, piece of equipment, reagent, or device for, among other things, any changes in the instructions or indication of usage and for added warnings and precautions. The fact that an organization or Website is referred to in this work as a citation and/or a potential source of further information does not mean that the author or the publisher endorses the information the organization or Website may provide or recommendations it may make. Further, readers should be aware that Internet Websites listed in this work may have changed or disappeared between when this work was written and when it is read. No warranty may be created or extended by any promotional statements for this work. Neither the publisher nor the author shall be liable for any damages arising herefrom.

Library of Congress Cataloging-in-Publication Data

Inorganic chemical biology : principles, techniques and applications / editor, Gilles Gasser.
     p. ; cm.
  Includes bibliographical references and index.
  ISBN 978-1-118-51002-5 (cloth)
I. Gasser, Gilles, editor of compilation.
  [DNLM: 1. Biochemical Phenomena. 2. Metals–chemistry. 3. Macromolecular Substances–chemistry. QU 130.2]
  QP606.D46
  572′.43–dc23

                                                                                                    2013049090

A catalogue record for this book is available from the British Library.

ISBN: 9781118510025

Typeset in 10/12pt TimesLTStd by Laserwords Private Limited, Chennai, India
Printed and bound in Singapore by Markono Print Media Pte Ltd

1   2014

Contents

11. Metal Complexes as Enzyme Inhibitors and Catalysts in Living Cells   341

Julien Furrer, Gregory S. Smith and Bruno Therrien

12. Other Applications of Metal Complexes in Chemical Biology   373

Tanmaya Joshi, Malay Patra and Gilles Gasser

About the Editor

Gilles Gasser was born in the French-speaking part of Switzerland in 1976 and obtained his PhD from the University of Neuchâtel (Switzerland) in 2004 in the field of supramolecular/coordination chemistry under the supervision Professor Helen Stoeckli-Evans. After post-docs on bioinorganic chemistry with Professor Leone Spiccia (Monash University, Australia), sponsored by the Swiss National Science Foundation (SNSF), and on medicinal organometallic chemistry, as an Alexander von Humboldt fellow in the group of Professor Nils Metzler-Nolte (Ruhr-University Bochum, Germany), Gilles was given the opportunity to return to Switzerland to start his independent research at the Department of Chemistry of the University of Zurich as an SNSF Ambizione fellow in 2010. Since March 2011, Gilles has been an assistant professor at the same institution endowed with an SNSF professorship. His current research interests cover various fields of inorganic chemical biology and medicinal inorganic chemistry, focusing on using metal complexes to understand cellular processes as well as to kill cancer cells and parasites.

List of Contributors

Christophe Biot, Unit of Structural and Functional Glycobiology, University of Lille1, France

Shawn C. Burdette, Department of Chemistry and Biochemistry, Worcester Polytechnic Institute, USA

Vivien M. Chen, Lowy Cancer Research Centre and Prince of Wales Clinical School, University of New South Wales, Australia

Rachel Codd, School of Medical Sciences (Pharmacology) and Bosch Institute, The University of Sydney, Australia

Najwa Ejje, School of Medical Sciences (Pharmacology) and Bosch Institute, The University of Sydney, Australia

Katherine J. Franz, Department of Chemistry, Duke University, USA

Julien Furrer, Department of Chemistry and Biochemistry, University of Berne, Switzerland

Gilles Gasser, Department of Chemistry, University of Zurich, Switzerland

Bim Graham, Medicinal Chemistry, Monash Institute of Pharmaceutical Sciences, Monash University, Australia

Jiesi Gu, School of Medical Sciences (Pharmacology) and Bosch Institute, The University of Sydney, Australia

Celina Gwizdala, Department of Chemistry, University of Connecticut, USA

Christian Hartinger, School of Chemical Sciences, The University of Auckland, New Zealand

Philip J. Hogg, Lowy Cancer Research Centre and Prince of Wales Clinical School, University of New South Wales, Australia

Minh Hua, Lowy Cancer Research Centre and Prince of Wales Clinical School, University of New South Wales, Australia

Tanmaya Joshi, Institute of Inorganic Chemistry, University of Zurich, Switzerland

Michael D. Lee, Medicinal Chemistry, Monash Institute of Pharmaceutical Sciences, Monash University, Australia

Tulip Lifa, School of Medical Sciences (Pharmacology) and Bosch Institute, The University of Sydney, Australia

Kenneth Kam-Wing Lo, Department of Biology and Chemistry, City University of Hong Kong, P.R. China

Andrée Kirsch-De Mesmaeker Organic Chemistry and Photochemistry, Faculty of Sciences, Free University of Brussels, Belgium

Lionel Marcélis, Organic Chemistry and Photochemistry, Free University of Brussels, Belgium

Ingo Ott, Institute of Medicinal and Pharmaceutical Chemistry, Technical University of Braunschweig, Germany

Danielle Park, Lowy Cancer Research Centre and Prince of Wales Clinical School, University of New South Wales, Australia

Malay Patra, Institute of Inorganic Chemistry, University of Zurich, Switzerland

Luca Quaroni, Swiss Light Source, Paul Scherrer Institute, Switzerland

Ulrich Schatzschneider, Institute for Inorganic Chemistry, Julius-Maximilians University of Würzburg, Germany

Ivan Ho Shon, Lowy Cancer Research Centre and Prince of Wales Clinical School, University of New South Wales, Australia

Peter V. Simpson, Institute for Inorganic Chemistry, Julius-Maximilians University of Würzburg, Germany

Gregory S. Smith, Department of Chemistry, University of Cape Town, South Africa

James D. Swarbrick, Medicinal Chemistry, Monash Institute of Pharmaceutical Sciences, Monash University, Australia

Bruno Therrien, Institute of Chemistry, University of Neuchatel, Switzerland

Willem Vanderlinden, Department of Chemistry, Laboratory of Photochemistry and Spectroscopy, Division of Molecular Imaging and Photonics, University of Leuven, Belgium

Qin Wang, Department of Chemistry, Duke University, USA

Kenneth Yin Zhang, Department of Biology and Chemistry, City University of Hong Kong, P.R. China

Fabio Zobi, Department of Chemistry, University of Fribourg, Switzerland

Preface

Chemical biology is a rapidly growing field. New chemical biology departments are being set up in universities across the globe. Also, to specifically match the emerging interest of numerous students in this research area in addition to inculcating an interest among others, new courses are being created at the Bachelor and Master levels. The vast majority of these courses are being taught by chemists with an organic chemistry background with only little reference to the influences of metal complexes. However, such compounds have a proven record of playing a pivotal role in this field of research, with their application expected to grow even further in the near future. However, for the time being, the resources available to lecturers to cover this area properly in a directed and specific manner are very limited, despite the importance of metal complexes having being quickly recognized [1–3]. The main goal behind the preparation of this book is to address this problem by providing a comprehensive overview of the current role played by metal complexes in chemical biology. At this stage, I must stress that the area of medicinal inorganic chemistry (e.g., anticancer, antimicrobial and antiparasitical agents) is not really covered in this book because this has recently been reviewed in detail in several other books/book chapters [4–8]. Also, since the typical definition of chemical biology is to understand, identify and/or influence biological systems using small molecules and/or chemical techniques [2, 9–12], the role of metal ions in biology is not covered in this book. Only the use of metal complexes for (molecular) biology purposes has been presented.

This book has been constructed in a manner that allows the subject matter to be easily taught by organic and inorganic lecturers alike and readily understood by students, thereby allowing this emerging research area to be appropriately covered in all chemical biology subjects. All chapters of this book are a mix of fundamental theoretical chemistry and of concrete examples to explain the concept presented. The first two chapters explain how metal complexes can help in purifying essential biomolecules (Chapter 1) and identifying their structures (Chapter 2). A chapter is then dedicated to a description of the analytical techniques that can help specifically in the detection of metals in living cells (Chapter 3). In this sense, the first part of the book directs interested readers to the use of such techniques for biological purposes and gives a good overall perspective of the coupling of the areas of inorganic chemistry and chemical biology. The second part of the book is then dedicated to the visualization of important organelles, molecules, and ions in living cells. More specifically, the imaging of particular cellular organelles using luminescent complexes (Chapter 4) and metal carbonyl ligand (Chapter 5) is first presented. The three subsequent chapters then explain how metal complexes can help in visualizing the different types of DNA (Chapter 6), proteins (Chapter 7) as well as metal ions, anions, and small molecules (Chapter 8). The third part of the book relates to the use of metal compounds to release biologically relevant metal ions (Chapter 9) and bioactive molecules (Chapter 10) in living cells as well as to inhibiting specific enzymes (Chapter 11). Finally, the editor's outlook on future potential applications of metal complexes in chemical biology is discussed in the last chapter (Chapter 12).

I hope this book will serve as a useful go-to reference for all new and experienced chemical biology professionals, to further encourage more biologists to use metal complexes in their research, and also to persuade more inorganic chemists to develop new metal-containing probes to further understand cellular processes.

Enjoy this book!

Gilles Gasser
Zurich
Switzerland

References

1. K.L. Haas and K.J. Franz (2009) Application of metal coordination chemistry to explore and manipulate cell biology, *Chem. Rev.*, **109**, 4921–4960, and references therein.
2. M. Patra and G. Gasser (2012) Organometallic compounds, an opportunity for chemical biology, *ChemBioChem*, **13**, 1232–1252.
3. S.J. Lippard (2006) The inorganic side of chemical biology, *Nat. Chem. Biol.*, **2**(10), 504–507.
4. E. Alessio (ed.) (2011) *Bioinorganic Medicinal Chemistry*, Wiley-VCH Verlag GmbH, Weinheim.
5. J.L. Sessler, S.R. Doctrow, T.J. McMurry and S.J. Lippard (2005) *Medicinal Inorganic Chemistry*, American Chemical Society, Washington, D.C.
6. M. Gielen and E.R.T. Tiekink (eds) (2005) *Metallotherapeutic Drugs & Metal-based Diagnostic Agents - The Use of Metals in Medicine*, John Wiley & Sons Ltd, Chichester.
7. J. C. Dabrowiak, (2009) *Metals in Medicine*, John Wiley & Sons, Ltd, Chichester.
8. G. Jaouen and N. Metzler-Nolte (eds) (2010) Medicinal Organometallic Chemistry, in *Topics in Organometallic Chemistry*, Springer-Verlag, Heidelberg.
9. S.L. Schreiber (2005) Small molecules: the missing link in the central dogma, *Nat. Chem. Biol.*, **1**, 64–66.
10. K.L. Morrison and G.A. Weiss (2006) The origins of chemical biology, *Nat. Chem. Biol.*, **2**(1), 3–6 (2006).
11. H. Waldmann and P. Janning, (2009) *Chemical Biology: Learning through Case Studies*, Wiley-VCH Verlag GmbH, Weinheim.
12. S.L. Schreiber, T.M. Kapoor and G. Wess (eds) (2007) *Chemical Biology: From Small Molecules to Systems Biology and Drug Design*, Wiley-VCH Verlag GmbH, Weinheim.

Acknowledgements

The editing of a book is a team effort and I have to admit that I have been extremely fortunate to be the "coach" of a world-class squad. Obviously, first of all I would like to thank all the (co-)authors for contributing to this book. It was my great pleasure to work with such knowledgeable scientists from five different continents. They have performed a tremendous job in a record-time. THANKS A LOT!

Special thanks also go to my current and past post-docs, PhD and Master students, Anna Leonidova, Jeannine Hess, Philipp Anstätt, Vanessa Pierroz, Cristina Mari, Dr Malay Patra, Sandro Konatschnig, Dr Tanmaya Joshi, Angelo Frei, and Dr Riccardo Rubbiani (in order of arrival in the group), who have not only proofread this book but importantly given me critical feedback on each chapter. I am very lucky to work with such a dedicated and bright group of young researchers. Thanks also to them for all the hard work they carry out each day in the labs in Zurich in such a pleasant atmosphere. I must not forget to gratefully acknowledge Dr Jacqui F. Young for her valuable help during the writing of the proposal of this book.

Finally, I would like to sincerely thank the publishers, Wiley, who have given me the opportunity to edit this book. Special thank goes to Sarah Higginbotham, Sarah Tilley, and Rebecca Ralf from the publishing team who have constantly tried to make the process of editing this book as smooth as possible.

Gilles Gasser
Zurich
Switzerland

1

New Applications of Immobilized Metal Ion Affinity Chromatography in Chemical Biology

Rachel Codd, Jiesi Gu, Najwa Ejje and Tulip Lifa
School of Medical Sciences (Pharmacology) and Bosch Institute, The University of Sydney, Australia

1.1 Introduction

Immobilized metal ion affinity chromatography (IMAC) was first introduced as a method for resolving native proteins with surface exposed histidine residues from a complex mixture of human serum [1]. IMAC has since become a routine method used in molecular biology for purifying recombinant proteins with histidine tags engineered at the *N*- or *C*-terminus. The success of IMAC for protein purification may have obscured its potential utility in other applications in biomolecular chemistry and chemical biology. Since there exists in nature a multitude of non-protein based low molecular weight compounds that have an inherent affinity towards metal ions, or that have a fundamental requirement for metal ion binding for activity, IMAC could be used to capture these targets from complex mixtures. This highly selective affinity-based separation method could facilitate the discovery of new anti-infective and anticancer compounds from bacteria, fungi, plants, and sponges. A recent body of work highlights new applications of IMAC for the isolation of known drugs and for drug discovery, metabolome profiling, and for preparing metal-specific molecular probes for chemical proteomics-based drug discovery. At its core, IMAC is a method underpinned by the fundamental tenets of coordination chemistry. This chapter will briefly focus on these aspects, before moving on to describe a number of recent innovations in IMAC. The ultimate intent of this chapter is to seed interest in other research groups for expanding the use of IMAC across chemical biology.

Inorganic Chemical Biology: Principles, Techniques and Applications, First Edition. Edited by Gilles Gasser.
© 2014 John Wiley & Sons, Ltd. Published 2014 by John Wiley & Sons, Ltd.

1.2   Principles and Traditional Use

An IMAC system comprises three variable elements (Fig. 1.1): the insoluble matrix (green), the immobilized chelate (depicted as iminodiacetic acid, IDA, red), and the metal ion (commonly Ni(II), blue). Critical to the veracity of IMAC as a separation technique is that the coordination sphere of the immobilized metal–chelate complex is unsaturated, which allows target compounds to reversibly bind to the resin via the formation and dissociation of coordinate bonds. Each element of the IMAC system can be varied independently or in combination, which, together with basic experimental conditions (buffer selection, pH value), will influence the outcome of a separation experiment. This modular type of experimental system allows a high level of control for optimization.

In accord with its original intended use, the majority of IMAC targets are proteins, which even as native molecules can bind to the immobilized metal–chelate complex with variable affinities, as determined by the presence of surface exposed histidine residues and, in some cases, more weakly binding cysteine residues (Fig. 1.1, protein shown at left). Compared with native proteins, recombinant proteins, which feature a hexameric histidine repeat unit (His-tag) engineered at the C- or N-terminus, are higher affinity IMAC targets (Fig. 1.1, protein shown at middle). In this case, the C-terminal histidine residues of the recombinant protein displace the three water ligands in the immobilized Ni(II)–IDA coordination sphere, with the majority of the components in the protein expression mixture not retained on the resin (Fig. 1.2). After washing the resin to remove these unbound components, the coordinate bonds between the Ni(II)–IDA complex and the C-terminal histidine residues are dissociated by competition upon washing the resin with a buffer containing a high concentration of imidazole.

Phosphorylated proteins (Fig. 1.1, protein at right) as studied in phosphoproteomics [2–4], are also isolable using an IMAC format, based upon the affinity between Fe(III) and phosphorylated proteins (Fe(III)–phosphoserine, $\log K \sim 13$ [5]). The IMAC-compatible metal ions most suited for phosphoproteomics include Fe(III), Ga(III), or Zr(IV), with

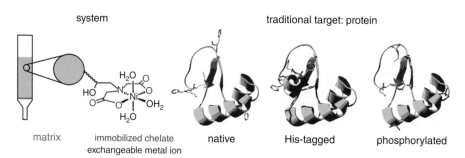

Figure 1.1   *The elements of an immobilized metal ion affinity chromatography (IMAC) experiment. The system (left-hand side) comprises an insoluble matrix (green) with a covalently bound chelate (iminodiacetic acid, IDA, red) which coordinates in a 1:1 fashion a metal ion (Ni(II), blue) to give a complex with vacant coordination sites available for the reversible binding of targets with metal binding groups. Traditional IMAC targets (right-hand side) include native proteins with surface exposed histidine residues, histidine-tagged proteins, and phosphorylated proteins*

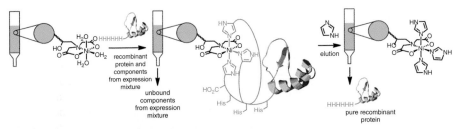

Figure 1.2 *The traditional use of IMAC for the purification of His-tagged recombinant proteins. The recombinant protein binds to the immobilized coordination complex upon the displacement of water ligands by the histidine residues engineered at the N- or C- (as shown) terminus. The resin is washed to remove unbound components from the expression mixture, and the purified protein is eluted from the resin by competition upon washing with a high concentration of imidazole buffer*

these hard acids having preferential binding affinities towards the hard base phosphate groups. This highlights that the IMAC technique is governed by key principles of coordination chemistry, including the hard and soft acids and bases (HSAB) theory [6], coordination number and geometry preferences, and thermodynamic and kinetic factors.

Because there is a significant market demand for IMAC-based separations, considerable research in the biotechnology sector has focused upon finding new and improved matrices and immobilized chelates. Common matrices include cross-linked agarose, cellulose, and sepharose. These polymers can be prepared with different degrees of cross-linking, branching, and different levels of activation, which affect the concentration of the immobilized chelate in the final matrix. There are several different types of immobilized chelates in use in IMAC applications (Fig. 1.3), with the most common being tridentate iminodiacetic acid (IDA, **A**) and tetradentate nitrilotriacetic acid (NTA, **B**). Immobilized tetradentate *N*-(carboxymethyl)aspartic acid (CM-Asp, **C**) and pentadentate *N,N,N'*-tris(carboxymethyl)ethylenediamine (TED, **D**) are used less frequently. These different *N*- and *O*-atom containing ligand types cover a range of degrees of coordinative unsaturation, which for a metal ion with an octahedral coordination preference would span: three available sites ($M(N_1O_2(OH_2)_3)$ (IDA)), two available sites ($M(N_1O_3(OH_2)_2)$ (NTA), $M(N_1O_3(OH_2)_2)$ (CM-Asp)), and one available site ($M(N_2O_3(OH_2))$ (TED)). A significant number of resins with non-traditional immobilized chelates, such as 1,4,7-triazocyclononane [7], 8-hydroxyquinoline [8] or *N*-(2-pyridylmethyl)aminoacetate [9] have been prepared, which have different performance characteristics with respect to protein purification, compared with the traditional IMAC resins.

The nature of the immobilized coordination complex, in terms of both chelate and metal ion, has a major influence on the outcome of an IMAC procedure. An example of the influence of the chelate is found in early studies, which focused on the development of IMAC for phosphoproteomics. Fractions of phosphoserine-containing ovalbumin were retained on an immobilized Fe(III)–IDA resin, but were not retained on an immobilized Fe(III)–TED resin [2]. While an explanation for this observation was not provided in the original work, we posit that this is most likely due to the difference between the number of available coordination sites in the Fe(III)–IDA complex (three sites) and the Fe(III)–TED complex (one site) (Fig. 1.3). This would suggest that retention of ovalbumin fractions via phosphoserine

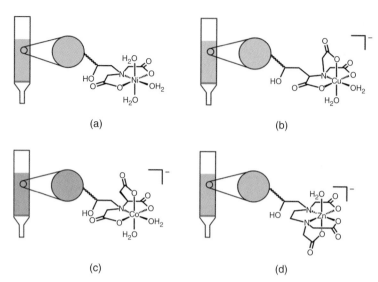

(a)   (b)

(c)   (d)

Figure 1.3 *Immobilized chelates used in IMAC applications. Chelates: iminodiacetic acid (IDA, a), nitrilotriacetic acid (NTA, b), N-(carboxymethyl)aspartic acid (CM-Asp, c) or N,N,N'-tris(carboxymethyl)ethylenediamine (TED, d). A range of metal ions, including Ni(II), Cu(II), Co(II) or Zn(II), are compatible with each type of immobilized chelate. The type of chelate and the coordination preferences of the metal ion will direct the degree of coordinative unsaturation of the immobilized complex*

residues involves at least a bidentate binding mode, and that the single coordination site at the Fe(III)–TED complex was insufficient for retaining the target.

1.3   A Brief History

As an enabling technology, IMAC has played a significant role in accelerating knowledge of molecular, cell, and human biology, through expediting access to significant quantities of pure proteins. For a technique that is conducted every day in many laboratories around the world, it is interesting to reflect briefly upon the history and acceptance of IMAC in its early phases of development. The many review articles available on the history of IMAC [10–14] warrants only a brief coverage of this topic here. The first description of IMAC for protein fractionation used Zn(II)- or Cu(II)-loaded IDA resins prepared in house, with the columns configured in series [1]. Processing of an aliquot of human serum showed that the Zn(II) column was enriched with transferrin, acid glycoprotein, and ceruloplasmin, while the Cu(II) column was enriched with albumin, haptoglobins and β-lipoprotein [1]. In the ten years following this initial report, aside from sporadic reports of IMAC formats using immobilized Cd(II), Ca(II) or Cu(II), the dominant IMAC format in use for protein separation was Zn(II)–IDA, which had mixed success. In its infancy, IMAC gathered only modest traction in the protein purification research community. The uptake of IMAC improved after a follow up study from the initial authors, which highlighted its broad utility [15]. In this second report, the use of Ni(II)- or Fe(III) loaded IDA resin was described, together with a TED-immobilized resin loaded with these same metal ions, which performed well in purifying native proteins with surface exposed histidine residues. Nickel(II)–IDA resin ultimately became the IMAC format of choice for many protein-based applications [16].

The exponential rise in the use of IMAC came with the demonstration of its utility for purifying recombinant proteins [17]. His-tagged protein constructs were isolated from complex mixtures using an Ni(II)–IDA (Fig. 1.2) or –NTA resin with extraordinary selectivity. The dovetailed techniques of recombinant molecular biology and IMAC guaranteed the rapid uptake of IMAC in life science research laboratories and spurred high activity in IMAC-related product development in biotechnology companies, which continues today. Both the lag time and eventual traction of IMAC is evident from a plot of the number of citations per year of the article first reporting its use for native proteins [1], and of the article that described its use for recombinant protein purification [17] (Fig. 1.4).

1.4   New Application 1: Non-protein Based Low Molecular Weight Compounds

Since many non-protein based low molecular weight compounds have an inherent affinity to metal ions, or have a fundamental requirement for metal ion binding for activity, IMAC could have potential for isolating these types of compounds. This could expedite natural products based drug discovery, because secondary metabolites in bacterial culture or extracts from plants and marine life are usually present in very small quantities and require careful purification to provide sufficient yields for downstream structural characterisation and biological screening [18]. As an affinity-based separation method, IMAC has particular value in this regard, since the target material can potentially be concentrated

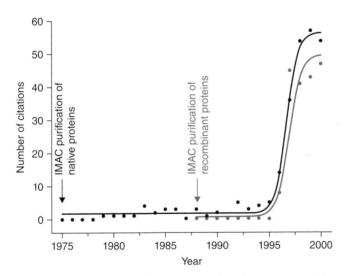

Figure 1.4   *Citations per year for the publications that first described IMAC for native protein purification (black) [1] and that first used IMAC for Hig-tag recombinant protein purification (grey) [17]. There was a 20-year lag time between the discovery of IMAC and its wide acceptance as a powerful method for the purification of recombinant proteins*

on the resin from large volumes of native dilute culture or extract, thereby circumventing concentration steps often necessary in more traditional purification protocols. To test the veracity of using IMAC for purifying non-protein based low molecular weight targets, our group selected bacterial siderophores as a test construct.

1.4.1   Siderophores

Our laboratory has a significant focus on research into the chemical biology of siderophores, which are low molecular weight ($M_r \sim 1000\,\mathrm{g\,mol^{-1}}$) organic chelates produced by bacteria for the purpose of Fe acquisition [19–25]. Under aqueous, pH neutral, and aerobic conditions, most Fe is present as insoluble Fe(III)–oxyhydroxide species, which restricts its availability to bacteria through passive uptake. In response to this environmental challenge, bacteria have evolved a number of mechanisms to guarantee supply of essential Fe, with the production of Fe(III)-specific siderophores as one of the most successful and widespread of these adaptations. Siderophores excreted by bacteria into the extracellular milieu solubilize local Fe(III) in soil, or in Fe-bound proteins such as transferrin in a mammalian host, to form stable Fe(III)–siderophore coordination complexes. The Fe(III) is returned to the bacterial cell through a protein-mediated cascade, which is initiated by an avid recognition event between the Fe(III)–siderophore complex and cell-surface receptors of the source bacterium. Ultimately, the Fe(III)–siderophore complex is dissociated in the cytoplasm to liberate Fe for incorporation into the multiple Fe-containing proteins, including cytochromes, ribonucleotide reductase and aconitase, which are fundamental for life [26, 27].

By virtue of their ability to coordinate Fe(III) with high affinity, and a range of other metal ions with lower affinity [28], siderophores have potential in treating metal ion mediated pathologies in humans, and have applications in environmental metal remediation [28–36]. Owing to the bacterial competition for limited Fe, it is often the case that a bacterial genome will code for the biosynthesis of a structurally unique siderophore that is recognized as the Fe(III)-loaded complex by a structurally unique cell receptor. Both the structural diversity and the breadth of metal binding applications in the areas of health and the environment fuel research towards building a comprehensive library of siderophores across the bacterial kingdom.

As is common to most targets in natural products, siderophores are produced in native cultures in very small quantities ($<1\,\mathrm{mg\,L^{-1}}$) [37, 38], which limits the ability to obtain sufficient amounts for structural characterisation using spectroscopy and X-ray crystallography. This limitation led us to consider more streamlined approaches to purifying siderophores from complex bacterial culture medium, with the aim of delivering a robust separation method that would facilitate siderophore profiling. Siderophores have been classified into three groups, based upon the nature of the Fe(III) binding groups: hydroxamic acids, catechols, or hydroxycarboxylic acids, with the last class being based on citric acid scaffolds [39]. The metal binding ability inherent to siderophores prompted us to consider whether IMAC could be used to select these ligands from a bacterial culture. Our first experiments in this regard focused upon the hydroxamic acid based siderophore, desferrioxamine B (DFOB), which is produced by the non-pathogenic soil bacterium *Streptomyces pilosus*. The mesylate salt of DFOB is used to treat, via sub-cutaneous infusion, secondary iron overload disease, which arises from the frequent blood transfusions necessary to prevent life-threatening anaemia symptomatic of genetic blood disorders, including beta-thalassaemia and myelodysplastic syndromes [40–42]. Despite the release of new orally-active synthetic iron chelating agents for these conditions, including deferasirox and deferiprone [40], DFOB remains the "gold standard" for iron overload disease, due to its high affinity towards Fe(III) and its low toxicity for chronic use.

The high affinity of siderophores towards Fe(III), prompted caution about using Fe(III)–IDA IMAC resins for their separation, since it would be likely that the siderophore (Fe(III)–DFOB, log K 30.6) would compete against the IDA (Fe(III)–IDA, log K 10.7) for Fe(III) and strip it from the resin, thereby providing no resolution from the bulk mixture (Table 1.1).

One report described the isolation of siderophores from *Alcaligenes eutrophus* CH34 using Fe(III)-based IMAC [46]. *A. eutrophus* CH34 was subsequently renamed *Ralstonia eutropha* CH34 and is known presently as *Cupriavidus metallidurans* CH34. The compound that gave a positive result in the universal siderophore chrome azurol sulfonate (CAS) detection assay, was identified in the original report as a hydroxamic acid based siderophore [46], subsequently as a novel phenolate-based siderophore (M_r 1.470 g mol^{-1}) that contained neither hydroxamic acid nor catecholate groups [47], and finally by other workers using chemical degradation and spectroscopy, as the citric acid based siderophore staphyloferrin B [48]. In the initial report, the reduced affinity between Fe(III) and a hydroxycarboxylate-based siderophore [49], as distinct from a hydroxamic acid based siderophore, may have enabled the use of Fe(III)-based IMAC purification, although this method was not used for the isolation of this siderophore beyond the first report.

In our studies, we chose to use Ni(II)-based IMAC to examine the ability of the method to select for DFOB from bacterial culture (Fig. 1.5). The Ni(II)–DFOB affinity constant

Table 1.1 Equilibrium constants for metal complexes formed with immobilized ligands relevant to IMAC and selected non-protein based low molecular weight molecular targets

Ligand	Abbrev.	Equilibrium	Log $K^{a,b}$					
			Fe(III)	Co(II)	Ni(II)	Cu(II)	Zn(II)	Yb(III)
Iminodiacetic acid	IDA	ML/(M·L)	10.7^c	6.9	8.1	10.6	7.2	7.4
N-methyliminodiacetic acid	MIDA	ML/(M·L)	NA^e	7.6	8.7	11.0	7.6	7.6
Nitrilotriacetic acid	NTA	ML/(M·L)	15.9	10.4	11.5	12.9	10.7	12.2
		ML$_2$/(M·L)^2	24.3^d	14.3	16.3	17.4	14.2	21.4^b
Ethylenediaminetetraacetic acid	EDTA	ML/(M·L)	25.0	16.3	18.5	18.7	16.4	19.5
Acetohydroxamic acid	AHA	ML/(M·L)	11.4^d	5.1^d	5.3^d	7.9^d	5.4^d	6.6^d
Desferrioxamine B	DFOB	MHL/(M·HL)	30.6^d	10.3^d	10.9^d	14.1^d	10.1^d	16.0^d
Desferrioxamine E	DFOE	ML/(M·L)	32.5^d	11.9^d	12.2^d	13.7^d	12.1^d	NA^e
Bleomycin A$_2$	BLMA$_2$	ML/(M·L)	NA^e	9.7^{d,f}	11.3^{d,f}	12.6^{d,f}	9.1^{d,f}	NA^e

^{a}25 °C, 0.1 M (unless specified otherwise).
^{b}From References 43 and 44 (unless specified otherwise).
^{c}25 °C, 0.5 M.
^{d}20 °C, 0.1 M.
^{e}NA, not available.
^{f}From Reference 45.

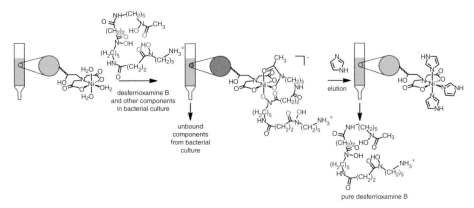

Figure 1.5 *A new use of IMAC for the purification of bacterial secondary metabolites with metal ion binding affinity, such as desferrioxamine B (DFOB). In a fashion similar to its traditional use (refer to Fig. 1.2), the target low molecular weight non-protein based metabolite binds to the immobilized coordination complex upon the displacement of water ligands by the metal binding functional groups (for DFOB, hydroxamic acid groups). The resin is washed to remove unbound components from the bacterial culture supernatant and the purified metabolite is eluted from the resin by competition upon washing with a high concentration of imidazole buffer, or by decreasing the pH value of the elution buffer to pH < pK_a (functional group). The figure shows a posited binding mode between analyte and resin*

(log K 10.9) foreshadowed that compared with Fe(III)–IDA based IMAC, DFOB would have a reduced propensity to leach Ni(II) from the Ni(II)–IDA complex. This would manifest as a binding event between DFOB and the immobilized Ni(II)–IDA complex, rather than elution of Ni(II)-loaded DFOB. Several reports of Ni(II)–hydroxamic acid coordination chemistry suggested that Ni(II) might be a judicious choice of metal ion [50–52].

These experiments showed that a 1 mL column of Ni(II)–IDA resin bound about 350 nmol of the monohydroxamic acid acetohydroxamic acid at pH 9.0, and, that under the same conditions, the binding capacity of the resin increased to about 3000 nmol for the dihydroxamic acid suberodihydroxamic acid and the trihydroxamic acid DFOB, which correlated with available stability constants [53], and reflected the potential ability of the latter two ligands to act as at least tridentate ligands towards the resin for improved binding. The optimal pH value for binding these hydroxamic acid standards was pH 9, which is close to the pK_a value of the N–OH proton in aqueous solvents [54]. At higher pH values, metal hydroxides can precipitate on the resin. Most striking about this study was the selection of native DFOB from a crude culture supernatant of *S. pilosus* [55]. This crude culture supernatant was not subject to any pre-treatment steps, aside from adjusting the pH value to 9.0. As evident from the HPLC (high-performance liquid chromatography) trace (Fig. 1.6, A), the crude mixture contained many components, including those from the bacteriological medium (amino acids, peptides, vitamins) and other secondary metabolites produced by *S. pilosus*. Single-step processing using Ni(II)-based IMAC selected five major components from the mixture, with two components co-eluting under the peak at $t_R = 8.9$ min (Fig. 1.6, B). Based on mass spectrometry and the use of samples spiked with authentic DFOB, the peaks at $t_R = 15.03$ and 15.45 min (B, boxed) were ascribed to $DFOA_1$ and DFOB, respectively,

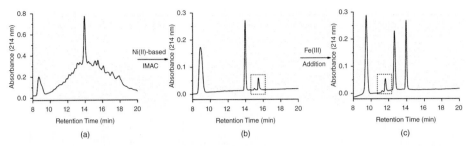

Figure 1.6 *Crude S. pilosus culture, which contained many components as measured by RP-HPLC (a), was processed using Ni(II)-based IMAC to select for DFOB (major) and DFOA₁ (minor) as metal free (b, boxed) and Fe(III)-loaded (c, boxed) species. The IMAC process selected five components from the complex mixture, which were more completely resolved upon the addition of Fe(III). Compared with the free ligands, Fe(III)-loaded DFOB and DFOA₁ eluted on the RP-HPLC column in a window described by increased water solubility, which is consistent with the role of siderophores to increase the water solubility of Fe(III). Adapted with permission from [55] © 2008 Royal Society of Chemistry*

from the ferrioxamine class of siderophores [19, 20, 56]. In the presence of added Fe(III), these peaks (C, boxed) shifted in a systematic fashion to a more hydrophilic region of the reverse-phase HPLC (RP-HPLC) trace, in accord with the role of siderophores to increase the water solubility of Fe(III). The LC–MS (liquid chromatography–mass spectrometry) data from the Fe(III)-loaded solution correlated with the identification of Fe(III)–DFOA₁ and Fe(III)–DFOB. The signals at $t_R = 9.45$, 12.63, and 13.93 min (C) were derived from components in the medium, and showed variable Fe(III) responsiveness.

This work demonstrated the potential of IMAC for selecting clinically valuable agents direct from bacterial culture. Nickel(II)-based IMAC would be predicted to be useful in the isolation of other types of siderophores, including but not limited to, the catecholate-based compounds enterobactin or salmochelin from *Escherichia* species, the 2-hydroxyphenyloxazoline ring-based siderophores from *Mycobacteria* [57], and marine siderophores, such as lystabactin [58]. Recent compendiums of the full structural diversity of siderophores [19, 20] provide a range of promising IMAC-compatible targets.

The metal ion selected for the IMAC procedure has a major influence on the experimental outcome. As predicted, Fe(III)-loaded IMAC resin was not suitable for the capture of high affinity Fe(III) binding hydroxamic acid based siderophores. In this case, the DFOB sequestered the Fe(III) from the immobilized Fe(III)–IDA complex (Equation 1.1, log K 19.9) and was eluted in the wash fraction as the Fe(III)-loaded complex (Fig. 1.7, A). In the case of Ni(II)-based IMAC, the respective Ni(II)–DFOB and Ni(II)–IDA affinity constants (Table 1.1) were better poised to enable successful product retention (Equation 1.2, log K 2.8) and elution of DFOB in a metal-free form (Fig. 1.7, B).

$$[\text{Fe(III)(IDA)(OH}_2)_3]^+ + \text{DFOBH}_4{}^+ \rightleftharpoons [\text{Fe(III)(DFOBH)}]^+ + \text{IDAH}_2 + 3\text{H}_2\text{O} + \text{H}^+$$
$$(1.1)$$

$$[\text{Ni(II)(IDA)(OH}_2)_3] + \text{DFOBH}_4{}^+ \rightleftharpoons [\text{Ni(II)(DFOBH)}] + \text{IDAH}_2 + 3\text{H}_2\text{O} + \text{H}^+ \quad (1.2)$$

In other experiments, V(IV)-loaded IDA resin was prepared with the intent of capturing DFOB. While DFOB and related hydroxamic acids have a rich coordination chemistry

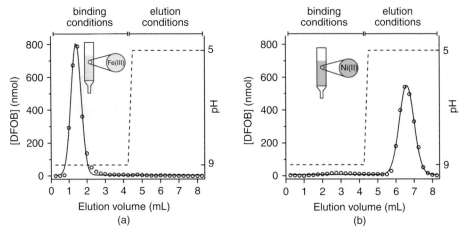

Figure 1.7 *The performance of Fe(III)- (a) or Ni(II)- (b) based IMAC for targeting desferriox-amine B (DFOB) purification. In the case of Fe(III)-based IMAC, DFOB out-competed IDA for Fe(III) (refer to Equation 1.1), and was eluted from the resin as unbound Fe(III)-loaded DFOB. In the case of Ni(II)-based IMAC, the competition between DFOB and IDA for Ni(II) (refer to Equation 1.2) was poised in a region that effected DFOB binding to the resin at pH 9, with its subsequent elution as the metal free ligand at pH 5*

with V(IV) and V(V) [38, 59–64], DFOB was not retained on this resin. This is probably attributable to the insufficient number of available coordination sites on the immobilized [V(IV)(O)(IDA)] complex. A similar rationale would suggest that V(V)-loaded IDA resins containing the immobilized $[V(V)(O)_2(IDA)]^-$ complex would be close to coordinatively saturated and prevent ligand binding.

1.4.2   Anticancer Agent: Trichostatin A

The enzyme-mediated modification of chromatin is at the forefront of cancer research, since the topology of this protein (histone)–polynucleotide structure directs transcrip-tional activity [65]. A common modification to the histone component of chromatin is the *N*-acetylation of selected lysine residues, with the acetylation status controlled by a two-enzyme system: the histone acetyltransferases (HATs) and the histone deacetylases (HDACs). Upregulated HDAC activity leads to a higher concentration of unmasked lysine residues, which causes chromatin to condense into a less transcriptionally active form, which in turn attenuates the expression of tumor suppressor genes. The gene silencing effects of upregulated HDAC activity is associated with the onset and progression of cancer, with this class of enzyme validated as an anticancer target [66, 67]. Three of the four classes of the 18 known HDAC isoforms are Zn(II)-containing enzymes, which render Zn(II) binding compounds as potential HDAC inhibitors [68–72].

One of the most potent inhibitors of Zn(II)-containing HDACs is the monohydrox-amic acid trichostatin A (TSA), which was discovered from cultures of *Streptomyces hygroscopicus* Y-50 [73]. TSA (HDAC1, IC_{50} ∼ 5 nM [74]) has served as a lead com-pound for designing inhibitors against HDACs and other disease relevant Zn(II)-containing enzymes, including matrix metalloproteinases, metallo-*β*-lactamases, carboxypeptidase A, and carbonic anhydrase [28, 30, 75–78]. The structurally simpler TSA analogue suberoylanilide hydroxamic acid (SAHA, vorinostat) inhibits HDACs through Zn(II) coordination [79] and is in clinical use for the treatment of cutaneous T cell lymphoma. Studies of the coordination chemistry of SAHA with metals including Ni(II) [80] and the experiments that showed acetohydroxamic acid bound to Ni(II)–IDA resin [55], prompted studies to establish the utility of IMAC for isolating TSA from culture. IMAC also had the potential to resolve mixtures of TSA and the β-glucosyl-substituted analogue trichostatin C (TSC). TSA is a costly chemical and can be prepared in its enantiomeric pure form (99% e.e.) using total synthesis only on a small scale [81]. The ability to access TSA and other analogues from culture would be valuable for drug discovery and for the fine chemicals industry.

A standard solution of TSA bound to Ni(II)–IDA resin at a loading of 0.5 μmol mL^{-1} with recovery at >95%. The capacity of the Ni(II)–IDA resin towards binding TSA from the culture of native TSA- and TSC-producing *Streptomyces hygroscopicus* MST-AS5346 was reduced, due to the presence of competing ligands, including those present in the culture medium and other secondary metabolites with inherent metal binding affinity [82]. In the native culture, there were at least 50 species detected by UV spectroscopy (ultraviolet) (Fig. 1.8, A) and 150–200 species detected using LC–MS in the total ion current (TIC) mode. The presence of TSA, TSC, and trichostatic acid, the last of which is both a TSA precursor and hydrolysis product, was confirmed from LC–MS measurements. The

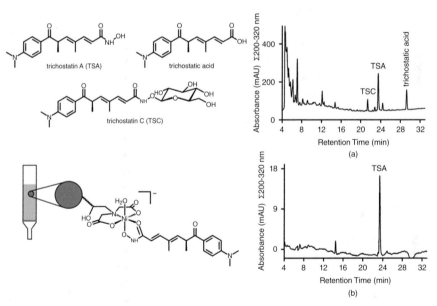

Figure 1.8 *Nickel(II)-based IMAC for the purification of trichostatin A (TSA) from Streptomyces hygroscopicus MST-AS5346 culture, with a posited binding mode between analyte and resin shown at the lower left. As determined by RP-HPLC, this method was effective at resolving TSA (B) from a complex mixture of components, including TSA, the −NH-O-glycosyl analogue TSC and trichostatic acid (A), present in the bacterial culture supernatant. Adapted with permission from [82] © 2012 Royal Society of Chemistry*

fraction containing unbound components was enriched with TSC and trichostatic acid, in addition to containing about 20% of the total TSA. The glucose masked NH−O group of TSC and the terminal carboxylic acid group of trichostatic acid prevented these compounds binding to the resin. The bound component was enriched with TSA (Fig. 1.8, B), with the method showing striking selectivity towards capturing TSA above TSC and trichostatic acid. A minor species in the bound fraction was detected at t_R 14.5 min (m/z_{obs} 432.89), which was not identified, but could be a potential inhibitor of Zn(II)- or Ni(II)-containing metalloproteins.

1.4.3   Anticancer Agent: Bleomycin

Bleomycins are a family of glycopeptide antibiotics produced by *Streptomyces verticillus* that are used clinically to treat a number of cancers [83–85]. Combination therapy of bleomycin, cisplatin, and etoposide has contributed to an increase to 90% in the cure rate of testicular cancer [86]. The total synthesis of bleomycin has been achieved [87], but for pharmaceutical-scale production, it is purified from large-scale *S. verticillus* fermentation broths. The structural elements of bleomycin include a metal-binding region, comprised of nitrogen donor atoms from imidazole, β-hydroxyhistidine (amide), pyrimidine, and β-aminoalanine (primary and secondary amines) [88–90], a bithiazole group with pendant groups that distinguish between different bleomycin congeners, and a disaccharide motif [84, 85]. The coordination of Fe(II) to the metal-binding region of bleomycin is integral to its mechanism of action, with the ultimate production of a low-spin ferric peroxide species (O_2^{2-}−Fe(III)−bleomycin) that cleaves DNA [91–93]. During fermentation, the metal binding region of bleomycin selects for Cu(II) (log K 12.6) (Table 1.1), which needs to be removed from the complex prior to clinical use.

Since bleomycin has a natural affinity towards Cu(II), we undertook to establish the efficacy of Cu(II)-based IMAC for bleomycin purification. Similar to the approaches taken for the isolation of DFOB and TSA, we began by optimizing the capture of a standard solution of bleomycin on Cu(II)−IDA resin, which showed a binding capacity of 300 nmol mL^{-1} [94]. In experiments aimed to select bleomycin directly from *S. verticillus* culture, the endogenous Cu(II) was first removed by adsorbing the culture onto a macroreticular XAD-2 resin and washing the resin with EDTA, similar to the process used industrially. After this process, the Cu(II)-free mixture was loaded at pH 9 onto the Cu(II)-loaded IMAC resin. Similar to the previous examples, compared with a standard bleomycin solution, the capture of bleomycin from culture was less efficient, with about 50% capture, due to the presence of competing ligands in the complex mixture. The crude XAD-2 treated culture contained at least 50 UV-active species (Fig. 1.9, A). The majority of these species appeared in the unbound fractions, with the two dominant bleomycin congeners, bleomycin A$_2$ (BLMA$_2$) and bleomycin B$_2$ (BLMB$_2$), appearing in the fractions collected under elution conditions (Fig. 1.9, B). The purification factor for the resolution of BLMA$_2$ and BLMB$_2$ from the bulk mixture was conservatively estimated as 25. The affinity of BLMB$_2$ towards the Cu(II)−IDA complex was somewhat greater than in the case of BLMA$_2$, with a higher proportion of the latter congener appearing in the wash fraction, and a higher proportion of the former congener appearing in the elution fraction. Since the primary metal binding region of BLMA$_2$ and BLMB$_2$ is the same, the difference in binding to the Cu(II)−IDA resin

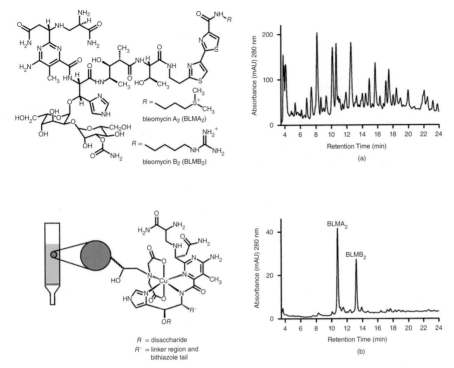

Figure 1.9   *Copper(II)-based IMAC for the purification of bleomycin from Streptomyces verti-cillus culture, with a posited binding mode between analyte and resin shown at the lower left. As determined by RP-HPLC, this method was effective at resolving the two major bleomycin congeners BLMA$_2$ and BLMB$_2$ (b) from a complex mixture of components (a) present in the bacterial culture supernatant. Reprinted with permission from [94] © 2012 Elsevier*

must be attributable to the bithiazole-substituted group. In the case of BLMB$_2$, a coordinate bond between Cu(II) and an agmatine-based nitrogen atom might be possible, similar to the observed binding between arginine or agmatine and the Mn(II)-containing enzymes arginase [95] or agmatinase [96], respectively. This additional interaction could result in a stronger interaction with the Cu(II)–IDA resin, as observed experimentally. Together with the studies of the selection of DFOB or TSA from crude culture medium, the experiments using bleomycin support the utility of IMAC for the streamlined capture of metal binding pharmaceutical agents.

1.4.4   Anti-infective Agents

The use of IMAC for isolating non-protein based pharmaceutical agents has been garnering increased research interest. A recent study determined the utility of Cu(II)- and Fe(III)-based IMAC for analyzing antibiotic drugs in veterinary and human use [97]. The study showed that solutions from standard samples of tetracycline-, fluoroquinoline-, macrolide-, aminoglycoside-, and β-lactam-based antibiotics were adsorbed with significant recoveries on both Cu(II)- and Fe(III)-loaded IMAC resins, and that sulfonamides and steroid- and non-steroid hormones were not retained. An examination of the structures

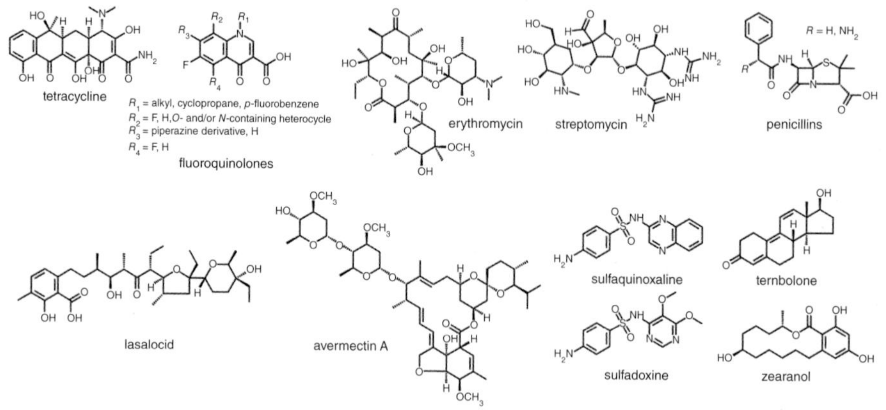

Figure 1.10 *Antibiotic compounds and veterinary anti-parasitic compounds and hormones examined as IMAC targets. The agents that bound to Cu(II)- or Fe(III)-loaded resins have functional groups configured to form coordination complexes compatible with IMAC processing*

of each of these compounds informs the experimental results (Fig. 1.10). The antibiotic components that were retained on IMAC resins have multiple oxygen and/or nitrogen donor atoms configured to form stable five- or six-membered chelate rings with the immobilized Cu(II)- or Fe(III)-IDA complex: tetracycline (keto, carboxamide, vicinal diol), fluoroquinolines (keto, carboxylate), erythromycin (vicinal diol), streptomycin (vicinal diol, α-hydroxycarboxylate, amine), and penicillins (amide, carboxylate). Each of these antibiotics has been discussed in the coordination chemistry literature [98, 99], which is a useful indicator of the veracity of using IMAC for their isolation. Components that showed minimal (lasalocid, avermectin A) or no retention (sulfonamides, trenbolone, zearanol) on the IMAC resin had insufficient and/or poorly configured donor groups for coordination. While close to 100% of the penicillin-based antibiotic ampicillin ($R = $ NH$_2$) was retained on both the Cu(II)- and Fe(III)-loaded IMAC resin, only 70% of benzylpenicillin ($R = $ H) was retained on the Fe(III)-loaded resin, and only 35% of the sample was retained on the Cu(II)-loaded resin. This is probably due to the amino nitrogen atom and carbonyl oxygen atom of the amide group of ampicillin being correctly configured to furnish a stable five-membered coordination ring with the immobilized Cu(II)- or Fe(III)–IDA complex. The significant difference in binding phenomenon between these two closely related compounds illustrates the influence of coordination chemistry in selecting suitable targets for IMAC purification. In the case of β-lactam antibiotics, the use of Zn(II)- or Co(II)-loaded resins could lead to the metal ion catalyzed hydrolysis of the β-lactam ring, which would yield ring opened structures upon elution [100].

1.4.5   Other Agents

Copper(II)-loaded IDA resins have been used to concentrate Cu(II)-specific ligands present in surface sea waters [101]. Because of low concentrations, the exact nature of these ligands was unable to be established from LC–MS measurements, but possible structures posited included small peptide units and/or fragments from degraded siderophores.

Copper(II)-based IMAC has also been used to study ligands present in soil samples [102]. The retention behavior of a series of model ligands with variable Cu(II) affinities was examined, which showed that pyridine-type ligands and salicylhydroxamic acid were strongly retained on the Cu(II)–IDA matrix, and ligands containing carboxylic acid groups as the predominant ligand type were generally not retained.

1.4.6   Selecting a Viable Target

A multitude of octahedral heteroleptic metal–IDA or –NTA complexes exist that have been characterized in the solid state and in solution, in which the remaining sites in the coordination sphere are filled by a different tri- or bidentate ligand, respectively [103–105]. The non-IDA- or NTA-based ligands in these cases, or more complex compounds that contain these ligand types as substructures, represent candidates for isolation using IMAC, since the IMAC resin can be considered representative of the metal–IDA or –NTA component of the mixed ligand complex. Examples include solution structures characterized between [Fe(III)(NTA)(OH$_2$)$_2$] and acetohydroxamic acid [104], which accord with the successful capture of hydroxamic acid compounds, including DFOB and TSA on IDA-based IMAC resins [55, 82]. An X-ray crystal structure of the Fe(III)–NTA based heteroleptic complex with the fluoroquinoline antibiotic ciprofloxacin (R_1 = cyclopropane, R_2 = H, R_3 = piperazine, R_4 = H) (Fig. 1.10) showed that the complex [Fe(ciprofloxacin)(NTA)]·3.5H$_2$O featured coordination from ciprofloxacin through the keto and carboxylate oxygen atoms [106]. Ciprofloxacin and other fluoroquinoline-based analogues are synthetic products, rather than products of bacterial fermentation, which would find IMAC being a possible step for purification during chemical synthesis. The structural characterization of [Fe(ciprofloxacin)(NTA)]·3.5H$_2$O is consistent with the successful capture of fluoroquinoline-based antibiotic compounds using Fe(III)-based IMAC [97]. X-ray crystal structure data from heteroleptic complexes between Cu(II)–*N*-methyl-IDA and monodentate coordinated adenine [107] or bidentate acyclovir (also known as aciclovir) [108] support the use of IMAC for polynucleotide purification [109].

The X-ray crystal structure of the heteroleptic complex [Cu(IDA)(α-picolinamide)]· 2H$_2$O [105] supports that Cu(II)–IDA based IMAC could be useful for the isolation of more complex natural products that feature α-picolinamide as a structural motif, including the vertilecanins, which were isolated and characterized from fermentations of the fungus *Verticillium lecanii*, and shown to have antibiotic properties [110]. The characterization of the heteroleptic complex [Ni(IDA)(DABT)(OH$_2$)] (DABT = 2,2′-diamino-4,4′-bi-1,3-thiazole) [111] supports that natural products containing the bithiazole motif would be candidates for IMAC-based purification. Examples of complex natural products with bithiazole groups include metabolites from the myxobacterium *Myxococcus fulvus*, including myxothiazole A [112], cystothiazoles [113], and melithiazoles [114], which have demonstrated various antimicrobial and cytotoxic effects [115]. Although selected members of this group of compounds can be produced from total synthesis [116, 117], the multi-step and complex syntheses give modest yields. As allowed by IMAC, a single-step isolation method direct from culture carries the advantages of both improved yield of the parent compound and the simultaneous capture of structural analogues for establishing chemical diversity. In addition to bithiazole-containing natural products, IMAC could be used for non-aromatic analogues, such as the siderophores yersiniabactin and pyochelin, which contain a 4,2-linked thiazoline-thiazolidine system [118–120]. While IDA- or NTA-based heteroleptic complexes are the best guide for IMAC compatible ligands, other heteroleptic complexes can also be used to inform the selection process. As one example, the characterization of the heteroleptic Cu(II) complex formed with 2,2′-bipyridine and the natural product plumbagin [121], suggests that related naphthoquinone-based compounds present in *Plumbaginaceae* extracts used in traditional Chinese medicine [122], could be isolable using Cu(II)-based IMAC.

Another strategy to determine targets that would be amenable to isolation using IMAC is to consider metalloprotein inhibitors. The metal–ligand active site of a metalloprotein is most often coordinatively unsaturated, to allow for substrate docking and subsequent chemical transformation. The coordinatively unsaturated metal–IDA or –NTA resin can in effect be viewed as a surrogate of a metalloprotein active site. This opens up a powerful means to select for potential metalloprotein inhibitors from a complex mixture. This has proven successful in the use of Ni(II)-based IMAC for selecting TSA from *S. hygroscopicus* MST-AS5346 culture, as one of the most potent documented inhibitors of the Zn(II)-containing histone deacetylases [72, 82]. The bengamides, which are natural products isolated from sponge [123], have been shown to coordinate to the bimetallic Ni(II), Mn(II) or Co(II) active site of human or *Mycobacterium tuberculosis* methionine aminopeptidase via bridged coordination from the trihydroxy-substituted amide extension region [124, 125]. These types of compounds could potentially be isolable from the extract using Ni(II)-, Mn(II)- or Co(II)-based IMAC. The stoichiometry of the ternary immobilized bengamide–metal–IDA/NTA complex would likely be different from the bridged structure observed from the protein X-ray data, but the IMAC method remains compatible with the selection of this type of compound. Documentation of all natural products based inhibitors of metalloproteins is beyond the scope of this chapter, but these few examples serve to illustrate the use of protein X-ray crystallography data to guide the selection of targets isolable using IMAC.

In the absence of available data on metal–IDA or –NTA or other heteroleptic complexes, or guiding knowledge of the metalloprotein inhibitory activity of a compound, it is sufficient to simply explore the coordination chemistry of a potential target to assess its viability for selection using IMAC. Bacteria, fungi, plants, and sponges produce a multitude of chemically diverse natural products that serve as drugs themselves or as scaffolds for chemical modification or the design of synthetic analogues [126–130]. Candidate compounds that have a native metal binding ability, such as siderophores, or a metal ion binding requirement for activity, such as bleomycins, and other metalloantibiotics [131], are cases in point. The IMAC-based purification protocol can be guided by the types and configuration of donor atoms and the most appropriate IMAC-compatible metal ion, according to the HSAB theory [6] and other considerations of coordination chemistry. The aminoglycoside antibiotics, which include streptomycin, gentamycin, tobramycin, amikacin, neomyxin, and paromomycin, contain an extensive array of amino and hydroxyl groups, which would bind to a Cu(II)- or Fe(III)-based IMAC resin [99, 132]. We are currently using IMAC to purify other anticancer compounds from complex fermentation mixtures, including the aminoglycoside-based doxorubicin, which has a documented coordination chemistry with Cu(II) and Fe(III). Solution complexes of Cu(II) and the antibiotic lincomycin have been established using ^{1}H and ^{13}C NMR (nuclear magnetic resonance) spectroscopy [133], which supports that Cu(II)-based IMAC would be useful in selecting for lincomycin and analogues from cultures of *Streptomyces lincolnensis*. The success of each of these examples will be dependent upon the water solubility of the target compound, since the IMAC procedure can generally tolerate only low levels of organic solvents, and is mostly incompatible with the use of coordinating solvents, such as methanol and dimethylformamide. Ultimately, it remains that experiment is the best guide to the efficacy of IMAC for the capture of non-protein based targets.

1.5   New Application 2: Multi-dimensional Immobilized Metal Ion Affinity Chromatography

Part of the driving force for developing IMAC for isolating non-protein based bacterial metabolites lies in the possibility of accessing more than one compound from a single culture. The ability to purify multiple components from a single mixture would increase production efficiency and streamline pharmaceutics processing. Two or more IMAC columns configured in series containing resins loaded with different metal ions could retain different classes of metabolites, as defined by differential metabolite–metal binding affinities. In our group, we have termed this IMAC mode, multi-dimensional IMAC (MD-IMAC). It could be predicted that each column would be enriched with a different class of metabolite, which would be recoverable upon disassembling the columns and eluting the bound components.

The original report on IMAC for the separation of serum proteins used an in-series configuration of Zn(II)- and Cu(II)-loaded IMAC resins [1]. Similar to the modest traction that IMAC itself garnered in its infancy (Fig. 1.4), the concept of using columns in series to gain additional resolving power essentially vanished from the literature, with only a few subsequent reports of this format, which were termed "tandem" IMAC [15] or "cascade" IMAC [134].

In our group, we are practising MD-IMAC for the resolution of more than one high-value clinical agent from a single culture. MD-IMAC has not previously been conducted to separate multiple bacterial secondary metabolites. Our preliminary experiments have focused upon the resolution of mixtures of standard solutions of DFOB and bleomycin. In single-dimensional IMAC, Ni(II)-based IMAC was effective for retaining a standard DFOB solution (Figs. 1.6 and 1.7) and Cu(II)-based IMAC was effective for retaining a standard bleomycin solution (Fig. 1.9). For the MD-IMAC format, significant resolution of these agents was achieved using Yb(III)-based IMAC for DFOB and Cu(II)-based IMAC for bleomycin (Fig. 1.11, panel on the left). The major difference between the results obtained in our work and the early results that used in-series columns for resolving serum proteins [1, 15], was that we observed a profound difference in the effect of the order of the columns. In the case where the column containing the Yb(III)-loaded resin was mounted above the column containing the Cu(II)-loaded resin, about 70% of the DFOB was retained on the Yb(III)-based resin, with about 30% DFOB, and the entire sample of bleomycin, retained on the Cu(II)-based resin. In the case where the column order was reversed, with the Cu(II)-loaded column mounted above the Yb(III)-loaded column, there was no resolution between DFOB and bleomycin, with both components retained on the upper Cu(II)-loaded resin (Fig. 1.11, panel on the right). This is an elegant

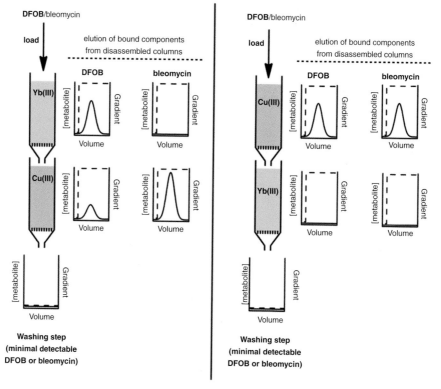

Figure 1.11 *Multi-dimensional immobilized metal ion affinity chromatography (MD-IMAC) for the separation of multiple bacterial secondary metabolites with differential metal ion affinities. In the MD-IMAC setup, two columns containing IMAC resins loaded with different metal ions are configured in series. After loading the mixture containing the analytes, the two columns are disassembled and the bound components are eluted from each column. The column order can affect the resolution: DFOB and bleomycin were well resolved using a Yb(III)-loaded resin mounted above a Cu(II)-loaded resin (left-hand side), but were not resolved using a Cu(II)-loaded resin mounted above a Yb(III)-loaded resin (right-hand side)*

example of the use of MD-IMAC and its underlying principle of separating non-protein based bacterial metabolites based upon differential metal ion affinities.

The successful partial resolution of DFOB and bleomycin was determined by the significant differential in binding affinities towards Yb(III). While DFOB has a significant affinity towards Yb(III) (log K 16.0), bleomycin, which has a combination of hard (O-based) and borderline (N-based) donor atoms in the metal-binding region (Fig. 1.9), has negligible affinity to the hard Yb(III) ion (log K not determined). In the case where the Cu(II) column was mounted above the Yb(III) column, the similarity in affinity constants of complexes between Cu(II) and DFOB (log K 14.1) or bleomycin (log K 12.6) resulted in both components binding to the Cu(II) column. In experiments that used in-series Cu(II)- and Fe(III)-IMAC formats to resolve serum proteins, there was no effect in the column order: the same protein profile was distributed across each column, independent of column order [15].

Multi-dimensional IMAC has significant potential for capturing multiple components from a single mixture. This could be used for pharmaceutics processing and also for targeted, analytical-scale separations of whole bacterial metabolomes, as discussed in the following section. In our work, we have used two in-series columns, but it would be possible to use additional columns, each loaded with a different metal ion. A hypothetical device that could advance MD-IMAC is shown in Fig. 1.12. In this device, multiple IMAC "cassettes" each containing a different metal ion are threaded in series in a Falcon tube-like housing. Following the binding step, which could be undertaken using gravity, pressure, or low centrifugal force, the cassettes would be disassembled and the components eluted from each individual cassette.

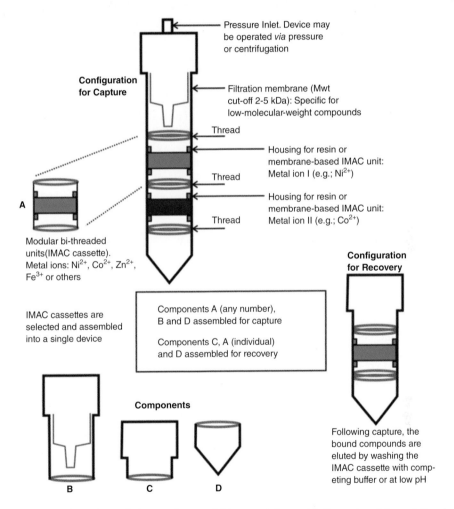

Figure 1.12 *A hypothetical device that could be used for multi-dimensional immobilized metal ion affinity chromatography (MD-IMAC) applications. This device shows individual IMAC "cassettes" threaded in series in a Falcon tube-like housing that could be subject to operation under low centrifugal force*

1.6   New Application 3: Metabolomics

Metabolomics is a recent addition to the "omics" family that describes the set of metabolites characteristic to a given species [135–138]. Each bacterial metabolome contains many hundreds of compounds and provides a rich resource for drug discovery [139, 140]. The bacterial metabolome also reports on the metabolic status of an organism, as modulated by the growth conditions local to the environment or host [141, 142]. While a given metabolome harbors a rich information content, its analysis is complicated by the sheer number of components, which are unable to be fully resolved for LC–MS analysis [142]. Separation technologies that reduce the complexity of a bacterial metabolome are needed to support metabolomics research across the health and environmental spheres.

IMAC offers the means to fractionate a bacterial metabolome into various pools of compounds with different metal ion affinities. By definition, the compounds that have no metal ion affinity will be removed from the whole metabolome. While this could be seen as a disadvantage, the concept does offer the potential to select for a significantly smaller population of compounds in a targeted fashion. The bacterial metabolome can be modulated *in situ* by external conditions, such as temperature and available nutrients, and also *in vitro* using precursor directed biosynthesis approaches, where non-native substrates are added to the growth medium for incorporation into native biosynthetic pathways to generate new compounds [143, 144]. IMAC would be well suited to profiling siderophore and bleomycin metabolomes, as well as the metabolomes of alternative IMAC-compatible compounds. As an example, precursor directed biosynthesis approaches have been used to generate new siderophore analogues of desferrioxamine E, pyochelin, and rhiozoferrin [145–147]. More recently, the siderophore metabolome of *Shewanella putrefaciens* has been modulated [148] by inhibiting endogenous levels of the 1,4-diaminobutane substrate used in the biosynthesis of its native siderophore putrebactin [149]. Each of these siderophore metabolomes would be amenable to profiling using IMAC.

Our group is currently undertaking studies to map the siderophore metabolome of the marine bacterium *Salinispora tropica*. Marine bacteria, including *S. tropica*, and other *Actinomycetes*, are under scrutiny in natural products research, given the wealth of clinical compounds discovered from their terrestrial counterparts [150, 151]. *Salinispora tropica* has particular notoriety with regard to chemical diversity, since almost 10% of its genome codes for secondary metabolites [152]. These studies, to be published shortly, underpin the utility of IMAC for the selective capture of siderophores from culture. New erythromycin analogues prepared using a precursor-directed biosynthesis approach with an engineered strain of *Escherichia coli* [153] could be isolable using Fe(III)- or Cu(II)-based IMAC [97].

1.7   New Application 4: Coordinate-bond Dependent Solid-phase Organic Synthesis

In a recent development, the IMAC format has been extended beyond the affinity-based separation of non-protein based low molecular weight bacterial secondary metabolites. In this application, a bed of IMAC resin containing a bound bacterial metabolite was subject to solid-phase synthesis on this same resin bed, to furnish a derivative of the metabolite [154]. This method is efficient, since the same resin bed is used to both capture the molecule and for the downstream semi-synthetic chemistry. Conducting solid-phase organic synthesis using a system where the primary molecule was bound to the IMAC resin via coordinate bonds carries all the advantages inherent to traditional solid-phase synthesis methods, in which the primary ligand is covalently bound to Merrifield resin, or Rink or Wang resins. The reaction was able to be driven to completion using an excess of reagents and these excess reagents and reaction byproducts were readily removed from the system via filtration. This method is distinct from other solid-phase organic synthesis methods by virtue of the immobilization of the primary ligand by multiple coordinate bonds rather than covalent bonds, and for this reason has been termed coordinate-bond dependent solid-phase organic synthesis (CBD-SPOS) [154].

In the first example of CBD-SPOS, a sample of DFOB was bound to Ni(II)-loaded IDA resin. At pH 9, DFOB is posited to bind to the IMAC resin via the hydroxamic acid functional groups with the pendant amine group remaining chemically reactive (Fig. 1.5), which was demonstrated by the orange color formed from reaction with 2,4,6,-trinitrobenzenesulfonic acid. Following DFOB binding, the water was exchanged for non- (or weakly) coordinating tetrahydrofuran (THF), which did not significantly compete with DFOB for the resin binding sites. Methanol and dimethylformamide were unsuitable solvents, since they acted as competitive ligands and displaced DFOB from the resin. The semi-synthetic target that was chosen was biotinylated DFOB. Biotin–DFOB was of interest, since it could be used as a molecular probe against the cognate cell proteome of the native DFOB producing bacterium *S. pilosus*. Reaction between DFOB-bound Ni(II)-loaded IDA resin and *N*-hydroxysuccinimide–biotin in THF at 40 °C overnight yielded biotin–DFOB, which remained bound to the resin (Fig. 1.13). Less than 1% of the DFOB was leached from the resin during this procedure. The resin was washed exhaustively with THF, which leached a minimal amount of product (<1%) from the resin.

The CBD-SPOS product was eluted from the resin with water at pH 6.0. This pH value was sufficiently acidic to displace the DFOB-based products from the resin (pH 6 < pK_a hydroxamic acid (pK_a 9.0)), but was not so acidic as to dissociate the immobilized Ni(II)–IDA coordination complex (pH 6 > pK_a IDA (pK_a 2.5)), which resulted in the liberation of biotin–DFOB as a metal-free adduct. The yield of biotin–DFOB was 76% and the product was significantly purer than the equivalent product prepared using traditional solution phase chemistry (Fig. 1.14). The veracity of the method depended upon the strength of the multiple coordinate bonds between DFOB and the immobilized Ni(II)–IDA chelate being functionally equivalent to a covalent bond, as used in traditional SPOS methods. Methods that use metabolites as molecular probes are keenly sought in metabolomics/proteomics applications [155–158], and CBD-SPOS delivers a new methodology to access these probes. We are currently using CBD-SPOS to prepare chemical probes of alternative amine-functionalized siderophores, including pyoverdine type II, which is produced by pathogenic *Pseudomonas aeruginosa* [31, 159, 160], and other siderophores native to bacterial pathogens, including ornibactin C-4 from *Burkholderia cepacia* [161], and exochelin MN from *Mycobacterium neoaurum* [162, 163].

1.8 Green Chemistry Technology

One of the major advantages of IMAC for metabolite capture is that the technique is water compatible. In experiments aimed to isolate DFOB from *S. pilosus* culture, the aqueous supernatant was adjusted to pH 9.0, as the only pre-treatment step before being loaded onto

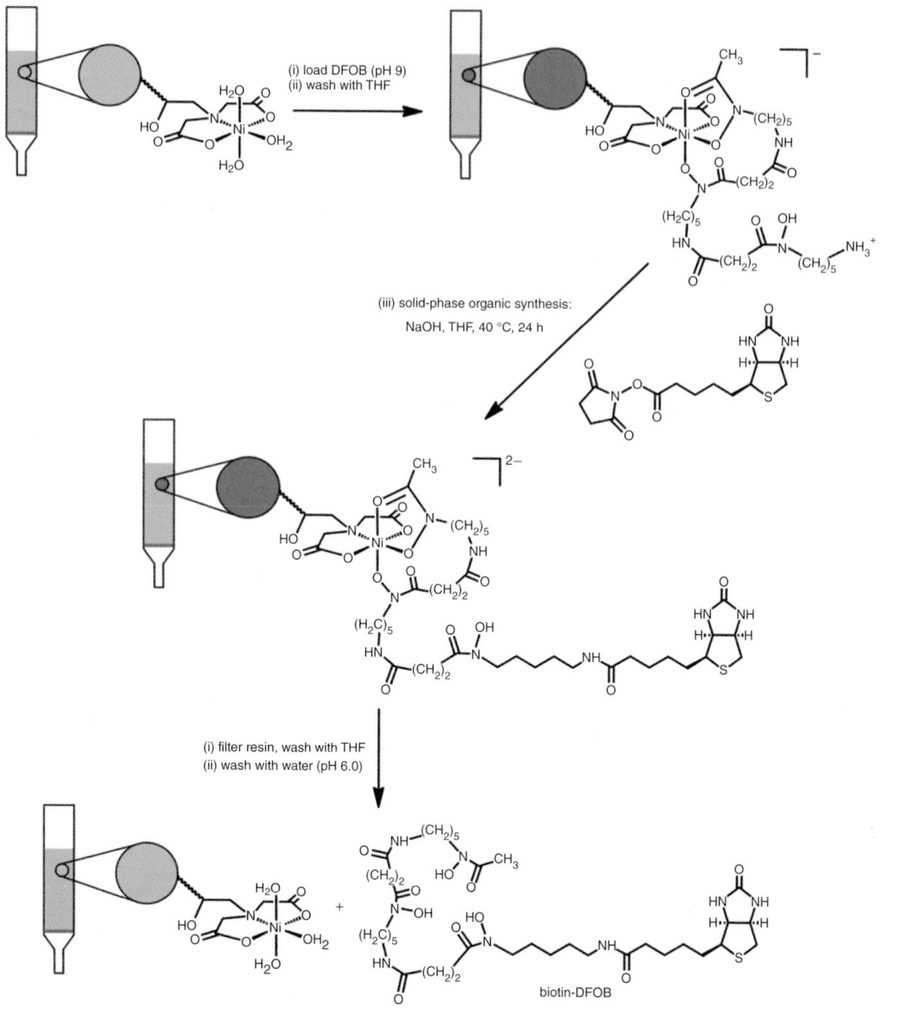

Figure 1.13 *Use of Ni(II)-loaded IMAC resin for the preparation of metal-specific probes and other semi-synthetic derivatives via coordinate-bond dependent solid-phase organic synthesis (CBD-SPOS). In CBD-SPOS, a metabolite with a chemically reactive group (e.g., amine) that is not involved in binding to the IMAC resin can be directly subjected to solid-phase chemical conversion on the same IMAC resin bed. This method carries all the advantages inherent to traditional solid-phase synthesis and has been used successfully to prepare biotin−DFOB under mild conditions. Adapted with permission from[154] © 2012 Royal Society of Chemistry*

an Ni(II)−IDA resin [55]. The retention of this compound, and of TSA in subsequent work [82], was undertaken entirely in the absence of organic solvents, and highlights the tenets of Green Chemistry inherent to IMAC processing. In other applications, components present in the bacteriological medium, including amino acids, peptides, and vitamins, which could compete for the metal-binding sites, can be removed using XAD-2 chromatography prior to IMAC processing [94]. This will improve the performance of the metal ion affinity step,

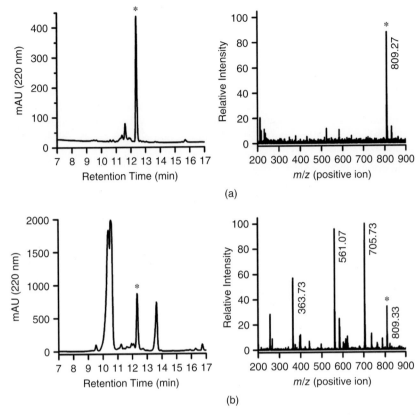

Figure 1.14 Biotin-DFOB ([M+Na$^+$]$^+$, m/z_{calc} 809.42, *) was prepared in 76% yield and high purity using coordinate-bond dependent solid-phase organic synthesis (CBD-SPOS) (a), compared with conventional solution methods (b). Adapted with permission from [154] © 2012 Royal Society of Chemistry

but does so at the expense of efficiency with the additional chromatographic step. One of the major goals in Green Chemistry is to reduce the use of organic solvents [164–166]. In pharmaceutics production, the extraction of clinical agents from fermentation mixtures may rely on the use of large volumes of organic solvents. This is the case for the current production of bleomycin [167] and also for DFOB, with 75% of the total production costs of the latter agent due to purification. Scaling IMAC processing to industrial scale could enable a significant reduction in the use of organic solvents in pharmaceutics processing. Further, the water compatibility of IMAC is particularly useful in natural products chemistry, since separating the polar components from an extract is generally more difficult than resolving the organic soluble components.

1.9 Conclusion

Based on the recent research from our laboratory and from other groups described in this chapter, it is evident that the applications of IMAC have not yet been exhausted. This technique, with regard to the isolation of single or multiple non-protein based low molecular weight bacterial metabolites direct from complex fermentation mixtures, has a rich future in sustainable pharmaceutics processing. The use of in-series columns to obtain resins that are enriched with different classes of metabolites can increase purification efficiency, which is often the major cost in producing agents for clinical use. This technique has utility for mapping the metabolomes of selected classes of non-protein based bacterial metabolites with native metal binding affinity, including, but not limited to, siderophores and bleomycins. This has implications for discovering new drugs and for metabolomics to gain an understanding of how the local environment of a given bacterial species modulates the type and/or relative concentrations of these metabolites. This provides experimental data to complement systems biology approaches to drug discovery and environmental research.

The types of IMAC formats developed by the biotechnology sector aimed at automating recombinant protein purification are ever expanding. These newer IMAC-based formats, which include 96-well plates, membranes, magnetic beads, and quantum dots [168] are all transferable to the applications described in this chapter, relating to non-protein based low molecular weight metal binding compounds. These formats herald a bright future for the use of IMAC in natural products discovery and in pharmaceutics processing.

One of the major advantages of IMAC is its water compatibility. In this age of Green Chemistry, there is a strong drive to reduce the use of large volumes of organic solvents for extracting new chemical entities from natural sources: IMAC meets this remit. The authors look forward to continuing their contributions to expanding the scope of IMAC and envisage a future where this easy-to-use method is more fully integrated into chemical biology research platforms.

Acknowledgments

Ms Douha Lozi is acknowledged for contributions to the early phase of this work. The work has been supported by an Australian Postgraduate Award (J.G.) and with funding from The University of Sydney (Bridging Support Grants (R.C.), University Postgraduate Awards (N.E., T.L.)) and the National Health and Medical Research Council of Australia (R.C.).

References

1. Porath, J., Carlsson, J., Olsson, I., Belfrage, G. (1975) Metal chelate affinity chromatography, a new approach to protein fractionation. Nature, 258: 598–599.
2. Andersson, L., Porath, J. (1986) Isolation of phosphoproteins by immobilized metal (Fe^{3+}) affinity chromatography. Anal. Biochem., 154: 250–254.
3. Nühse, T. S., Stensballe, A., Jensen, O. N., Peck, S. C. (2003) Large-scale analysis of *in vivo* phosphorylated membrane proteins by immobilized metal ion affinity chromatography and mass spectrometry. Mol. Cell. Proteomics, 2: 1234–1243.
4. Imam-Sghiouar, N., Joubert-Caron, R., Caron, M. (2005) Application of metal-chelate affinity chromatography to the study of the phosphoproteome. Amino Acids, 28: 105–109.

5. Österberg, R. (1957) Metal and hydrogen-ion binding properties of *O*-phosphoserine. Nature, 179: 476–477.

6. Pearson, R. G. (1963) Hard and soft acids and bases. J. Am. Chem. Soc., 85: 3533–3539.

7. Jiang, W., Graham, B., Spiccia, L., Hearn, M. T. W. (1998) Protein selectivity with immobilized metal ion-tacn sorbents: chromatographic studies with human serum proteins and several other globular proteins. Anal. Biochem., 255: 47–58.

8. Zachariou, M., Hearn, M. T. W. (2000) Adsorption and selectivity characteristics of several human serum proteins with immobilized hard Lewis metal ion-chelate adsorbents. J. Chromatogr. A, 890: 95–116.

9. Chaouk, H., Hearn, M. T. W. (1999) New ligand, *N*-(2-pyridylmethyl)aminoacetate, for use in the immobilized metal ion affinity chromatographic separation of proteins. J. Chromatogr. A, 852: 105–115.

10. Cheung, R. C. F., Wong, J. H., Ng, T. B. (2012) Immobilized metal ion affinity chromatography: a review on its applications. Appl. Microbiol. Biotechnol., 96: 1411–1420.

11. Block, H., Maertens, B., Spriestersbach, A., Brinker, N., Kubicek, J., Fabis, R., Labahn, J., Schäfer, F. (2009) Immobilized-metal affinity chromatography: a review. Methods Enzymol., 463: 439–473.

12. Gutierrez, R., Martin del Valle, E. M., Galan, M. A. (2007) Immobilized metal-ion affinity chromatography: Status and trends. Sep. Purif. Rev., 36: 71–111.

13. Chaga, G. S. (2001) Twenty-five years of immobilized metal ion affinity chromatography: past, present and future. J. Biochem. Biophys. Methods, 49: 313–334.

14. Gaberc-Porekar, V., Menart, V. (2001) Perspectives of immobilized-metal affinity chromatography. J. Biochem. Biophys. Methods, 49: 335–360.

15. Porath, J., Olin, B. (1983) Immobilized metal ion affinity adsoption and immobilized metal ion affnity chromatography of biomaterials. Serum protein affinities for gel-immobilized iron and nickel ions. Biochemistry, 22: 1621–1630.

16. Priestman, D. A., Butterworth, J. (1985) Prolinase and non-specific dipeptidase of human kidney. Biochem. J., 231: 689–694.

17. Hochuli, E., Bannwarth, W., Döbeli, H., Gentz, R., Stüber, D. (1988) Genetic approach to facilitate purification of recombinant proteins with a novel metal chelate adsorbent. Bio/Technology, 6: 1321–1325.

18. Bucar, F., Wube, A., Schmid, M. (2013) Natural product isolation - how to get from biological material to pure compounds. Nat. Prod. Rep., 30: 525–545.

19. Budzikiewicz, H. (2010) Microbial siderophores, in: Kinghorn, A. D., Falk, H. and Kobayashi, J. (eds) *Progress in the Chemistry of Organic Natural Products*, pp. 1–75 (New York, Springer-Verlag).

20. Hider, R. C., Kong, X. (2010) Chemistry and biology of siderophores. Nat. Prod. Rep., 27: 637–657.

21. Neilands, J. B. (1995) Siderophores: structure and function of microbial iron transport compounds. J. Biol. Chem., 270: 26723–26726.

22. Raymond, K. N., Dertz, E. A. (2004) Biochemical and physical properties of siderophores, in: Crosa, J. H., Mey, A. R. and Payne, S. M. (eds) *Iron Transport in Bacteria*, pp. 3–17 (Washington, DC, ASM Press).

23. Crumbliss, A. L., Harrington, J. M. (2009) Iron sequestration by small molecules: thermodynamic and kinetic studies of natural siderophores and synthetic model complexes. Adv. Inorg. Chem., 61: 179–250.

24. Butler, A., Theisen, R. M. (2010) Iron(III)-siderophore coordination chemistry: Reactivity of marine siderophores. Coord. Chem. Rev., 254: 288–296.

25. Boukhalfa, H., Crumbliss, A. L. (2002) Chemical aspects of siderophore mediated iron transport. BioMetals, 15: 325–339.

26. Miethke, M., Marahiel, M. A. (2007) Siderophore-based iron acquisition and pathogen control. Microbiol. Mol. Biol. Rev., 71: 413–451.

27. Miethke, M. (2013) Molecular strategies of microbial iron assimilation: from high-affinity complexes to cofactor assembly systems. Metallomics, 5: 15–28.

28. Codd, R. (2008) Traversing the coordination chemistry and chemical biology of hydroxamic acids. Coord. Chem. Rev., 252: 1387–1408.

29. Scott, L. E., Orvig, C. (2009) Medicinal inorganic chemistry approaches to passivation and removal of aberrant metal ions in disease. Chem. Rev., 109: 4885–4910.

30. Marmion, C. J., Griffith, D., Nolan, K. B. (2004) Hydroxamic acids. An intriguing family of enzyme inhibitors and biomedical ligands. Eur. J. Inorg. Chem.: 3003–3016.

31. Schalk, I. J., Hannauer, M., Braud, A. (2011) New roles for bacterial siderophores in metal transport and tolerance. Environ. Microbiol., 13: 2844–2854.

32. Telpoukhovskaia, M. A., Orvig, C. (2013) Werner coordination chemistry and neurodegeneration. Chem. Soc. Rev., 42: 1836–1846.

33. Casentini, B., Pettine, M. (2010) Effects of desferrioxamine-B on the release of arsenic from volcanic rocks. Appl. Geochem., 25: 1688–1698.

34. Duckworth, O. W., Bargar, J. R., Sposito, G. (2009) Quantitative structure-activity relationships for aqueous metal-siderophore complexes. Environ. Sci. Technol., 43: 343–349.

35. Liddell, J. R., Obando, D., Liu, J., Ganio, G., Volitakis, I., Mok, S. S., Crouch, P. J., White, A. R., Codd, R. (2013) Lipophilic adamantyl- or deferasirox-based conjugates of desferrioxamine B have enhanced neuroprotective capacity: implications for Parkinson disease. Free Radic. Biol. Med., 60: 147–156.

36. Gez, S., Luxenhofer, R., Levina, A., Codd, R., Lay, P. A. (2005) Chromium(V) complexes of hydroxamic acids: Formation, structures, and reactivities. Inorg. Chem., 44: 2934–2943.

37. Pakchung, A. A. H., Soe, C. Z., Codd, R. (2008) Studies of iron-uptake mechanisms in two bacterial species of the *Shewanella* genus adapted to middle-range (*Shewanella putrefaciens*) or Antarctic (*Shewanella gelidimarina*) temperatures. Chem. Biodivers., 5: 2113–2123.

38. Pakchung, A. A. H., Soe, C. Z., Lifa, T., Codd, R. (2011) Complexes formed in solution between vanadium(IV)/(V) and the cyclic dihydroxamic acid putrebactin or linear suberodihydroxamic acid. Inorg. Chem., 50: 5978–5989.

39. Drechsel, H., Winkelmann, G. (1997) Iron Chelation and Siderophores, in: Winkelmann, G. and Carrano, C. J. (eds) *Transition Metals in Microbial Metabolism*, pp. 1–49 (Amsterdam, Harwood Academic).

40. Bernhardt, P. V. (2007) Coordination chemistry and biology of chelators for the treatment of iron overload disorders. Dalton Trans., 3214–3220.

41. Liu, J., Obando, D., Schipanski, L. G., Groebler, L. K., Witting, P. K., Kalinowski, D. S., Richardson, D. R., Codd, R. (2010) Conjugates of desferrioxamine B (DFOB) with derivatives of adamantane or with orally available chelators as potential agents for treating iron overload. J. Med. Chem., 53: 1370–1382.

42. Kalinowski, D. S., Richardson, D. R. (2005) The evolution of iron chelators for the treatment of iron overload disease and cancer. Pharmacol. Rev., 57: 547–583.

43. Martell, A. E., Smith, R. M. (1974) Critical Stability Constants. Vol. 1. (New York, Plenum Press).

44. Martell, A. E., Smith, R. M. (1977) Critical Stability Constants. Vol. 3. (New York, Plenum Press).

45. Sugiura, Y., Ishizu, K., Miyoshi, K. (1979) Studies of metallobleomycins by electronic spectroscopy, electron spin resonance spectroscopy, and potentiometric titration. J. Antibiot., 32: 453–461.

46. Khan, M. A., Lelie, D., Cornelis, P., Mergeay, M. (1994) Purification and characterization of "alcaligin E", a hydroxamate-type siderophore produced by *Alcaligenes eutrophus* CH 34, Conference Proceedings of the *Plant Pathogenic Bacteria: 8th International Conference*, Versailles, France, 1992, pp. 591–597.

47. Gilis, A., Khan, M. A., Cornelis, P., Meyer, J.-M., Mergeay, M., van der Lelie, D. (1996) Siderophore-mediated iron uptake in *Alcaligenes eutrophus* CH34 and identificatoin of *ale*B encoding the ferric iron-alcaligin E receptor. J. Bacteriol., 178: 5499–5507.

48. Münzinger, M., Taraz, K., Budzikiewicz, H. (1999) Staphyloferrin B, a citrate siderophore of *Ralstonia eutropha*. Z. Naturforsch. C., 54: 867–875.

49. Harris, W. R., Carrano, C. J., Raymond, K. N. (1979) Coordination chemistry of microbial iron transport compounds. 16. Isolation, characterization and formation constants of ferric aerobactin. J. Am. Chem. Soc., 101: 2722–2727.

50. Stemmler, A. J., Kampf, J. W., Kirk, M. L., Pecoraro, V. L. (1995) A model for the inhibition of urease by hydroxamates. J. Am. Chem. Soc., 117: 6368–6369.

51. Benini, S., Rypniewski, W. R., Wilson, K. S., Miletti, S., Ciurli, S., Mangani, S. (2000) The complex of *Bacillus pasteurii* urease with acetohydroxamate anion from X-ray data at 1.55 Å resolution. J. Biol. Inorg. Chem., 5: 110–118.

52. Gaynor, D., Starikova, Z. A., Ostrovsky, S., Haase, W., Nolan, K. B. (2002) Synthesis and structure of a heptanuclear nickel(II) complex uniquely exhibiting four distinct binding modes, two of which are novel, for a hydroxamate ligand. Chem. Commun., 506–507.

53. Brown, D. A., Geraty, R., Glennon, J. D., Choileain, N. N. (1986) Design of metal chelates with biological activity. 5. Complexation behavior of dihydroxamic acids with metal ions. Inorg. Chem., 25: 3792–3796.

54. Fazary, A. E. (2005) Thermodynamic studies on the protonation equilibria of some hydroxamic acids in $NaNO_3$ solutions in water and in mixtures of water and dioxane. J. Chem. Eng. Data, 50: 888–895.

55. Braich, N., Codd, R. (2008) Immobilized metal affinity chromatography for the capture of hydroxamate-containing siderophores and other Fe(III)-binding metabolites from bacterial culture supernatants. Analyst, 133: 877–880.

56. Keller-Schierlein, W., Mertens, P., Prelog, V., Wasler, A. (1965) Metabolic products of microorganisms. XLIX. Ferrioxamines A1, A2, and D2. Helv. Chim. Acta, 48: 710–723.

57. De Voss, J. J., Rutter, K., Schroeder, B. G., Su, H., Zhu, Y., Barry, C. E. I. (2000) The salicylate-derived mycobactin siderophores of *Mycobacterium tuberculosis* are essential for growth in macrophages. Proc. Natl. Acad. Sci. USA, 97: 1252–1257.

58. Zane, H. K., Butler, A. (2013) Isolation, structure elucidation, and iron-binding properties of lystabactins, siderophores isolated from a marine *Pseudoaltermonas* sp. J. Nat. Prod., 76: 648–654.

59. Rehder, D. (1999) The coordination chemistry of vanadium as related to its biological functions. Coord. Chem. Rev., 182: 297–322.

60. Butler, A., Parsons, S. M., Yamagata, S. K., de la Rosa, R. I. (1989) Reactivation of vanadate-inhibited enzymes with desferrioxamine B, a vanadium(V) chelator. Inorg. Chim. Acta, 163: 1–3.

61. Bell, J. H., Pratt, R. F. (2002) Mechanism of inhibition of the β-lactamase of *Enterobacter cloacae* P99 by 1:1 complexes of vanadate with hydroxamic acids. Biochemistry, 41: 4329–4338.

62. Goldwaser, I., Li, J., Gershonov, E., Armoni, M., Karnieli, E., Fridkin, M., Shechter, Y. (1999) L-Glutamic acid γ-monohydroxamate. A potentiator of vanadium-evoked glucose metabolism in vitro and in vivo. J. Biol. Chem., 274: 26617–26624.

63. Haratake, M., Fukunaga, M., Ono, M., Nakayama, M. (2005) Synthesis of vanadium(IV,V) hydroxamic acid complexes and *in vivo* assessment of their insulin-like activity. J. Biol. Inorg. Chem., 10: 250–258.

64. Luterotti, S., Grdinic, V. (1986) Spectrophotometric determination of vanadium(V) with desferrioxamine B. Analyst, 111: 1163–1165.

65. Minucci, S., Pelicci, P. G. (2006) Histone deacetylase inhibitors and the promise of epigenetic (and more) treatments for cancer. Nat. Rev. Cancer, 6: 38–51.

66. Bolden, J. E., Peart, M. J., Johnstone, R. W. (2006) Anticancer activities of histone deacetylase inhibitors. Nat. Rev. Drug Discov., 5: 769–784.

67. Liu, T., Kuljaca, S., Tee, A., Marshall, G. M. (2006) Histone deacetylase inhibitors: multifunctional anticancer agents. Cancer Treat. Rev., 32: 157–165.

68. Bertrand, P. (2010) Inside HDAC with HDAC inhibitors. Eur. J. Med. Chem., 45: 2095–2116.

69. Marks, P. A., Breslow, R. (2007) Dimethylsulfoxide to vorinostat: Development of this histone deacetylase inhibitor as an anticancer drug. Nat. Biotechnol., 25: 84–90.

70. Codd, R., Braich, N., Liu, J., Soe, C. Z., Pakchung, A. A. H. (2009) Zn(II)-dependent histone deacetylase inhibitors: suberoylanilide hydroxamic acid and trichostatin A. Int. J. Biochem. Cell Biol., 41: 736–739.

71. Liao, V., Liu, T., Codd, R. (2012) Amide-based derivatives of β-alanine hydroxamic acid as histone deacetylases inhibitors: Attenuation of potency through resonance effects. Bioorg. Med. Chem. Lett., 22: 6200–6204.

72. Bieliauskas, A. V., Pflum, M. K. H. (2008) Isoform-selective histone deacetylase inhibitors. Chem. Soc. Rev., 37: 1402–1413.

73. Tsuji, N., Kobayashi, M., Nagashima, K., Wakisaka, Y., Koizumi, K. (1976) A new antifungal antibiotic, trichostatin. J. Antibiot., 29: 1–6.

74. Woo, S. H., Frechette, S., Khalil, E. A., Bouchain, G., Vaisburg, A., Bernstein, N., Moradei, O., Leit, S., Allan, M., Fournel, M., Trachy-Bourget, M.-C., Li, Z., Besterman, J. M., Delorme, D. (2002) Structurally simple trichostatin A-like straight chain hydroxamates as potent histone deacetylase inhibitors. J. Med. Chem., 45: 2877–2885.

75. Anzellotti, A. I., Farrell, N. P. (2008) Zinc metalloproteins as medicinal targets. Chem. Soc. Rev., 37: 1629–1651.

76. Puerta, D. T., Cohen, S. M. (2004) A bioinorganic perspective on matrix metalloproteinase inhibition. Curr. Top. Med. Chem., 4: 1551–1573.

77. Whittaker, M., Floyd, C. D., Brown, P., Gearing, A. J. H. (1999) Design and therapeutic application of matrix metalloproteinase inhibitors. Chem. Rev., 99: 2735–2776.

78. Hu, J., Van den Steen, P. E., Sang, Q.-X. A., Opdenakker, G. (2007) Matrix metalloproteinase inhibitors as therapy for inflammatory and vascular diseases. Nat. Rev. Drug. Discov., 6: 480–498.

79. Finnin, M. S., Donigian, J. R., Cohen, A., Richon, V. M., Rifkind, R. A., Marks, P. A., Breslow, R., Pavletich, N. P. (1999) Structures of a histone deacetylase homologue bound to the TSA and SAHA inhibitors. Nature, 401: 188–193.

80. Griffith, D. M., Szöcs, B., Keogh, T., Suponitsky, K. Y., Farkas, E., Buglyó, P., Marmion, C. J. (2011) Suberoylanilide hydroxamic acid, a potent histone deacetylase inhibitor; its X-ray crystal structure and solid state and solution studies of its Zn(II), Ni(II), Cu(II) and Fe(III) complexes. J. Inorg. Biochem., 105: 763–769.

81. Zhang, S., Duan, W., Wang, W. (2006) Efficient, enantioselective organocatalytic synthesis of trichostatin A. Adv. Synth. Catal., 348: 1228–1234.

82. Ejje, N., Lacey, E., Codd, R. (2012) Analytical-scale purification of trichostatin A from bacterial culture in a single step and with high selectivity using immobilised metal affinity chromatography. RSC Adv., 2: 333–337.

83. Umezawa, H., Takita, T. (1980) The bleomycins: Antitumor copper-binding antibiotics. Struct. Bond., 40: 73–99.

84. Chen, J., Stubbe, J. (2005) Bleomycins: Towards better therapeutics. Nat. Rev. Cancer, 5: 102–112.

85. Galm, U., Hager, M. H., Van Lanen, S. G., Ju, J., Thorson, J. S., Shen, B. (2005) Antitumor antibiotics: Bleomycin, enediyenes, and mitomycin. Chem. Rev., 105: 739–758.

86. Einhorn, L. H. (2002) Curing metastatic testicular cancer. Proc. Natl. Acad. Sci. USA, 99: 4592–4595.

87. Aoyagi, Y., Katano, K., Suguna, H., Primeau, J., Chang, L.-H., Hecht, S. M. (1982) Total synthesis of bleomycin. J. Am. Chem. Soc., 104: 5537–5538.

88. Sugiyama, M., Kumagai, T., Hayashida, M., Maruyama, M. (2002) The 1.6-Å crystal structure of the copper(II)-bound bleomycin complexed with the bleomycin-binding protein from bleomycin-producing Streptomycin verticillus. J. Biol. Chem., 277: 2311–2320.

89. Iitaka, Y., Nakamura, H., Nakatani, T., Muraoka, Y., Fujii, A., Takita, T., Umezawa, H. (1978) Chemistry of bleomycin. XX The X-ray structure determination of P-3A Cu(II)-complex, a biosynthetic intermediate of bleomycin. J. Antibiot., 31: 1070–1072.

90. Decker, A., Chow, M. S., Kemsley, J. N., Lehnert, N., Solomon, E. I. (2006) Direct hydrogen-atom abstraction by activated bleomycin: An experimental and computational study. J. Am. Chem. Soc., 128: 4719–4733.
91. Sam, J. W., Tang, X.-J., Peisach, J. (1994) Electrospray mass spectrometry of iron bleomycin: Demonstration that activated bleomycin is a ferric peroxide complex. J. Am. Chem. Soc., 116: 5250–5256.
92. Burger, R. M., Peisach, J., Horwitz, S. B. (1981) Activated bleomycin. A transient complex of drug, iron, and oxygen that degrades DNA. J. Biol. Chem., 256: 11636–11644.
93. Westre, T. E., Loeb, K. E., Zaleski, J. M., Hedman, B., Hodgson, K. O., Solomon, E. I. (1995) Determination of the geometric and electronic structure of activated bleomycin using X-ray absorption spectroscopy. J. Am. Chem. Soc., 117: 1309–1313.
94. Gu, J., Codd, R. (2012) Copper(II)-based metal affinity chromatography for the isolation of the anticancer agent bleomycin from *Streptomyces verticillus* culture. J. Inorg. Biochem., 115: 198–203.
95. Bewley, M. C., Jeffrey, P. D., Patchett, M. L., Kanyo, Z. F., Baker, E. N. (1999) Crystal structure of *Bacillus caldovelox* arginase in complex with substrate and inhibitors reveal new insights into activation, inhibition and catalysis in the arginase superfamily. Structure, 7: 435–448.
96. Ahn, H.-J., Kim, K. H., Lee, J. K., Ha, J.-Y., Lee, H. H., Kim, D., Yoon, H.-J., Kwon, A.-R., Suh, S. W. (2004) Crystal structure of agmatinase reveals structural conservation and inhibition mechanism of the ureohydrolase superfamily. J. Biol. Chem., 279: 50505–50513.
97. Takeda, N., Matsuoka, T., Gotoh, M. (2010) Potentiality of IMAC as sample pretreatment tool in food analysis for veterinary drugs. Chromatographica, 72: 127–131.
98. Drechsel, H., Fiallo, M., Garnier-Suillerot, A., Matzanke, B. F., Schünemann, V. (2001) Spectroscopic studies on iron complexes of different anthracyclines in aprotic solvent systems. Inorg. Chem., 40: 5324–5333.
99. Gokhale, N., Patwardhan, A., Cowan, J. A. (2007) Metalloaminoglycosides: Chemistry and biological relevance, in: Arya, D. P. (ed.) *Aminoglycoside Antibiotics: From Chemical Biology to Drug Discovery*, pp. 235-254 (Hoboken, NJ, Wiley-Interscience).
100. Chen, Z.-F., Tang, Y.-Z., Liang, H., Zhong, X.-X., Li, Y. (2006) Cobalt(II)-promoted hydrolysis of cephalexin: crystal structure of the cephalosporate-cobalt(II) complex. Inorg. Chem. Commun., 9: 322–325.
101. Ross, A. R. S., Ikonomou, M. G., Orians, K. J. (2003) Characterization of copper-complexing ligands in seawater using immobilized copper(II)-ion affinity chromatography and electrospray ionization mass spectrometry. Mar. Chem., 83: 47–58.
102. Paunovic, I., Schulin, R., Nowack, B. (2005) Evaluation of immobilized metal-ion affinity chromatography for the fractionation of natural Cu complexing ligands. J. Chromatogr., Sect. A., 1100: 176–184.
103. Kruppa, M., König, B. (2006) Reversible coordinative bonds in molecular recognition. Chem. Rev., 106: 3520–3560.

104. Gabričević, M., Crumbliss, A. L. (2003) Kinetics and mechanism of iron(III)-nitrilotriacetate complex reactions with phosphate and acetohydroxamic acid. Inorg. Chem., 42: 4098–4101.

105. Bugella-Altamirano, E., González-Pérez, J. M., Choquesillo-Lazarte, D., Niclós-Gutiérrez, J., Castiñeiras-Campos, A. (2000) Structural relationships obtained from the coordination of α-picolinamide to the (iminodiacetato)copper(II) chleate: synthesis, crystal structure, and properties of (α-picolinamide)(iminodiacetato) copper(II) dihydrate. Z. Anorg. Allg. Chem., 626: 930–936.

106. Wallis, S. C., Gahan, L. R., Charles, B. G., Hambley, T. W. (1995) Synthesis and X-ray structural characterisation of an iron(III) complex of the fluoroquinoline antimicrobial ciprofloxacin, [Fe(CIP)(NTA)]3.5H$_2$O (NTA = nitrilotriacetato). Polyhedron, 14: 2835–2840.

107. Bugella-Altamirano, E., Choquesillo-Lazarte, D., González-Pérez, J. M., Sánchez-Moreno, M. J., Marín-Sánchez, R., Martín-Ramoa, J. D., Covelo, B., Carballo, R., Castiñeiras, A., Niclós-Gutiérrez, J. (2002) Three new modes of adenine-copper(II) coordination: interligand interactions controlling the selective N3-, N7- and bridging μ-N3,N7-metal-bonding of adenine to different N-substituted iminodiacetato-copper(II) chelates. Inorg. Chim. Acta, 339: 160–170.

108. del Pilar Brandi-Blanco, M., Choquesillo-Lazarte, D., Domínguez-Martín, A., González-Pérez, J. M., Castiñeiras, A., Niclós-Gutiérrez, J. (2011) Metal ion binding patterns of acyclovir: Molecular recognition between this antiviral agent and copper(II) chelates with iminodiacetate or glycylglycinate. J. Inorg. Biochem., 105: 616–623.

109. Kanakaraj, I., Jewell, D. L., Murphy, J. C., Fox, G. E., Wilson, R. C. (2011) Removal of PCR error products and unincorporated primers by metal-chelate affinity chromatography. PLoS ONE, 6: e14512. doi:10.1371/journal.pone.0014512.

110. Soman, A. G., Gloer, J. B., Angawi, R. F., Wicklow, D. T., Dowd, P. F. (2001) Vertilecanins: new phenopicolinic acid analogues from *Verticillium lecanii*. J. Nat. Prod., 64: 189–192.

111. Liu, J.-G., Xu, D.-J. (2005) Synthesis and crystal structure of aqua(diaminobithiazole) (iminodiacetato)nickel(II) hydrate. J. Coord. Chem., 58: 735–740.

112. Ahn, J.-W., Jang, K. H., Yang, H.-C., Oh, K.-B., Lee, H.-S., Shin, J. (2007) Bithiazole metabolites from the Myxobacterium *Myxococcus fulvus*. Chem. Pharm. Bull., 55: 477–479.

113. Suzuki, Y., Ojika, M., Sakagami, Y., Fudou, R., Yamanaka, S. (1998) Cystothiazoles C-F, new bithiazole-type antibiotics from the myxobacterium *Cystobacter fuscus*. Tetrahedron, 54: 11399–11404.

114. Sasse, F., Böhlendorf, B., Hermann, M., Kunze, B., Forche, E., Steinmetz, H., Höfle, G., Reichenbach, H. (1999) Melithiazols, new β-methoxyacrylate inhibitors of the respiratory chain isolated from Myxobacteria. J. Antibiot., 52: 721–729.

115. Weissman, K. J., Müller, R. (2010) Myxobacterial secondary metabolites: bioactives and modes-of-action. Nat. Prod. Rep., 27: 1276–1295.

116. Colon, A., Hoffman, T. J., Gebauer, J., Dash, J., Rigby, J. H., Arseniyadis, S., Cossy, J. (2012) Catalysis-based enantioselective total synthesis of myxothiazole Z, (14S)-melithiazole G and (14S)-cystothiazole F. Chem. Commun., 48: 10508–10510.

117. Williams, D. R., Patnaik, S., Clark, M. P. (2001) Total synthesis of cystothiazoles A and C. J. Org. Chem., 66: 8463–8469.
118. Quadri, L. E. N., Keating, T. A., Patel, H. M., Walsh, C. T. (1999) Assembly of the *Pseudomonas aeruginosa* nonribosomal peptide siderophore pyochelin: *in vitro* reconstitution of aryl-4,2-bisthiazoline synthetase activity from PchD, PchE, and PchF. Biochemistry, 38: 14941–14954.
119. Brandel, J., Humbert, N., Elhabiri, M., Schalk, I. J., Mislin, G. L. A., Albrecht-Gary, A.-M. (2012) Pyochelin, a siderophore of *Pseudomonas aeruginosa*: Physiochemical characterization of the iron(III), copper(II) and zinc(II) complexes. Dalton Trans., 41: 2820–2834.
120. Hare, N. J., Soe, C. Z., Rose, B., Harbour, C., Codd, R., Manos, J., Cordwell, S. J. (2012) Proteomics of *Pseudomonas aeruginosa* Australian epidemic strain 1 (AES-1) cultured under conditions mimicking the cystic fibrosis lung reveals increased iron acquisition via the siderophore pyochelin. J. Proteome Res., 11: 776–795.
121. Chen, Z.-F., Tan, M.-X., Liu, L.-M., Liu, Y.-C., Wang, H.-S., Yang, B., Peng, Y., Liu, H.-G., Liang, H., Orvig, C. (2009) Cytotoxicity of the traditional chinese medicine (TCM) plumbagin in its copper chemistry. Dalton Trans., 10824–10833.
122. Padhye, S., Dandawate, P., Yusufi, M., Ahmad, A., Sarkar, F. H. (2012) Perspectives on medicinal properties of plumbagin and its analogs. Med. Res. Rev., 32: 1131–1158.
123. Quinoa, E., Adamczeski, M., Crews, P., Bakus, G. J. (1986) Bengamides, heterocyclic anthelmintics from a Jaspidae marine sponge. J. Org. Chem., 51: 4494–4497.
124. Towbin, H., Bair, K. W., DeCaprio, J. A., Eck, M. J., Kim, S., Kinder, F. R., Morollo, A., Mueller, D. R., Schindler, P., Song, H. K., van Oostrum, J., Versace, R. W., Voshol, H., Wood, J., Zabludoff, S., Phillips, P. E. (2003) Proteomics-based target identification: bengamides as a new class of methionine aminopeptidase inhibitors. J. Biol. Chem., 278: 52964–52971.
125. Lu, J.-P., Yuan, X.-H., Yuan, H., Wang, W.-L., Wan, B., Franzblau, S. G., Ye, Q.-Z. (2011) Inhibition of *Mycobacterium tuberculosis* methionine aminopeptidase by bengamide derivatives. ChemMedChem, 6: 1041–1048.
126. Newman, D. J., Cragg, G. M. (2012) Natural products as sources of new drugs over the 30 years from 1981–2010. J. Nat. Prod., 75: 311–335.
127. Lam, K. S. (2007) New aspects of natural products in drug discovery. Trends Microbiol., 15: 279–289.
128. Garson, M. J. (1993) The biosynthesis of marine natural products. Chem. Rev., 93: 1699–1733.
129. Demain, A. L. (1999) Pharmaceutically active secondary metabolites of microorganisms. Appl. Microbiol. Biotechnol., 52: 455–463.
130. Ganesan, A. (2008) The impact of natural products upon modern drug discovery. Curr. Opin. Chem. Biol., 12: 306–317.
131. Ming, L.-J. (2003) Structure and function of metalloantibiotics. Med. Res. Rev., 23: 697–762.
132. Priuska, E. M., Clark-Baldwin, K., Pecoraro, V. L., Schacht, J. (1998) NMR studies of iron-gentamycin complexes and the implications for aminoglycoside toxicity. Inorg. Chim. Acta, 273: 85–91.

133. Gaggelli, E., Gaggelli, N., Valensin, D., Valensin, G., Jeżowska-Bojczuk, M., Kozłowski, H. (2002) Structure and dynamics of the lincomycin-copper(II) complex in water solution by ^{1}H and ^{13}C NMR studies. Inorg. Chem., 41: 1518–1522.

134. Porath, J., Hansen, P. (1991) Cascade-mode multiaffinity chromatography: fractionation of human serum proteins. J. Chromatogr., 550: 751–764.

135. Kuehnbaum, N. L., Britz-McKibbin, P. (2013) New advances in separation science for metabolomics: resolving chemical diversity in a post genomic era. Chem. Rev., 113: 2437–2468.

136. Patti, G. J., Yanes, O., Siuzdak, G. (2012) Innovation metabolomics: the apogee of the omics trilogy. Nat. Rev. Mol. Cell Biol., 13: 263–269.

137. Johnson, C. H., Gonzalez, F. J. (2012) Challenges and opportunities of metabolomics. J. Cell. Physiol., 227: 2975–2981.

138. Mounicou, S., Szpunar, J., Lobinski, R. (2009) Metallomics: the concept and methodology. Chem. Soc. Rev., 38: 1119–1138.

139. Kersten, R. D., Dorrestein, P. C. (2009) Secondary metabolomics: Natural products mass spectrometry goes global. ACS Chem. Biol., 4: 599–601.

140. Rochfort, S. (2005) Metabolomics reviewed: a new "omics" platform technology for systems biology and implications for natural products research. J. Nat. Prod., 68: 1813–1820.

141. Phelan, V. V., Liu, W.-T., Pogliano, K., Dorrestein, P. C. (2012) Microbial metabolic exchange - the chemotype-to-phenotype link. Nat. Chem. Biol., 8: 26–35.

142. Ryan, D., Robards, K. (2006) Metabolomics: The greatest omics of them all? Anal. Chem., 78: 7954–7958.

143. Thiericke, R., Rohr, J. (1993) Biological variation of microbial metabolites by precursor-directed biosynthesis. Nat. Prod. Rep., 10: 265–289.

144. Bode, H. B., Bethe, B., Hofs, R., Zeeck, A. (2002) Big effects from small changes: possible ways to explore nature's chemical diversity. ChemBioChem, 3: 619–627.

145. Meiwes, J., Fiedler, H.-P., Zähner, H., Konetschny-Rapp, S., Jung, G. (1990) Production of desferrioxamine E and new analogues by directed fermentation and feeding fermentation. Appl. Microbiol. Biotechnol., 32: 505–510.

146. Ankenbauer, R. G., Staley, A. L., Rinehart, K. L., Cox, C. D. (1991) Mutasynthesis of siderophore analogues by *Pseudomonas aeruginosa*. Proc. Natl. Acad. Sci. USA, 88: 1878–1882.

147. Tschierske, M., Drechsel, H., Jung, G., Zähner, H. (1996) Production of rhizoferrin and new analogues obtained by directed fermentation. Appl. Microbiol. Biotechnol., 45: 664–670.

148. Soe, C. Z., Pakchung, A. A. H., Codd, R. (2012) Directing the biosynthesis of putrebactin or desferrioxamine B in *Shewanella putrefaciens* through the upstream inhibition of ornithine decarboxylase. Chem. Biodivers., 9: 1880–1890.

149. Ledyard, K. M., Butler, A. (1997) Structure of putrebactin, a new dihydroxamate siderophore produced by *Shewanella putrefaciens*. J. Biol. Inorg. Chem., 2: 93–97.

150. Capon, R. J. (2012) Biologically active natural products from Australian marine organisms, in: Tringali, C. (ed.) *Bioactive Compounds from Natural Sources*, pp. 579–602 (Boca Raton, FL, CRC Press).

151. Kim, T. K., Garson, M. J., Fuerst, J. A. (2005) Marine actinomycetes related to the 'Salinospora' group from the Great Barrier Reef sponge *Pseudoceratina clavata*. Environ. Microbiol., 7: 509–518.

152. Udwary, D. W., Zeigler, L., Asolkar, R. N., Singan, V., Lapidus, A., Fenical, W., Jensen, P. R., Moore, B. S. (2007) Genome sequencing reveals complex secondary metabolome in the marine actinomycete *Salinispora tropica*. Proc. Natl. Acad. Sci., USA, 104: 10376–10381.

153. Harvey, C. J. B., Publisi, J. D., Pande, V. S., Cane, D. E., Khosla, C. (2012) Precursor-directed biosynthesis of an orthogonally functional erythromycin analog: selectivity in the ribosome macrolide binding pocket. J. Am. Chem. Soc., 134: 12259–12265.

154. Lifa, T., Ejje, N., Codd, R. (2012) Coordinate-bond-dependent solid-phase organic synthesis of biotinylated desferrioxamine B: A new route for metal-specific probes. Chem. Commun., 48: 2003–2005.

155. Hou, Y., Braun, D. R., Michel, C. R., Klassen, J. L., Adnani, N., Wyche, T. P., Bugni, T. S. (2012) Microbial strain prioritization using metabolomics tools for the discovery of natural products. Anal. Chem., 84: 4277–4283.

156. Boughton, B. A., Callahan, D. L., Silva, C., Bowne, J., Nahid, A., Rupasinghe, T., Tull, D. L., McConville, M. J., Bacic, A., Roessner, U. (2011) Comprehensive profiling and quantitation of amine group containing metabolites. Anal. Chem., 83: 7523–7530.

157. Boettcher, T., Pitscheider, M., Sieber, S. A. (2010) Natural products and their biological targets: Proteomic and metabolomic labeling strategies. Angew. Chem., Int. Ed., 49: 2680–2698.

158. Carlson, E. E. (2010) Natural products as chemical probes. ACS Chem. Biol., 5: 639–653.

159. Visca, P., Imperi, F., Lamont, I. L. (2007) Pyoverdine siderophores: From biogenesis to biosignificance. Trends Microbiol., 15: 22–30.

160. Lamont, I. L., Beare, P. A., Ochsner, U., Vasil, A. I., Vasil, M. L. (2002) Siderophore-mediated signaling regulates virulence factor production in *Pseudomonas aeruginosa*. Proc. Natl. Acad. Sci. USA, 99: 7072–7077.

161. Meyer, J.-M., Van, V. T., Stintzi, A., Berge, O., Winkelmann, G. (1995) Ornibactin production and transport properties in strains of *Burkholderia vietnamiensis* and *Burkholderia cepacia* (formerly *Pseudomonas cepacia*). BioMetals, 8: 309–317.

162. Sharman, G. J., Williams, D. H., Ewing, D. F., Ratledge, C. (1995) Determination of the structure of exochelin MN, the extraceullular siderophore from *Mycobacterium neoaurum*. Chem. & Biol., 2: 553–561.

163. Dhungana, S., Miller, M. J., Dong, L., Ratledge, C., Crumbliss, A. L. (2003) Iron chelation properties of an extracellular siderophore exochelin MN. J. Am. Chem. Soc., 125: 7654–7663.

164. Sheldon, R. A. (2005) Green solvents for sustainable organic synthesis: state of the art. Green Chem., 7: 267–278.

165. Cue, B. W., Zhang, J. (2009) Green process chemistry in the pharmaceutical industry. Green Chem. Lett. Rev., 2: 193–211.

166. Sheldon, R. A. (2007) The E factor: fifteen years on. Green Chem., 9: 1273–1283.
167. Umezawa, H., Suhara, Y., Takita, T., Maeda, K. (1966) Purification of bleomycins. J. Antibiot., 19: 210–215.
168. Gupta, M., Caniard, A., Touceda-Varela, A., Campopiano, D. J., Mareque-Rivas, J. C. (2008) Nitrilotriacetic acid-derivatized quantum dots for simple purification and site-selective fluorescent labeling of active proteins in a single step. Bioconjugate Chem., 19: 1964–1967.

2

Metal Complexes as Tools for Structural Biology

Michael D. Lee, Bim Graham and James D. Swarbrick
Monash Institute of Pharmaceutical Sciences, Monash University, Australia

2.1   Structural Biological Studies and the Major Techniques Employed

With genomic data being annotated at an ever-increasing rate, structural analyses of proteins and nucleic acids, as well as their complexes and multimeric assemblies, are necessary to provide definitive insights into their *ex vivo* functions, which can ultimately be used to elucidate their roles in complex biological systems. Additionally, many of the major advancements in medical science and drug development have and will continue to depend critically on our ability to determine the three-dimensional (3D) structures of bio-macromolecules, most notably rational (structure-based) drug design.

Structural biologists currently have at their disposal a diverse and powerful range of tools and techniques that can be drawn upon to generate structural models. X-ray crystallography [1] and nuclear magnetic resonance (NMR) spectroscopy [2] provide high-resolution structures of proteins and nucleic acids with atomic level of detail. In conjunction with the structures of individual domains/molecules obtained via these methods, small-angle X-ray scattering (SAXS) [3], small-angle neutron scattering (SANS) [4], and cryo-electron microscopy [5] can yield lower resolution structures of larger multimeric assemblies. Similarly, several "sparse" data techniques exist that deliver few, but important, long-range insights into the interaction and arrangement of subunits within such assemblies. These include chemical cross-linking combined with mass spectrometry (CX-MS) [6], electron paramagnetic resonance (EPR) spectroscopy [7], and luminescence resonance energy transfer (LRET) analysis [8].

Inorganic Chemical Biology: Principles, Techniques and Applications, First Edition. Edited by Gilles Gasser.
© 2014 John Wiley & Sons, Ltd. Published 2014 by John Wiley & Sons, Ltd.

Central to the success of all of these techniques has been the development of a range of computational algorithms and tools to generate 3D models that are consistent with the various types of experimental data. In the absence of experimental data, *in silico* simulations (e.g., homology modeling, molecular dynamics, small-molecule docking, and protein folding) can also be effective in determining likely structures [9]. Moreover, great advances in computational power, data processing and analysis, in conjunction with synchrotron light sources and higher field NMR spectrometers, are increasing the speed of structure determination [10]. *Ab initio* computational simulations currently reach beyond the millisecond timescale and are working alongside, for example, NMR relaxation studies to reveal key insights into the important role of dynamics in bio-macromolecular assembly and function, the most spectacular example being elucidation of the pathways via which proteins fold [11].

Despite the huge number of protein and nucleic acid structures that have been determined (with the Protein Data Bank (PDB) approaching 100 000 entries) [12], there are several important classes of proteins, in particular, that remain notoriously difficult to structurally characterize with respect to NMR and/or X-ray methods. Aside from the molecular weight limitation for NMR, there are many factors contributing to this, ranging from low expression levels of recombinant proteins in heterologous cell lines or strains, through to intrinsic disorder or difficulties in crystallization due to the presence of highly dynamic regions within the structures. A class of protein that poses challenges at all levels – expression, solubilization, purification, and crystallization – are those that require a membranous environment to fold and function. Perhaps the most prominent amongst these are the G protein-coupled receptors (GPCRs), which, together with kinases, are the most heavily investigated drug targets in the pharmaceutical industry. GPCR-targeting drugs account for the majority of best-selling drugs and about 40% of all prescription pharmaceuticals on the market [13]. It is known that binding of drugs and endogenous ligands to GPCRs can induce complex changes in receptor topology for coupling to signal transduction. The structural characterization of GPCRs (and other integral membrane-bound proteins), including establishment of the structural basis behind their function, therefore presents researchers with one of today's most important and difficult challenges. There is a clear need for new technologies, or improvements in existing technologies, to address this challenge, as well as many other important problems in biology and medicine requiring a detailed knowledge of bio-macromolecular structure.

2.2   What do Metal Complexes have to Offer the Field of Structural Biology?

Metal complexes have, and will continue to find use as tools to facilitate structural analyses of proteins and nucleic acids via a number of the techniques mentioned above. In some cases, this has seen a dramatic "step-change" in terms of both the *range* of biological systems amenable to study, as well as the *quality* of the structural information that may be obtained. Within X-ray crystallographic studies, metal complexes have been employed for heavy-atom derivatization of proteins, enabling extraction of phase information, whilst NMR spectroscopic analysis of proteins and other macromolecules has benefited greatly from the introduction of so-called "lanthanide-binding tags". Such tags have proven particularly valuable for the collection of long-rang distance and angular restraints, as well as resolving heavily overlapped protein NMR spectra. Selected luminescent lanthanide complexes have found application as "donors" in fluorescence resonance energy transfer (FRET) spectroscopic measurements, greatly extending the upper limit and accuracy of the distances that can be measured via this technique. Most recently, the use of half-integer high-spin metal complexes as alternative spin labels to the traditionally used nitroxide radicals has seen a marked enhancement in the sensitivity and range of EPR-based distance measurements. The background behind each of these applications is discussed in the following sections, together with representative examples that serve to illustrate the benefits that may derive from employing metal complexes as tools for structural studies of biomolecules.

2.3   Metal Complexes for Phasing in X-ray Crystallography

The highest resolution structures of bio-macromolecules and their assemblies are obtained using X-ray crystallography, in which a crystalline sample is exposed to a focused beam of X-rays to generate a 3D map of electron density that is then used to infer the coordinates of the individual atoms. The electron density, $\rho(x, y, z)$, at a particular point, (x, y, z), in the unit cell of a crystal is given by Equation 2.1:

$$\rho(x, y, z) = \frac{1}{v} \sum_{h} \sum_{k} \sum_{l} |F(h, k, l)| \cos 2\pi \{hx + ky + lz - \phi(h, k, l)\} \qquad (2.1)$$

where V is the crystal volume, $F(h, k, l)$ is the amplitude of the wave which is proportional to the square root of the intensity (measured during the crystallographic experiment as a reflection or "spot", and referred to as a "structure factor"), h, k, l are the Miller indices of the diffracted beams (the reciprocal points, which specify the measured location of the reflection on the detector), and $\phi(h, k, l)$ represents the phases of the structure factors [1]. Unfortunately, whilst the spot positions and intensities, as well as the crystal volume, can be determined, the phases of the different diffracted beams are not measurable during data collection. Thus, in order to calculate the electron density, and therefore to determine the coordinates of the atoms inside the unit cell, the so-called "phase problem" must be addressed.

For crystals composed of bio-macromolecules, the phase problem can be solved in one of three ways: (i) by introducing atoms in the structure with high scattering power (referred to as multiple isomorphous replacement, MIR); (ii) by introducing atoms that scatter X-rays anomalously (single-wavelength anomalous diffraction, SAD, and multi-wavelength anomalous diffraction, MAD); and (iii) by using the known structure of a homologous molecule as an initial model for the macromolecule under investigation, which is subsequently refined (molecular replacement, MR) [1]. With regards to the first two methods, biosynthetic incorporation of selenomethionine has proven an invaluable strategy for introduction of the necessary "heavy atoms" [14], however metal complexes have also featured prominently amongst the classes of heavy-atom reagents employed for phasing [15, 16]. The high scattering power of the metal ions within these complexes compared with that of the atoms within a bio-macromolecule changes the intensities of the diffraction pattern when compared with that of the native macromolecule. These differences in intensity between the two spectra are used to generate a map of interatomic vectors between the heavy atom positions (a Patterson map), from which it is relatively easy to determine their coordinates within the unit cell.

The range of metal complexes employed as phasing agents in X-ray crystallography include simple aquo complexes of metal ions that interact electrostatically with bio-macromolecules, complexes that coordinate to specific groups within proteins or nucleic acids, multi-metal cluster complexes, and complexes that can be covalently linked to bio-macromolecules via coordinating ligands [15]. Depending on the complex, these may be introduced through either soaking crystals of a protein/nucleic acid with a solution of the complex, co-crystallization of the two species, or covalent reaction between the complex and bio-macromolecule prior to crystallization. In the past, K_2PtCl_4, $KAu(CN)_2$, K_2HgI_4, $UO_2(OAc)_2$, $HgCl_2$, $K_3UO_2F_5$, *p*-chloromercury benzoic sulfate, trimethyllead acetate, and ethylmercury thiosalicylate (thiomersal) have proven particularly popular as phasing agents [15]. These complexes preferentially bind to the *S*-containing side-chains of cysteine and methionine residues, and the imidazole side-chain of histidine residues, within proteins.

The intense and sharp "white lines" of the L_{II} and L_{III} absorption edges of lanthanides (associated with the absorption of radiation of sufficient energy to promote electrons from the $2p(^2P_{1/2})$ and $2p(^2P_{3/2})$ orbital states to the continuum state), combined with the development of third-generation synchrotron radiation sources, have also proven useful for the derivation of phase information. Kahn and coworkers, for example, reported successful *de novo* phasing via SAD using a series of gadolinium(III) complexes that pack within the crystal lattice of a protein (e.g., **1** in Fig. 2.1) [17]. These complexes have since been employed by a number of other groups. Hermoso and coworkers, for example, employed **1** to help solve the structure of CbpF, a bifunctional choline-binding protein and autolysis regulator from *Streptococcus pneumonia* [18], whilst Mayer and coworkers determined the crystal structure of the *Pyrococcus abyssi* Pab87 peptidase protein with the aid of the anomalous signal of the corresponding dysprosium(III) complex [19]. Wiener and coworkers have described two thiol-reactive chelates, **2** and **3** (Fig. 2.1), suitable for incorporation of lanthanides into protein crystals and SAD-based phasing [20], whilst Imperiali and coworkers have developed genetically-encodable peptide-based lanthanide-binding tags that may be used for this purpose [21].

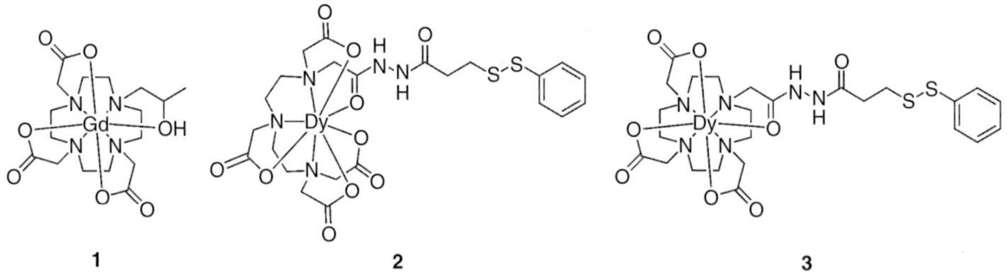

Figure 2.1   *Examples of lanthanide(III) complexes that have been used for phasing protein X-ray diffraction data*

A particularly novel strategy for the introduction of well-ordered heavy metal ions into proteins has been recently proposed and demonstrated by Schultz and coworkers, involving the genetic incorporation of metal-chelating non-natural amino acids into recombinantly-expressed proteins [22]. This approach makes use of an orthogonal amber suppressor tRNA/tyrosyl-tRNA synthetase pair evolved from *Methanococcus jannaschii* to install non-natural amino acids in response to the amber stop codon [23]. Using *O*-acetylserine sulfhydrylase as a test protein and 2-amino-3-(8-hydroxyquinolin-3-yl) propanoic acid (HQ-Ala, **4**) as a non-natural amino acid, Schultz and coworkers were able to successfully immobilize a zinc(II) ion onto the surface of the protein and to employ this for SAD phasing during crystallographic determination of the protein's structure (Fig. 2.2).

2.4 Metal Complexes for Derivation of Structural Restraints via Paramagnetic NMR Spectroscopy

Many bio-macromolecules and their complexes with other biomolecules or ligands can be difficult or impossible to crystallize in a form suitable for the collection of high-quality diffraction data, hindering or precluding analysis by X-ray crystallography. The dynamic nature of proteins, protein complexes, and other bio-macromolecular assemblies in solution can also significantly hamper efforts to study these systems by crystallography. It is partly for these reasons that NMR spectroscopy holds such a pre-eminent position amongst the arsenal of techniques available to structural biologists.

NMR spectroscopy, as applied to structural biology, typically involves the running of several multi-dimensional (2D–4D) experiments to assign the chemical shifts of all atoms, together with nuclear Overhauser effect spectroscopy (NOESY) experiments to identify all protons that are in close spatial proximity (<5–6 Å) [2]. The NMR structure of a macro-molecule is then "folded" from a randomly extended strand by calculating an ensemble

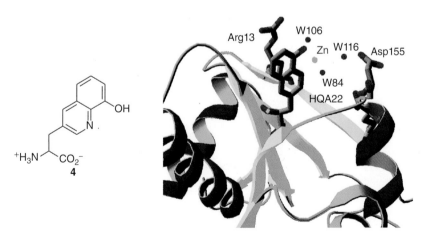

Figure 2.2 *The metal ion-chelating non-natural amino acid, HQ-Ala (**4**), and the structure of an O-acetylserine sulfhydrylase mutant containing a Zn(II) center coordinated by a genetically incorporated HQ-Ala residue. The Zn(II) is coordinated by the N and deprotonated O of the HQ-Ala side-chain. Adapted with permission from [22] © 2009 American Chemical Society*

of models that conform to these distance restraints. Included in these calculations are any other restraints derived from a variety of parameters recorded in a wide range of different experiments, most commonly angle restraints, particularly those of the backbone *Phi* and *Psi* torsional angles. A simplified atomistic force field, supplemented with a number of potential energy terms for each restraint type, is required that maintains the correct covalent geometry of the atoms (bond lengths and angles, e.g., tetrahedral sp^3 carbons) and van der Waals repulsion during the calculation procedure. Improvements in computational algorithms and hardware technology have advanced biological NMR spectroscopy considerably over the years. Nevertheless, in its "traditional" form, it is still generally limited to the analysis of systems of less than 35 kDa in molecular weight (notwithstanding the molecular weight limitation of high-resolution NMR structures, backbone atoms of systems in excess of 100 kDa can be readily assigned and the side-chain methyl atoms in large symmetric assemblies over 300 kDa have been achieved; both give rise to unrivalled site-specific "fingerprint" maps containing many spy nuclei to probe changes in structure and/or dynamics of large biomolecules) [24]. Beyond this limit, NMR signals broaden substantially and the sensitivity of most experiments rapidly diminishes. Furthermore, heavily overlapping signals, typically found in spectra of high molecular weight or inherently unstructured bio-macromolecules, are challenging to assign, as is unraveling intermolecular distance data for symmetric oligomers.

The recent advent of "paramagnetic NMR spectroscopy", which exploits the unique magnetic properties of paramagnetic metal ions, particularly those from the lanthanide series, can potentially extend the scope of the NMR technique in structural biological studies [25–28]. The unpaired electrons in the 4f orbitals of paramagnetic lanthanide ions have gyromagnetic ratios almost 1000 times higher than that of a proton, which leads to a large magnetic susceptibility [29]. When bound to a protein or other macromolecule in a suitable fashion (*vide infra*), this can give rise to a number of effects in the NMR spectrum. Extraction and quantification of these often quite spectacular effects provides additional sets of restraints that can be used for structure refinement and in studies of protein–protein interactions [30, 31] and protein–small molecule docking (including fragment-based lead generation) [32, 33]. The real power of paramagnetic NMR spectroscopy, however, stems from the long-range nature of these effects (up to 40 Å or more from the metal ion) [34], which enables the structural analysis of truly large and/or previously intractable systems.

There are three main effects that can be observed upon the attachment of a paramagnetic lanthanide ion to a biomolecule: paramagnetic relaxation enhancement (PRE), residual dipolar coupling (RDC), and pseudo-contact shifts (PCS) [25–28]. The impacts of these effects on the NMR spectrum are governed by the magnetic susceptibility tensor of the lanthanide ion, which may be either isotropic (as for gadolinium) or anisotropic (as for all other paramagnetic lanthanides), depending on the number and arrangement of the unpaired electrons within the ion.

2.4.1 Paramagnetic Relaxation Enhancement (PRE)

PRE refers to the enhanced rate of relaxation of nuclear spins caused by dipolar coupling with the unpaired electrons of a paramagnetic lanthanide ion [35]. It results in a distance-dependent broadening of NMR resonances, with the degree of broadening proportional to $1/r^6$, where r is the distance between the nuclei and paramagnetic center. The half-filled 4f orbitals of gadolinium provide the largest isotropic magnetic susceptibility and largest PRE effect of all lanthanides. The quantification of PREs can provide short-to-moderate distance restraints (<15 Å) between nuclei and the paramagnetic center.

2.4.2 Residual Dipolar Coupling (RDC)

When a paramagnetic lanthanide ion with an anisotropic magnetic susceptibility is bound to a bio-macromolecule, rather than tumbling randomly in solution, the molecule may slightly align within the magnetic field. This anisotropic tumbling, characterized by the alignment tensor, results in measurable dipolar couplings between nuclear spins (in the order of a few-to-10s of Hz) [36]. The RDCs manifest as a change in the total coupling constant in the presence (J + RDC) and absence (J) of a paramagnetic-induced alignment, and can be positive, negative or zero. The size and sign of RDCs are easily determined between a pair of heteronuclear spins, typically from an amide bond (in the case of proteins) with large J coupling (e.g., $J^1_{HN-}{}^{15}_N = 93\,Hz$). Furthermore, as the ^{15}N and ^1HN distance is fixed, the $^1H–^{15}N$ RDCs are dependent only on the angle that the N–H bond vector makes with the alignment tensor, and thus yield long-range angular restraints for any amide bond that is not otherwise broadened by PRE due to proximity to the lanthanide metal ion.

2.4.3 Pseudo-Contact Shifts (PCS)

PCS are produced by paramagnetic lanthanides with anisotropic magnetic susceptibilities and manifest as a change in the chemical shift of nuclei from the paramagnetic sample compared to that in the diamagnetic sample. As the magnitude is proportional to $1/r^3$, quantification of PCS can yield long-range distance information, much greater than that derived from PRE ($1/r^6$ dependence). Thus, for strongly paramagnetic lanthanide ions with large anisotropic susceptibility tensors, such as Dy^{3+}, Tb^{3+}, and Tm^{3+}, PCS can be observed for nuclei over 40–50 Å from the metal center [34]. PRE-broadening effects are considerable for these inherently strong paramagnetic lanthanides, for example, for Dy^{3+} ions, 2D ^{15}N HSQC peaks from amides within about 15 Å of the metal are broadened beyond detection, leading to a characteristic PRE "invisible zone" in the spectra. PCS from nuclei close to the lanthanide can be obtained, however, from weakly paramagnetic lanthanide ions with smaller inherent anisotropic susceptibilities, such as Eu^{3+} or Ce^{3+}. Unlike PRE, the size and sign of the PCS depends on both the distance and the *angle* the nuclei make with respect to the susceptibility tensor. The distance and angular dependence of the PCS can be conveniently described by calculating an isosurface representation of the anisotropic magnetic susceptibility tensor showing regions in 3D space of equal PCS (Fig. 2.3).

In practice, PCS trace out parallel lines in the 2D spectra of a lanthanide ion-tagged protein (Fig. 2.3) because the ^{15}N and the attached ^1HN spin have very similar individual PCS owing to their close proximity. The assignment of PCSs can be challenging, but is greatly facilitated by the superposition of several ^{15}N HSQC spectra, each recorded from the attachment of a different lanthanide ion. Calculation of the anisotropic susceptibility tensor parameters from analysis of at least eight PCS for each of these metal ions thus gives access to a unique metal-centered coordinate system to characterize the 3D position of the spin by virtue of its measured PCS. This is the basis behind the structural studies referred to in the following section.

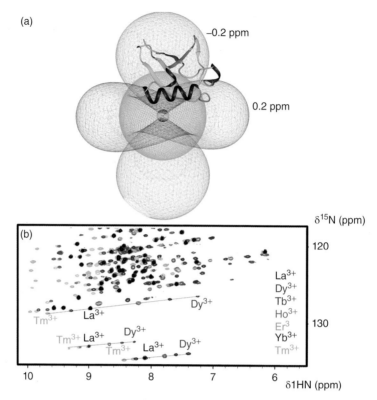

Figure 2.3 *(a) Isosurface representation of the anisotropic magnetic susceptibility tensor (pink and blue) and PRE-broadened zone (green) for a lanthanide ion (Dy³⁺; yellow) bound to a protein (ubiquitin). The NMR signals of nuclei whose coordinates fall on the tensor isosurfaces experience PCS of ±0.2 ppm, with the signals of those nuclei on the pink isosurface shifted in the opposite direction to those on the blue one. (b) Superimposed ¹H–¹⁵N HSQC NMR spectra for a protein (ubiquitin) tagged with various paramagnetic lanthanide(III) ions or lanthanum(III) as a diamagnetic reference*

2.4.4 Strategies for Introducing Lanthanide Ions into Bio-Macromolecules

The first examples of the exploitation of paramagnetic lanthanide ions in protein NMR spectroscopy were limited to proteins featuring a natural metal ion binding site, in which the native metal ion could be substituted for a paramagnetic lanthanide ion. Lee and Sykes, for example, exchanged the Ca^{2+} ions from the EF-hand calcium-binding domains of a number of parvalbumin proteins with paramagnetic Yb^{3+} ions and, using NMR spectroscopy, were able to reveal small structural differences in the positions of the nuclei within the EF domains, below the level of the resolution of the X-ray data available at the time [37]. Otting and coworkers substituted lanthanide ions into the Mg^{2+}/Mn^{2+}-binding site of the ϵ subunit of *E. coli* polymerase III to rapidly derive the structure of this multi-protein complex using PCS data in combination with rigid-body docking (Fig. 2.4) [38].

Most proteins, however, do not contain natural metal ion binding sites. In order to harness the potential of paramagnetic lanthanide ions and generalize their usage in structural

Figure 2.4 *Anisotropic magnetic susceptibility tensor isosurfaces calculated for the ε (left) and θ (middle) subunits of E. coli polymerase III and the corresponding complex of the ε and θ subunits (right), determined by PCS-based rigid body docking. Reprinted with permission from [38] © 2006 American Chemical Society*

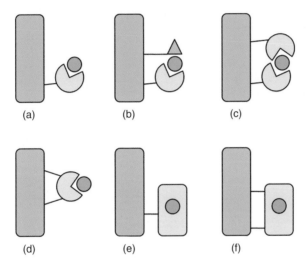

Figure 2.5 *Different strategies for introducing lanthanide ions into proteins (blue rectangles represent protein, yellow shapes chelating ligands/peptides, and green circles lanthanide ions): a) single-point conjugation of a stable lanthanide complex; b) single-point conjugation of a stable lanthanide chelate combined with coordination of an amino acid residue side-chain (blue triangle); c) two-point conjugation with a pair of lanthanide chelators bonded to a single metal ion; d) two-point conjugation of a stable lanthanide complex; e) single-point conjugation of a lanthanide-binding peptide; and f) two-point conjugation of a lanthanide-binding peptide*

NMR studies, a number of different strategies have been developed for introducing these ions into proteins in a site-specific fashion (Fig. 2.5). The most widely applicable of these involve the use of synthetic lanthanide complexes or chelating agents (lanthanide-binding tags, LBTs) that may be covalently linked (conjugated) to a bio-macromolecule's surface. Fig. 2.6 shows some representative examples.

In order to position a lanthanide ion at a specific site on a protein's surface, the ion must be bound by an appropriate set of donor atoms. In some scenarios, the ability to exchange a diamagnetic and paramagnetic metal ion in the same NMR experiment can be advantageous. For example, with the assistance of a nearby aspartic acid residue, Swarbrick

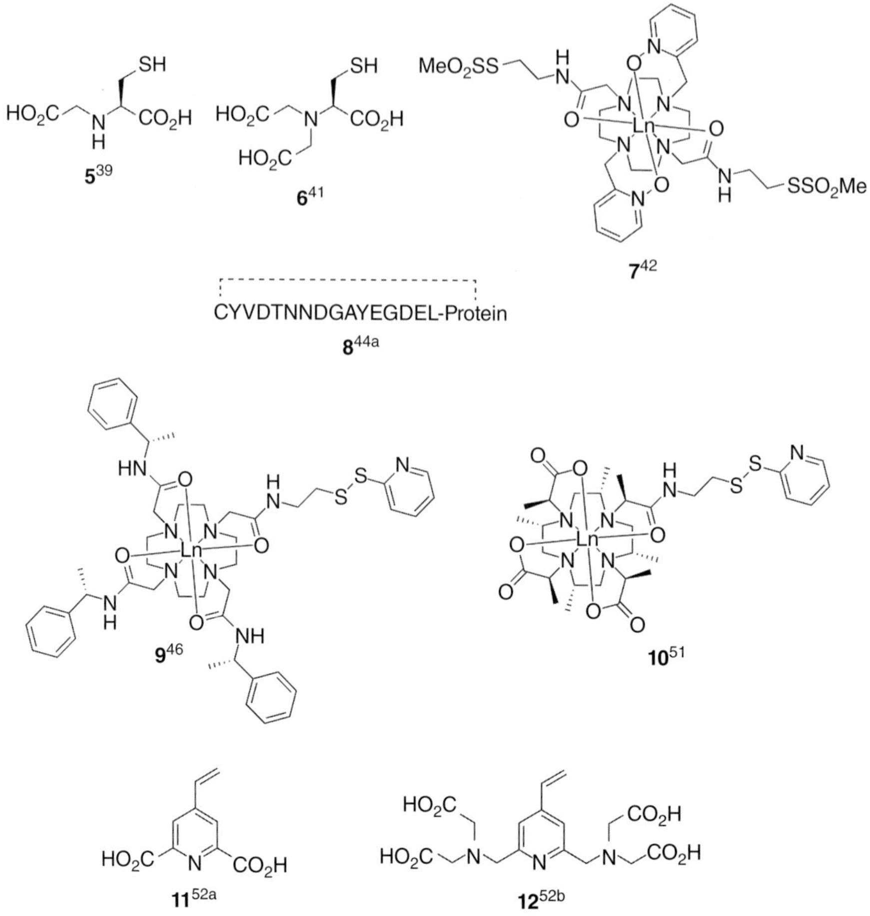

Figure 2.6 *Examples of synthetic lanthanide complexes and chelating agents (lanthanide-binding tags, LBTs) employed in paramagnetic NMR studies*

et al's LBT **5** (Fig. 2.6) captured and released lanthanide ions at an appropriate rate to allow exchange spectroscopy, which enabled PCS to be readily assigned with the aid of exchange cross-peaks (Fig. 2.7), as well as the measurement of PCS of slower-relaxing heteronuclei in positions close to the metal (where the proton PCS are broadened by PRE) [39]. More commonly, however, a highly stable lanthanide complex is desirable, to minimize the level of free paramagnetic metal ion in solution (a number of proteins do not tolerate free metal ions in solution or need to be studied in the presence of their own metal ion cofactors).

An important aspect of LBT design is that the lanthanide ion should be held as *rigidly* as possible with respect to the bio-macromolecule surface, particularly for the observation of sizeable PCSs and even more so for RDCs. This is because, unlike PRE, both RDC and PCS can take positive or negative values and increased motion of the bound lanthanide ion relative to the bio-macromolecule averages these values towards zero. Different strategies for achieving rigid lanthanide ion binding have been demonstrated (see Fig. 2.5).

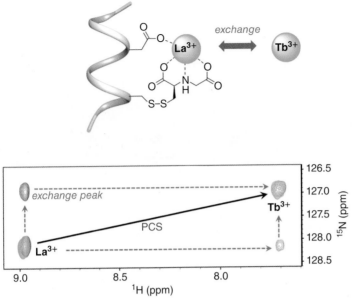

Figure 2.7 *Positioning of LBT **5** and an aspartic acid residue in an i, i + 4 arrangement within an α-helix leads to rigid, yet kinetically-labile binding of lanthanide ions (top). This allows exchange NMR spectroscopy to be carried out, in which protein spectra are recorded in the presence of a mixture of diamagnetic (e.g., La³⁺) and paramagnetic (e.g., Tb³⁺) ions. Signals for both the diagmagnetically- and paramagnetically-labeled protein are observed in the NMR spectrum, together with exchange cross-peaks (bottom). Reproduced with permission from [39] © 2011 Wiley-VCH Verlag GmbH & Co. KGaA, Weinheim*

As mentioned previously, the binding of lanthanides to LBT **5** is reliant on an aspartic acid in an $i + 4$ position relative to the tag, to cooperatively complex the lanthanide between the two coordinating groups [39, 40]. Similarly, two copies of LBT **6** (Fig. 2.6) in an i, $i + 4$ α-helical arrangement can complex lanthanides in a rigid fashion (Fig. 2.8) [41]. Alternatively, a single tag, that forms a tight complex with the lanthanide and is attached to the protein via two residues, can be used. A particularly noteworthy example is the 1,4,7,10-tetraazacyclododecane (cyclen)-based LBT **7** (known as CLaNP-5) of Ubbink and coworkers, which binds to proximal cysteine residues (Fig. 2.9) [42]. This tag has been successfully employed in the analysis of protein dynamics [43], and the structural analysis of a small molecule protein [32] and protein–protein complexes [30, 31].

Lanthanide binding peptides can be used to coordinate lanthanide ions in a well-defined geometry, which can then be linked to the protein of interest via a single or two points of conjugation to introduce rigidity (LBT **8** in Fig. 2.6) [44, 45]. By introducing a lanthanide binding peptide to galectin-3, for example, Prestegard and coworkers were able to measure and utilize PCSs and RDCs, as well as a single intermolecular NOE restraint, to determine the structure of the galectin-3-lactose complex (Fig. 2.10) [45].

Although the methods of lanthanide immobilization discussed here have proved successful, they require at least some structural knowledge of the target protein and potentially multiple mutations to the native protein sequence. Tags that bind lanthanides tightly and

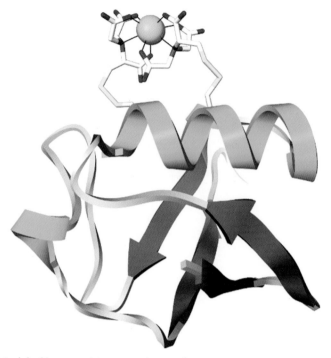

Figure 2.8 *Model of human ubiquitin with an Ln³⁺ ion immobilized via a pair of LBT **6**. The protein is shown as a ribbon, the side-chains of Cys24 and Cys28 with the attached LBTs as sticks, and the lanthanide ion as a light blue sphere. Reprinted with permission from [41] © 2011 Royal Society of Chemistry*

only require single-point attachment (e.g., LBTs **9–12** in Fig. 2.6) are simpler to use and potentially applicable to a greater range of proteins. The cyclen-based LBT **9**, for example, features bulky phenyl groups within its pedant arms, which impart rigidity once conjugated to a protein (Fig. 2.11) [46]. It has been successfully utilized in a number of studies, including PCS-based investigation of the conformational changes accompanying binding of inhibitors to the drug target, dengue virus NS2B-NS3 protease [47], and the binding of deoxyguanosine 5'-monophosphate to the R3H domain of the nucleic acid-binding protein, Sμbp-2 [48].

Several of the LBTs that have been developed are based on the well-known chelator, 1,4,7,10-tetraazacyclododecane-N,N',N'',N'''-tetraacetic acid (DOTA) (e.g., LBTs **7**, **9**, **10** in Fig. 2.6). DOTA forms extremely stable complexes with lanthanides, with dissociation constants ranging from 10^{-23} to 10^{-27} M. However, it is itself unsuitable for use in paramagnetic NMR spectroscopy since, in solution, its Ln(III) complexes exist as a mixture of interconverting stereoisomers (Fig. 2.12) [49, 50]. These differ in terms of the conformation of the cyclen ring (defined by the four NCCN torsion angles, which may be either δδδδ or λλλλ) and the helicity of the pendant arms (defined by the NCCO torsion angle as either Δ or Λ). When Ln(III)–DOTA complexes interact with chiral molecules such as proteins, the result is up to four distinct diastereomers, resulting in the generation of multiple PCS peaks in NMR spectra, which complicates PCS assignment. The chiral elements in LBTs such as LBT **9** and in LBT **10** of Häussinger *et al.* [51] serve to "freeze" these conformational

Figure 2.9 *Two-point attachment of LBT **7** to pseudoazurin via disulfide linkages to two prox-imal cysteine residues. The LBT is shown in stick form and the lanthanide ion as a pink sphere. Reprinted with permission from [25] © 2011 Elsevier*

exchange processes and preferentially favor the formation of a single isomer, resulting in single sets of PCSs. In the case of LBT **7**, the C_2 symmetry of the tag and two-point method of attachment restrict the LBT–protein to a single diastereomeric form [42].

To date, the vast majority of LBTs have been designed for conjugation to target proteins via cysteine-based chemistry. Most commonly, this involves formation of a disulfide bond between the tag and protein (LBTs **5–7**, **9**, and **10** in Fig. 2.6), though formation of a thioether linkage via a Michael addition reaction between an alkene and cysteine has also been used (LBTs **11** and **12**) [52]. These methods are reliant on the presence of a single solvent-exposed cysteine residue (or a pair in the case of two-point attachment), which may require mutations to introduce a cysteine(s) at an appropriate site, or the removal of cysteines for proteins that contain multiple cysteines. This may be problematic if another cysteine is required for catalytic activity. Furthermore, if the LBT is attached by a disulfide bond, problems arise if the protein must be studied under reducing conditions, or if the LBT severs any of the structure-stabilizing disulfide bonds within a protein by thiol–disulfide exchange. Screening libraries can contain ligands with disulfide-reactive groups (including thiols) that could potentially displace the LBT in a similar manner.

In view of the above issues, progress has been made in the development of other non-cysteine-based methods of introducing LBTs to proteins. Very recently, Otting and coworkers reported two alkyne-bearing tags, LBTs **13** and **14**, that can be attached to

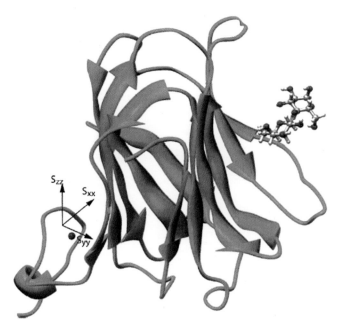

Figure 2.10   *Structure of galectin-3-lactose complex. A lanthanide-binding peptide has being introduced to galectin-3 with the lanthanide ion position indicated by a pink–purple sphere at the bottom left. Reprinted with permission from [45] © 2008 The Protein Society*

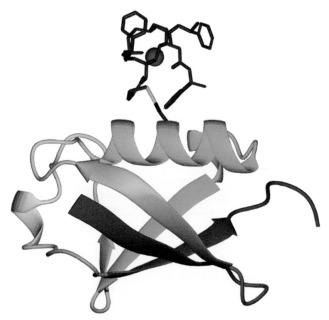

Figure 2.11   *Model of the LBT **9** bound to Cys28 in the ubiquitin A28C mutant. The lanthanide ion is represented by a magenta sphere, the disulfide bond of the linker is highlighted in yellow, and the side-chain of Cys28 and of the LBT are shown in stick form. Reprinted with permission from [46] © 2011 American Chemical Society*

Twisted Square Antiprism
Δ(δδδδ)

Square Antiprism
Λ(δδδδ)

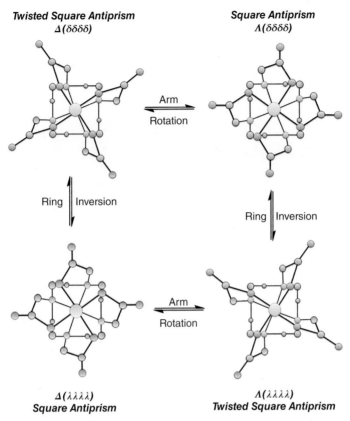

Arm
Rotation

Ring | Inversion

Ring | Inversion

Arm
Rotation

Δ(λλλλ)
Square Antiprism

Λ(λλλλ)
Twisted Square Antiprism

Figure 2.12 The four stereoisomeric forms of Ln(III)–DOTA. These interconvert by ring flipping or arm rotation. Adapted with permission from [50] © 2005 Royal Society of Chemistry

proteins in a site-specific manner via a "bio-orthogonal" Cu(I)-catalyzed azide-alkyne "click" reaction with a genetically-encoded p-azido-L-phenylalanine residue [53]. Although the resulting tethers to the protein are longer than the disulfide tethers afforded by LBTs such as **9** and **10** (Fig. 2.6), tags **13** and **14** were found to generate sizable PCSs with remarkable reliability. The use of a Cu(I)–stabilizing ligand, such as 2-[4-{(bis[(1-tert-butyl-1H-1,2,3-tri-azol-4-yl)methyl]amino)methyl}-1H-1,2,3-triazol-1-yl]acetic acid (BTTAA), is required to ensure good ligation yields and prevent binding of the Cu(I) by the protein.

Another alternative approach that has been explored is the use of lanthanide complexes as "shift reagents" that bind to proteins through specific, yet non-covalent interactions with proteins. Otting and coworkers, for example, have shown that the complex formed between three dipicolinic acid ligands (DPA) and a lanthanide ion (**15** in Fig. 2.14) can bind to a pair of adjacent and positively-charged amino acid residues whose charges are not compensated by negatively-charged residues nearby [54, 55]. Moreover, binding is, in effect, selective for a single site if a 1:1 complex-to-protein ratio is used, allowing determination of the magnetic susceptibility tensor and interpretation of the measured PCSs for structure analysis. The PCSs delivered by this approach are significantly smaller than those produced by many of the tags in Fig. 2.6, however the approach does have the advantage of

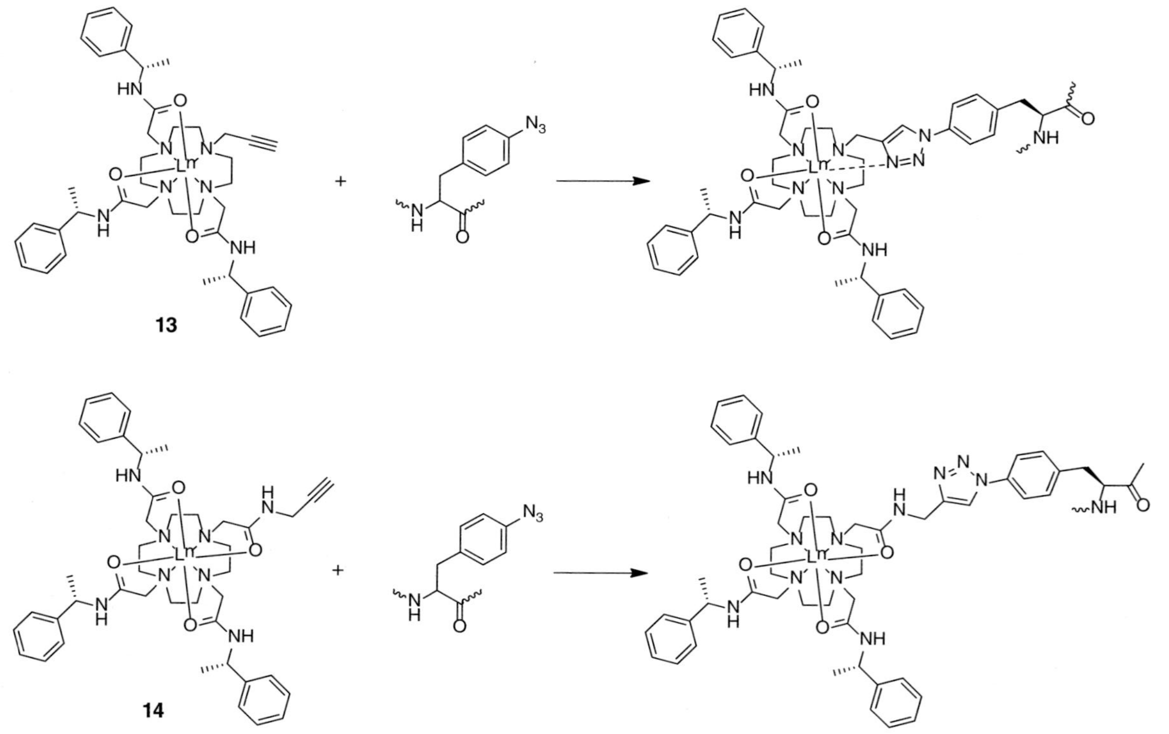

Figure 2.13 *"Clickable" LBTs that have been successfully incorporated into proteins via reaction with a genetically encoded p-azi-do-L-phenylalanine residue and used to generate PCS data* [53]

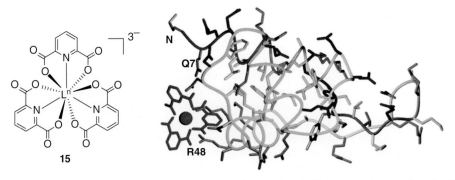

Figure 2.14 *The [Ln(DPA)₃]³⁻ complex (**15**) and a model of **15** (magenta) bound to ArgN. Adapted with permission from [54] © 2009 American Chemical Society*

not requiring any chemical manipulation of the protein prior to analysis and, unlike with other LBTs, PCS appear on the fast, rather than slow, chemical shift timescale and are thus easily assigned.

Complexes of some paramagnetic transition metal ions have also been employed in biological NMR studies, but to a lesser extent than LBTs. For example, Kojima and coworkers showed that the copper(II) complex of iminodiacetic acid coordinates to histidine residues on the surface of proteins and can be used to provide long-range distance information [56], whilst Bertini [57] and Jaroniec and coworkers [58] have demonstrated that long-range, up to about 20 Å, electron–nucleus distance restraints can be obtained using spinning solid-state nuclear magnetic resonance (SSNMR) techniques and proteins doped with paramagnetic transition metal ions or modified with covalently-attached paramagnetic transition metal–EDTA complexes. Recently, Otting and coworkers reported the successful generation of PCS in protein NMR spectra using a genetically-encoded cobalt(II)-binding non-natural amino acid featuring a bipyridyl side-chain [59].

2.5 Metal Complexes as Spin Labels for Distance Measurements via EPR Spectroscopy

Structural characterization of homo-dimers, homo-oligomers, and bio-macromolecular complexes via X-ray crystallography or NMR spectroscopy becomes increasingly harder as the systems become larger and more complex. For this reason, there has been increasing interest in the development of distinct, yet complementary methodologies to obtain additional structural restraints. Long-range distance restraints provided by EPR-based techniques such as double electron–electron resonance (DEER, also known as pulsed electron double resonance, PELDOR) are particularly useful for structure validation, for filtering candidate structures of macromolecular assemblies, or as additional constraints for *de novo* structure determination [7, 60]. DEER experiments involve the detection of the dipolar interaction between unpaired electrons through perturbation of the EPR spectrum at one microwave frequency (the "pumping" frequency) and observation of the effects on the EPR spectrum at a second frequency. This allows the distance between the unpaired electrons to be accurately determined.

DEER experiments have traditionally employed proteins tagged with pairs of nitroxide spin labels [7, 60]. Recently, however, a new class of spin labels based on half-integer high-spin gadolinium(III) and manganese(II) chelates have been proposed, and their use demonstrated in a number of model systems [61–65]. The advantage of these metal complex-based labels is that they allow nanometer-scale measurements (up to 8 nm or more) to be carried out at W-band (95 GHz), Q-band (about 34 GHz) and K-band (about 30 GHz) frequencies, at which the intrinsic sensitivity of DEER measurements is greatly enhanced (measurements with nitroxide labels are most commonly carried out at X-band (about. 9.5 GHz, 0.35 T) frequency, since anisotropy effects hamper their use at the higher frequencies). Thus, for example, a Gd(III)–Gd(III) distance of 5.73 nm has been measured at K-band frequency within a DNA duplex bearing a tag at the end of each strand [62]. W-band DEER experiments have also been reported using LBT **5** (Fig. 2.6), which demonstrated that outstandingly accurate Gd(III)–Gd(III) distance measurements in the 6 nm range may be made within protein assemblies available in sub-nanomole amounts (using the homodimeric protein ERp29 as a model system) (Fig. 2.15) [64].

2.6 Metal Complexes as Donors for Distance Measurements via Luminescence Resonance Energy Transfer (LRET)

One of the most commonly used spectroscopic methods for measuring distances in bio-macromolecules and their assemblies is FRET, in which a donor chromophore, initially in its electronic excited state, transfers energy to an acceptor chromophore

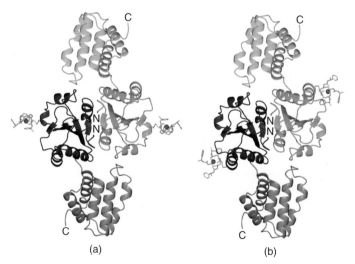

(a)   (b)

Figure 2.15   *The dimer structure of ERp29 with the Gd(III)–**9** tag modeled at (a) position 114 and (b) position 147. The protein monomers are shown in blue and green, using darker colors for the N-terminal domain, and the tags are drawn in stick form (gray) with magenta balls identifying the Gd^{3+} ions. The modeled Gd(III)–Gd(III) distances agree exceptionally well with those measured via DEER experiments. Reprinted with permission from [64] © 2011 American Chemical Society*

through non-radiative dipole–dipole coupling [66]. In order for FRET to occur, the emission spectrum of the donor must overlap with the excitation spectrum of the acceptor. The rate of energy transfer, k_T, is then given by Equation 2.2:

$$k_\tau = \frac{1}{\tau_D}\left(\frac{R_0}{r}\right)^6 \tag{2.2}$$

where τ_D is the decay time of the donor in the absence of the acceptor, r is the distance between the donor and acceptor, and R_0 is the distance between the donor and acceptor for which the energy transfer is 50% of the maximum value, the so-called Förster radius [67], given by Equation 2.3:

$$R_0 = 0.211[k^2 n^{-4} Q_D J(\lambda)]^{1/6} \tag{2.3}$$

where n is the refractive index of the medium, Q_D is the quantum yield of the donor in the absence of the acceptor, $J(\lambda)$ is the spectral overlap of the donor–acceptor pair, and k^2 is the dipole orientation factor, usually assumed to be 2/3 [66]. The dependence of k_T on the inverse sixth power of the intermolecular separation is what renders FRET suitable for measuring distances on the bio-macromolecular scale (1–7 nm). FRET may be detected as a decrease in the fluorescence intensity, lifetime or anisotropy of the donor, or as an increase in the fluorescence intensity of the acceptor (if it is fluorescent).

A modification of conventional FRET that allows the measurement of longer distances with greater accuracy is LRET, which involves the use of luminescent terbium(III) or europium(III) complexes as donors [8]. Such complexes are characterized by their exceedingly long luminescence lifetimes (of the order of milliseconds), their large Stokes' shifts (difference between absorption and emission maxima) and narrow emission bands (<10 nm at half-maximum), which offer a number of advantages over the organic fluorophores classically employed as FRET donors [8]. Firstly, the long luminescence lifetime of lanthanide donors greatly extends the lifetime of sensitized emission from the acceptor, enabling temporal discrimination of this emission from nanosecond timescale background fluorescence (sample auto-fluorescence and directly excited acceptor fluorescence) via introduction of a short delay (≥100 ns) between sample excitation and detection. This drastically improves signal-to-noise (by 50-fold or more), enabling much larger distances to be measured, and also renders FRET lifetime measurements much more accurate. Secondly, the long Stokes' shifts of lanthanide complexes reduces "bleeding" of the excitation light into the measurement, and their sharp emission bands provide several appropriate wavelengths for performing measurements with minimal contamination from donor emission. Finally, the orientation dependence of energy transfer is minimized by the donor's multiple electronic transitions and long lifetime, limiting uncertainty in the measured distance due to orientation effects. By way of example, illustrating the utility of LRET in complex macromolecular structural studies, the thiol-reactive complex **16** (Fig. 2.16) has been successfully employed in LRET experiments to measure voltage-dependent distance changes in a potassium channel protein [68], conformational changes in myosin [69] and in AE1 protein homodimer from red blood cells [70], and distances within a DNA double helix [8] and an RNA polymerase complex [71].

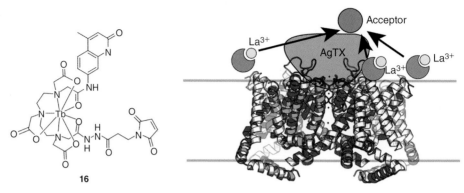

Figure 2.16   *A luminescent terbium(III) complex,* **16**, *employed as an LRET donor for distance measurements within biomolecules and their complexes. Reprinted with permission from [72] © 2010 Elsevier*

2.7   Concluding Statements and Future Outlook

As illustrated in this chapter, metal complexes have been utilized in conjunction with a number of the major techniques employed in structural biology, either to facilitate data analysis (X-ray crystallography) or to extend the capabilities of these techniques beyond their existing limits (NMR and EPR spectroscopy and FRET). In the case of X-ray crystallography and LRET, the use of metal complexes has now become commonplace. The utility of paramagnetic metal complexes in NMR and EPR spectroscopy has not quite yet gained the same level of familiarity amongst the structural biology community, however this situation is expected to change fairly dramatically within the next few years as LBTs become increasingly available. Akin to the emergence of supramolecular chemistry ("chemistry beyond the molecule") in the later half of the 20th century, we are now witnessing structural biology take its first definitive steps "beyond the bio-macromolecule" and towards the study of the structural changes that occur as groups of proteins and/or nucleic acids interact together to achieve particular functions. It is almost certain that paramagnetic metal complex-based reagents, with their unique ability to provide accurate long-range restraints, will have a major role to play in this exciting new era of structural biology.

References

1. a) Albrecht Messerschmidt. *X-Ray Crystallography of Biomacromolecules: A Practical Guide*, Wiley-VCH Verlag GmbH: Weinheim, 2007 b) Jan Drenth, *Principles of Protein X-ray Crystallography*, 3rd edition, Springer: New York, 2007.
2. a) Gerhard Wider. Structure determination of biological macromolecules in solution using nuclear magnetic resonance spectroscopy. *Biotechniques*, **2000**, *29*, 1278–1282 b) Michael Bieri, Ann H. Kwan, Mehdi Mobli, Glenn F. King, Joel P. Mackay, Paul R. Gooley. Macromolecular NMR spectroscopy for the non-spectroscopist: beyond macromolecular solution structure determination. *FEBS Journal*, **2011**, *278*, 704–715.

3. Alexander Grishaev, Justin Wu, Jill Trewhella, and Ad Bax. Refinement of multido-main protein structures by combination of solution small-angle X-ray scattering and NMR data. *J. Am. Chem. Soc.*, **2005**, *127*, 16621–16628.

4. Maxim V. Petoukhov, and Dmitri I. Svergun. Analysis of X-ray and neutron scattering from biomacromolecular solutions. *Curr. Opin. Struct. Biol.*, **2007**, *17*, 562–571.

5. Joachim Frank. *Three-Dimensional Electron Microscopy of Macromolecular Assemblies*. Oxford University Press: New York, 2006.

6. David L. Tabb. Evaluating protein interactions through cross-linking mass spectrometry. *Nat. Methods*, **2012**, *9*, 879–881.

7. Gunnar Jeschke, and Yevhen Polyhach. Distance measurements on spin-labelled biomacromolecules by pulsed electron paramagnetic resonance. *Phys. Chem. Chem. Phys.*, **2007**, *9*, 1895–1910.

8. a) Paul R. Selvin, and John E. Hearst. Luminescence energy transfer using a terbium chelate: Improvements on fluorescence energy transfer. *Proc. Natl. Acad. Sci. USA*, **1994**, *91*, 10024–10028 b) Paul R. Selvin, Tariq M. Rana, and John E. Hearst. Luminescence resonance energy transfer. *J. Am. Chem. Soc.*, **1994**, *116*, 6029–6030.

9. a) Narayanan Eswar, Ben Webb, Marc A. Marti-Renom, M.S. Madhusudhan, David Eramian, Min-yi Shen, Ursula Pieper, and Andrej Sali. Comparative Protein Structure Modeling with MODELLER. In *Current Protocols in Bioinformatics*, Hoboken: John Wiley & Sons, Inc., Supplement 15, 5.6.1–5.6.30, **2006**; b) James C. Phillips, Rosemary Braun, Wei Wang, James Gumbart, Emad Tajkhorshid, Elizabeth Villa, Christophe Chipot, Robert D. Skeel, Laxmikant Kale, and Klaus Schulten. Scalable molecular dynamics with NAMD. *J. Comp. Chem.*, **2005**, *26*, 1781–1802; c) Douglas B. Kitchen, Hélène Decornez, John R. Furr, and Jürgen Bajorath. Docking and scoring in virtual screening for drug discovery: methods and applications. *Nat. Rev. Drug Discov.*, **2004**, *3*, 935–949.

10. a) Janet L. Smith, Robert F.Fischetti, and Masaki Yamamoto. Micro-crystallography comes of age. *Curr. Opin. Struct. Biol.*, **2012**, *22*, 602–612; b) Torsten Herrmann, Peter Güntert, and Kurt Wüthrich. Protein NMR structure determination with automated NOE assignment using the new software CANDID and the torsion angle dynamics algorithm DYANA. *J. Mol. Biol.*, **2002**, *319*, 209–227.

11. a) Kresten Lindorff-Larsen, Paul Maragakis, Stefano Piana, Michael P. Eastwood, Ron O. Dror, and David E. Shaw. Systematic validation of protein force fields against experimental data. *PLoS ONE*, **2012**, *7*, e32131; b) David E. Shaw, Paul Maragakis, Kresten Lindorff-Larsen, Stefano Piana, Ron O. Dror, Michael P. Eastwood, Joseph A. Bank, John M. Jumper, John K. Salmon, Yibing Shan, and Willy Wriggers. Atomic-level characterization of the structural dynamics of proteins. *Science*, **2010**, *330*, 341–346.

12. RCSB Protein Data Bank; see http://www.rcsb.org/pdb/home.home.do (accessed 12 November 2013).

13. a) John P. Overington, Bissan Al-Lazikani, and Andrew L. Hopkins. How many drug targets are there? *Nat. Rev. Drug Discov.*, **2006**, *5*, 993–996; b) David Filmore. It's a GPCR world. *Mod. Drug Discov.*, **2004**, *7*, 24–28.

14. Sylvie Doublié. Preparation of selenomethionyl proteins for phase determination. *Methods Enzymol.*, **1997**, *276*, 523–530.

15. Elspeth Garman, and James W. Murray. Heavy-atom derivatization. *Acta Crystallogr.*, **2003**, *D59*, 1903–1913.

16. The Heavy Atom Databank provides extensive information on heavy-atom use in protein crystallography; see http://www.sbg.bio.ic.ac.uk/had/ (accessed 12 November 2013).

17. Éric Girard, Meike Stelter, Pier L. Anelli, Jean Vicata, and Richard Kahn. A new class of gadolinium complexes employed to obtain high-phasing-power heavy-atom derivatives: results from SAD experiments with hen egg-white lysozyme and urate oxidase from *Aspergillusflavus*. *Acta Crystallogr.*, **2003**, *D59*, 118–126.

18. Rafael Molina, Ana González, MeikeStelter, Inmaculada Pérez-Dorado, Richard Kahn, María Morales, Susana Campuzano, Nuria E. Campillo, Shahriar Mobashery, José L. García, Pedro García, and Juan A. Hermoso. Crystal structure of CbpF, a bifunctional choline-binding protein and autolysis regulator from *Streptococcus pneumonia*. *EMBO Rep.*, **2009**, *10*, 246–251.

19. Vanessa Delfosse, Eric Girard, Catherine Birck, Michaël Delmarcelle, Marc Delarue, Olivier Poch, Patrick Schultz, and Claudine Mayer. Structure of the archaeal Pab87 peptidase reveals a novel self-compartmentalizing protease family. *PLoS ONE*, **2009**, *4*, e4712.

20. Michael D. Purdy, Pinghua Ge, Jiyan Chen, Paul R. Selvin, and Michael C. Wiener. Thiol-reactive lanthanide chelates for phasing protein X-ray diffraction data. *Acta Crystallogr.*, **2002**, *D58*, 1111–1117.

21. Nicholas R. Silvaggi, Langdon J. Martin, Harald Schwalbe, Barbara Imperiali, and Karen N. Allen. Double-lanthanide-binding tags for macromolecular crystallographic structure determination. *J. Am. Chem. Soc.*, **2007**, *129*, 7114–7120.

22. Hyun Soo Lee, Glen Spraggon, Peter G. Schultz, and Feng Wang. Genetic incorporation of a metal-ion chelating amino acid into proteins as a biophysical probe. *J. Am. Chem. Soc.*, **2009**, *131*, 2481–2483.

23. Chang C. Liu, and Peter G. Schultz. Adding new chemistries to the genetic code. *Annu. Rev. Biochem.*, **2010**, *79*, 413–444.

24. Mark P. Foster, Craig A. McElroy, and Carlos D. Amero. Solution NMR of large molecules and assemblies. *Biochemistry*, **2007**, *46*, 331–340.

25. Peter H. J. Keizers, and Marcellus Ubbink. Paramagnetic tagging for protein structure and dynamics analysis. *Prog. NMR Spectrosc.*, **2011**, *58*, 88–96.

26. Julia Koehler, and Jens Meiler. Expanding the utility of NMR restraints with paramagnetic compounds: background and practical aspects. *Prog. NMR Spectrosc.*, **2011**, *59*, 360–389.

27. Gottfried Otting. Prospects for lanthanides in structural biology by NMR. *J. Biomol. NMR*, **2008**, *42*, 1–9.

28. Michael John, and Gottfried Otting. Strategies for measurements of pseudocontact shifts in protein NMR spectroscopy. *ChemPhysChem*, **2007**, *8*, 2309–2313.

29. CRC *Handbook of Chemistry and Physics*, 93rd edition, CRC Press: Boca Raton, FL, 2012–2013.

30. Xingfu X, Peter H. J. Keizers, Wolfgang Reinle, Frank Hanneman, Rita Bernhardt, and Marcellus Ubbink. Molecular dynamics studied by paramagnetic tagging. *J. Biomol. NMR*, **2009**, *43*, 247–254.

31. Peter H. J. Keizers, Berna Mersinli, Wolfgang Reinle, Julia Donauer, Yoshitaka Hiruma, Frank Hannemann, Mark Overhand, Rita Bernhardt, and Marcellus Ubbink.

A solution model of the complex formed by adrenodoxin and adrenodoxinreductase determined by paramagnetic NMR spectroscopy. *Biochemistry*, **2010**, *49*, 6846–6855.

32. Tomohide Saio, Kenji Ogura, Kazumi Shimizu, Masashi Yokochi, Terrence R. Burke, Jr., and Fuyuhiko Inagaki. An NMR strategy for fragment-based ligand screening utilizing a paramagnetic lanthanide probe. *J. Biomol. NMR*, **2000**, *51*, 395–408.

33. Jia-Ying Guan, Peter H.J. Keizers, Wei-Min Liu, Frank Loehr, Simon Peter Skinner, Edwin A. Heeneman, Harald Schwalbe, Marcellus Ubbink, and Gregg David Siegal. Small molecule binding sites on proteins established by paramagnetic NMR spectroscopy. *J. Am. Chem. Soc.*, **2013**, *135*, 5859-5868.

34. Marco Allegrozzi, Ivano Bertini, Matthias B. L. Janik, Yong-Min Lee, Gaohua Liu, and Claudio Luchinat. Lanthanide-induced pseudocontact shifts for solution structure refinements of macromolecules in shells up to 40 Å from the metal ion. *J. Am. Chem. Soc.*, **2000**, *122*, 4154–4161.

35. G. Marius Clore, and Junji Iwahara. Applications of paramagnetic relaxation enhancement for the characterization of transient low-population states of biological macromolecules and their complexes. *Chem. Rev.*, **2009**, *109*, 4108–4139.

36. Rebecca S. Lipsitz, and Nico Tjandra. Residual dipolar couplings in NMR structure analysis. *Annu. Rev. Biophys. Biomol. Struct.*, **2004**, *33*, 387–413.

37. a) Lana Lee, and Brian D. Sykes. Strategies for the uses of lanthanide NMR shift probes in the determination of protein structure in solution. Application to the EF calcium binding site of carp parvalbumin. *Biophys. J.*, **1980**, *32*, 193–210; b) Lana Lee, David C. Corson, and Brian D. Sykes. Structural studies of calcium-binding proteins using nuclear magnetic resonance. *Biophys. J.*, **1985**, *47*, 139–142.

38. Guido Pintacuda, Ah Y. Park, Max A. Keniry, Nicholas E. Dixon, and Gottfried Otting. Lanthanide labeling offers fast NMR approach to 3D structure determinations of protein-protein complexes. *J. Am. Chem. Soc.*, **2006**, *128*, 3696–3702.

39. James D. Swarbrick, Phuc Ung, Sandeep Chhabra, and Bim Graham. An iminodiacetic acid based lanthanide binding tag for paramagnetic exchange NMR spectroscopy. *Angew. Chem. Int. Ed.*, **2011**, *50*, 4403–4406.

40. Hiromasa Tagi, Ansis Maleckis, and Gottfried Otting. A systematic study of labelling an α-helix in a protein with a lanthanide using IDA-SH or NTA-SH tags. *J. Biomol. NMR*, **2013**, *55*, 157–166.

41. James D. Swarbrick, PhucUng, Xun-Cheng Su, Ansis Maleckis, Sandeep Chhabra, Thomas Huber, Gottfried Otting, and Bim Graham. Engineering of a bis-chelator motif into a protein α-helix for rigid lanthanide binding and paramagnetic NMR spectroscopy. *Chem. Commun.*, **2011**, *47*, 7368–7370.

42. Peter H. J. Keizers, Athanasios Saragliadis, Yoshitaka Hiruma, Mark Overhand, and Marcellus Ubbink. Design, synthesis, and evaluation of a lanthanide chelating protein probe: CLaNP-5 yields predictable paramagnetic effects independent of environment. *J. Am. Chem. Soc.*, **2008**, *130*, 14802–14812.

43. Mathias A. S. Hass, Peter H. J. Keizers, Anneloes Blok, Yoshitaka Hiruma, and Marcellus Ubbink. Validation of a lanthanide tag for the analysis of protein dynamics by paramagnetic NMR spectroscopy. *J. Am. Chem. Soc.*, **2010**, *132*, 9952–9953.

44. a) Tomohide Saio, Kenji Ogura, Masashi Yokochi, Yoshihiro Kobashigawa, and Fuyuhiko Inagaki. Two-point anchoring of a lanthanide-binding peptide to a target protein enhances the paramagnetic anisotropic effect. *J. Biolmol. NMR*, **2009**, *44*,

157–166; b) Katja Barthelmes, Anne M. Reynolds, Ezra Peisach, Hendrik R. A. Jonker, Nicholas J. De Nunzio, Karen N. Allen, Barbara Imperiali, and Harald Schwalbe. Engineering encodable lanthanide-binding tags into loop regions of proteins. *J. Am. Chem. Soc.*, **2011**, *133*, 808–819c) Xun-Cheng Su, Kerry McAndrew, Thomas Huber, and Gottfried Otting. Lanthanide-binding peptides for NMR measurements of residual dipolar couplings and paramagnetic effects from multiple angles. *J. Am. Chem. Soc.*, **2008**, *130*, 1681–1687.

45. Tiandi Zhuang, Han-Seung Lee, Barbara Imperiali, and James H. Prestgard. Structure determination of a galectin-3-carboyhydrae complex using paramagnetism-based NMR constraints. *Protein Sci.*, **2008**, *17*, 1220–1231.

46. Bim Graham, Choy Theng Loh, James David Swarbrick, PhucUng, James Shin, Hiromasa Yagi, Xinying Jia, Sandeep Chhabra, Nicholas Barlow, Guido Pintacuda, Thomas Huber, and Gottfried Otting. DOTA-amide lanthanide tag for reliable generation of pseudocontact shifts in protein NMR spectra. *Bioconjugate Chem.*, **2011**, *22*, 2118–2125.

47. Laura de la Cruz, Thi Hoang Duong Nguyen, Kiyoshi Ozawa, James Shin, Bim Graham, Thomas Huber, and Gottfried Otting. Binding of low molecular weight inhibitors promotes large conformational changes in the dengue virus NS2B-NS3 protease: fold analysis by pseudocontact shifts. *J. Am. Chem. Soc.*, **2011**, *133*, 19205–19215.

48. Kristaps Jaudzems, Xinying Jia, Hiromasa Yagi, Dmitry Zhulenkovs, Bim Graham, Gottfried Otting, and Edvards Liepinsh. Structural basis for 5'-end-specific recognition of single-stranded DNA by the R3H domain from human $S\mu bp$-2. *J. Mol. Biol.*, **2012**, *424*, 42–53.

49. Lorenzo Di Bari, and Piero Salvadori. Static and dynamic stereochemistry of chiral Ln DOTA analogues. *Chem. Phys. Chem.*, **2011**, *12*, 1490–1497.

50. Mark Woods, Mauro Botta, Stefano Avedano, Jing Wang, and A. Dean Sherry. Towards the rational design of MRI contrast agents: a practical approach to the synthesis of gadolinium complexes that exhibit optimal water exchange. *Dalton Trans.*, **2005**, 3829–3837.

51. Daniel Häussinger, Jie-rong Huang, and Stephan Grzesiek. DOTA-M8: an extremely rigid, high-affinity lanthanide chelating tag for PCS NMR spectroscopy. *Bioconjugate Chem.*, **2011**, *22*, 2118–2125.

52. a) Qin-Feng Li, Yin Tang, Ansis Maleckis, Gottfired Otting, and Xun-Cheng Su. Thiol-ene reaction: a versatile tool in site-specific labelling of proteins with chemically inert tags for paramagnetic NMR., *Chem. Commun.*, **2012**, *48*, 2704–2706; b) Yin Yang, Qing-Feng Li, Chan Cao, Feng Huang, and Xun-Cheng Su. Site-specific labelling of proteins with a chemically stable, high-affinity tag for protein study. *Chem. Eur. J.*, **2013**, *19*, 1097–1103.

53. Choy T. Loh, Kiyoshi Ozawa, Kellie L. Tuck, Nicholas Barlow, Thomas Huber, Gottfried Otting, and Bim Graham. Lanthanide tags for site-specific ligation to an unnatural amino acid and generation of pseudocontact shifts in proteins. *Biocon. Chem.*, **2013**, *24*, 260–268.

54. Xun-Cheng Su, Haobo Liang, Karin V. Loscha, and Gottfried Otting. Ln(DPA)$_3$]$^{3-}$ is a convenient paramagnetic shift reagent for protein NMR studies. *J. Am. Chem. Soc.*, **2009**, *131*, 10352–10353.

55. Xinying Jia, Hiromasa Yagi, Xun-Cheng Su, Mitchell Stanton-Cook, Thomas Huber, and Gottfried Otting. Engineering $[Ln(DPA)_3]^{3-}$ binding sites in proteins: a widely applicable method for tagging proteins with lanthanide ions. *J. Biomol. NMR*, **2011**, *50*, 411–420.

56. Makoto Nomuraa, Toshitatsu Kobayashia, Toshiyuki Kohnob, Kenichiro Fujiwarac, Takeshi Tennod, Masahiro Shirakawac, ItsukoIshizakia, Kazuo Yamamotoe, Toshifumi Matsuyamae, Masaki Mishimaa, and Chojiro Kojima. Paramagnetic NMR study of Cu^{2+}-IDA complex localization on a protein surface and its application to elucidate long distance information. *FEBS Lett.*, **2004**, *566*, 157–161.

57. Stéphane Balayssac, Ivano Bertini, Anusarka Bhaumik, Moreno Lelli, and Claudio Luchinat. Paramagnetic shifts in solid-state NMR of proteins to elicit structural information. *Proc. Natl. Acad. Sci. USA*, **2012**, *109*, 11095–11100.

58. Philippe S. Nadaud, Jonathan J. Helmus, Stefanie L. Kall and Christopher P. Jaroniec. Paramagnetic ions enable tuning of nuclear relaxation rates and provide long-range structural restraints in solid-state NMR of proteins. *J. Am. Chem. Soc.*, **2009**, *131*, 8108–8120; b) Ishita Sengupta, Philippe S. Nadaud, Jonathan J. Helmus, Charles D. Schwieters, and Christopher P. Jaroniec. Protein fold determined by paramagnetic magic-angle spinning solid-state NMR spectroscopy. *Nat. Chem.*, **2012**, *4*, 410–417.

59. Thi Hoang Duong Nguyen, Kiyoshi Ozawa, Mitchell Stanton-Cook, Russell Barrow, Thomas Huber, and Gottfried Otting. Generation of pseudocontact shifts in protein NMR spectra with a genetically encoded cobalt(II)-binding amino acid. *Angew. Chem. Int. Ed.*, **2010**, *49*, 1–3.

60. Janet E. Banham, Christiane R. Timmel, Rachel J. M. Abbott, Susan M. Lea, and Gunnar Jeschke. The characterization of weak protein-protein interactions: evidence from DEER for the trimerization of a von Willebrand Factor A domain in solution. *Angew. Chem. Int. Ed.*, **2006**, *45*, 1058–1061.

61. Arnold M. Raitsimring, Chidambaram Gunanathan, Alexey Potapov, Irena Efremenko, Jan M. L. Martin, David Milstein, and Daniella Goldfarb. Gd^{3+} complexes as potential spin labels for high field pulsed EPR distance measurements. *J. Am. Chem. Soc.*, **2007**, *129*, 14138–14140.

62. Alexey Potapov, Hiromasa Yagi, Thomas Huber, Slobodan Jergic, Nicholas E. Dixon, Gottfried Otting, and Daniella Goldfarb. Nanometer-scale distance measurements in proteins using Gd^{3+} spin labeling. *J. Am. Chem. Soc.*, **2010**, *132*, 9040–9048.

63. Ying Song, Thomas J. Meade, Andrei V. Astashkin, Eric L. Klein, John H. Enemark, and Arnold Raitsimring. Pulsed dipolar spectroscopy distance measurements in biomacromolecules labeled with Gd(III) markers. *J. Magn. Reson.*, **2011**, *210*, 59–68.

64. Hiromasa Yagi, Debamalya Banerjee, Bim Graham, Thomas Huber, Daniella Goldfarb, and Gottfried Otting. Gadolinium tagging for high-precision measurements of 6 nm distances in protein assemblies by EPR. *J. Am. Chem. Soc.*, **2011**, *133*, 10418–10421.

65. Debamalya Banerjee, Hiromasa Yagi, Thomas Huber, Gottfried Otting, and Daniella Goldfarb. Nanometer-range distance measurement in a protein using Mn^{2+} tags. *J. Phys. Chem. Lett.*, **2012**, *3*, 157–160.

66. Joseph R. Lakowicz, *Principles of Fluorescence Spectroscopy*, New York: Kluwer Academic/Plenum Publishers, **1999**.

67. Theodor Förster, Zwischenmolekulare Energiewanderung und Fluoreszenz. *Ann. Phys.*, **1948**, *6*, 55–74.
68. Albert Cha, Gregory E. Snyder, Paul R. Selvin, and Francisco Bezanilla. Atomic scale movement of the voltage-sensing region in a potassium channel measured via spectroscopy. *Nature*, **1999**, *402*, 809–813.
69. Ming Xiao, Handong Li, Gregory E. Snyder, Roger Cooke, Ralph G. Yount, and Paul R. Selvin. Conformational changes between the active-site and regulatory light chain of myosin as determined by luminescence resonance energy transfer: The effect of nucleotides and actin. *Proc. Natl. Acad. Sci. USA*, **1998**, *95*, 15309–15314.
70. Phillip A. Knauf, and Prithwish Pal. Use of luminescence resonance energy transfer to measure distances in the AE1 anion exchange protein dimer. *Blood Cells, Mol. Dis.*, **2004**, *32*, 360–365.
71. Tomasz Heyduk. Luminescence resonance energy transfer analysis of RNA polymerase complexes. *Methods*, **2001**, *25*, 44–53.
72. Justin W. Taraska, and William N. Zagotta. Fluorescence applications in molecular neurobiology. *Neuron*, **2010**, *66*, 170–189.

3

AAS, XRF, and MS Methods in Chemical Biology of Metal Complexes

Ingo Ott,[a] Christophe Biot[b] and Christian Hartinger[c]
[a]Institute of Medicinal and Pharmaceutical Chemistry, Technical University of Braunschweig, Germany
[b]Unit of Structural and Functional Glycobiology 47, University of Lille1, France
[c]School of Chemical Sciences, The University of Auckland, New Zealand

3.1 Introduction

Understanding of the chemical biology of metal complexes and their involvement in biological processes has been acquired, to a considerable extent, through advanced instrumental analytical procedures, which have also facilitated the development of sophisticated metal-based agents as well as metallodrugs. Owing to their unique spectroscopic features, metal complexes offer interesting opportunities for bioanalytical chemistry, which can be exploited via appropriate element specific methods. In general, either the metal itself can be detected by trace element quantification procedures or features of the coordinated ligands can be used for detection purposes. With respect to applications in chemical biology, the sensitive and selective detection of metal complexes in biological matrices, such as cell or tissue suspensions, are critical and challenging analytical tasks.

In this chapter, metal specific bioanalytical methods for qualitative and quantitative measurement will be addressed. These include, among others, atomic spectroscopic techniques, such as atomic absorption spectroscopy (AAS) or inductively coupled plasma mass spectrometry (ICP-MS), as well as X-ray fluorescence (XRF)-based technologies requiring synchrotron radiation. The basic principles of the respective techniques will be discussed and examples of practical applications in inorganic chemical biology will be given.

Inorganic Chemical Biology: Principles, Techniques and Applications, First Edition. Edited by Gilles Gasser.
© 2014 John Wiley & Sons, Ltd. Published 2014 by John Wiley & Sons, Ltd.

While AAS has been the most widely used trace elemental technique for many decades, in recent years ICP-based methods have progressively entered the area of atomic spectroscopy. Nowadays both techniques are well established for the sensitive quantification of metal traces in biological samples. Coupled techniques involving ICP-MS and chromatographic/electrophoretic separation methods have emerged lately and represent powerful combinations of advanced separation methods with a highly sensitive element-specific quantification detector. XRF methods have evolved for both quantitative and qualitative metal determinations. In particular, microscopic XRF methods have experienced a steep increase in technological improvements, which has enabled a significant advance in information on the spatial biodistribution of metal complexes in tissues, as has the introduction of laser ablation ICP-MS.

The mentioned techniques are becoming increasingly applied in the emerging field of *metallomics* [1]. In analogy to genomics or proteomics, in metallomics, the distribution and properties of metals and metal complexes in the biological environment and in organisms are investigated [2]. For obvious reasons, metal specific analytical methods are crucial in this area of research.

3.2   Atomic Absorption Spectroscopy (AAS)

3.2.1   Fundamentals and Basic Principles of AAS

Atomic absorption spectroscopy (AAS) is based on the absorption of radiation by free gaseous atoms, which are usually in the ground state [3–6]. The underlying principles were established by Bunsen and Kirchoff and rely on the ability of an atom to specifically absorb the radiation that the atom is also able to emit. The effect of the element-specific radiation absorption had been observed earlier in the form of Fraunhofer lines in the optical spectrum of the sun. These dark absorption lines in the solar spectrum are the consequence of the presence of gaseous atoms in colder gases between the observer and the sun. AAS makes use of these fundamental principles of physics for the selective and sensitive spectroscopic detection of many metal atoms.

For AAS analysis, an appropriate wavelength, which corresponds to the transition of the atom in its ground state to the respective excited level, is chosen for absorption readings. As a consequence of the findings of Kirchoff and Bunsen, this particular wavelength is highly specific for the selected element. According to the Lambert–Beers law, the obtained absorption value is proportional to the concentration of the atoms in the ground state and thereby reflects the concentration of the respective element. With appropriate instrumentation the absorption can be registered over time and the absorbance of the peak-shaped signal can be obtained by integration as the area under the curve (AUC). In general, both the signal height and the AUC can be used for data analysis.

Since the number of atomic absorption lines is usually small, spectral overlap with lines from other elements in the sample is rather rare. In contrast, background interferences are often the result of undissociated molecules that lead to molecular absorption over a broader spectral range or the presence of particles, which cause scattering of the primary irradiation beam. As outlined in more detail in the following discussions, several techniques and

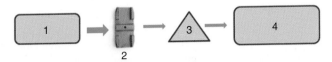

Figure 3.1 *Simplified schematic instrumental setup of an atomic absorption spectrometer: (1) primary irradiation source, (2) atomization unit (e.g., graphite tube), (3) monochromator, and (4) detector*

sample preparation procedures are available to help to effectively diminish these unwanted interferences.

Overall, AAS offers high analytical selectivity and sensitivity and enables the convenient quantification of metals in trace amounts even in complex matrices of biological origin. These advantages have made AAS a widely used instrumental analytical technique for quantitative bioanalysis of metals in trace concentrations.

3.2.2 Instrumental and Technical Aspects of AAS

Basic instrumental elements of atomic absorption spectrometers include the irradiation source, an atomizer, a monochromator, a signal amplifier, and a detector (Fig. 3.1) [3–6].

3.2.2.1 *Primary irradiation source*

The primary irradiation source provides the appropriate measurement line, which is specific for the measured element and according to the law of Kirchoff and Bunsen also produced by the very same element. An important requirement for this device is a high intensity emission at the respective atomic absorption line. The line should be well separated from other lines that might cause background absorptions. The most commonly used irradiation sources are hollow-cathode lamps, which contain the element of interest in the cathode and are filled with an inert gas, or electrodeless discharge lamps, which contain the element of interest in a sealed argon filled silica tube.

A more recent development is advanced continuum radiation source based spectroscopy using a Xe short-arc lamp as the light source. In contrast to hollow-cathode lamps or electrodeless discharge lamps, which emit specific lines for the element of interest, the Xe short-arc lamp is a continuum emitter. The use of continuum emitters had previously been limited due to the too weak emission of available lamps but has been overcome by the radiation source working in the so-called hot-spot mode. Together with other technical improvements concerning the monochromator and detector, this has led to high resolution continuum source AAS (HR-CS AAS) as a modern AAS technique, which allows the use of wavelength as a third dimension (in addition to the absorption intensity and time) for analytical data processing [7–9]. This presents additional opportunities for detecting molecular absorptions resulting, for example, from certain diatomic species such as PO or CS.

3.2.2.2 *Atomizers*

An atomizer is a technical device that is used to provide gaseous free atoms from the analyte metal in the sample. The atomization is commonly achieved by heating the probes in a flame or furnace to temperatures in the range of from 1500 to 3000 °C and ideally results in a complete atomization of the analyte.

The flame AAS (FAAS) technique initially converts the liquid sample into an aerosol in a nebulizer. The aerosol is then aspirated through a slot in the burner into the flame, which is most commonly fuelled by air–acetylene or N_2O–acetylene mixtures. Metal concentrations in the low milligram per liter dimension (mg l^{-1}, ppm) can be determined by the FAAS technique.

Substantially lower detection limits, mainly in the low microgram per liter to the nanogram per liter (μg l^{-1} to ng l^{-1}) range can be conveniently reached by the so-called graphite furnace AAS (GFAAS), which is also known as electrothermal AAS (ETAAS). The increased sensitivity of this technique compared with flame FAAS is in part the consequence of a longer residence time of the gaseous free atoms in the irradiation beam. Besides its lower detection limits, the low sample consumption is another advantage of this technique. Only a few microliters are necessary to perform accurate analyses. The sample volume is injected through a small hole into a graphite tube, which itself is located in a furnace. After injection, the furnace is heated up electrothermally in a stepwise manner according to a furnace temperature program to pyrolyze the sample and to atomize the analyte metal. The irradiation beam of the light source is directed through the tube for the absorption reading. During heating and pyrolysis, a constant flow of argon is passed through the graphite tube to remove volatile sample components. This allows the removal of most of the sample matrix and prevents oxidative corrosion of the tube at elevated temperatures. For atomization of the analyte metal, the argon flow is stopped. In a final high temperature cleaning step the remaining sample material is vaporized again and transported out of the tube by restarting the argon flow. After cooling of the tube the next sample can be injected into the same tube without memory effects caused by the preceding sample. An improvement of the conventional standard graphite tubes was the introduction of a thin graphite platform in the tube. The sample is then deposited onto this platform, which reaches the atomization temperature later than the wall of the tube. In this way, atomization occurs when the wall of the tube has already reached a constant temperature plateau and this leads to a more effective sample dissociation and reduces interferences.

3.2.2.3 Monochromator and detector

Since most irradiation sources emit more than one line, a monochromator has to be placed before the detector to eliminate unwanted emissions. Usually, interference filters, plane gratings or echelle gratings are used for this purpose. The spectral bandwidths are typically in the range of 0.2–1.0 nm. When the flame atomization technique is used, a stable signal is recorded for a few seconds. In GFAAS a peak-shaped signal can be determined by measuring the absorption over time. For HR-CS AAS a high resolution double monochromator, which consists of a prism pre-monochromator and an echelle grating monochromator, is used in combination with a modern linear CCD array detector that contains hundreds of individual detectors. Overall this enables a simultaneous background correction and a three-dimensional imaging of the analyte signal as well as its spectral environment (Fig. 3.2). Since the read-out rates of the available CCD systems are still limited, an area of within less than 1 nm can be monitored. This very narrow spectral window so far still prevents strictly simultaneous multi-element measurements over the whole spectral range [7–9].

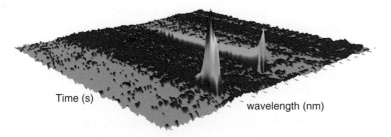

Time (s)   wavelength (nm)

Figure 3.2 *HR-CS AAS measurement of a sample containing both iron and ruthenium. Selected elements with atomic lines in very close spectral proximity can be detected by a single measurement*

3.2.2.4 *Options for background correction*

Background absorption is the sum of interferences that cause unspecific loss of intensity at the atomic absorption line. Modern AAS spectrometers contain sophisticated devices to identify and handle such phenomena. In conventional instruments the so-called Deuterium and Zeeman correction are most frequently used, while HR-CS AAS uses a simultaneous background correction.

Deuterium correction. Besides the element-specific lines, a continuum light beam can be directed through the atomization device. For the continuum emission a broader bandwidth is used and the absorption is not measured at the same time as the specific atomic absorption. Therefore, this method does not correct strictly at the absorption line and is not simultaneous. For background correction purposes, deuterium lamps as a continuum emitter are most frequently used.

Zeeman correction. The Zeeman correction is based on the Zeeman effect, which describes the splitting of spectral lines in a magnetic field. A pulsating magnetic field is applied and when the field is active the resonance line is split with two additional lines appearing approximately ±0.01 nm next to the resonance line. The split lines are polarized and either the central line eliminated by the polarizer can be used for background correction at the same wavelength or alternatively the additional lines can be used for background measurements near the analytical wavelength.

Simultaneous background correction. Based on the combined use of a continuum irradiation source, a high resolution double monochromator and a CCD detector, in HR-CS AAS spectrometers the background around the analyte signal can be measured efficiently and strictly at the same time as the analyte signal. This represents to date the most advanced option for background correction in AAS.

3.2.3 Method Development and Aspects of Practical Application

Method development of AAS analytical procedures includes the optimization of the instrumental setup according to the needs of the respective atomization technique as well as sample preparation involving, for example, the addition of chemical modifiers. In particular, of general importance is the removal of matrix effects, the suppressing of spectral interferences, and an efficient analyte atomization with high recovery rates. Spectral interferences are not very common in AAS due to the high element specificity of atomic absorption lines, whereas matrix effects (e.g., related to unspecific absorption by volatile components of the probe) and incomplete atomization of the analyte metal have to be carefully considered, especially for samples of biological origin.

Calibration can ideally be carried out by the simple use of external standards and data analysis can be performed after linear regression employing external standardization. However, for many practical applications, accurate calibration involves the use of more time consuming methods, such as matrix-matched calibration or the standard addition method.

3.2.3.1   *Parameters for FAAS and GFAAS method development*

With respect to the FAAS technique, where the gaseous atoms are obtained by heating the sample in a flame, the type of gas mixture is very important. Most commonly air–acetylene and N_2O–acetylene combinations are used to achieve temperatures in the range of from 2000 to 3000 °C. For optimal atomization, the ratio of both gas components is important as well as the position of the sample irradiation beam within the burning flame, which can be adjusted by moving the burner up or down vertically.

When GF-AAS is used as the atomization technique, method development focuses primarily on the setup of an appropriate temperature program for the furnace, which contains the graphite tube with the sample. Initial heating steps aim at removing solvents and drying the sample before thermal pre-treatment and pyrolysis take place (Fig. 3.3). Ideally all non-analyte components are removed from the tube by an argon flow and the analyte metal is reduced to its elemental state. Atomization takes place typically at temperatures between 2000 and 3000 °C and a final heating step after measurement is used to clean the tube. Parameters that can be modified, besides the respective temperatures, are the heating ramps and the temperature holding time. Drying steps are usually optimized empirically whereas appropriate temperatures for pyrolysis and atomization can be identified in a methodical manner. In this procedure, a constant atomization temperature is initially set and temperatures for pyrolysis are increased in stepwise fashion. For each pyrolysis temperature, a sample of the same concentration is measured. At an optimal pyrolysis temperature, a high analyte signal is obtained with a low unspecific background absorbance, which is measured

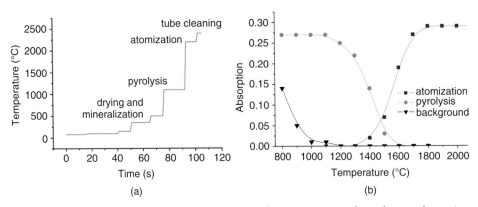

Figure 3.3   *(a) Temperature program for GFAAS; (b) optimization of pyrolysis and atomization Temperatures*

at the same time. Too low pyrolysis temperatures will result in high background absorption and too high pyrolysis temperatures will cause vaporization and loss of analyte mass before atomization. In the second step, atomization temperatures are varied at a constant pyrolysis temperature and thereby a suitable temperature plateau can usually be identified where efficient atomization takes place. Too high atomization temperatures can lead to lower signals, for example due to thermal ionization of the metal.

3.2.3.2   Interferences and the use of modifiers

In atomic absorption spectroscopy, interferences can have a spectral or a non-spectral nature. Non-spectral interferences can, for example, be the result of the formation of salts that are difficult to atomize. In chemical biology, this has to be considered in particular for probes containing high anion concentrations. For example, a well-known interference is that caused by phosphate in the determination of calcium. Spectral interferences can be the consequence of non-specific absorption and of light scattering phenomena by components of the sample matrix or result from overlapping of resonance lines. Spectral interferences caused by lines from other elements are not particularly frequent and are generally well characterized. Usually this type of interference can be easily managed by switching to another analytical line of the element under investigation or by background correction. Non-specific absorptions or light scattering can be caused by components of the sample matrix and lead to elevated background noise. Many of these interferences can be conveniently handled by an appropriate method parameter setup (e.g., effective removal of matrix components by heating in GF-AAS) or by appropriate background correction techniques (see earlier discussion).

If interferences continue to be a problem, chemical modifiers offer some options for improvement. Chemical modifiers are additives that can suppress interferences and lead to a stable atomization signal. Many types of chemical modifiers have been described and are used routinely. For example, La^{3+} can form thermally stable salts with phosphate. Therefore, it can be used to suppress interferences of phosphate in calcium determinations as calcium will be released and be able to form free atoms. Ionization interference occurs when an element is easy to ionize and the number of atoms in the elemental state is decreased. This can be the case particularly for alkali metals. In this case ionization buffers such as CsCl can be used. The ionization buffer is added in excess and forms a high number of electrons upon ionization. This shifts the equilibrium of the alkali metal under study towards the elemental state. In GFAAS, modifiers can be used specifically to increase the difference in the volatilization of the matrix components in comparison with the element of interest or to reduce reactions with the graphite surface. Mixtures of nitrates of Mn and Pd have evolved as widely used "universal modifiers". Also, organic additives such as Triton, EDTA or ascorbic acid are often applied to improve solubilization or facilitate reduction of the analyte.

3.2.4   Selected Application Examples

3.2.4.1   Cellular uptake of ruthenium polypyridyl complexes

In recent years, ruthenium polypyridyl complexes have been shown to display promising *in vitro* data as novel anticancer drug candidates. The mode of action of these species is not yet completely understood, however, recent data strongly hint at the contribution of DNA interactions and/or antimitochondrial properties [10, 11].

Ruthenium can be measured in cell preparations by GFAAS with good sensitivity. For structurally closely related complexes of the type $[(\eta^6\text{-}C_6Me_6)Ru(L)(\text{polypyridyl})]^{n+}$ cellular uptake quantification in HT-29 and MCF-7 cells provided clear correlations between size, overall charge and lipophilicity of the species (Fig. 3.4) [10]. The highest uptake was generally observed with complexes containing the large dppn (4,5,9,16-tetraaza-dibenzo[a,c]naphthacene) ligand. Comparing the data obtained for dppn complexes **a, b** and **d** (Fig. 3.4) shows that the uptake decreases if the overall charge of the complexes switches from +1 to +2 and that tetramethylthiourea containing complexes experience a slightly higher accumulation than the less lipophilic thiourea congeners. With decreasing size of the dppn ligand, the cellular ruthenium levels decrease. However, it should be noted in this context that such stringent correlations between chemical/physico-chemical properties of metal complexes and their cellular accumulation levels are rarely observed.

3.2.4.2   *Cellular uptake and intracellular distribution of gold(I) complexes*

Biomedical applications of gold complexes have played an important role since ancient times. Nowadays some gold species are used for the treatment of symptoms of rheumatoid arthritis and increasing research efforts clearly confirm a considerable potential for use as anti-infectives or anticancer drugs [12–14].

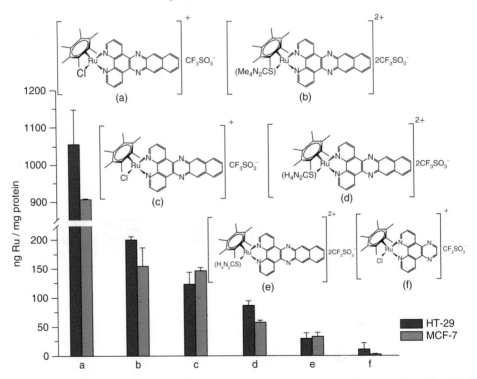

Figure 3.4   *Cellular ruthenium levels in HT-29 and MCF-7 cancer cells exposed to 10 μM of the complexes **a–f** for 4 h*

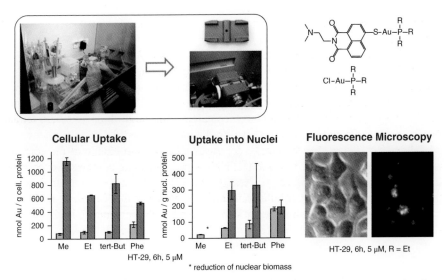

Figure 3.5 *Quantification of cellular and nuclear gold levels. A comparative study of complexes containing a naphthalimide-thiolato or a chlorido ligand. The accumulation of the naphthalimides in the nuclei was additionally visualized by fluorescence microscopy. Reprinted with permission from [17] © 2009, American Chemical Society*

Gold can be conveniently quantified in biological tissues by GFAAS, however, due to matrix effects calibration should be carried out by matrix matching or the standard addition method [15].

Cellular gold levels of several gold(I)–phosphane complexes containing either a chlorido or a naphthalimide-thiolato ligand were studied in a comparative manner in tumor tissue (Fig. 3.5) [16, 17]. Interestingly, the naphthalimide-derived ligand caused an elevated gold uptake in comparison with the chlorido derivatives. The complexes feature different substituents on the phosphorus atom, which, however, did not translate into an obvious effect on cellular accumulation. Naphthalimides are well known DNA intercalators [18]. Accordingly, with the naphthalimide derivatives, high gold concentrations were measured in the nuclei isolated from the cells. The enhanced uptake into the nuclei could be further confirmed by microscopic registration of the strong blue fluorescence of the naphthalimide core.

Another class of gold complexes that has recently gained attention in anticancer drug research is based on *N*-heterocyclic carbene (NHC) ligands. The interest in this type of organometallic compounds is related among other effects to their efficient triggering of apoptosis, their cytotoxicity against tumor cells, and their potential to inhibit the activity of the enzyme thioredoxin reductase [19–21].

Depending on the type of coordinated ligands, neutral or cationic gold NHC complexes can be obtained. For the Au(I)–NHC complexes depicted in Fig. 3.6 it was shown that the cationic (and lipophilic) derivatives led to much higher cellular gold levels than the respective neutral complex [22]. This also correlated with enhanced tumor cell growth inhibition of the positively charged complexes. Investigation of mitochondrial fractions isolated from drug candidate-exposed tumor cells confirmed an elevated gold content especially for an

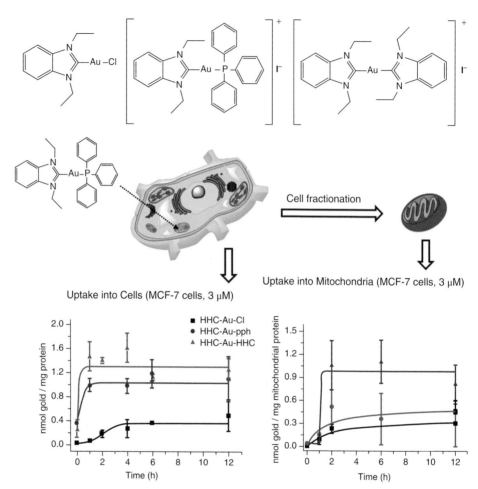

Figure 3.6   *Au(I)–NHC complexes and their biodistribution. Reprinted with permission from [22] © 2011, American Chemical Society*

Au(I)–(triphenylphosphane) derivative. The biodistribution behavior of this and related compounds is supposedly related to its lipophilic cationic nature. Delocalized lipophilic cations are in general known to accumulate inside mitochondria and to trigger antimitochondrial effects, which is related to the larger transmembrane potential of these cell organelles [23].

3.3   Total Reflection X-Ray Fluorescence Spectroscopy (TXRF)

3.3.1   Fundamentals and Basic Principles of TXRF

From a general viewpoint, X-ray spectroscopy is based on the excitation of samples with an X-ray beam [24, 25]. As a consequence of the high-energy irradiation, core electrons of the

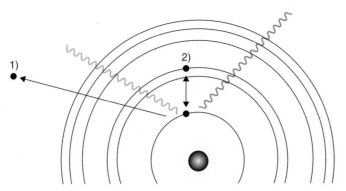

Figure 3.7 *Principle of X-ray fluorescence: X-ray radiation (orange beam) excites an inner shell electron, which can be ejected (1) from the atom or transfer to an excited level (2). When the excited electron relaxes by returning to the inner shell, an element specific fluorescence emission (blue beam) occurs*

element under study can be ejected from the K, L or M layer of the atom (Fig. 3.7). When these excited electrons relax again, an element specific fluorescence emission occurs. This secondary emission is called X-ray fluorescence.

X-ray fluorescence techniques are widely used in materials chemistry and surface analysis. With the development of total reflection X-ray fluorescence (TXRF) spectroscopy, the technique has progressed to the field of trace metal quantification. In TXRF, the sample is irradiated with the X-ray beam at a very low angle (e.g., 0.1°) and total-reflection occurs. This reduces absorption and scattering phenomena with the consequence of a reduced background noise. As only the elements on the sample surface are irradiated, TXRF offers a strongly increased sensitivity compared with conventional X-ray fluorescence methods.

The intensity of the fluorescence emission relates to the amount/concentration of the studied element and the energy (usually indicated in keV) where the emission occurs is specific for the respective element. Consequently, TXRF enables both quantitative and qualitative analysis of many elements in a simultaneous manner.

3.3.2 Instrumental/Methodical Aspects of TXRF and Applications

For analysis, solutions or suspensions of a few microliters of the samples are deposited on a carrier disc and dried, which leads to the formation of a thin layer with diameters in the low micrometer range. In the TXRF instrument, the discs are irradiated with the X-ray beam and the emitted fluorescence intensity is recorded. The most common irradiation sources contain molybdenum or tungsten anodes.

Advantages, which make TXRF an "easy-to-use" technique for trace element analysis, include the simple sample pre-treatment procedures, the use of internal standard calibration, and the low number of instrumental parameters that influence the measurement outcome.

Parameters that have to be considered in method development include sample pre-treatment procedures (e.g., concerning homogenization and drying), measurement time, and the choice of internal standard. Since TXRF is still not a routine technology,

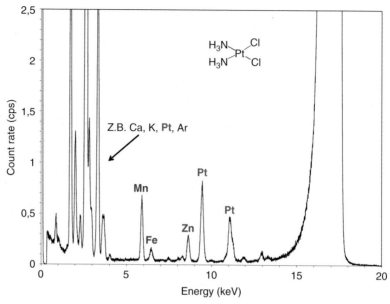

Figure 3.8 *Example of a TXRF spectrum: cisplatin (5.1 μM) in an HT-29 cell suspension; two specific platinum signals are shown among other elements of the sample matrix and the internal standard Mn*

only a limited number of applications have been reported. Therefore, comparison with established methods of trace metal quantification is difficult. However, recent literature reports suggest that this technique has high potential in pharmaceutical impurity analysis as well as for the bioanalysis of metal-based drugs (Fig. 3.8) [26–29].

3.4 Subcellular X-ray Fluorescence Imaging of a Ruthenium Analogue of the Malaria Drug Candidate Ferroquine Using Synchrotron Radiation

Synchrotron XRF imaging is a powerful technique that exploits the spectrally pure and finely focused X-ray beam from a synchrotron. It has opened up new application modes such as trace element analysis, surface analysis, chemical state analysis, and microanalysis. Digital images of microscopic or nanoscopic samples are built, pixel by pixel, by scanning the sample through the beam (Fig. 3.9).

The resulting X-ray fluorescence radiation is characteristic of the chemical elements at each pixel. Mathematical deconvolution of the fluorescence spectrum reveals the chemical composition, from which quantitative elemental images of the sample are assembled. They are often displayed as false-color maps. The combination of high flux and low emittance provided by synchrotron sources has proved to be crucial for the enormous success of experiments in the field of synchrotron-based X-ray fluorescence.

In biological applications, the maps may give a direct and clear observation of element occurrences in different regions of the sample. Elemental maps depicting the subcellular distribution and concentration of a certain element have great potential in biomedical

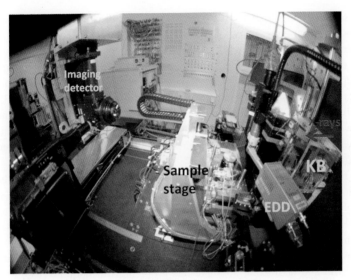

Figure 3.9 *Photography of the experimental setup of an XRF instrument. The imaging detector, the sample stage, the Kirkpatrick–Baez mirrors (KB) and the energy dispersive diffractometer (EDD) are depicted*

research, thanks to the low detection limit and the high spatial resolution of synchrotron X-ray fluorescence microscopy [30].

3.4.1 Application of X-ray Fluorescence in Drug Development Using Ferroquine as an Example

While the choice of drugs against malaria is limited, the organometallic compound ferroquine (FQ or SSR97193, Fig. 3.10) is currently providing new hope in the fight against this deadly parasitic disease. FQ was designed by starting from the chemical structure of the established antimalarial drug chloroquine (CQ, Fig. 3.10) [31].

CQ was used successfully for 50 years. However, the human malaria parasite *Plasmodium falciparum* became resistant to CQ [32]. In order to restore the susceptibility of CQ, a ferrocenyl moiety, which is stable, lipophilic, redox active, and is usually devoid of serious toxicity, was introduced into the structure [33]. FQ is active against both CQ-susceptible

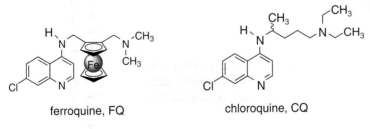

Figure 3.10 *Chemical structures of ferroquine (FQ) and chloroquine (CQ)*

and CQ-resistant *P. falciparum* and *P. vivax* strains and/or isolates [34]. FQ was tested in clinical studies in humans (phases I and II), alone or in combination with another antimalarial drug, that is, artesunate [35]. FQ is one of the most important antimalarial drug candidates currently in the pipeline. In future periods, focus should be set on the identification of an appropriate partner drug allowing the development of a single-dose antimalarial. Recent data concerning the mechanisms of action of FQ have been published [36–38]. The properties linked to the specificity of the metallocene, such as the lipophilicity and the redox behavior, were discussed [36]. Nevertheless, it was mandatory to obtain a detailed picture of the subcellular distribution of FQ within infected red blood cells in order to elucidate the mechanism of action of FQ.

In small red blood cells (RBC, size around 6–8 μm), *Plasmodium* digests hemoglobin in a specific organelle called the digestive vacuole (DV, Fig. 3.11). In the DV (size around 1–2 μm), the degraded compounds, which are potentially toxic to the parasite itself, are crystallized into hemozoin, an insoluble, brown pigment (also called the malaria pigment) that contains different iron species [39].

In this environment, it is clearly impossible to distinguish the different iron atoms of FQ, heme, hematin, and hemozoin. To solve this problem, ruthenoquine (RQ, Fig. 3.12), an analogue of FQ in which the iron atom was substituted by a ruthenium atom, was used as

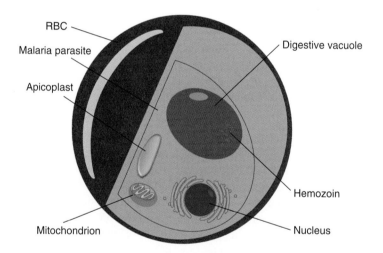

Figure 3.11   *Schematic representation of a red blood cell infected by Plasmodium falciparum*

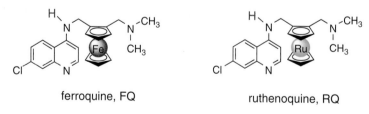

Figure 3.12   *Chemical structures of ferroquine (FQ) and ruthenoquine (RQ)*

a cold tracer [40]. Ruthenium was used due to its location just below iron in the Periodic Table of Elements. RQ has similar physico-chemical and antimalarial activities to FQ.

One of the main problems in tracing FQ/RQ under biological conditions was the sensitivity of the biophysical techniques available. An RBC with an estimated volume of about 87 μm^3 contains approximately 10^{14} atoms. If one assumes an estimated ruthenium concentration over the whole RBC of around 4 μM, 1 atom in 500 000 000 has to be detected. Therefore, the cellular uptake of RQ (Fig. 3.13) was determined by means of inductively coupled plasma mass spectrometry (ICP-MS) analyses via measurement of Ru in the CQ-susceptible HB3 strain of *P. falciparum* infected RBCs (iRBCs) exposed to a clinically relevant dose (IC_{90} value, 40 nM) as a function of time (5, 10, and 15 min) [40].

RQ accumulates quickly (10 min) and preferentially in iRBCs compared with RBCs with a ratio of 15:1 (iRBCs:RBCs) and to culture medium with a ratio of 180:1 (iRBCs:medium) (Table 3.1). This confirms to some extent the data of Sullivan *et al.* who showed that ^{3}H-labelled CQ ([^{3}H]CQ), accumulates preferentially in HB3 iRBCs compared with RBCs with a similar ratio of 23:1 (iRBCs:RBCs) but for a longer period of time (20–24 h) and a 30-fold higher concentration (1.2 μM) [41].

Total Ru concentration determination does not allow localizing the exact accumulation site in iRBCs. Owing to the presence of the ruthenium atom, RQ was used as staining agent in transmission electron microscopy (TEM) to provide contrast with the image. In untreated iRBCs, no particular contrast was detected. On the contrary, the DV membranes of parasites (HB3) in iRBCs treated with RQ (40 nM) after only 15 min developed a sharp contrast, suggesting that the compound became associated with the DV membranes [40].

After 30 min of treatment, the contrast increases for the DV membranes suggesting a higher concentration of ruthenium in the DV membranes. Moreover, a sharp contrast was now observed inside the DV close to the hemozoin crystals. Of note, some opacity also

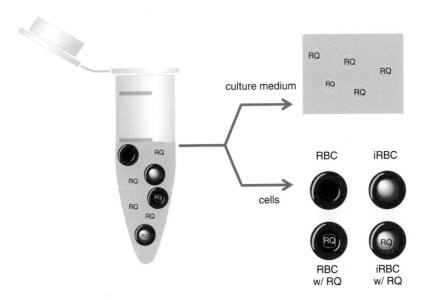

Figure 3.13 *Schematic representation of the samples analyzed by ICP-MS*

Table 3.1   *ICP-MS determination of Ru accumulation ratio between uninfected and infected erythrocytes treated with RQ as a function of time*

t (min)	Ru uptake in infected erythrocytes (%)	Ru uptake in uninfected erythrocytes (%)	Accumulation ratio
5	10.98	0.65	17
10	11.06	0.67	16
15	13.07	0.62	21
t (min)	RQ accumulation (%) in 1 μl of media	RQ accumulation (%) in 1 μl of infected erythrocytes	Relative concentration in infected erythrocytes
5	0.06	10.99	184
10	0.06	11.07	187
15	0.06	13.07	215

defined clear structures in the cytosol. Those structures were identified as lipid droplets, which are believed to catalyze the biocrystallization of heme into hemozoin.

To prove that the black dots observed by TEM are indeed ruthenium atoms, the X-ray emission characteristics of ruthenium were used, employing synchrotron chemical nano-imaging based on X-ray fluorescence (Fig. 3.14). The European Synchrotron Radiation Facility (ESRF) nano-imaging end-station can obtain a 60–100 nm X-ray spot range at 29 keV excitation energy with very high flux. This extraordinarily fine and bright focus offers the unique opportunity to rely on high sensitivity for K-shell excitation not

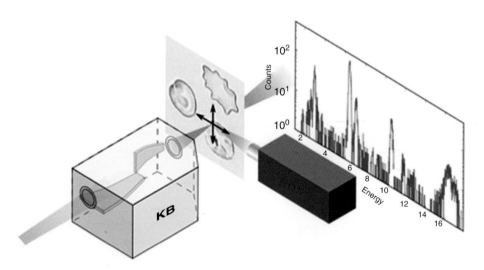

Figure 3.14   *X-ray fluorescence microscopy of individual red blood cells. The sample was scanned through the nanofocus produced by the KB optics of the ID22NI station while the X-ray fluorescence signal was collected*

only of ruthenium but also of iron, sulfur, and chlorine atoms, which are among the main elements of biological interest.

The experiments were performed with RBCs infected with the HB3 CQ-susceptible strain and treated (or not) with RQ (40 nM, 30 min). The iRBCs were then deposited on 500 nm thick silicon nitride windows, fixed in methanol, and washed with distilled water. Optical micrographs revealed an apparent normal morphology of iRBCs after the treatment and fixation procedures. Fig. 3.15 features a typical series of images both for control and RQ-treated iRBCs.

In untreated and RQ-treated iRBCs, the DV was located thanks to the intense iron signal, as a high iron content is characteristic of the hemozoin crystals within the DV. In RQ-treated iRBCs, no ruthenium signal could be detected for the whole cell as the background noise was higher than the total amount of ruthenium detected. However, an intense ruthenium signal was detected in the DV, clearly proving specific accumulation of ruthenium close to the malarial pigment.

Although such a nanoprobe is unique for X-ray methods, the high elemental sensitivity does not allow visualization of the membranes (thickness between 10 and 20 nm). Combination of these results with those previously obtained with TEM (spatial resolution:

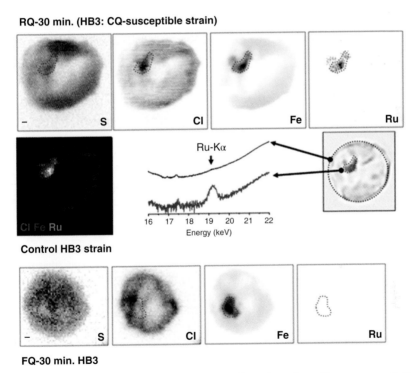

Figure 3.15 Synchrotron chemical nano-imaging based on X-ray fluorescence. Sulfur (S), chlorine (Cl), iron (Fe), and ruthenium (Ru) Kα X-ray fluorescence intensity maps determined in iRBCs exposed for 30 min to RQ (40 nM) and of unexposed control RBCs. The corresponding optical image of the analyzed iRBC shows that the integrity of chemically fixed cells is preserved and that hemozoin is well delineated (clear match with the iron distribution in iRBCs)

2 nm) on ultra-thin sections are complementary for all qualitative information. Nevertheless, X-ray emission characteristics of the chlorine atoms can also be used for chlorine mapping as RQ, FQ, and even CQ contain a chlorine atom at position 7 in the quinoline ring. Interestingly, the co-localization of ruthenium and chlorine atoms clearly evidences the specific accumulation of RQ (or of RQ derivative(s)) within the DV.

Additional experiments were thus performed with RBCs infected with the HB3 CQ-susceptible strain and treated by FQ or CQ (40 nM, 30 min) using chlorine as a tracer for the compounds. Whereas similar data were obtained for FQ and RQ, chlorine was distributed equally in CQ-treated iRBCs even if the ratio of total chlorine atoms in the DV is significantly higher than in the control. This kinetic difference is in agreement with data previously obtained by ICP-MS.

EXAFS (extended X-ray absorption fine structure) is the oscillating part of the X-ray absorption spectrum (XAS) that extends to about 1000 eV above the absorption edge of a particular element in a sample [42]. Different samples of RQ (neutral or diprotonated) were analyzed in the solid state and in aqueous solutions under oxidizing conditions (H_2O_2 or hemine) mimicking the environment of the DV. EXAFS spectroscopy provides information on the coordination number, the nature of the scattering atoms surrounding the absorbing atom, the interatomic distance between the absorbing atom and the backscattering atoms, and the Debye–Waller factor, which accounts for the disorder due to the static displacements and thermal vibrations. Using a combination of statistical methods, the detection of very subtle changes in the $Ru-C_{Cp}$ distances suggests the implication of non-binding doublets of the nitrogen atom of the aminoquinoline ring. It also suggests that RQ may be partially oxidized *in vitro* [43].

3.5   Mass Spectrometric Methods in Inorganic Chemical Biology

Mass spectrometry (MS) has found application at all stages of drug development, be it as a characterization method for new small molecules or biologicals, in the identification of metabolites or in target identification, pharmacokinetics, and proteomics. For years, research in inorganic chemical biology focused on the detection of reaction products between inorganic compounds and biological molecules. More recently, with the development of improved instrumentation, more sophisticated methods and techniques have been implemented. Most of the inorganic compounds used in these studies are related to the development of new cancer chemotherapeutics and contain elements that usually do not occur naturally in biological samples [44]. In this section, the most important mass spectrometric techniques will be summarized. Their strengths and disadvantages will be highlighted and discussed using case studies from the literature with a main focus on anticancer drug development.

Much of the research involving mass spectrometry was initially focused on the anticancer drug cisplatin and related second and third generation compounds. Platinum-based anticancer agents are estimated to be used in more than 50% of the chemotherapy regimens [45]. The modes of action of these antitumor agents are related to binding to the biological macromolecule DNA and its structural modifications, thereby hindering replication/transcription of tumor cells. Newer developments involve non-platinum anticancer compounds with biomolecular targets other than DNA.

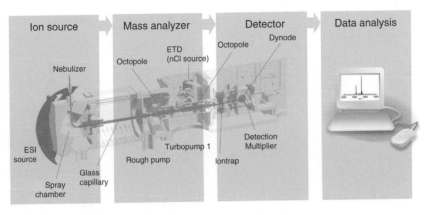

Figure 3.16 *The main components of a mass spectrometer*

The last decades have brought enormous progress in the development of new mass spectrometry instrumentation. This is true for virtually all aspects of mass spectrometers [46], from ion sources and mass analyzers to software solutions for data analysis (Fig. 3.16). Many metals have properties that make them well suited for analysis by MS: they exhibit characteristic masses and isotopic distributions, and the metals used as drug compounds in therapy and diagnosis are normally not found in biological samples. These properties allow unambigous identification of species containing metals without high risk of interference from other sources. Furthermore, high sensitivity instruments with high sample throughput and minute sample amounts as compared with other methods, make MS a powerful tool at many stages of drug development and research in inorganic chemical biology.

Among the major advances in mass spectrometry was the introduction of new ion sources that eventually allowed the analysis of high molecular weight molecules and of compounds with weak bonds. Electrospray ionization (ESI), matrix assisted laser desorption/ionization (MALDI) and inductively-coupled plasma (ICP) are currently the most frequently applied ionization methods in inorganic chemical biology (Fig. 3.17). The soft ionization methods ESI and MALDI allow the detection of intact biomolecules and also after adduct formation with metal moieties and they are often referred to as molecular mass spectrometry methods. ESI normally yields multiply charged ions, whereas in MALDI-MS mostly species are detected carrying a single charge. In both cases the charge stems from adduct formation with protons and alkali metal ions such as Na^+ or deprotonation. In contrast, ICP-MS is an element-specific method with unrivaled sensitivity. Given the fact that in this field of research the elements used are often not found in biological systems, this element-specific method is of great importance and is often also referred to as inorganic mass spectrometry.

In a mass spectrometer, after ionization, the ions are transferred into a mass analyzer. A variety of mass analyzers (and combinations thereof) are used in modern mass spectrometers. Mass analyzers such as quadrupoles (q), 3D and linear ion traps (ITs) as well as time-of-flight analyzers (ToFs) have been on the market for quite some time and many applications in inorganic chemical biology have been reported. More recently, although already known for some time, Fourier transform (FT) ion cyclotron resonance (ICR) has been used to solve problems where higher resolutions are needed. However,

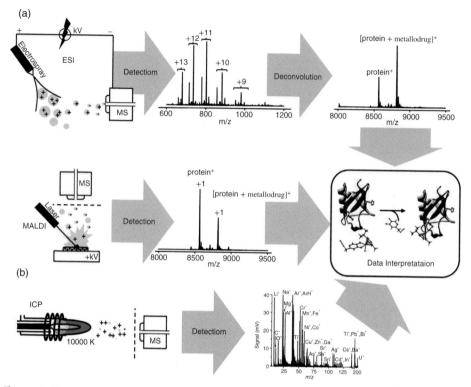

Figure 3.17 *From sample ionization to detection and data interpretation in (a) ESI- and MALDI-MS as compared with (b) element specific ICP-MS. Both ionization methods are extensively used in metallodrug research in combination with various mass analyzers. Parts of the figure were adapted from [47] with permission of Samuel M. Meier and reproduced with permission from [48] © 2013 Wiley-VCH Verlag GmbH & Co. KGaA, Weinheim*

the major breakthrough in recent years was the development of the orbitrap analyzer, another FT-based approach. The latter type of instruments delivers high resolution and therewith high mass accuracy, which is required in particular in proteomics and in life science-related applications.

Whereas with ICP quadrupoles are most frequently used, the softer ionization methods are usually linked to ion traps, ToFs, and FT-based analyzers. Both ionization method and mass analyzer choice contribute to the price of a mass spectrometer.

In order to gain additional information on a particular ion, such as on its structure and composition, tandem mass spectrometry methods are employed. Fragmentation is achieved by different methods. Collision induced dissociation (CID) is the standard method used and is found in most commercially available instruments. With the development of new mass analyzers, alternative fragmentation methods have been developed and nowadays infrared multiphoton dissociation (IRMPD), higher energy C-trap dissociation (HCD), electron transfer dissociation (ETD), and electron capture dissociation (ECD), among others, are available options in mass spectrometers. The underlying differences in the fragmentation mechanisms cause the techniques to often yield complementary data and thereby enhance the information gained [49]. This makes these interesting features when considering the purchase of new MS instruments.

Another important factor that will influence the quality of data gained from mass spectrometry is the coupling to a separation device before mass spectrometric analysis. Separation of biological sample mixtures in particular using liquid (or to a smaller extent gas) chromatography or electrophoresis allows the complexity of mass spectra to be reduced and leads to improved data sets.

3.5.1   Mass Spectrometry and Inorganic Chemical Biology: Selected Applications

Many of the mass spectrometric methods developed for application in inorganic chemical biology are related to the discovery of the anticancer activity of relatively simple, low molecular weight metal complexes [44]. These comprise cisplatin (Fig. 3.18), the first platinum anticancer agent, and its next generation derivatives carboplatin and oxaliplatin, which are widely used in the treatment of tumorigenic diseases [45]. Their modes of action are related to DNA binding, while serum protein interaction after intravenous administration is considered to lead to side effects of chemotherapy. In addition to platinum complexes, mainly Ru- but also Ga-based anticancer agents are currently being investigated (Fig. 3.18). Several examples have been translated into ongoing clinical trials (KP1339, NAMI-A, KP46) with the aim of reducing the side effects and resistance experienced with classic chemotherapeutics and to increase the number of treatable tumors. Many new metal-containing pharmacophores, especially those based on Ru (compare RAPTA-C and [Ru(arene)(en)Cl]$^+$ in Fig. 3.18), are currently under investigation as potential anticancer drugs.

3.5.1.1   *Interaction of metallodrugs with proteins*

Most of the registered anticancer metallodrugs as well as those in preclinical development are designed for intravenous administration. In the bloodstream the first binding partners are proteins, with human serum albumin (HSA) and transferrin (Tf) in particular as high

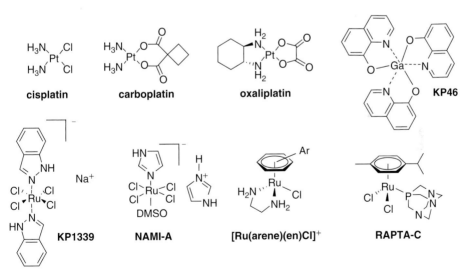

Figure 3.18   *The chemical structures of the anticancer agents cisplatin, carboplatin, and oxaliplatin, together with the formulae of a selection of metal-based anticancer drugs currently undergoing clinical trials or at advanced preclinical development*

abundance proteins being major targets [50]. Both proteins have been suggested to contribute to the delivery of anticancer active moieties to the tumor. HSA might act as a reservoir and contribute to the uptake into the tumor tissue through the enhanced permeability and retention effect (EPR) that allows macromolecules such as proteins to penetrate leaky blood vessels and thereby accumulate in the neoplasm. Transferrin receptors are overexpressed on tumor cells and loading Tf with cytotoxics may therefore provide a means for accumulation of anticancer drugs in the tumor with minor impact on healthy tissue.

Mass spectrometric methods have been used extensively to study the reactions of metallodrugs with blood proteins. The binding of anticancer Pt, Ru, and Ga complexes to serum proteins has been studied by CE- and HPLC-ICP-MS and also offline molecular MS. Since neither of these metals are normally found in organisms, the element selectivity of ICP-MS can be exploited to determine the fate in complex biological matrices (e.g., see the analysis of clinical blood plasma samples of a patient treated with the Ru(III) compound KP1019 in Fig. 3.19). In general, these metal complexes show high affinity for donor atoms of biomolecules and, for example, the Ru fragment of KP1019 was found mainly attached to HSA, whereas binding to transferrin was confirmed for the Ga complex KP46. For both of these complexes, the transport through serum protein binding has been thought to be an essential part of their modes of action [50] and the low side effects observed during the treatment of cancer patients in clinical trials may be related to this property. On the other hand, platinum anticancer agents are thought to be deactivated through protein binding, which is, however, in contrast with clinical trials of protein conjugates of cisplatin in the 1980s, which showed similar activity to cisplatin *i.v.* at reduced side effects [51, 52]. The positively charged multinuclear Pt complex BBR3464 is another platinum complex which was studied in clinical trials and because of its cationic nature, non-covalent interactions with HSA were proposed based on ESI-MS studies [53].

The exact binding site of the metal fragment from metallodrugs on proteins such as HSA and Tf has been the subject of several studies. Recently, the role of Tf in the delivery of the organometallic drug candidate $[Ru(\eta^6\text{-biphenyl})(en)Cl]^+$ was investigated [55]. Tryptic digestion of the protein conjugates and analysis by HPLC-ESI-MS revealed that the complex coordinated to the N-donors His242, His273, His578, and His606, whereas cisplatin showed, under the same conditions, very different binding pattern. However, neither compound had an effect on the Fe(III) binding ability of Tf, nor interfered with the transferrin receptor interaction or endocytosis of Tf. Notably the Ru complex retained its biological activity whereas cisplatin was deactivated.

Furthermore, the role of the structure on the biological activity can be studied by MS. Cisplatin and its anticancer inactive trans isomer transplatin show differing reactivity and adduct formation with proteins. Whereas cisplatin tends to form bi- or even trifunctional adducts (binding to the Pt center via two or three coordinative bonds), transplatin tends to bind in a monofunctional manner, in both cases retaining some of the original ammonia/chlorido ligands. This behavior was studied, for example, by ESI-MS and MALDI-MS employing the model protein ubiquitin (Ub) [56]. These experiments also provided preliminary insight into the binding sites of the platinum complexes on the protein. The ammine is trans-activated in the case of cisplatin through methionine-1 binding at Ub, which was not observed for transplatin. This demonstrates how MS can contribute to the understanding of the mechanisms of action of metallodrugs through protein binding studies.

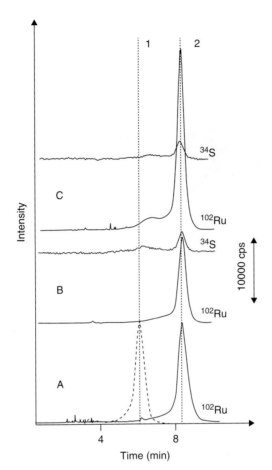

Figure 3.19 The plasma protein binding of the Ru compound KP1019 in a clinical setting: analysis of clinical samples (A, Tf and HSA standards after incubation with KP1019; B, 24 h after first infusion; C, after sixth infusion). Peak identification: 1, Tf adduct, 2, HSA adduct. Reproduced with permission from [54] © 2008 Wiley-VCH Verlag GmbH & Co. KGaA, Weinheim

Identification of protein targets in the cell is challenging due to the multitude of potential binding partners, the high affinity of metal ions to biological ligands and their donor atoms. However, several approaches have been applied to characterize the molecular targets of metal-based drugs in cells. The potential of using proteomic studies in the evaluation of cancer cell response to Pt, Au, As, and Ru complexes has been demonstrated by various groups [57–59]. Wolters *et al.* employed MS-based multi-dimensional protein identification technology (MudPIT) [59] and identified histones as playing an important role in the mode of action of RAPTA-T [59–60].

Most protein binding studies were conducted on anticancer metal complexes. More recently, Sun and colleagues reported an intriguing approach utilizing continuous-flow gel electrophoresis with ICP-MS to identify proteins in cells that contain metal ions [48]. One

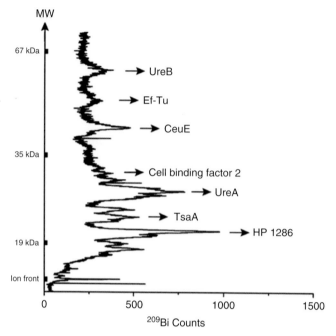

Figure 3.20 *Bi-binding protein profile in* H. pylori *analyzed by gel electrophoresis with ICP-MS detection of ^{209}Bi after treatment with Bi subcitrate. Reproduced with permission from [48] © 2013 Wiley-VCH Verlag GmbH & Co. KGaA, Weinheim*

example reported was the binding of Bi from Bi subcitrate as a treatment of *H. pylori*, which is responsible for the formation of ulcers and is also related to stomach cancer [61]. The combination with peptide mass fingerprinting by MALDI-ToF-MS allowed identification of the Bi-containing peptides as shown in Fig. 3.20 [48]. This is another example of where the combination of complementary methods has provided important insight into the mode of action of biologically active metal systems.

3.5.1.2   *Binding of metallodrugs to DNA and DNA model compounds*

Anticancer platinum drugs exhibit their tumor-inhibiting activity through binding to DNA and in particular adjacent guanine bases to form bifunctional adducts [45]. Such interaction with DNA induces structural changes and eventually leads to induction of apoptosis of the cancer cell. Many analytical methods have been used to investigate the binding of metal complexes with DNA or its components, which are often referred to as DNA models, with the most prominent being guanosine 5'-monophosphate (5'-GMP and also the 2-deoxy variant 5'-dGMP). The employed techniques include NMR spectroscopy, capillary electrophoresis (CE), high-performance liquid chromatography (HPLC), X-ray diffraction analysis and also mass spectrometry [62–68]. NMR spectroscopy, MS, and X-ray diffraction allow the structural characterization [69] of adducts formed between metal complexes and (oligo)nucleotides and other DNA models [44, 68, 70–73]. In contrast, the separation techniques CE and HPLC (also hyphenated to MS detectors) have been mainly used to study the binding kinetics of such reactions [66, 74–80].

Hyphenation of separation methods with MS gives complementary information. For example, it was shown that the cisplatin interaction with the DNA model 5'-GMP is dependent on the incubation buffer [81, 82]. By employing CE with ESI-MS, not only could the kinetics of the reaction be elucidated but also a variety of adducts containing buffer constituents were identified and characterized [82]. Such studies provide basic information for the development of new anticancer complexes, since the screening of Pt anticancer drug candidates in particular towards DNA models has become a routine step in the early development stages. In analogy, similar approaches were also taken for non-platinum complexes. The reactivity of different Ru-, Rh-, and Os-PTA complexes towards guanine was monitored by ESI-MS [83], whereas the adduct formation between the Ru compounds KP1019, its imidazole analogue KP418 and RAPTA-C and dGMP was analysed by CE-MS techniques [84, 85].

The major disadvantage of using single DNA nucleotides, nucleosides or even only the nucleobases is the lack of influence of the 3D structure of the macromolecule DNA on the binding event. For more realistic conditions, short oligonucelotides of typically less than 15 bases have been used. By careful selection of the sequence of the bases, important information on the binding behavior and preference of metal complexes can be elucidated. ESI- and MALDI-MS were combined early on with enzymatic digestion for elucidation of the binding sites of a cisplatin, multinuclear Pt as well as Rh compounds on oligonucleotides, also confirming guanosine as a preferred binding partner [86–89]. An LC-ESI-MS method for the detection and quantification of GG and AG intrastrand cross-links induced by oxaliplatin was successfully applied to mouse tissue samples [90]. In an FT-ICR-MS study with different fragmentation techniques, preferential cross-linking of adjacent guanosine residues by cisplatin was demonstrated [91], confirming the major binding mode of anticancer platinum complexes to DNA. Similarly, the mechanism of dissociation of platinated oligonucleotides in the gas phase was analyzed by ESI-qToF-MS^2 [92]. In an elegant approach, the kinetics of the adduct formation between cisplatin and oligonucleotides was studied by coupling continuous gel electrophoresis (GE) to ICP- and MALDI-MS. By the combination of these techniques, quantitative as well as structural information could be gained in a complementary way (Fig. 3.21) [93]. MALDI-MS was also used to study the adduct formation between a 14-mer oligonucleotide and RAPTA-C [94], HPLC-ESI-MS was employed to probe the competition reactions of an Ru-arene complex with histidine, cytochrome c, and a 14-mer oligonucleotide [95].

3.5.1.3   Metal-distribution in living systems

ICP-MS with its high sensitivity and element specificity has been used for the quantification of metallodrugs in biological tissue after digestion. However, by digesting the sample, no spatial resolution of the metal distribution can be achieved, for example, in organs. Laser ablation (LA)-ICP-MS provides a means to overcome this problem and allows the scanning of different materials for the metal distribution [96]. The method has, for example, been successfully applied to study the distribution of Pt in kidneys from cisplatin-treated rats (Fig. 3.22) [97]. Exploiting the high spatial resolution (down to 8 μm) for metal mapping in renal substructures and combination of the results with histological studies allowed correlation of the Pt accumulation in the cortex and corticomedullary regions with renal damage. Furthermore, ICP-MS allows parallel measurement of elements and in this case

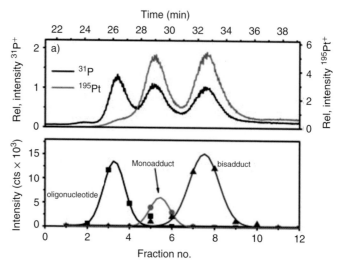

Figure 3.21 *Complementary GE-ICP-MS (top) and GE-MALDI-MS (bottom) data obtained from the reaction of 5′-TCCGGTCC-3′ and cisplatin. Reproduced with permission from [94] © 2008 Wiley-VCH Verlag GmbH & Co. KGaA, Weinheim*

determination of Zn and Cu distributions indicated their displacement by Pt in renal cells. Similar approaches were also followed for a variety of different applications, such as for analyses involving Gd-based contrast agents [98], or allowing quantitative imaging of Cu, Fe, Mn, and Zn in a mouse model with unilateral 6-hydroxydopamine (6-OHDA) lesions [99]. The major drawbacks of LA-ICP-MS are, however, long acquisition times as well as difficult quantification and low sensitivity [96].

3.5.1.4 Tagging of proteins with metal complexes

Proteins are considered as potential carrier systems for some metallodrugs. However, by selectively functionalizing proteins with metal-based tags, it is also possible to monitor unambiguously their fate in complex biological systems (see also Chapters 1 and 2 for additional information about the use of metal-based tags to purify proteins or to determine structures of biomolecules). Whetstone *et al.* reported different element-coded metal chelates to afford identification of a tagged peptide from a complex mixture [100]. In a proof-of-principle approach, a synthetic peptide was modified at a Cys residue with polydentate ligands coordinated to terbium or yttrium. Digestion by trypsin, affinity column chromatography, and LC-MS/MS demonstrated the suitability of the approach. DOTA–rare earth element complexes have been found advantageous for use as affinity tags because of their high polarity and water-solubility and as many rare earth elements are monoisotopic. Tanner and co-workers applied a related approach to the quantification of multiple proteins in complex biological samples [101]. The use of distinguishable element-tagged antibodies allows their detection of proteins by ICP-MS in a concentration range as low as $2-100 \, ng \, ml^{-1}$.

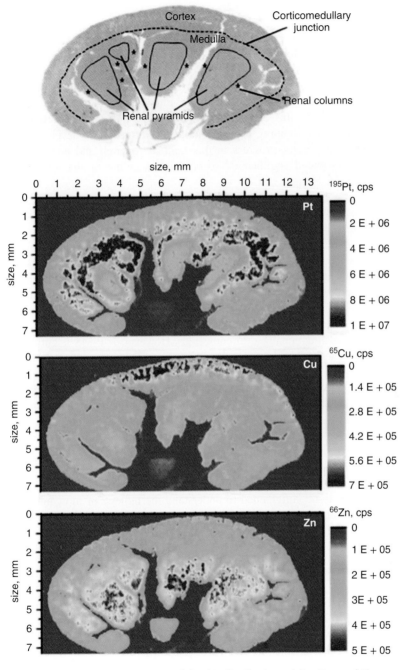

Figure 3.22 *Simultaneous monitoring of the biodistribution of Pt, Cu, and Zn on a sagittal section from a rat kidney after treatment with a single dose of 16 mg kg^{-1} cisplatin by LA-ICP-MS. Histological staining indicates the cortex, medulla, and corticomedullary junction (dotted lines). Reprinted with permission from [97] © 2011 American Chemical Society*

3.6 Conclusions

Modern analytical techniques have contributed significantly to the development of novel metal-based therapeutics for treatment of a wide variety of diseases. However, it soon became obvious that employment of complementary methods is required to obtain unambiguous information on the biological activity of metal-containing species, as each method has its own strengths and disadvantages. Element-specific methods play a major role in such studies, and this element specificity offers a unique opportunity to learn about the speciation, distribution, and even molecular targets. The development of such techniques has led to major progress in this field of research and novel bioanalytical methods have contributed significantly to our understanding of the role of metals in biological processes.

A multitude of applications exploiting the specific properties of metals in bioanalysis have been reported. Mass spectrometry with its diversity of instruments, ionization methods, and analyzers, each of them with its own strength, has proven to be one of the most powerful analysis techniques. In particular, the modern instruments with extremely high resolutions allow determination of small differences in the molecular weights and therefore accurate elemental composition. Small inorganic modifications to biomolecules based on mass shifts and isotope distributions are highly indicative of many elements and allow unambiguous speciation. With the emergence of LA-ICP-MS and MALDI imaging, MS has also obtained a spatial dimension and now allows determination of the distribution of metals and metal-functionalized biomolecules. However, ICP-MS is currently mainly used in inorganic chemical biology as a detector hyphenated to separation methods or after digestion of biological tissue for metal quantification. The latter is also the strength of AAS methods and new methodology allowing the reduction of interferences has boosted its use in this research area in recent years. Among the available complementary tools for the study of metals in biology and medicine, synchrotron-based X-ray spectroscopy enables sub-cellular chemical imaging and therefore new opportunities. However, this comes with the disadvantage of requiring a synchrotron radiation source.

Within this book chapter, selected examples demonstrating the advantages and applicability of the various methods have been discussed. However, we still lack fundamental knowledge, which needs to be enlightened to understand the biology of inorganic substances. Application of novel methodology in creative approaches and development of even more sophisticated instrumentation will help us manage this task.

Acknowledgements

C.B. would like to thank Dr Sylvain Bohic (Inserm, U836, ESRF) for helpful discussion. C.G.H. is grateful for support from the University of Auckland and thanks all the students and collaborators involved in the MS research area. The authors acknowledge financial support from the COST actions D39, CM1105 and CM0902.

References

1. H. Haraguchi, Metallomics as integrated biometal science, *J. Anal. At. Spectrom.*, **19**(1), 5–14 (2004).

2. M. Groessl and C. G. Hartinger, Anticancer metallodrug research analytically painting the "omics" picture-current developments and future trends, *Anal. Bioanal. Chem.*, **405**(6), 1791–1808 (2013).

3. D. A. Skoog and J. L. Leary, Instrumentelle Analytik, *Springer-Verlag (Berlin)* (1996).

4. K. Cammann, Instrumentelle Analytische Chemie, *Spektrum Akademischer Verlag (Heidelberg)* (2001).

5. R. A. Kellner, J.-M. Mermet, M. Otto, M. Valcarel and H. M. Widmer, Analytical Chemistry, *Wiley-VCH Verlag GmbH (Weinheim)* (2004).

6. B. Welz and M. Sperling, Atomabsorptionsspektrometrie, *Wiley-VCH Verlag, GmbH (Weinheim)* (1997).

7. B. Welz, H. Becker-Ross, S. Florek and U. Heitmann, High-Resolution Continuum Source AAS, *Wiley-VCH Verlag GmbH (Weinheim)* (2005).

8. B. Welz, High-resolution continuum source AAS: the better way to perform atomic absorption spectrometry, *Anal. Bioanal. Chem.*, **381**, 69–71 (2005).

9. M. Resano and E. García-Ruiz, High-resolution continuum source graphite furnace atomic absorption spectrometry: Is it as good as it sounds? A critical review, *Anal. Bioanal. Chem.*, **399**, 323–330 (2011).

10. S. Schäfer, I. Ott, R. Gust and W. S. Sheldrick, Influence of the polypyridyl (pp) ligand size on the DNA binding properties, cytotoxicity and cellular uptake of organoruthenium(II) complexes of the type $[(\eta^6\text{-}C_6Me_6)Ru(L)(pp)]^{n+}$ [L = Cl, n = 1; L = $(NH_2)_2CS$, n = 2], *Eur. J. Inorg. Chem.*, 3034–3046 (2007).

11. V. Pierroz, T. Joshi, A. Leonidova, C. Mari, J. Schur, I. Ott, L. Spiccia, S. Ferrari and G. Gasser, Molecular and cellular characterization of the biological effects of ruthenium(II) complexes incorporating 2-pyridyl-2-pyrimidine-4-carboxylic acid, *J. Am. Chem Soc.*, **134**, 20376–20387 (2012).

12. I. Ott, On the medicinal chemistry of gold complexes as anticancer drugs, *Coord. Chem. Rev.*, **253**, 1670–1681 (2009).

13. F. Magherini, A. Modesti, L. Bini, M. Puglia, I. Landini, S. Nobili, E. Mini, M. A. Cinellu, C. Gabbiani and L. Messori, Exploring the biochemical mechanisms of cytotoxic gold compounds: a proteomic study, *J. Biol. Inorg. Chem.*, **15**, 573–582 (2010).

14. A. Casini and L. Messori, Molecular mechanisms and proposed targets for selected anticancer gold compounds, *Curr. Top. Med. Chem.*, **11**(21), 2647–2660 (2011).

15. I. Ott, H. Scheffler and R. Gust, Development of a method for the quantification of the molar gold concentration in tumour cells exposed to gold-containing drugs, *ChemMedChem*, **2**, 702–707 (2007).

16. C. P. Bagowski, Y. You, H. Scheffler, D. H. Vlecken, D. J. Schmitz and I. Ott, Naphthalimide gold(I) phosphine complexes as anticancer metallodrugs, *Dalton Trans.*, 10799–10805 (2009).

17. I. Ott, X. Qian, Y. Xu, D. H. W. Vlecken, I. J. Marques, D. Kubutat, J. Will, W. S. Sheldrick, P. Jesse, A. Prokop and C. P. Bagowski, A gold(I) phosphine complex containing a naphthalimide ligand functions as a TrxR inhibiting antiproliferative agent and angiogenesis inhibitor, *J. Med. Chem.*, **52**, 763–770 (2009).

18. M. F. Brana and A. Ramos, Naphthalimides as anticancer agents: synthesis and biological activity, *Curr. Med. Chem. Anti-Cancer Agents*, **1**, 237–255 (2001).

19. L. Oehninger, R. Rubbiani and I. Ott, *N*-Heterocyclic carbene metal complexes in medicinal chemistry, *Dalton Trans.*, **42**, 3269–3284 (2013).

20. A. Gautier and F. Cisnetti, Advances in metal–carbene complexes as potent anti-cancer agents, *Metallomics*, **4**, 23–32 (2012).

21. W. Liu and R. Gust, Metal *N*-heterocyclic carbene complexes as potential antitumor metallodrugs, *Chem.Soc. Rev.*, **42**, 755–773 (2013).

22. R. Rubbiani, S. Can, I. Kitanovic, H. Alborzinia, M. Stefanopoulou, M. Kokoschka, S. Mönchgesang, W. S. Sheldrick, S. Wölfl and I. Ott, Comparative in vitro evaluation of *N*-heterocyclic carbene gold(I) complexes of the benzimidazolylidene type, *J. Med. Chem.*, **54**, 8646–8657 (2011).

23. J. S. Modica-Napolitano and J. R. Aprille, Delocalized lipophilic cations selectively target the mitochondria of carcinoma cells, *Adv. Drug Delivery Rev.*, **69**, 63–70 (2001).

24. P. Wobrauschek, Total reflection X-ray fluorescence analysis - a review, *X-ray Spectrom.*, **36**, 289–300 (2007).

25. M. West, A. T. Ellis, P. J. Potts, C. Streli, C. Vanhoof, D. Wegrzynekf and P. Wobrauschek, Atomic spectrometry update–X-ray fluorescence spectrometry, *J. Anal. At. Spectrom.*, **25**, 1503–1545 (2010).

26. B. J. Shaw, D. J. Semin, M. E. Rider and M. R. Beebe, Applicability of total reflection X-ray fluorescence (TXRF) as a screening platform for pharmaceutical inorganic impurity analysis, *J. Pharm. Biomed. Anal.*, **63**, 151–159 (2012).

27. F. J. Antosz, Y. Xiang, A. R. Diaz and A. J. Jensen, The use of total reflectance X-ray fluorescence (TXRF) for the determination of metals in the pharmaceutical industry, *J. Pharm. Biomed. Anal.*, **62**, 17–22 (2012).

28. A. Meyer, S. Grotefend, A. Gross, H. Wätzig and I. Ott, Total reflection X-ray fluorescence spectrometry as a tool for the quantification of gold and platinum metallodrugs: Determination of recovery rates and precision in the ppb concentration range, *J. Pharm. Biomed. Anal.*, **70**, 713–717 (2012).

29. N. Szoboszlai, Z. Polgári, V. G. Mihucz and G. Záray, Recent trends in total reflection X-ray fluorescence spectrometry for biological applications, *Anal. Chim. Acta*, **633**, 1–18 (2009).

30. S. Bohic, M. Cotte, M. Salome, B. Fayard, M. Kuehbacher, P. Cloetens, G. Martinez-Criado, R. Tucoulou and J. Susini, Biomedical applications of the ESRF synchrotron-based microspectroscopy platform, *J. Struct. Biol.*, **177**(2), 248–258 (2012).

31. D. Dive and C. Biot, Ferrocene conjugates of chloroquine and other antimalarials: the development of ferroquine, a new antimalarial, *ChemMedChem*, **3**(3), 383–391 (2008).

32. P. G. Bray and S. A. Ward, Malaria chemotherapy - resistance to quinoline containing drugs in plasmodium-falciparum, *FEMS Microbiol. Lett.*, **113**(1), 1–7 (1993).

33. D. R. van Staveren and N. Metzler-Nolte, Bioorganometallic chemistry of ferrocene, *Chem. Rev.*, **104**(12), 5931–5985 (2004).

34. C. Biot, F. Nosten, L. Fraisse, D. Ter-Minassian, J. Khalife and D. Dive, The antimalarial ferroquine: from bench to clinic, *Parasite*, **18**(3), 207–214 (2011).

35. U.S. National Institutes of Health, Dose Ranging Study of Ferroquine With Artesunate in African Adults and Children With Uncomplicated Plasmodium Falciparum

Malaria (FARM). http://clinicaltrials.gov/ct2/show/NCT00988507 (accessed 13 November 2013).

36. F. Dubar, T. J. Egan, B. Pradines, D. Kuter, K. K. Ncokazi, D. Forge, J. F. Paul, C. Pierrot, H. Kalamou, J. Khalife, E. Buisine, C. Rogier, H. Vezin, I. Forfar, C. Slomianny, X. Trivelli, S. Kapishnikov, L. Leiserowitz, D. Dive and C. Biot, The antimalarial ferroquine: role of the metal and intramolecular hydrogen bond in activity and resistance, *ACS Chem. Biol.*, **6**(3), 275–287 (2011).

37. C. Biot, D. Taramelli, I. Forfar-Bares, L. A. Maciejewski, M. Boyce, G. Nowogrocki, J. S. Brocard, N. Basilico, P. Olliaro and T. J. Egan, Insights into the mechanism of action of ferroquine. Relationship between physicochemical properties and antiplasmodial activity, *Mol Pharmaceut*, **2**(3), 185–193 (2005).

38. F. Dubar, C. Slomianny, J. Khalife, D. Dive, H. Kalamou, Y. Guérardel, P. Grellier and C. Biot, The ferroquine antimalarial conundrum : redox activation and reinvasion inhibition, *Angew. Chem. Int. Ed.*, **52**, 7690–7693 (2013).

39. S. Pagola, P. W. Stephens, D. S. Bohle, A. D. Kosar and S. K. Madsen, The structure of malaria pigment beta-haematin, *Nature*, **404**(6775), 307–310 (2000).

40. F. Dubar, S. Bohic, C. Slomianny, J. C. Morin, P. Thomas, H. Kalamou, Y. Guerardel, P. Cloetens, J. Khalife and C. Biot, In situ nanochemical imaging of label-free drugs: a case study of antimalarials in *Plasmodium falciparum*-infected erythrocytes, *Chem. Commun.*, **48**(6), 910–912 (2012).

41. D. J. Sullivan, I. Y. Gluzman, D. G. Russell and D. E. Goldberg, On the molecular mechanism of chloroquine's antimalarial action, *Proc. Natl. Acad. Sci. USA*, **93**(21), 11865–11870 (1996).

42. F. de Groot, High resolution X-ray emission and X-ray absorption spectroscopy, *Chem. Rev.*, **101**(6), 1779–1808 (2001).

43. E. Curis, F. Dubar, I. Nicolis, S. Benazeth and C. Biot, Statistical methodology for the detection of small changes in distances by EXAFS: Application to the antimalarial ruthenoquine, *J. Phys. Chem. A*, **116**(23), 5577–5585 (2012).

44. C. G. Hartinger, M. Groessl, S. M. Meier, A. Casini and P. J. Dyson, Application of mass spectrometric techniques to delineate the modes-of-action of anticancer metallodrugs, *Chem. Soc. Rev.*, **42**, 6186–6199 (2013).

45. M. A. Jakupec, M. Galanski, V. B. Arion, C. G. Hartinger and B. K. Keppler, Antitumour metal compounds: more than theme and variations, *Dalton Trans.*, (2), 183–194 (2008).

46. R. B. Cole (ed.) Electrospray and MALDI Mass Spectrometry: Fundamentals, Instrumentation, Practicalities, and Biological Applications: Fundamentals, Instrumentation, and Applications, 2nd edn, *John Wiley & Sons, Inc. (Hoboken)* 2010.

47. J. Pelka, H. Gehrke, A. Rechel, M. Kappes, F. Hennrich, C. G. Hartinger and D. Marko, DNA damaging properties of single walled carbon nanotubes in human colon carcinoma cells, *Nanotoxicology*, **7**(1), 2–20 (2013).

48. L. Hu, T. Cheng, B. He, L. Li, Y. Wang, Y.-T. Lai, G. Jiang and H. Sun, Identification of metal-associated proteins in cells by using continuous-flow gel electrophoresis and inductively coupled plasma mass spectrometry, *Angew. Chem. Int. Ed.*, **52**(18), 4916–4920 (2013).

49. M. L. Nielsen, M. M. Savitski and R. A. Zubarev, Improving protein identification using complementary fragmentation techniques in Fourier transform mass spectrometry, *Mol. Cell. Proteomics*, **4**(6), 835–845 (2005).

50. A. R. Timerbaev, C. G. Hartinger, S. S. Aleksenko and B. K. Keppler, Interactions of antitumor metallodrugs with serum proteins: Advances in characterization using modern analytical methodology, *Chem. Rev.*, **106**(6), 2224–2248 (2006).

51. J. D. Holding, W. E. Lindup, C. van Laer, G. C. Vreeburg, V. Schilling, J. A. Wilson and P. M. Stell, Phase I trial of a cisplatin-albumin complex for the treatment of cancer of the head and neck, *Br. J. Clin. Pharmacol.*, **33**(1), 75–81 (1992).

52. F. Kratz, Drug conjugates with albumin and transferrin, *Expert Opin. Therap. Pat.*, **12**(3), 433–439 (2002).

53. E. I. Montero, B. T. Benedetti, J. B. Mangrum, M. J. Oehlsen, Y. Qu and N. P. Farrell, Pre-association of polynuclear platinum anticancer agents on a protein, human serum albumin. Implications for drug design, *Dalton Trans.*, (43), 4938–4942 (2007).

54. M. Groessl, C. G. Hartinger, K. Polec-Pawlak, M. Jarosz and B. K. Keppler, Capillary electrophoresis hyphenated to inductively coupled plasma-mass spectrometry: a novel approach for the analysis of anticancer metallodrugs in human serum and plasma, *Electrophoresis*, **29**(10), 2224–2232 (2008).

55. W. Guo, W. Zheng, Q. Luo, X. Li, Y. Zhao, S. Xiong and F. Wang, Transferrin serves as a mediator to deliver organometallic ruthenium(II) anticancer complexes into cells, *Inorg. Chem.*, **52**(9), 5328–5338 (2013).

56. C. Scolaro, A. B. Chaplin, C. G. Hartinger, A. Bergamo, M. Cocchietto, B. K. Keppler, G. Sava and P. J. Dyson, Tuning the hydrophobicity of ruthenium(II)-arene (RAPTA) drugs to modify uptake, biomolecular interactions and efficacy, *Dalton Trans.*, (**43**), 5065–5072 (2007).

57. F. Guidi, A. Modesti, I. Landini, S. Nobili, E. Mini, L. Bini, M. Puglia, A. Casini, P. J. Dyson, C. Gabbiani and L. Messori, The molecular mechanisms of antimetastatic ruthenium compounds explored through DIGE proteomics, *J. Inorg. Biochem.*, **118**, 94–99 (2013).

58. X. Sun, C.-N. Tsang and H. Sun, Identification and characterization of metallodrug binding proteins by (metallo)proteomics, *Metallomics*, **1**(1), 25–31 (2009).

59. D. A. Wolters, M. Stefanopoulou, P. J. Dyson and M. Groessl, Combination of metallomics and proteomics to study the effects of the metallodrug RAPTA-T on human cancer cells, *Metallomics*, **4**(11), 1185–1196 (2012).

60. B. Wu, M. S. Ong, M. Groessl, Z. Adhireksan, C. G. Hartinger, P. J. Dyson and C. A. Davey, A ruthenium antimetastasis agent forms specific histone protein adducts in the nucleosome core, *Chem-Eur J*, **17**(13), 3562–3566 (2011).

61. B. J. Marshall and J. R. Warren, Unidentified curved bacilli in the stomach of patients with gastritis and peptic ulceration, *The Lancet*, **323**(8390), 1311–1315 (1984).

62. A. M. Fichtinger-Schepman, A. T. van Oosterom, P. H. Lohman and F. Berends, *cis*-Diamminedichloroplatinum(II)-induced DNA adducts in peripheral leukocytes from seven cancer patients: quantitative immunochemical detection of the adduct induction and removal after a single dose of *cis*-diamminedichloroplatinum(II), *Cancer Res.*, **47**(11), 3000–3004 (1987).

63. J. L. Beck, M. L. Colgrave, A. Kapur, P. Iannitti-Tito, S. F. Ralph, M. M. Sheil, A. Weimann and G. Wickham, Electrospray and tandem mass spectrometry of drug-DNA complexes, *Adv. Mass Spectrom.*, **15**, 175–192 (2001).

64. J. L. Beck, M. L. Colgrave, S. F. Ralph and M. M. Sheil, Electrospray ionization mass spectrometry of oligonucleotide complexes with drugs, metals, and proteins, *Mass Spectrom. Rev.*, **20**(2), 61–87 (2001).

65. J. M. Koomen, B. T. Ruotolo, K. J. Gillig, J. A. McLean, D. H. Russell, M. Kang, K. R. Dunbar, K. Fuhrer, M. Gonin and J. A. Schultz, Oligonucleotide analysis with MALDI-ion-mobility-TOFMS, *Anal. Bioanal. Chem.*, **373**(7), 612–617 (2002).

66. C. G. Hartinger and B. K. Keppler, Capillary electrophoresis in anticancer metallo-drug research - an update, *Electrophoresis*, **28**(19), 3436–3446 (2007).

67. C. G. Hartinger, W. H. Ang, A. Casini, L. Messori, B. K. Keppler and P. J. Dyson, Mass spectrometric analysis of ubiquitin-platinum interactions of leading anticancer drugs: MALDI versus ESI, *J. Anal. At. Spectrom.*, **22**(8), 960–967 (2007).

68. S. L. Kerr, T. Shoeib and B. L. Sharp, A study of oxaliplatin-nucleobase interactions using ion trap electrospray mass spectrometry, *Anal. Bioanal. Chem.*, **391**(6), 2339–2348 (2008).

69. E. R. Jamieson and S. J. Lippard, Structure, recognition, and processing of cisplatin-DNA adducts, *Chem. Rev.*, **99**(9), 2467–2498 (1999).

70. S. E. Sherman, D. Gibson, A. H. J. Wang and S. J. Lippard, Crystal and molecular structure of *cis*-[Pt(NH3)2[d(pGpG)]], the principal adduct formed by *cis*-diamminedichloroplatinum(II) with DNA, *J. Am. Chem. Soc.*, **110**(22), 7368–7381 (1988).

71. H. Huang, L. Zhu, B. R. Reid, G. P. Drobny and P. B. Hopkins, Solution structure of a cisplatin-induced DNA interstrand cross-link, *Science*, **270**(5243), 1842–1845 (1995).

72. S. O. Ano, F. P. Intini, G. Natile and L. G. Marzilli, A novel head-to-head conformer of d(GpG) cross-linked by Pt: New light on the conformation of such cross-links formed by Pt anticancer drugs, *J. Am. Chem. Soc.*, **120**(46), 12017–12022 (1998).

73. S. Komeda, T. Moulaei, K. K. Woods, M. Chikuma, N. P. Farrell and L. D. Williams, A third mode of DNA binding: Phosphate clamps by a polynuclear platinum complex, *J. Am. Chem. Soc.*, **128**(50), 16092–16103 (2006).

74. R. Da Col, L. Silvestro, C. Baiocchi, D. Giacosa and I. Viano, High-performance liquid chromatographic-mass spectrometric analysis of cis-dichlorodiamineplatinum-DNA complexes using an ionspray interface, *J. Chromatogr.*, **633**(1–2), 119–128 (1993).

75. F. Reeder, Z. Guo, P. D. S. Murdoch, A. Corazza, T. W. Hambley, S. J. Berners-Price, J.-C. Chottard and P. J. Sadler, Platination of a GG site on single-stranded and double-stranded forms of a 14-base oligonucleotide with diaqua cisplatin followed by NMR and HPLC. Influence of the platinum ligands and base sequence on 5'-G versus 3'-G platination selectivity, *Eur. J. Biochem.*, **249**(2), 370–382 (1997).

76. A. Zenker, M. Galanski, T. L. Bereuter, B. K. Keppler and W. Lindner, Capillary electrophoretic study of cisplatin interaction with nucleoside monophosphates, di- and trinucleotides, *J. Chromatogr. A*, **852**(1), 337–346 (1999).

77. D. B. Strickmann, A. Küng and B. K. Keppler, Application of capillary electrophoresis-mass spectrometry for the investigation of the binding behavior of oxaliplatin to 5'-GMP in the presence of the sulfur-containing amino acid L-methionine, *Electrophoresis*, **23**(1), 74–80 (2002).

78. E. Volckova, L. P. Dudones and R. N. Bose, HPLC determination of binding of cisplatin to DNA in the presence of biological thiols: implications of dominant platinum-thiol binding for its anticancer action, *Pharm. Res.*, **19**(2), 124–131 (2002).

79. U. Warnke, C. Rappel, H. Meier, C. Kloft, M. Galanski, C. G. Hartinger, B. K. Keppler and U. Jaehde, Analysis of platinum adducts with DNA nucleotides and nucleosides by capillary electrophoresis coupled to ESI-MS: Indications of guanosine 5'-monophosphate O6-N7 chelation, *ChemBioChem*, **5**(11), 1543–1549 (2004).

80. M. Groessl, C. G. Hartinger, P. J. Dyson and B. K. Keppler, CZE-ICP-MS as a tool for studying the hydrolysis of ruthenium anticancer drug candidates and their reactivity towards the DNA model compound dGMP, *J. Inorg. Biochem.*, **102**(5–6), 1060–1065 (2008).

81. R. C. Todd, K. S. Lovejoy and S. J. Lippard, Understanding the effect of carbonate ion on cisplatin binding to DNA, *J. Am. Chem. Soc.*, **129**(20), 6370–6371 (2007).

82. G. Grabmann, B. Keppler and C. Hartinger, A systematic capillary electrophoresis study on the effect of the buffer composition on the reactivity of the anticancer drug cisplatin to the DNA model 2'-deoxyguanosine 5'-monophosphate (dGMP), *Analytical and Bioanalytical Chemistry.*, **405**(20), 6417–6424 (2013).

83. A. Dorcier, C. G. Hartinger, R. Scopelliti, R. H. Fish, B. K. Keppler and P. J. Dyson, Studies on the reactivity of organometallic Ru-, Rh- and Os-pta complexes with DNA model compounds, *J. Inorg. Biochem.*, **102**(5–6), 1066–1076 (2008).

84. M. Groessl, C. G. Hartinger, P. J. Dyson and B. K. Keppler, CZE–ICP-MS as a tool for studying the hydrolysis of ruthenium anticancer drug candidates and their reactivity towards the DNA model compound dGMP, *J. Inorg. Biochem.*, **102**, 1060–1065 (2008).

85. P. Schluga, C. G. Hartinger, A. Egger, E. Reisner, M. Galanski, M. A. Jakupec and B. K. Keppler, Redox behavior of tumor-inhibiting ruthenium(III) complexes and effects of physiological reductants on their binding to GMP, *Dalton Trans.*, (14), 1796–1802 (2006).

86. F. Gonnet, F. Kocher, J. C. Blais, G. Bolbach, J. C. Tabet and J. C. Chottard, Kinetic analysis of the reaction between d(TTGGCCAA) and [Pt(NH$_3$)$_3$(H$_2$O]$^{2+}$ by enzymic degradation of the products and ESI and MALDI mass spectrometries, *J. Mass Spectrom.*, **31**(7), 802–809 (1996).

87. H. T. Chifotides, J. M. Koomen, M. Kang, S. E. Tichy, K. R. Dunbar and D. H. Russell, Binding of DNA purine sites to dirhodium compounds probed by mass spectrometry, *Inorg. Chem.*, **43**(20), 6177–6187 (2004).

88. R. Gupta, J. L. Beck, M. M. Sheil and S. F. Ralph, Identification of bifunctional GA and AG intrastrand crosslinks formed between cisplatin and DNA, *J. Inorg. Biochem.*, **99**(2), 552–559 (2005).

89. J. Zhu, Y. Zhao, Y. Zhu, Z. Wu, M. Lin, W. He, Y. Wang, G. Chen, L. Dong, J. Zhang, Y. Lu and Z. Guo, DNA cross-linking patterns induced by an antitumor-active trinuclear platinum complex and comparison with its dinuclear analogue, *Chem. Eur. J.*, **15**(21), 5245–5253 (2009).

90. R. C. Le Pla, K. J. Ritchie, C. J. Henderson, C. R. Wolf, C. F. Harrington and P. B. Farmer, Development of a liquid chromatography-electrospray ionization tandem mass spectrometry method for detecting oxaliplatin-DNA intrastrand cross-links in biological samples, *Chem. Res. Toxicol.*, **20**(8), 1177–1182 (2007).

91. A. E. Egger, C. G. Hartinger, H. Ben Hamidane, Y. O. Tsybin, B. K. Keppler and P. J. Dyson, High resolution mass spectrometry for studying the interactions of cisplatin with oligonucleotides, *Inorg. Chem.*, **47**(22), 10626–10633 (2008).

92. A. Nyakas, M. Eymann and S. Schuerch, The influence of cisplatin on the gas-phase dissociation of oligonucleotides studied by electrospray ionization tandem mass spectrometry, *J. Am. Soc. Mass Spectrom.*, **20**(5), 792–804 (2009).

93. W. Bruechert, R. Krueger, A. Tholey, M. Montes-Bayon and J. Bettmer, A novel approach for analysis of oligonucleotide-cisplatin interactions by continuous elution gel electrophoresis coupled to isotope dilution inductively coupled plasma mass spectrometry and matrix-assisted laser desorption/ionization mass spectrometry, *Electrophoresis*, **29**(7), 1451–1459 (2008).

94. W. H. Ang, E. Daldini, C. Scolaro, R. Scopelliti, L. Juillerat-Jeannerat and P. J. Dyson, Development of organometallic ruthenium-arene anticancer drugs that resist hydrolysis, *Inorg. Chem.*, **45**(22), 9006–9013 (2006).

95. F. Wang, J. Bella, J. A. Parkinson and P. J. Sadler, Competitive reactions of a ruthenium arene anticancer complex with histidine, cytochrome c and an oligonucleotide, *J. Biol. Inorg. Chem.*, **10**(2), 147–155 (2005).

96. I. Konz, B. Fernández, M. Fernández, R. Pereiro and A. Sanz-Medel, Laser ablation ICP-MS for quantitative biomedical applications, *Analytical and Bioanalytical Chemistry*, **403**(8), 2113–2125 (2012).

97. E. Moreno-Gordaliza, C. Giesen, A. Lazaro, D. Esteban-Fernandez, B. Humanes, B. Canas, U. Panne, A. Tejedor, N. Jakubowski and M. M. Gomez-Gomez, Elemental bioimaging in kidney by LA-ICP-MS as a tool to study nephrotoxicity and renal protective strategies in cisplatin therapies, *Anal. Chem.*, **83**(20), 7933–7940 (2011).

98. A. Sussulini, E. Wiener, T. Marnitz, B. Wu, B. Müller, B. Hamm and J. Sabine Becker, Quantitative imaging of the tissue contrast agent [Gd(DTPA)]$^{2-}$ in articular cartilage by laser ablation inductively coupled plasma mass spectrometry, *Contrast Media Mol. Imaging*, **8**(2), 204–209 (2013).

99. A. Sussulini, A. Matusch, M. Klietz, A. Bauer, C. Depboylu and J. S. Becker, Quantitative imaging of Cu, Fe, Mn and Zn in the L-DOPA-treated unilateral 6-hydroxydopamine Parkinson's disease mouse model by LA-ICP-MS, *Biomed. Spectrosc. Imaging, Chem. Eur. J*, **1**(2), 125–136 (2012).

100. P. A. Whetstone, N. G. Butlin, T. M. Corneillie and C. F. Meares, Element-coded affinity tags for peptides and proteins, *Bioconjug. Chem.*, **15**(1), 3–6 (2004).

101. Z. A. Quinn, V. I. Baranov, S. D. Tanner and J. L. Wrana, Simultaneous determination of proteins using an element-tagged immunoassay coupled with ICP-MS detection, *J. Anal. At. Spectrom.*, **17**(8), 892–896 (2002).

4

Metal Complexes for Cell and Organism Imaging

Kenneth Yin Zhang[a] and Kenneth Kam-Wing Lo[b]
[a]Key Laboratory for Organic Electronics & Information Displays and Institute of Advanced Materials, Nanjing University of Posts & Telecommunications, P.R. China
[b]Department of Biology and Chemistry, City University of Hong Kong, P.R. China

4.1 Introduction

Organic fluorophores [1], quantum dots [2], and genetically encoded fluorescent proteins [3] have been widely used in cell imaging to understand cellular structures, intracellular molecular interactions, and biological processes. In view of their rich and useful photophysical properties [4], luminescent inorganic complexes, including transition metal complexes and lanthanide chelates, have attracted increasing attention in the development of imaging reagents for cells and organisms [5]. In this chapter we describe the fundamental photophysical properties of luminescent inorganic complexes, common techniques to detect complexes in an intracellular environment, and summarize their applications in cell and organism imaging. We distinguish between the terms fluorescence and phosphorescence, and introduce two-photon absorption (TPA) and upconversion luminescence, in which low-energy excitation is converted into high-energy emission. The detection of complexes taken up by cells by confocal laser-scanning microscopy (CLSM), fluorescence lifetime imaging microscopy (FLIM), and flow cytometry through their luminescence is also described. In the section on cell and organism imaging, we explain the effects of the formal charge, lipophilicity, molecular size, and substrate on the cellular uptake properties of the complexes. The strategies of designing organelle-specific imaging reagents are also discussed. Additionally, we summarize recent applications of

Inorganic Chemical Biology: Principles, Techniques and Applications, First Edition. Edited by Gilles Gasser.
© 2014 John Wiley & Sons, Ltd. Published 2014 by John Wiley & Sons, Ltd.

two-photon and upconversion luminescence in cell and organism imaging. Furthermore, we review the strategies for sensing and labeling intracellular ions and molecules using inorganic complexes equipped with a recognition unit and/or a reactive functional group.

4.2   Photophysical Properties

4.2.1   Fluorescence and Phosphorescence

Luminescence occurs when an excited molecule undergoes relaxation by releasing a photon. Its efficiency is described as the quantum yield (Φ), which is the ratio of the number of photons emitted to the number of photons absorbed. Luminescence is mainly divided into fluorescence and phosphorescence, depending on whether the transitions are spin-allowed or not; emission generated from a singlet-to-singlet state is fluorescence while that from a triplet excited state to a singlet ground state is phosphorescence in nature. The electronic states and photophysical processes between these states are illustrated in a Jablonski diagram (Fig. 4.1). The ground, first, and second singlet, and first triplet excited states are depicted by S_0, S_1, S_2, and T_1, respectively. Each of these states has many vibrational energy levels. Absorption of a photon promotes a molecule to a singlet excited state, S_n ($n > 0$). Before luminescence occurs, the excited molecule undergoes internal conversion, where the electron either maintains its spin orientation, or swaps it through intersystem crossing. In the latter case, there is a transition from a singlet to a triplet state. In both cases, the electron ultimately finishes at the lowest excited state, that is, S_1 or T_1. In singlet excited states, the electron in the excited-state orbital is paired with the one in the ground-state orbital, and hence the fluorescence process is spin-allowed and takes place rapidly. In triplet excited states, the excited electron has the same spin orientation as the ground-state electron, and its transition to the ground state, that is, phosphorescence, is spin-forbidden, and takes a much longer time than fluorescence. Since triplet states are lower in energy compared with their corresponding singlet states, phosphorescence is always accompanied with a larger Stokes' shift (the difference in wavelength between the maxima of the absorption and emission spectra). In theory, transitions cannot occur directly from the ground state S_0 to triplet excited states, T_n. In practice, the possibility of such singlet–triplet processes is about 10^{-6} less than that of singlet–singlet absorption [4a–e].

In contrast to most organic molecules, which display fluorescence originating from the ($\pi \rightarrow \pi^*$) transitions with lifetimes in the nanosecond scale, luminescent transition

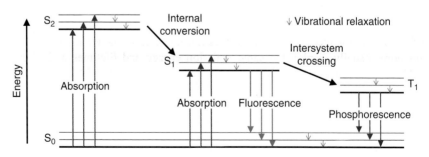

Figure 4.1   *Jablonski diagram illustrating the various electronic states and photophysical processes pathways between them*

metal complexes, including those of ruthenium(II), osmium(II), rhenium(I), iridium(III), rhodium(III), platinum(II), and gold(I), show long-lived phosphorescence in the micro- or sub-microsecond timescale [4f–h, 5c,j,l]. Such phosphorescence processes can involve both metal- and ligand-based orbitals, resulting in a diversity of emissive-state characters. Common emissive states include metal-to-ligand charge-transfer (MLCT), ligand-to-metal charge-transfer (LMCT), intraligand (IL), and ligand-to-ligand charge-transfer (LLCT). Complexes involving a breakage of covalent metal–ligand bonds upon excitation may exhibit high metal–ligand sigma-bond (σ-bond) character in the highest occupied molecular orbital (HOMO) and a ligand-based lowest unoccupied molecular orbital (LUMO). The phosphorescence of these complexes is probably dominated by sigma-bond-to-ligand charge-transfer (SBLCT) character. Excited states of metal-metal-to-ligand charge-transfer (MMLCT) and ligand-to-metal-metal charge transfer (LMMCT) characters are commonly found in planar polynuclear d^8 or d^{10} metal complexes, in which metal–metal interaction is observed. Lanthanides are unique luminescent metal ions that display emission from transitions involving only a redistribution of electrons within the 4f orbitals, because the effective shielding by 5s and 5p orbitals minimizes the influence of the external ligand field [4e,f, 5a,b]. As the transitions are spin-forbidden, lanthanides display very long emission lifetimes, usually in the millisecond timescale. Because their extinction coefficients are very small ($<10\,\mathrm{M}^{-1}\,\mathrm{cm}^{-1}$), the emissive states of lanthanides are populated through energy transfer from a sensitizing chelator instead of direct excitation. Therefore, the excitation spectra of lanthanide chelates reflect the absorption of their chelators.

Organic fluorophores have been used extensively as bioimaging reagents due to their high absorption and emission quantum yields, providing excellent detectability. The interesting and rich photophysical properties of inorganic transition metal complexes and lanthanide chelates render them attractive candidates for cell and organism imaging reagents. Advantages of using these complexes include: (1) they are photochemically inert, causing negligible photobleaching; (2) their high emission intensity results in high sensitivity; (3) their long emission lifetimes allow time-resolved applications; (4) their large Stokes' shifts minimize self-quenching; and (5) they show variable emission energy and can be used as multicolor probes.

4.2.2 Two-photon Absorption

Two-photon absorption (TPA) is simultaneous absorption of two photons by a single molecule resulting in an excited state with energy equal to the sum of the energy of the incident photons (Fig. 4.2) [4e]. There is no intermediate state for the molecule to reach before achieving the final excited state. TPA can be accomplished using low-energy excitation such as NIR (near-infrared) and this process is not constrained by selection rules for one-photon spectroscopy. TPA does not interfere in the emission process, and luminophores usually display the same emission energy and lifetimes as if they were excited by one-photon absorption with anti-Stokes' shifts. Since the NIR excitation allows deeper penetration, weaker autofluorescence, less photobleaching, and lower phototoxicity than the ultraviolet and visible light excitation, TPA technology has very high application potential in bioimaging [6]. Examples include kidney-dynamics visualization [7], neurobiology and brain studies [8], cardiovascular imaging [9], chemical and structural imaging of eyes [10], architecture phenomena in the skin [11], imaging and analysis of dynamic processes in living cells and tissues [12], clinical two-photon microendoscopy [13], noninvasive diagnostic procedures for the detection of malignancy in organs [14],

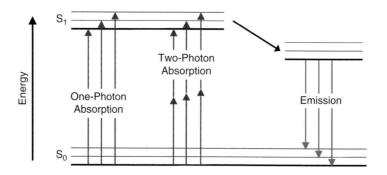

Figure 4.2   *Jablonski diagram illustrating the TPA process*

cancer imaging [15], lysosome and vascular imaging [16], and diagnostic tools for tissue engineering and drug delivery [17]. However, TPA suffers from two main limitations, which are: (1) the high peak powers required to increase the possibility that two photons are available for simultaneous absorption and (2) small TPA cross-sections (the analogue of extinction coefficients in linear absorption) that result in low emission efficiency. These would present a challenge to the design of equipment and luminescent reagents.

4.2.3   Upconversion Luminescence

Upconversion luminescence is a unique process in which sequential absorption of multiple low-energy photons leads to high-energy emission, for example, conversion from NIR into visible light. Similar to TPA, upconversion luminescence benefits from NIR excitation in the cell and organism imaging. It is different from TPA in that intermediate states are allowed for the occurrence of further photon absorption or energy transfer before the final emissive state is reached. Upconversion luminescence mainly occurs with lanthanide-doped nanoparticles [18]. There are three main types of processes that cause upconversion of lanthanide-doped materials, which are illustrated in Fig. 4.3: (a) excited-state absorption (ESA), (b) energy-transfer upconversion (ETU), and (c) photon avalanche (PA). In ESA, ground-state absorption (GSA) leads to the population of an intermediate metastable and long-lived excited state E1, which is followed by the second absorption that promotes the lanthanide ion from E1 to a higher excited state E2, where upconversion luminescence corresponding to the E2 → G transition occurs. The ETU involves energy transfer between two lanthanide ions at the E1 state. The donor undergoes nonradiative relaxation to the ground state while the acceptor receives the required energy to be excited to the E2 state, which undergoes luminescence. In the PA process, non-resonant GSA is followed by resonant ESA to excite one lanthanide ion to the E2 state. Cross-relaxation energy transfer between the excited ion at E2 and another ground-state ion results in the occupation of both ions at the intermediate state E1. Further ESA and cross-relaxation energy transfer take place circularly, promoting exponentially the population of the E2 state, where upconversion luminescence occurs like an avalanche. The upconversion efficiency of these processes follows the order ESA < ETU < PA. ETU is instant and pump-power independent, while PA depends on the pump power and shows a slow response to excitation because of numerous looping cycles of the ESA and cross-relaxation processes.

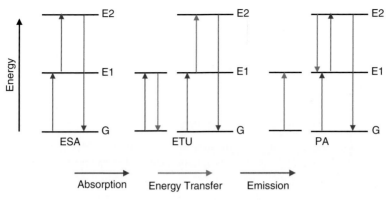

Figure 4.3 *Jablonski diagrams illustrating the three processes that cause upconversion luminescence*

Triplet–triplet annihilation (TTA) is another useful upconversion approach. It requires a triplet sensitizer and an acceptor. In a typical TTA process, the sensitizer is firstly excited to its singlet excited state (^{1}S*), which is followed by intersystem crossing (ISC) to the triplet state (^{3}S*), where triplet–triplet energy transfer (TTET) occurs to promote the acceptor to its triplet state (^{3}A*) (Fig. 4.4) [19]. TTA of two triplet acceptors eventually populates a ground state and a singlet excited state (^{1}A*), the latter of which decays radiatively, giving rise to upconversion luminescence. The upconversion quantum yield (Φ_{uc}) depends on the efficiency of all the processes including ISC, TTET, TTA, and the luminescence of the acceptor. Transition metal complexes are usually employed as the sensitizers since they undergo very high efficient ISC (almost 100%). TTET is, in most cases, a Dexter process, for which the rate constant sharply decreases with increasing distance between the donor and acceptor. Thus, covalent linking of the sensitizer and acceptor facilitates the energy transfer. The triplet energy levels of the acceptor and sensitizer should closely match to enhance the TTET efficiency. Additionally, the acceptor must primarily exhibit a high luminescence quantum yield to maximize the upconversion luminescence. Compared with lanthanide-based upconversion, TTA requires much lower excitation power density,

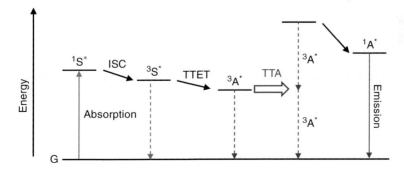

Figure 4.4 *Jablonski diagram illustrating the TTA upconversion process*

and exhibits higher overall upconversion capability ($\eta = \Phi_{uc} \times \epsilon$, where ϵ is the extinction coefficient), because the absorption of lanthanide materials is usually very weak.

4.3   Detection of Luminescent Metal Complexes in an Intracellular Environment

4.3.1   Confocal Laser-scanning Microscopy

Fluorescence microscopy is a powerful tool in various biological sciences since optical signals from a wide range of high-quality luminescent probes can provide important information with respect to the biochemical, biophysical, and structural status of cells and tissues [20]. In a typical fluorescence microscopy measurement, a sample is illuminated with light that excites luminescent probes contained in it and the luminescence is then imaged through a microscope objective (Fig. 4.5). As the entire sample is flooded evenly by the light source and all parts of the sample in the optical path are excited at the same time, the detected luminescence includes a considerable contribution from the unfocused background.

Confocal laser-scanning microscopy (CLSM) has gained popularity in the scientific and industrial communities, and has now become the most commonly used technique for optical slice imaging. It takes advantage of two pinholes to offer high-resolution optical images with depth selectivity. A light source passes through the excitation pinhole and is focused by an objective onto a small spot in the sample (Fig. 4.6). The luminescence (red line) from the focal plane is then focused to pass through the pinhole in front of the detector, whereas the light (green and blue lines) from below or above the focal plane is selectively rejected and cannot reach the detector. For a given objective, increasing the diameter of the pinhole allows more light to pass through, which improves sensitivity but decreases spatial resolution. As a general rule, a smaller pinhole gives a thinner focal plane and hence a higher resolution. However, once the pinhole size is decreased beyond the diffraction-limited Airy disk, the resolution is no longer improved but is reduced.

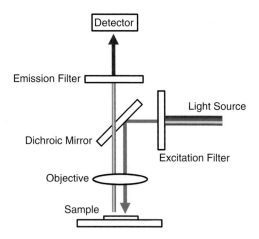

Figure 4.5   *Scheme of a fluorescence microscope*

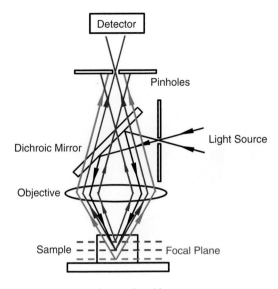

Figure 4.6 *Scheme of a confocal laser-scanning microscope*

4.3.2 Fluorescence Lifetime Imaging Microscopy

The long emission lifetimes of luminescent transition metal complexes and lanthanide chelates (in the microsecond and millisecond scales, respectively) facilitate their applications in fluorescence lifetime imaging microscopy (FLIM). Since the emission lifetimes of luminophores are independent of their concentrations, FLIM is particularly useful in imaging of cells and tissues, in which the concentration of a luminophore cannot be easily measured or controlled [21]. Also, the effect of photon scattering in thick layers of samples is minimized. Importantly, FLIM allows luminophores with different emission decay rates to be distinguished even if they absorb or emit at the same wavelength. Similar to intensity-based imaging microscopy, FLIM also produces images that provide information such as pH values, ion concentrations, oxygen contents, and structural status inside cells and tissues. At present there are three common methods widely used for FLIM. The first is frequency domain phase modulation measurements. In this method, the sample is excited with high frequency sinusoidally modulated light. The sinusoidally modulated emission is at the same frequency as the excitation, but is shifted in a phase ($\Delta\phi$) and reduced in modulation depth (Fig. 4.7), from which the lifetime can be calculated.

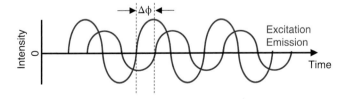

Figure 4.7 *Basic principle of FLIM in the frequency domain phase modulation measurements*

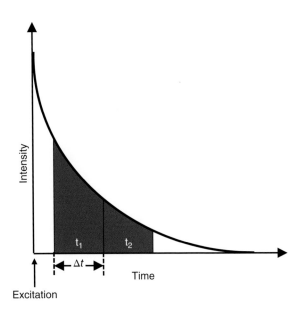

Figure 4.8   *Scheme of two-gate lifetime detection*

The second method is based on time-gated detection of luminescence after pulsed excitation of the sample. The luminescence detected sequentially in multiple delay times is used for lifetime determination. For example, in two-gate detection, the lifetime is given by the equation $\tau_0 = \Delta t / \log(I_{t1}/I_{t2})$, where Δt is the time difference between the start of the two time windows, and I_{t1} and I_{t2} are integrated luminescence intensities of time windows t_1 and t_2, respectively (Fig. 4.8). The third method is time-correlated single photon counting (TCSPC), which is a histogramming technique that records the time when individual photons are detected after single-pulse excitation. The avoidance of time gating and wavelength switching results in very high counting efficiency.

4.3.3   Flow Cytometry

Flow cytometry is a laser-based biophysical technology, which does not give direct images of cells, but provides analysis of multiple parameters of individual cells within heterogeneous populations [22]. Flow cytometry allows simultaneous detection of the physical and chemical characteristics of up to thousands of cells per second. It is a valuable tool for cell counting, sorting, and biomarker detection, and protein engineering. In typical flow cytometry, a stream of a fluid suspension of cells is illuminated with a light source, and a number of detectors are employed to collect certain information (Fig. 4.9). The direct light beam is arrested after passing through sample cells. The scattered light, which is of the same energy as the excitation light source, is collected by forward scatter (FSC) and side scatter (SSC)

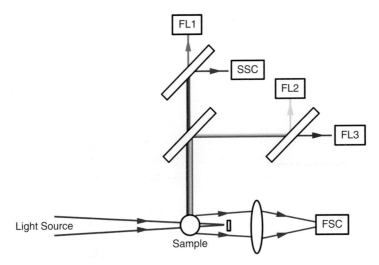

Figure 4.9   *Scheme of the optics of a flow cytometer: SSC, side scatter detector; FSC, forward scatter detector; FL1, FL2, and FL3 are fluorescence detectors for different emission wavelengths*

detectors, which are in line with and perpendicular to the light source, respectively, and correlate with the volume and inner complexity of the cells. The emission of luminophores inside the cells or attached to the cell surfaces is collected simultaneously by fluorescence detector(s) (FL1, FL2, or FL3 in Fig 4.9).

4.4   Cell and Organism Imaging

4.4.1   Factors Affecting Cellular Uptake

As the building block of life, a cell is the basic structure and functional unit of all living organisms. Live cell imaging is useful not only in visualizing cell structures, but also in understanding biological interactions and processes at both the molecular and cellular levels. Cell imaging reagents must possess reasonably good membrane permeability. In the past few years, the cellular uptake of inorganic complexes has been extensively studied using CLSM, FLIM, and flow cytometry [5]. Because of the involvement of transition metal and lanthanide ions, the cellular uptake of these complexes can be directly and conveniently measured by inductively coupled plasma-mass spectrometry (ICP-MS), inductively coupled plasma-atomic emission spectroscopy (ICP-AES), and atomic absorption spectroscopy (AAS). In many cases, transition metal complexes enter cells by an energy-independent diffusion-like pathway and/or an energy-dependent endocytosis-like pathway [5c,j,l]. However, most reported lanthanide chelates are taken up by cells via macropinocytosis [5a,b]. Since the cell membrane potentials are normally maintained negatively charged inside, positively-charged complexes enter living cells more efficiently than neutral or anionic ones; for example, while the neutral trinuclear rhenium(I) complex [Re(CO)$_3$(N^N-py)]$_3$ (**1**) is membrane-impermeable, sequestration of Ag$^+$ through the three uncoordinated pyridines yields the cationic Re–Ag adduct, which is able to cross membranes [23]. In another study, the zinc(II) complexes [Zn(ATSM)] (**2**) are found to exhibit much more rapid internalization throughout the cytoplasm than the copper(II) analogues [Cu(ATSM)], because the zinc(II) complexes are more easily protonated and become cationic at physiological pH values [24, 25].

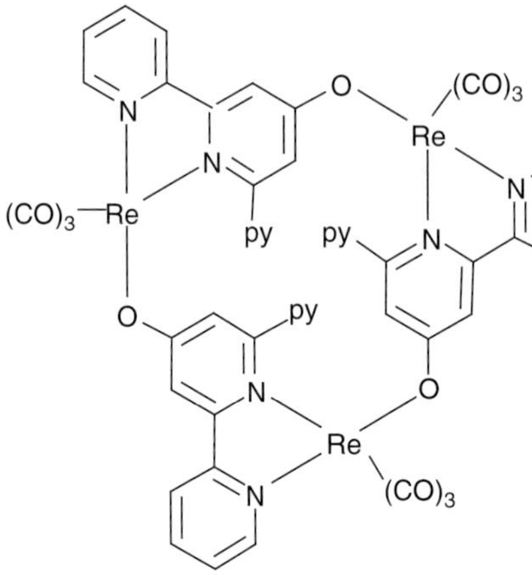

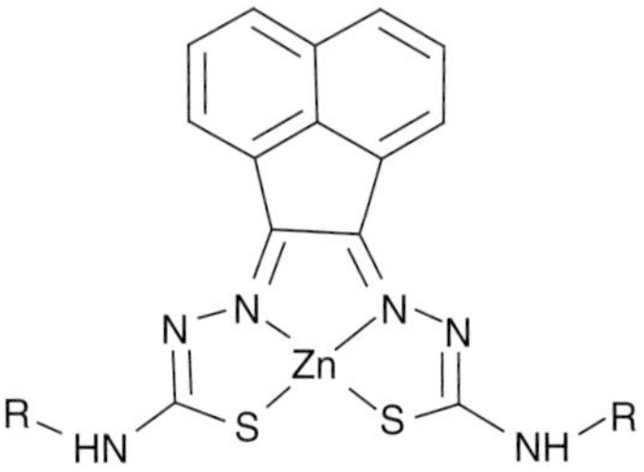

R = CH$_3$, C$_2$H$_5$

1   **2**

The lipophilicity of a compound refers to its affinity for dissolving in fats, oils or lipids, and is quantified as the partition coefficient (log $P_{o/w}$) in *n*-octanol–water. It is an important parameter with regard to not only the ability of the compound to permeate biological membranes, but also to its intracellular distribution and transportation [26]. Many studies have shown that inorganic complexes with higher $P_{o/w}$ values display higher cellular uptake efficiency; for example, among the ruthenium(II) complexes [Ru(N^N)$_2$(dppz)]$^{2+}$ (**3**) (lipophilicity ranging from −1.48 to 1.30), the most hydrophobic Ph$_2$-phen complex (log $P_{o/w}$ = 1.30) exhibits substantial cellular uptake as revealed by the most intense emission from HeLa cells treated by the complex [27]. In another example, the emission intensity of the cytoplasm of KB cells incubated with the iridium(III) complexes [Ir(N^C)$_2$(Hdcbpy)] (**4**) is found to increase monotonically with their lipophilicity (Fig. 4.10) [28]. This highlights the importance of lipophilicity on the uptake efficiency of a probe.

The molecular size of an inorganic complex also contributes to its cellular uptake rate. As expected, larger molecules enter cells less efficiently compared with smaller ones; for

3

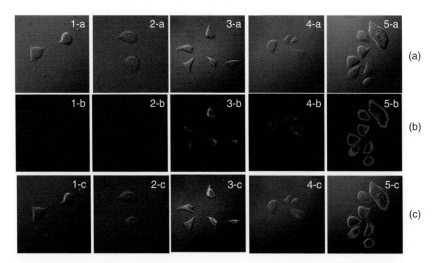

4

example, although the iridium(III) alkyl complex $[Ir(pq)_2(bpy-C_{18}H_{37})]^+$ (**5a**) is more lipophilic (log $P_{o/w}$ = 9.89) than $[Ir(pq)_2(bpy-C_{10}H_{21})]^+$ (**5b**) (log $P_{o/w}$ = 5.34), its larger molecular size reduces its cellular uptake rate [29]. The lipophilicity of the octanuclear dendritic iridium(III) complex $[\{Ir(ppy)_2\}_8(bpy-8)]^{8+}$ (**6a**) and the mononuclear complex $[Ir(ppy)_2(bpy-Et)]^+$ (**6b**) are 1.66 and 0.44, respectively [30]. Interestingly, ICP-MS analysis shows that the cellular uptake of complex **6a** is much less efficient than that of complex **6b**, probably because of its much larger molecular size.

Figure 4.10 *Bright-field (a), fluorescence (b), and overlaid (c) laser-scanning confocal microscopy images of KB cells treated with complexes **4**. The lipophilicity of these complexes increases from left to right. Reprinted with permission from [28] © 2010 American Chemical Society*

$$n = 18, \textbf{5a}$$
$$n = 10, \textbf{5b}$$

Complexes containing biologically active substrates can be specifically transported into cells by proteins located in the cell membrane through a receptor-mediated uptake pathway; for example, folate receptors (FRs) mediate delivery of folate (vitamin B_9) and reduced folic acid derivatives into cells. The rhenium(I) folate complex $[Re(N^N^N-B_9)(CO)_3]^+$ (**7**) is efficiently internalized by A2780/AD ovarian cancer cells that overexpress FRs [31]. No internalization of the complex is observed when the incubation medium is added with excess folic acid or when non-FR-expressing CHO cells are used. Cobalamin binds to the glycosylated protein intrinsic factor (IF) specifically, and the cellular uptake of the formed B_{12}–IF conjugate is mediated by cubilin, a 460 kDa protein. With the assistance of IF, the rhenium(I) cobalamin (vitamin B_{12}) complex $[Re(N^N^N-B_{12})(CO)_3]^+$ (**8**) enters cells in a similar cubilin-mediated pathway [32]. Glucose is the most important sugar in cellular metabolism and energy source for the growth of cells [33]. The uptake of glucose is mediated by a family of facilitative transporters known as glucose transporters (GLUTs) [34].

The iridium(III) glucose complex $[Ir(pq)_2(bpy-Glu)]^+$ (**9**) [35] and the ruthenium(II) glucose complex $[Ru(bpy-\{CH_2S-Glu\}_2)_3]^{2+}$ (**10**) [36] enter cells much more efficiently than their galactose, lactose, mannose, and maltose counterparts (Fig. 4.11, as an example), which has been ascribed to the GLUTs in the cell membrane.

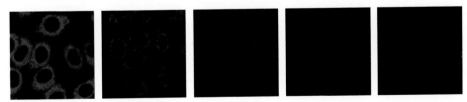

Figure 4.11   *Laser-scanning confocal microscopy images of HeLa cells incubated with complex **9** (5 µM) and other sugar analogues (from left to right: complexes with β-glucose, α-galactose, β-galactose, β-lactose, and β-maltose) for 2 h at 37 °C. Reprinted with permission from [35] © 2010 American Chemical Society*

6b

6a

7

8

9

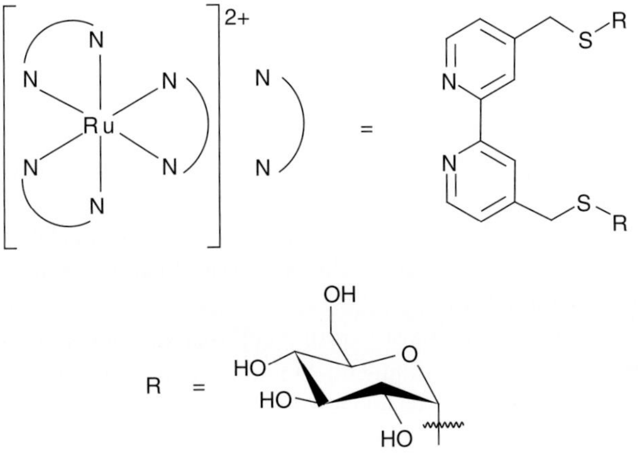

10

The rhenium(I) glucose complex [Re(Ph$_2$-phen)(CO)$_3$(py-glucose)]$^+$ (**11**) is found to display cell-selective uptake [37]. The internalization of this complex by transformed cells (HeLa and MCF-7, which overexpress GLUTs) is much more efficient than non-transformed cells (HEK293T and NIH-3T3). This illustrates the potential of the complexes as probes for cancer cells with high selectivity.

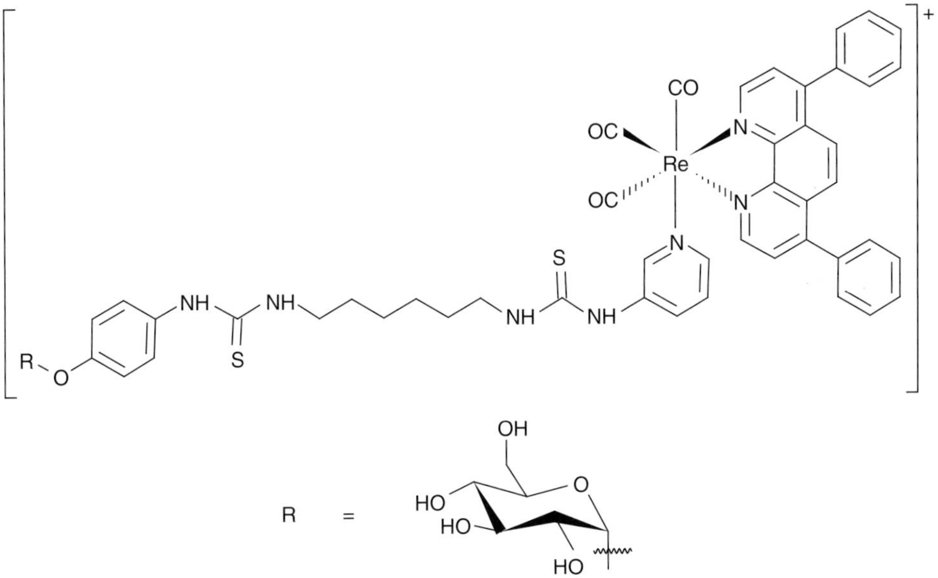

11

4.4.2   Organelle Imaging

Many luminescent transition metal complexes and lanthanide chelates with simple ligands and chelators undergo efficient internalization into different types of cells. Although their intracellular distributions are uneven inside the cytoplasm, most of the complexes do not show organelle specificity. Since many diseases have been found to have a close link with disrupted organelles, imaging of various organelles is of great significance in acquiring information on their morphology and ultrastructural dynamics in cells. Specific organelle imaging can be achieved by a judicious choice of coordinating ligands and chelators of luminescent inorganic complexes.

The cell membrane is the outermost part of a cell that surrounds the cell and separates the interior from the outside environment. Staining of the cell membrane is important for cell-surface demarcation and is integral to the study of cellular functions such as cell fusion and division. However, dyes that exclusively stain the cell membrane are not common because most compounds either undergo efficient internalization or do not show any affinity towards the membrane at all. Thus, the design of membrane-specific staining reagents is mainly based on the interactions between the dye and membrane components, that is, lipid bilayer and embedded proteins. Thus, two main types of staining reagents for plasma membranes are available: (1) compounds with a charged hydrophilic head group and lipophilic tails such as hydrophobic alkyl chains, which are structurally similar to phospholipids and favor the fusion with the lipid bilayer; and (2) compounds that show specific affinity to proteins embedded in the plasma membrane. For the design of the first type of reagents, transition metal polypyridine cores may serve as a positively-charged hydrophilic head, and the introduction of lipophilic tails can afford efficient membrane staining dyes; for example, the octadecyl cyclometalated platinum(II) complex $[Pt(N^{\wedge}N^{\wedge}C\text{-}C18)(P\{C_6H_4\text{-}SO_3\}_3)]^{2-}$ (**12**) [38] and the bis-butylaminomethyl substituted iridium(III) complex $[Ir(pppy\text{-}C4)_2(bpy\text{-}et)]^+$ (**13**) (K. Y. Zhang, H.-W. Liu, and K. K.-W. Lo, University of Hong Kong, unpublished work) exhibit rapid and efficient staining of the plasma membrane of live HeLa cells. Regarding the development of the second type of reagents, the iridium(III) complex $[Ir(ppy)_3]$ has been conjugated to Ac-TZ14011 peptides affording the complex $[Ir(ppy)_2(ppy\text{-}Ac\text{-}TZ14011)]$ (**14**), which can specifically target the membrane protein chemokine receptor 4 (CXCR4) [39]. The resulting conjugates specifically stain the cell membrane of MDAMB231 cells that overexpress CXCR4.

Mitochondria are the power stations of a cell as they produce ATP molecules that are the energy currencies that drive cellular reactions and mechanisms. The inner mitochondrial membrane, similar to the cell membrane, is negatively charged inside. Its potential ($\Delta\Psi = -120$ to $-180\,mV$) is more negative than that of the cell membrane ($\Delta\Psi = -30$ to $-90\,mV$) [40]. At equilibrium, the accumulation of a monovalent cationic compound in the mitochondria is calculated by the Nernst equation: $\Delta\Psi = -59\cdot\log C_{in}/C_{out}$, where C_{in} and C_{out} are concentrations of the compound inside and outside the mitochondria, respectively. Thus, theoretically speaking, mitochondria can accumulate monovalent cationic compounds to a concentration ratio of up to 1000:1 relative to other parts of the cell and of as high as 30000:1 to extracellular space. Thus, the design of inorganic dyes for mitochondria has been focused on membrane permeable positively charged lipophilic complexes. Examples include the rhenium(I) glucose complex $[Re(Ph_2\text{-}phen)(CO)_3(py\text{-}glucose)]^+$ (**11**) (Fig. 4.12) [37] and rhenium(I)

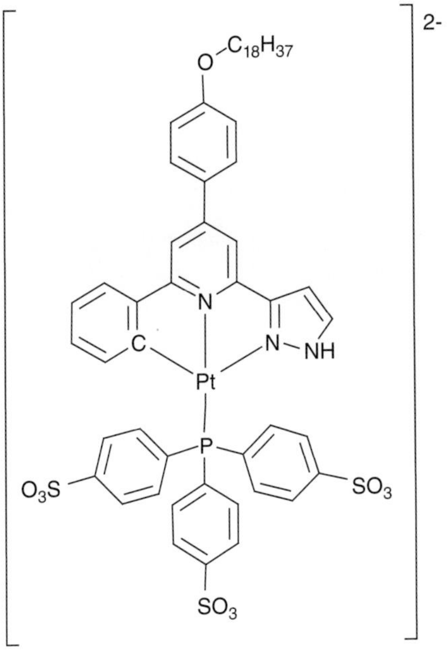

12

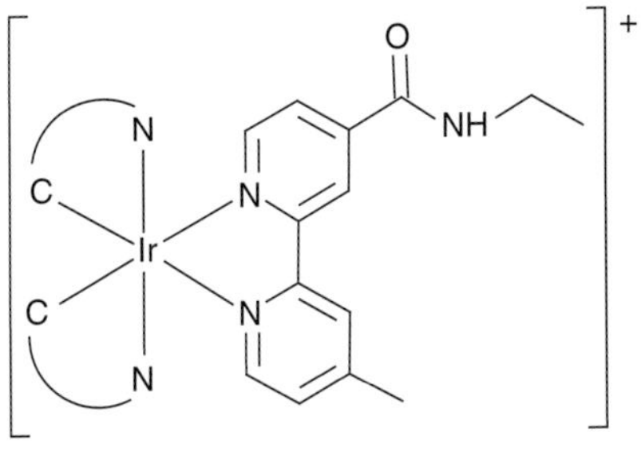

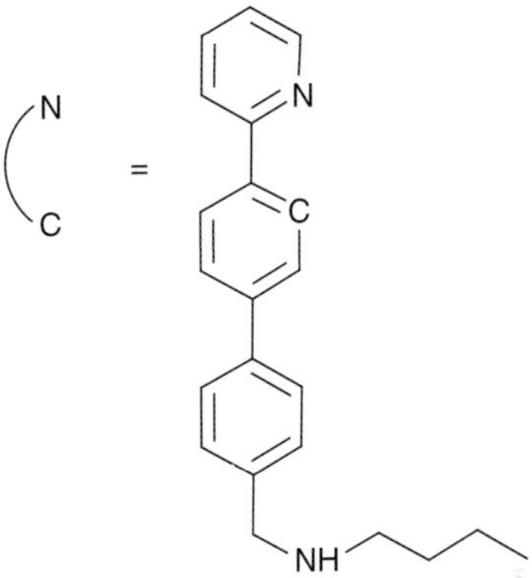

13

chloromethyl and hydroxylmethyl complexes $[Re(CO)_3(bpy)(py-CH_2R)]^+$ (**15**) [41], platinum(II) *N*-heterocarbene complex $[Pt(C\wedge N\wedge N)(^nBu_2NHC)]^+$ (**16**) [42], ruthenium(II) polypyridine complexes $[Ru(phen)_2(MOPIP)]^{2+}$ (**17a**) [43a] and $[Ru(dppz)_2(CppH)]^{2+}$ (**17b**) [43b], iridium(III) biphenylpyridine complex $[Ir(pppy)_2(Ph_2\text{-}phen)]^+$ (**18**) (K. Y. Zhang, H.-W. Liu, K. K.-W. Lo, City University of Hong Kong, unpublished work), europium(III) chelate (**19**) [44], and ytterbium(III) chelate (**20**) [45]. Most of these complexes exhibit a positive formal charge, and all of their ligands and chelators are considerably lipophilic in nature. Although these complexes are efficiently and selectively localized in the mitochondria, their accumulation induces depolarization of the mitochondrial membrane potential, which may activate mitochondria-mediated apoptosis [43].

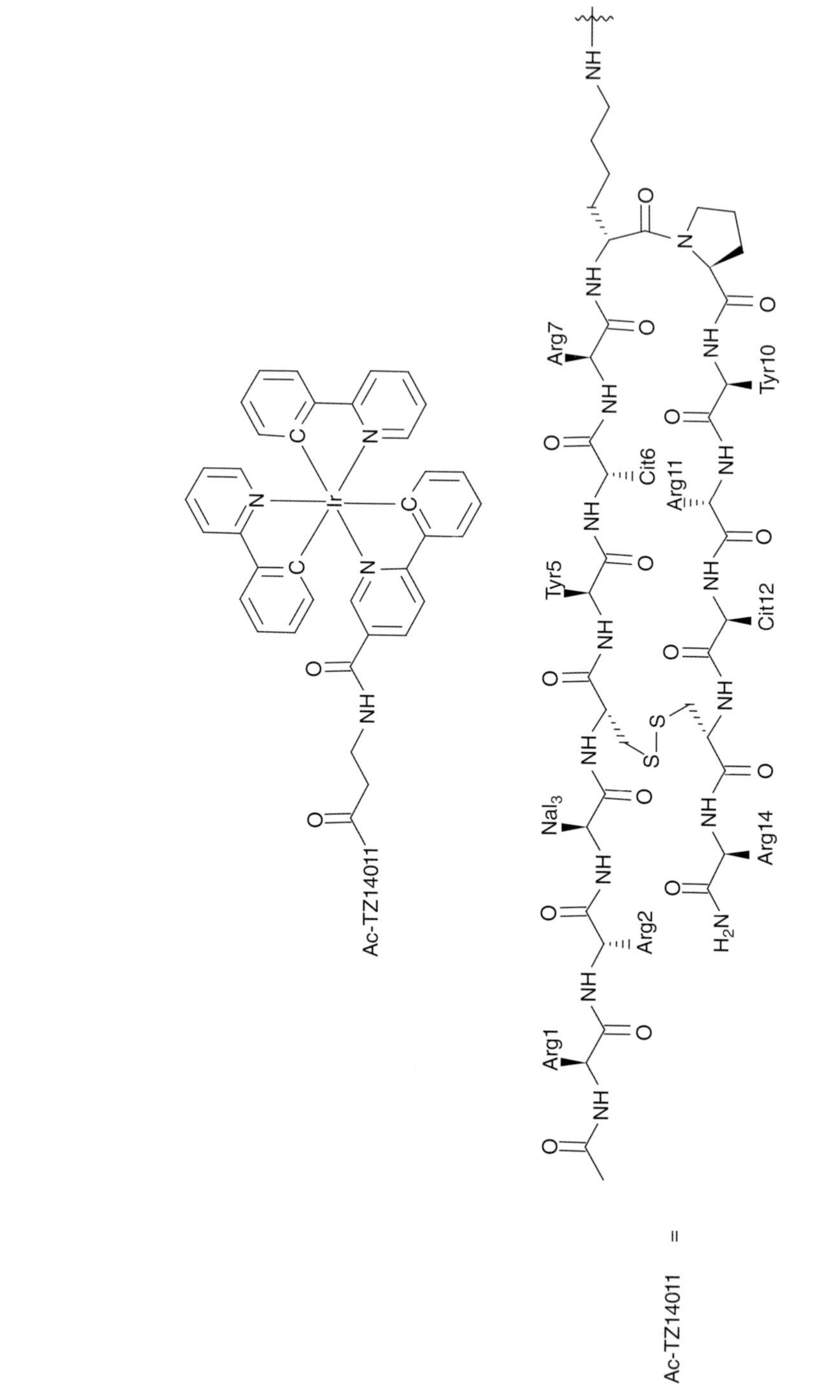
Ac-TZ14011
Ac-TZ14011 =
Arg1
Arg2
Nal3
Tyr5
Cit6
Arg7
Tyr10
Arg11
Cit12
Arg14

14

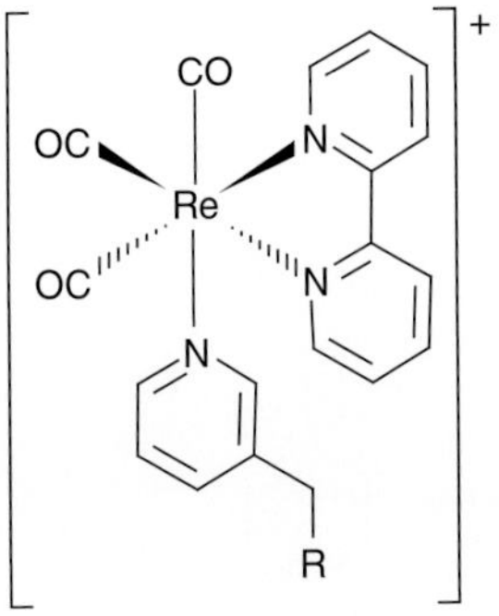

R = Cl, OH

15

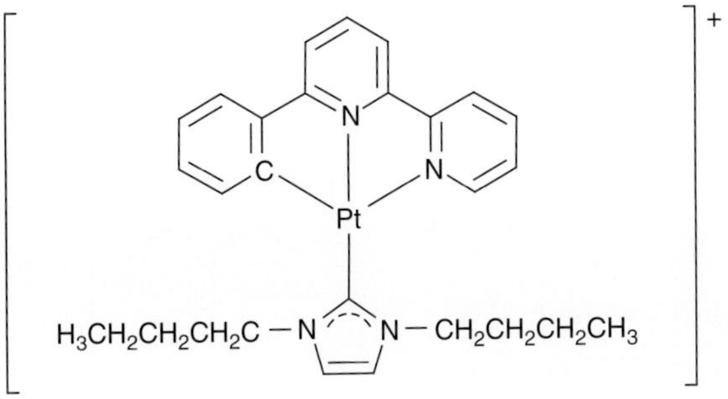

16

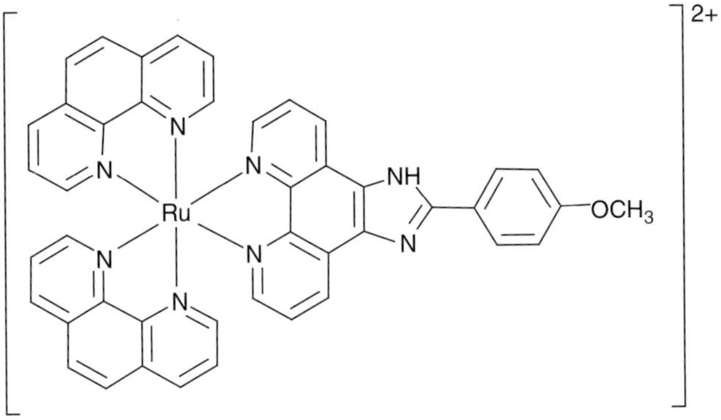

17a

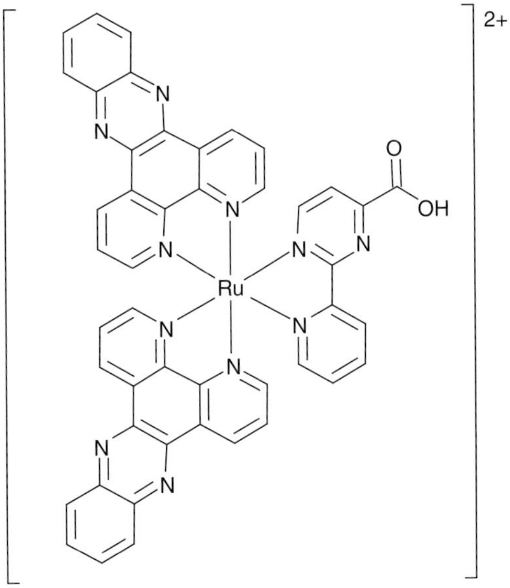

17b

18

19

20

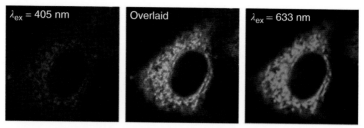

Figure 4.12 *Laser-scanning confocal microscopy images of HeLa cells treated successively with MitoTracker Deep Red FM (100 nM, 20 min, λ_{ex} = 633 nm) and complex **11** (100 μM, 5 min, λ_{ex} = 405 nm) in a glucose-free medium at 37 °C. Reproduced with permission from [37] © 2011 Wiley-VCH Verlag GmbH & Co. KGaA, Weinheim*

Endoplasmic reticulum (ER) is a network of channels enclosed by a single membrane throughout the cell from the outer nuclear membrane to the inner cell membrane. There are two types of ERs: (1) the smooth ER is used for the storage and sudden release of calcium ions and (2) the rough ER is covered with ribosomes and is responsible for protein synthesis. The synthesized proteins are processed in the Golgi apparatus (a stack of cisternae and the distribution point of the cells), and then directed to their destinations. Both ER and the Golgi apparatus are very important organelles of a cell and a number of inorganic complexes have been found to specifically stain these functional cellular compartments. For example, the zinc(II) complexes [Zn(salen)] (**21**) [46] and europium(III) chelate (**22a**) [47] are specifically co-localized with an endoplasmic reticulum marker in live cells, while the rhenium(I) complexes [Re(N^N)(CO)$_3$(py-biotin-NCS)]$^+$ (**23**) [48] and [Re(N^N)(CO)$_3$(Cl)]$^+$ (**24**) [49] and the iridium(III) complexes [{Ir(ppy)$_2$}$_8$(bpy-8)]$^{8+}$ (**6a**) [30] and [Ir(ppy-N)$_2$(ppy-B)] (**25a**) [50] bind to the Golgi apparatus (Fig. 4.13).

Additionally, the ytterbium(III) chelate (**26**) exhibits anticancer activity through apoptosis associated with the ER stress pathway [51]. It is believed that the chemical structures of these inorganic dyes play a very important role in their staining properties. Slight modification on the ligand or chelator affects the staining preference. For example, the chelate (**22b**) [47] and the iridium(III) complexes [Ir(ppy)$_2$(bpy-Et)]$^+$ (**6b**) [30] and [Ir(ppy-N)$_3$] (**25b**) [50], which are the tris(thenoyltrifluoroacetone) counterpart of **22a**, the mononuclear counterpart of **6a**, and the homoleptic counterpart of **25a**, respectively, are all localized in the cytoplasm and bind to different organelles with no selectivity.

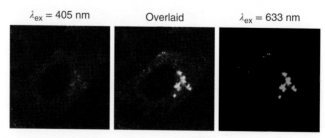

Figure 4.13 *Laser-scanning confocal microscopy images of HeLa cells treated successively with complex **6a** ([Ir] = 2 μM) at 37 °C for 2 h, PBS containing 3% paraformaldehyde, anti–golgin-97 (human) mouse IgG$_1$ (1 μg ml^{-1}, 1 h), and Alexa 635 goat anti-mouse IgG (H+L) (10 μg ml^{-1}, 30 min). Reprinted with permission from [30] © 2010 American Chemical Society*

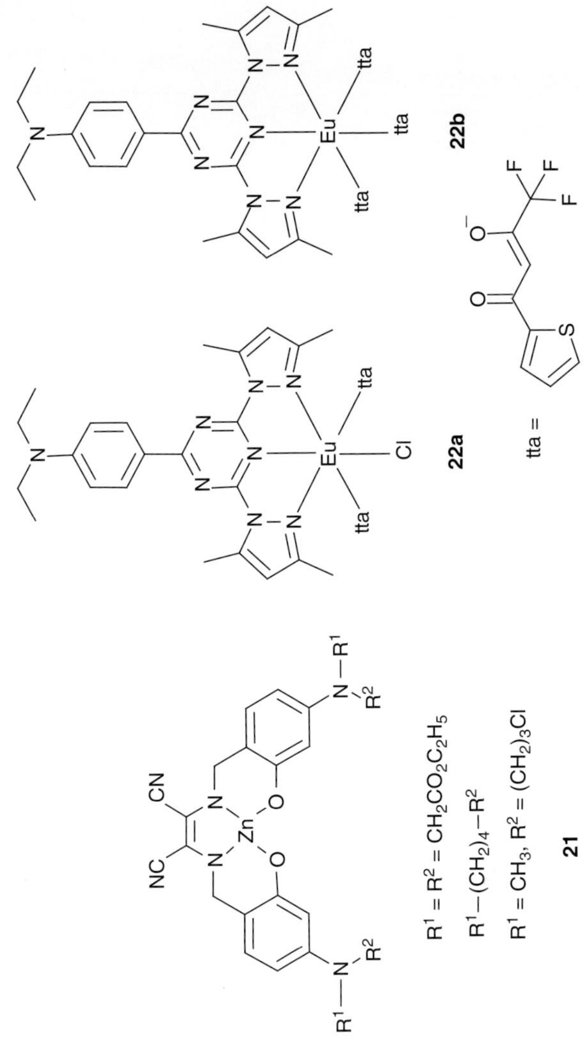

$R^1 = R^2 = CH_2CO_2C_2H_5$

$R^1–(CH_2)_4–R^2$

$R^1 = CH_3$, $R^2 = (CH_2)_3Cl$

21

22a

22b

tta =

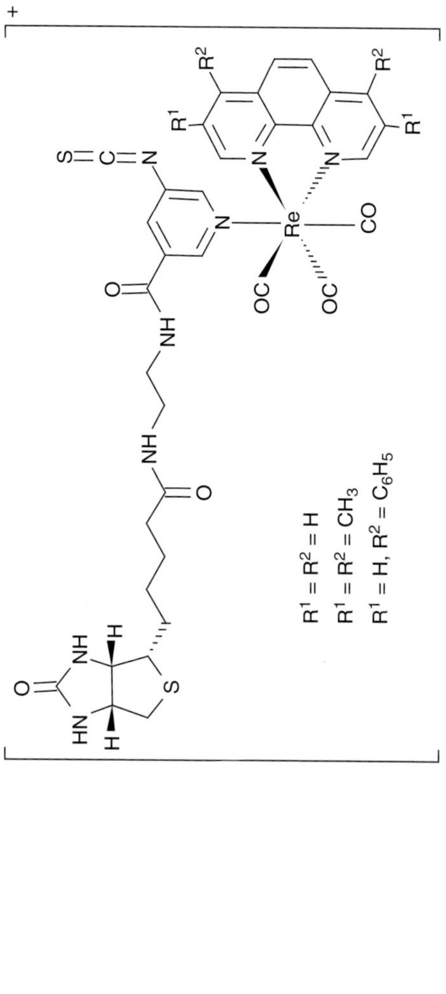

R¹ = R² = H
R¹ = R² = CH₃
R¹ = H, R² = C₆H₅

23

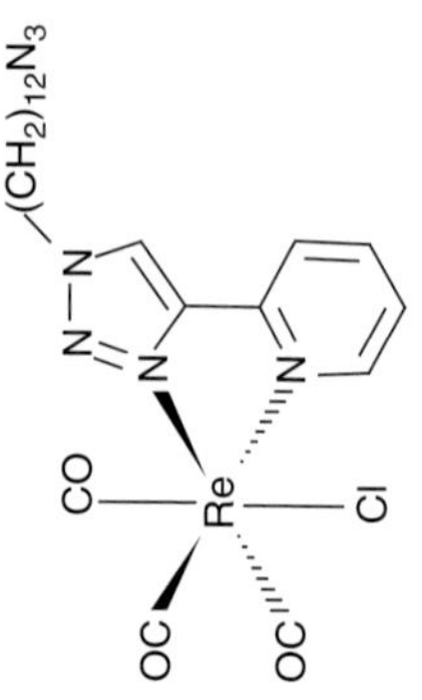

24

The cell nucleus is the largest and most important organelle in a cell. It is bounded by a double membrane nuclear envelope that encloses the genetic material and separates the contents of the nucleus from the cytoplasm. Materials exchange between the nucleus and the cytoplasm is realized with the aid of nuclear pores, which are large protein complexes crossing the nuclear envelope [52]. Linear gold(I) phosphine complexes are much smaller in molecular size compared with square planar and octahedral transition metal complexes and lanthanide chelates of different structures, which probably explain the observation that they enter the nucleus more efficiently. For example, the gold(I) complexes [Au(Cl)(PR$_3$)] (**27**) show substantial nuclear uptake, which can be detected and quantified inside the nuclei of HT-29 cells by electrothermal atomic absorption spectrometry [53]. Also, the efficiency of nuclear uptake is related to the lipophilicity of the complexes; for example, the accumulation of the more lipophilic naphthalimide complexes [Au(Naphth)(PR$_3$)] (**28**) in the nucleus is much more efficient than the chloro analogues **27** [54].

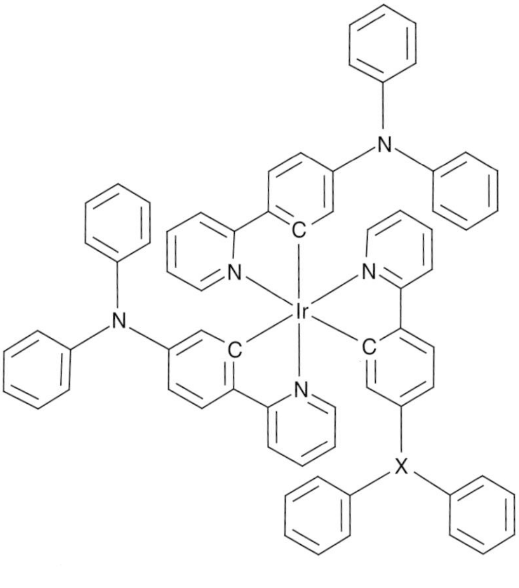

X = B, 25a
N, 25b

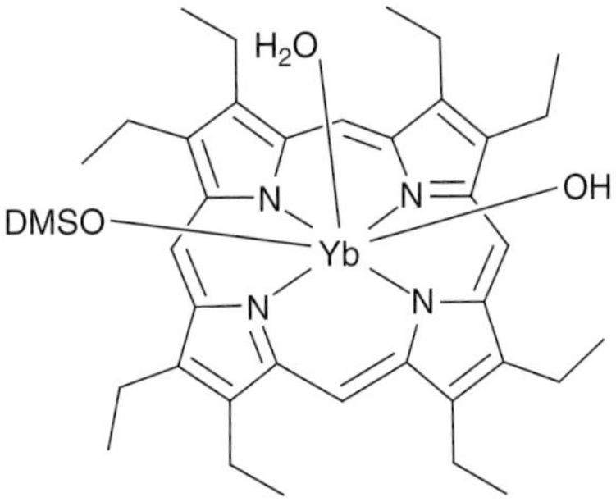

26

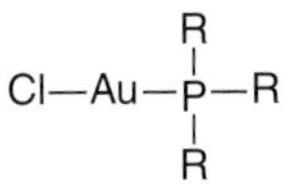

R = CH$_3$, C$_2$H$_5$, C(CH$_3$)$_3$, C$_6$H$_5$

27

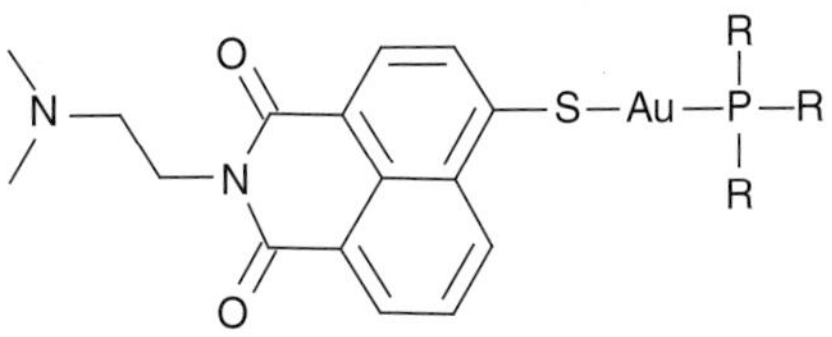

R = CH$_3$, C$_2$H$_5$, C(CH$_3$)$_3$, C$_6$H$_5$

28

Most octahedral polypyridine metal complexes do not show nuclear uptake properties because of their large and rigid structure skeleton. Strategies that enhance nuclear uptake usually involve physically or chemically increasing the nuclear envelope penetration

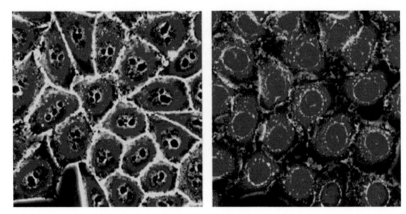

Figure 4.14 *Laser-scanning confocal microscopy images of HeLa cells treated with complex* ***30*** *(left) after pre-incubation of NEM (right) at 37 °C for 20 min. Reprinted with permission from [56] © 2009 American Chemical Society*

and reducing the nuclear export. For example, physical electroporation of V79 Chinese hamster cells significantly increases the nuclear uptake of the dinuclear ruthenium(II) complex $[Ru(phen)_2(cpdppz-C4-cpdppz)Ru(phen)_2]^{4+}$ (**29**) [55]. Treatment of cells with *N*-ethylmaleimide (NEM) induces nuclear pore malformation and the inhibitory effect on nuclear export pathways, leading to an increased nuclear entry of the iridium(III) indole complex (**30**), which is originally localized in the perinuclear region of live cells (Fig. 4.14) [56]. The design of nuclear dyes has been mainly focused on complexes that exhibit a strong affinity towards nucleic acids; for example, the ruthenium(II) complex $[Ru(phen)_2(tpphz)Ru(phen)_2]^{4+}$ (**31**) is non-emissive in water but displays intense luminescence when bound to DNA grooves [57]. After cellular uptake, it is localized in the nucleoplasm of MCF-7 cells, and interacts with nuclear DNA. At this stage, we also invite the readers to check Chapter 6 for more information on the probing of DNA using metal complexes.

Peptide nucleic acid (PNA) is a synthetic polymer that binds to complementary nucleic acid strands [58]. HEK-239 cells treated with the PNA-modified rhenium(I) complex $[(CO)_3Re(pyridazine-PNA)(Cl)_2Re(CO)_3]$ (**32**) reveal interesting nuclear staining due to the binding of the PNA pendant to intranuclear nucleic acid strands [59]. The introduction of cell-penetrating peptides to transition metal complexes not only increases their cellular uptake, but also facilitates entry of the metal complexes to the nucleus of live cells; for example, the oligoarginine linkers in the rhodium(III) complex $[Rh(chrysi)(phen)(bpy-R_8-fluor)]^{3+}$ (**33**) [60] and ruthenium(II) complex $[Ru(phen)(dppz)(bpy-R_8-fluor)]^{2+}$ (**34**) [61] increase the non-specific binding affinity of the complexes to matched and mismatched DNA molecules. The complexes show substantial nuclear localization after cellular uptake. Interestingly, the terbium(III) chelates (**35**) and (**36**) show concentration-dependent intracellular localization [62]. HeLa cells treated with these complexes at relatively high concentrations (10–100 μM) display endosomal/lysosomal staining. In contrast, incubation of cells with the complexes of a much lower concentration (<1 μM) leads to a clear nuclear localization profile, especially in cells undergoing division in the M phase, where membrane integrity is somewhat compromised. Most importantly, these complexes allow visualization of mitotic chromosomes of dividing cells.

29

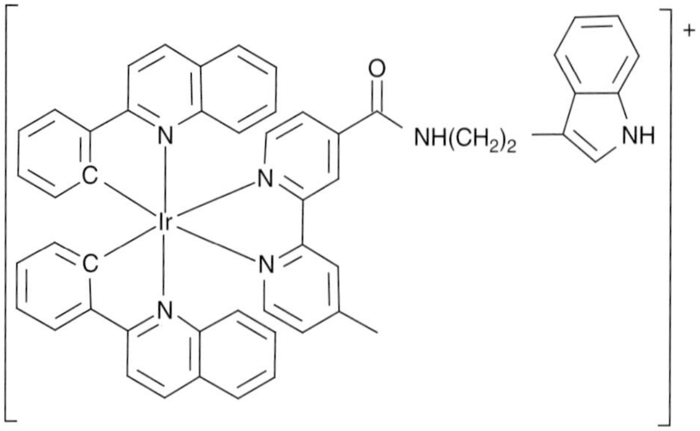

30

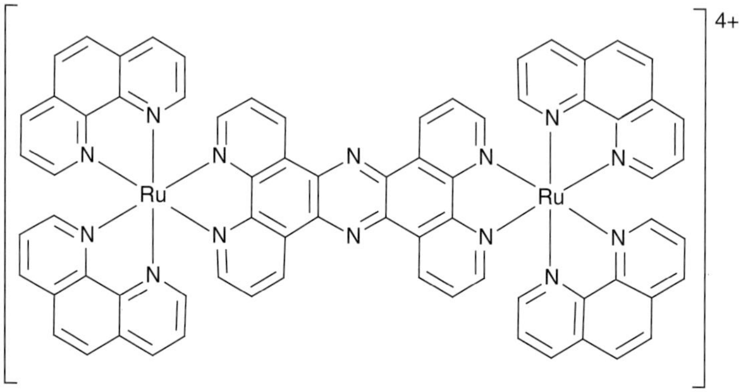

31

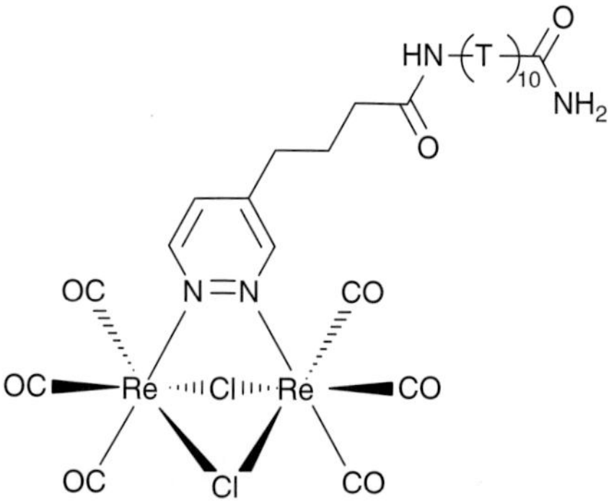

T = Thymine PNA monomer

32

33

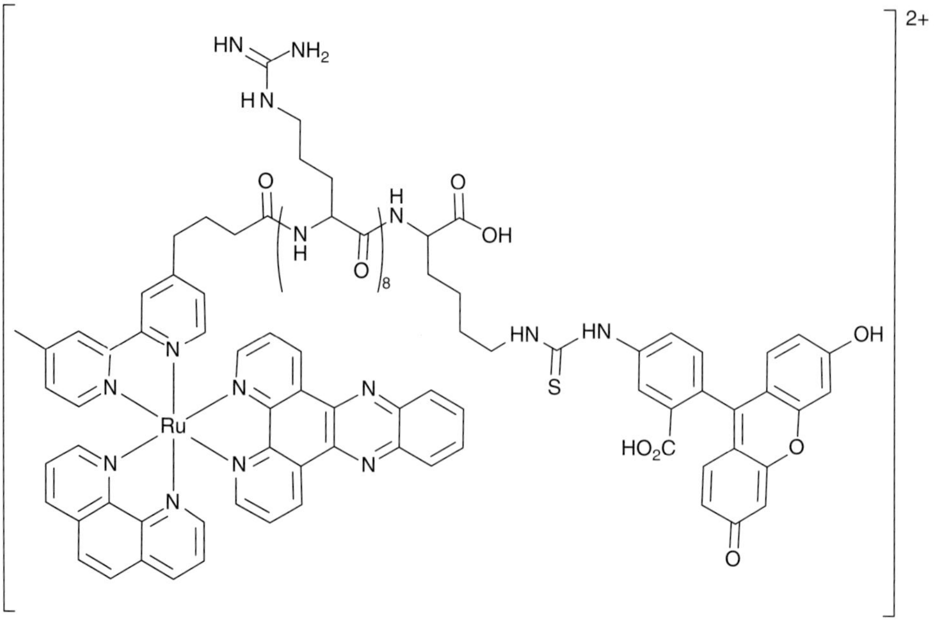

34

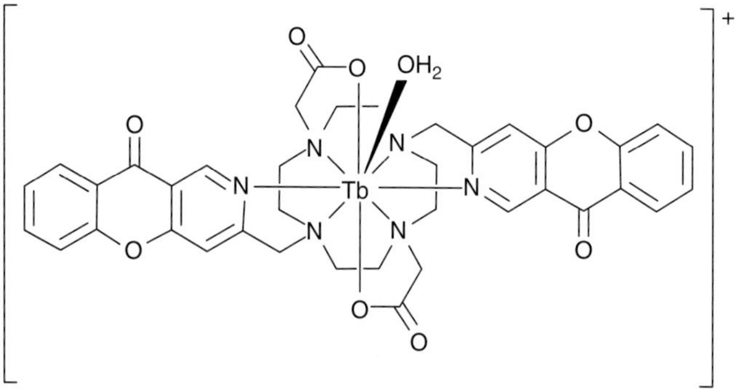

35

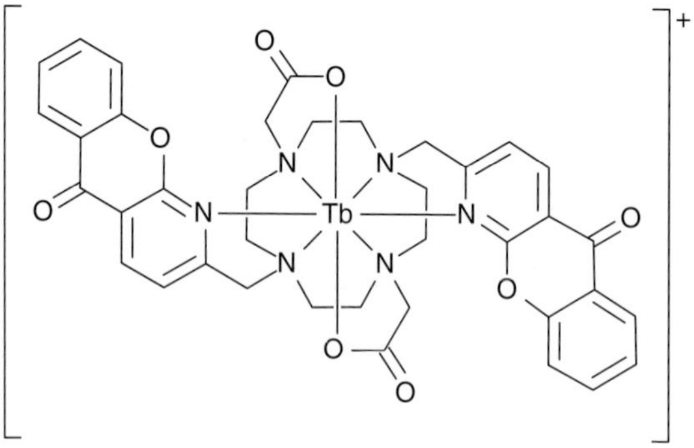

36

The nucleolus is a subnuclear non-membrane bound structure, and is composed of proteins and nucleic acids. It is the localization where ribosomal RNA is transcribed and processed. The design of inorganic dyes for the nucleolus is based on complexes that target nucleolar proteins or RNA; for example, the ruthenium(II) complex [Ru(bpy)$_2$(phen-ethidium)]$^{3+}$ (**37**) [63] and platinum(II) complex [Pt(N^C^N-COOMe)Cl] (**38**) [64] bind strongly to RNA and show substantial nucleoli staining in CHO, fibroblast, C8161, and MDA-MB-231 cells. Although bioinformatic analysis of the amino acid usage of nucleolar proteins does not reveal a preference for hydrophobic amino acids [65], cell biology experiments indicate that the nucleolus can be imaged using nucleus penetrating hydrophobic probes [66]. The iridium(III) complexes [Ir(N^C)$_2$(N^N)]$^+$ (**39**) accumulate efficiently in the nucleoli of cells (Fig. 4.15) [67]. Their hydrophobic DNA intercalating diimine ligands facilitate the nuclear uptake of these complexes, and lead to the binding of **39** to the hydrophobic pockets of the nucleolar proteins. Similarly, the platinum(II) complex [Pt(C^N^N-PPh$_3$)Cl]$^+$ (**40**) [68] and europium(III) chelate **41** [69] are also able to enter the nucleus and are subsequently localized in the nucleoli of NIH-3T3, HeLa, and HDF cells.

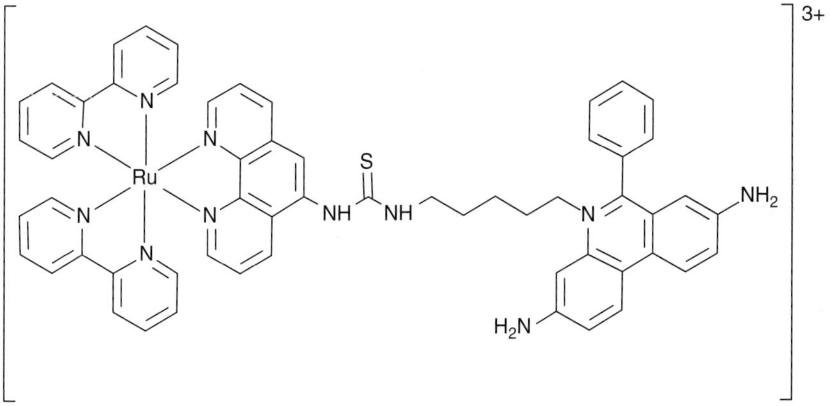

37

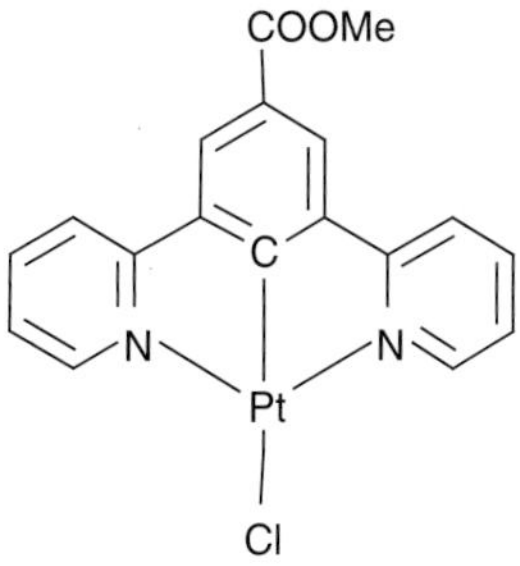

38

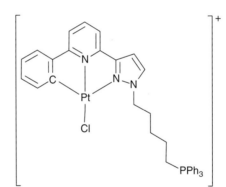

R = H, CONH(CH₂)₃CH₃

39

40

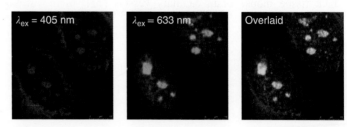

λ_{ex} = 405 nm λ_{ex} = 633 nm   Overlaid

Figure 4.15 *Laser-scanning confocal microscopy images of fixed MDCK cells treated successively with fibrillarin antibody (20 µl ml⁻¹, 1 h), Alexa 633 antirabbit IgG antibody (20 µg ml⁻¹, 30 min), and one of complex **39** [Ir(ppz)₂(dpq)](PF₆) (5 µM, 30 min). Reprinted with permission from [67] © 2010 American Chemical Society*

41

4.4.3 Two-photon and Upconversion Emission Imaging for Cells and Organisms

Recently, interest in using a low-energy excitation source, such as NIR, in performing live cell imaging has been increasing. Compared with the ultraviolet and visible light excitation, NIR excitation allows deeper tissue penetration, weaker autofluorescence, less photobleaching, and lower phototoxicity. Luminescent transition metal complexes and lanthanide chelates that can undergo two-photon excitation have been developed as cellular imaging reagents. For example, the aforementioned cell membrane-staining platinum(II) complex [Pt(N^N^C-C18)(P{C$_6$H$_4$-SO$_3$}$_3$)]$^{2-}$ (**12**) [38], ER-staining zinc(II) complex [Zn(salen)] (**21**) [46], and europium(III) chelate **22a** [47], the Golgi apparatus staining iridium(III) complex [Ir(ppy-N)$_2$(ppy-B)] (**25a**) [50], and the nucleolus-targeting platinum(II) complex [Pt(C^N^N-PPh$_3$)Cl]$^+$ (**40**) [68] exhibit two-photon cross-sections of 28–350 GM (Goeppert–Mayer units) in the NIR and IR regions, where 1 GM is 10^{-50} cm$^4 \cdot$s$\cdot$photon^{-1}. Their two-photon cross-sections are much larger than the minimum value (0.1 GM) for optical imaging applications in live specimens [70]. The ytterbium(III) chelate **42** has been designed to demonstrate the two-photon emission imaging of thick tissue mouse brain capillary vessels [71]. In this strongly scattering sample, the images clearly show that blood vessels of the mouse brain slices are stained with a reasonable signal-to-noise ratio up to a depth of 80 μm. In another study, two-photon emission imaging has been combined with time-resolved emission imaging to further minimize autofluorescence. The inert platinum(II) complex [Pt(N^C^N)Cl] (**43**) with a two-photon cross-section of 4 GM displays a preference for accumulation in nucleoli [72].

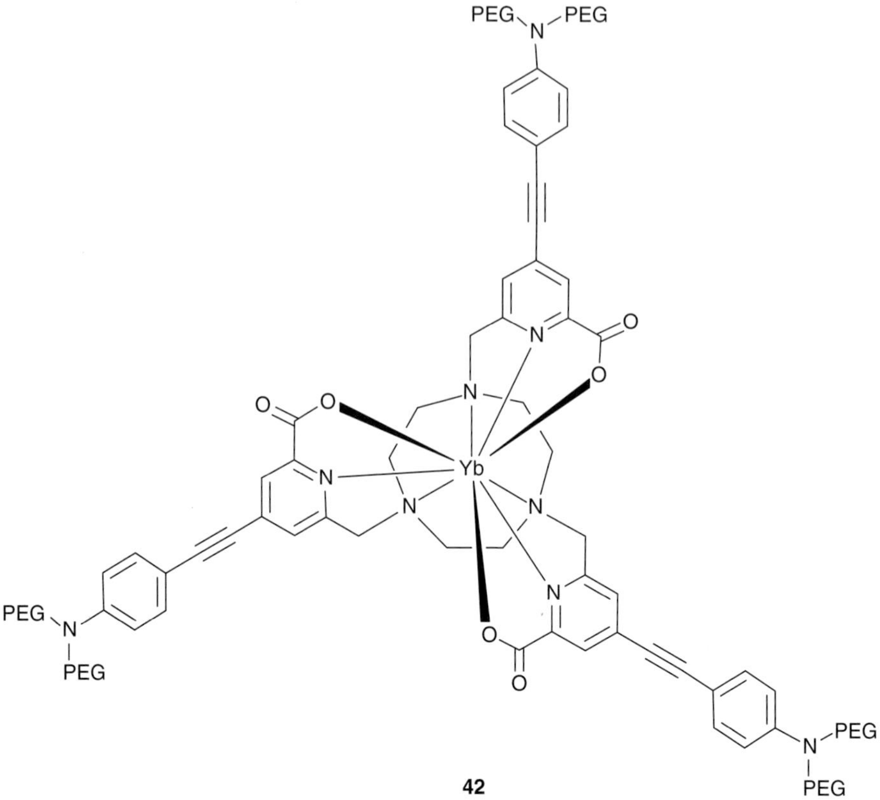

42

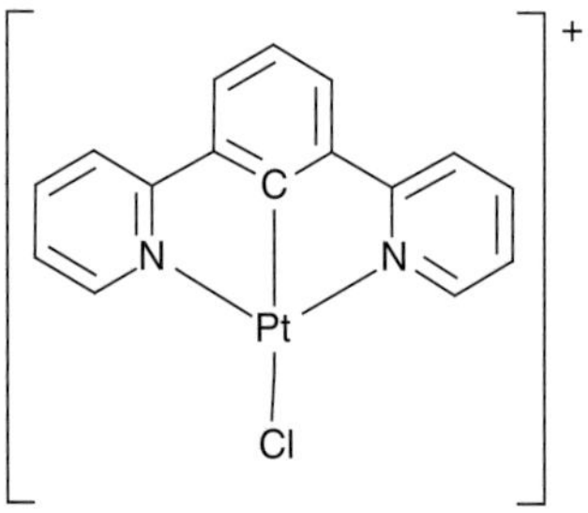

43

Sharing the same advantages as two-photon emission imaging, upconversion materials have been used in cell and organism imaging. Lanthanide-doped upconversion nanomaterials exhibit unique emission upon continuous NIR excitation. Their emission colors can be fine-tuned through the dopant compositions and doping level, as well as the morphology, crystallinity, size, and surface ligands [73]. In addition to the advantage of low-energy excitation, these nanomaterials can be imaged in the NIR and IR regions because many lanthanide-doped particles show downconversion upon NIR excitation. Furthermore, their cathodoluminescence (the emission of photons caused by the impact of electrons) allows ultra-high spatial resolution imaging using scanning electron microscopy. Several recent examples of their applications in cell and organism imaging are given below.

The water-soluble silica coated $NaYF_4$:Yb,Er upconversion nanoparticles have been used to image bone marrow derived stem cells and skeletal myoblasts [74]. Intense luminescence from intracellular nanoparticles with a high signal-to-background ratio is observed upon excitation at 980 nm. The biodistribution study using normal Wistar rats as a model with an injection at a dose of $10\,mg\cdot kg^{-1}$ reveals that the nanoparticles are accumulated in the lung and heart at 30 min post-injection, and are then gradually excreted by urine or faeces in 7 days. All the rats maintained their health status and behavior throughout the study.

Coating of nanoparticles with different molecules can alter their cellular uptake properties; for example, the oleic acid-capped $NaYF_4$:Yb,Er has been coated with folic acid labeled *N*-succinyl-*N'*-octyl chitosan, and then loaded with a zinc(II) phthalocyanine photosensitizer to yield nanoparticles that exhibit green luminescence upon NIR excitation at 980 nm [75]. These nanoparticles have been applied in both NIR confocal microscopy and small animal imaging (Fig. 4.16). The folic acid units on the surface enhance the tumor-selectivity of the nanoparticles to cancer cells that overexpress FRs. In another study, polyethylene glycol (PEG) functionalized multifunctional nanoparticles were prepared from $NaYF_4$:Yb,Er (Y:Yb:Er = 69:30:1), poly(acrylic acid), magnetic Fe_3O_4, and dopamine, and coated with a thin layer of gold shell [76]. These nanoparticles allow both upconversion optical and magnetic resonance imaging. Mouse mesenchymal stem cells treated with these nanoparticles show intense luminescence. To investigate the *in vivo* stem cell tracking, cells stained with these nanoparticles were subcutaneously injected into the back of a nude mouse, which was then imaged under excitation at 980 nm. A detection limit was determined to be as low as ten cells.

TTA-based upconversion luminescence, which requires lower excitation power density and exhibits higher upconversion capability compared with lanthanide-based upconversion, is another promising low-energy excitation approach for bioimaging [19]. Transition metal complexes, especially those absorbing strongly at low energy, are excellent triplet sensitizers in this system; for example, water-soluble TTA-based upconversion silica

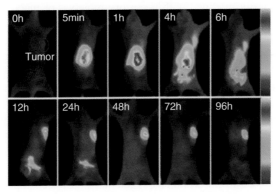

Figure 4.16 *In vivo tumor-targeting of the nanoparticles. Fluorescence images of nude mice bearing Bel-7402 tumors with intravenous injection of the nanoparticles. Reprinted with permission from [75] © 2013 American Chemical Society*

nanoparticles have been prepared with the palladium(II) octaethylporphyrin **44** as the sensitizer and 9,10-diphenylanthracene **45** as the acceptor [77]. Upon excitation with green light, the nanoparticles display blue emission with an overall upconversion luminescence quantum yield of 4.5% in pure water. The use of these nanoparticles to image live cells and the lymph node of Kunming mice has been demonstrated. The signal-to-noise ratio was quantified as high as 25 in the highly scattering biosamples. In another study, the platinum(II) tetraphenyl-tetrabenzoporphyrin **46** (sensitizer) and BODIPY dyes **47** (acceptors) have been loaded into BSA−dextran stabilized soybean oil droplets to yield TTA-based upconversion nanoparticles [78]. The use of soybean oil in this system significantly decreases the aggregation- and oxygen-induced quenching of TTA-based upconversion emission.

44	**45**

4.4.4   Intracellular Sensing and Labeling

Owing to their sensitive luminescence towards the coordinated ligands or chelators and surrounding microenvironment [4, 5], luminescent transition metal polypyridine complexes and lanthanide chelates have been used in the development of sensors for pH, metal cations, anions, gas vapors, small biomolecules such as sugars, nucleotides, and amino acids, and macromolecules such as nucleic acids and proteins. The ligands and chelators of inorganic complexes can be easily modified with a sensory moiety. For macromolecular sensing, spacer-arms are employed to reduce steric hindrance between the probe and the analyte. The detection of intracellular ions and molecules has recently emerged as an important research topic because of their roles in biology and neuroscience. Intracellular analysts of interest include oxygen, protons, essential metal ions, free radicals, and biomolecules. In this chapter, we will focus our attention only on oxygen and protons since the detection of essential metal ions, free radicals, and biomolecules is treated in detail in Chapter 8.

Intracellular oxygen concentration is of primary importance in the determination of physiological and pathological processes in biological systems. Luminescent transition metal polypyridine complexes have been commonly used in oxygen sensing applications due to the efficient quenching of their triplet emissive states by the oxygen molecules. Recent work has been extended to intracellular oxygen sensing: for example, live cells treated with the iridium(III) complex [Ir(btp)$_2$(acac)] (**48**) emit much more strongly when cultured at a 5% oxygen concentration compared with those at a 20% oxygen concentration [79]. As tumor tissues are known to be hypoxic, this complex has been used for

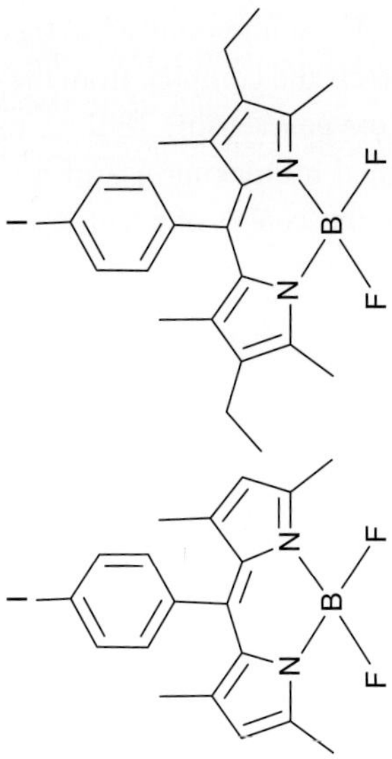

47

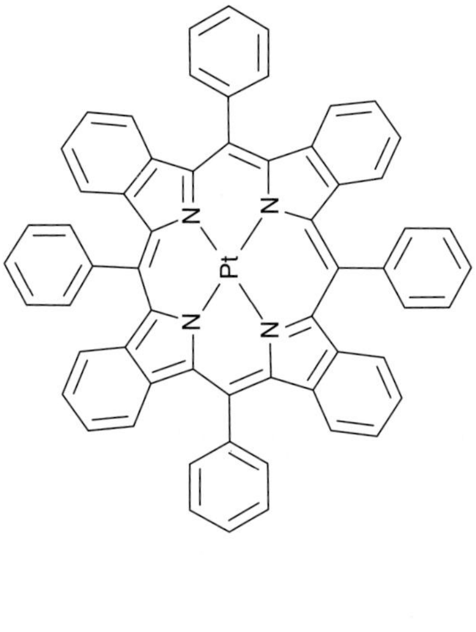

46

tumor imaging in living animals. Female athymic nude mice were implanted with five tumor cell lines, mouse oral squamous carcinoma derived SCC-7, human glioma U87, human lymphoma derived RAMOS, human colon carcinoma HT-29, and mouse lung cancer LL-2. All these five tumors exhibit intense emission at 5 min post-injection of complex **48**. Since the luminescence of transition metal complexes is sensitive not only towards oxygen molecules, but also many other microenvironmental parameters, the ruthenium(II) complex $[Ru(bpy)_2(bpy-pyr)]^{2+}$ (**49**) is immobilized to the surface of phospholipid coated polystyrene particles, which protects the complex from the complicated intracellular environment, but allows molecular oxygen sensing [80]. A ratiometric intracellular oxygen sensing strategy has been designed by making use of a mixture of the terbium(III) and europium(III) chelates **50**, the luminescences of which are sensitive and insensitive towards oxygen quenching, respectively [81].

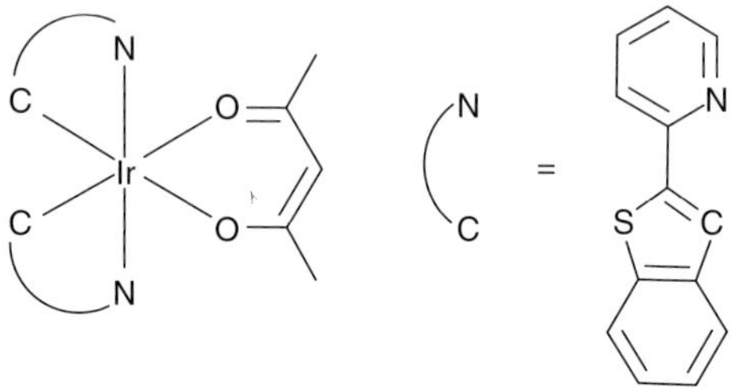

48

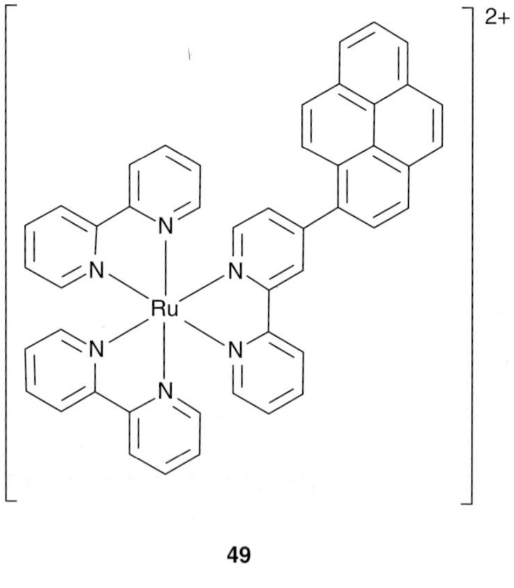

49

Ln = Tb, Eu

50

The determination of pH values is one of the most common measurements in industrial processing, pollution control, food manufacture, and clinical diagnosis. Intracellular pH measurements are also important, since acidic environments are known to be related to certain diseases including poorly vascularized tumors [82], cystic fibrosis [83], and asthma [84]. Luminescent transition metal polypyridine complexes and lanthanide chelates have high potential to function as pH sensors. A general approach to the design of pH-sensitive dyes is to modify their ligands or chelators with a proton donor or acceptor. The acid–base equilibrium acts as a means to control the electronic structures and hence the absorption and/or emission spectra. However, in heterogeneous media, such as intracellular environments, where the dye concentration or visual observation is limited, pH measurements require more sensitive indicators and more precise methods. pH sensors displaying profile changes in the emission spectrum upon interaction with protons utilize the advantages of ratiometric methodology. The proton concentrations can be related to the ratio of intensities at two emission wavelengths. The europium(III) chelate **51** functionalized with an *N*-methylsulfonamide moiety allows ratiometric pH measurements in the range of from 6 to 8 [85]. Protonation of the sulfonamide group increases the inner-sphere hydration number of the europium(III) center, which results in luminescence enhancement at 587 nm and reduction at 680 nm. This chelate has been used for mapping the pH within the NIH 3T3 cells. By measuring the luminescence intensity ratio, a global pH value within the cell has been estimated to be 7.4 [85].

In addition to intracellular sensing, luminescent transition metal complexes have been designed to label cellular components. In fact, metal complexes containing a reactive functional group for protein labeling have attracted attention since the 1990s [5g,i]. Upon internalization into live cells, these complexes maintain their reactivity and label intracellular molecules and structures; for example, the iridium(III) bis-aldehyde complexes $[Ir(qba)_2(Ph_2\text{-}phen)]^+$ (**52**) efficiently label intracellular proteins of HeLa cells through the reaction between the aldehyde groups and amino groups of amino acids [86]. The luminescence intensity of the stained cells is retained through fixation and washing with methanol (Fig. 4.17).

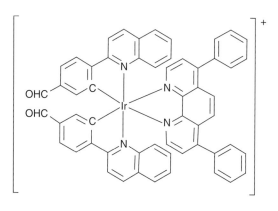

51

52

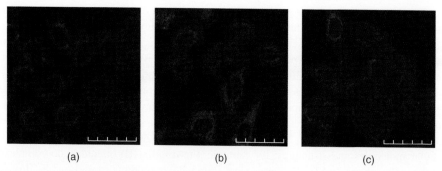

(a)   (b)   (c)

Figure 4.17   *Fluorescence images of HeLa cells incubated with complex **52** (5 μM) at 37 °C for 1 h (a) before and (b) after subsequent fixation with MeOH, or (c) fixation with paraformaldehyde, followed by extensive washing with MeOH. Reprinted with permission from [86] © 2011 Royal Society of Chemistry*

In another study, the iridium(III) complex [Ir(ppy)$_2$(DMSO)$_2$]$^+$ (**53**) with the labile DMSO (dimethyl sulfoxide-) ligands underwent substitution of the weak-field solvent ligands with strong-field π-accepting imidazole of histidine (His), resulting in labeling of intracellular His-containing proteins [87]. Live cells treated with this complex reveal histidine-mediated nucleus staining. Despite these studies, the intracellular targets of the reactive complexes and the biological processes perturbed by these complexes have not been identified. The iridium(III) isothiocyanate complex [Ir(pq)$_2$(phen-NCS)]$^+$ (**54**) has been used to identify the exact protein targets [88]. Through a series of experiments, including protein enrichment by fractional precipitation with ammonium sulfate, protein separation by polyacrylamide gel electrophoresis, and MALDI-TOF/TOF mass spectrometry analysis, one of the targets has been found to be the mitochondrial protein, voltage-dependent anion channel 1 (VDAC1). This is consistent with co-staining experiments, which show that the complex has mitochondrial-localization properties.

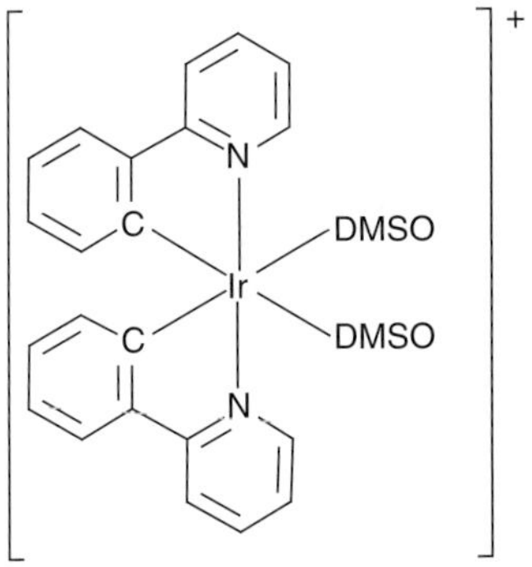

53

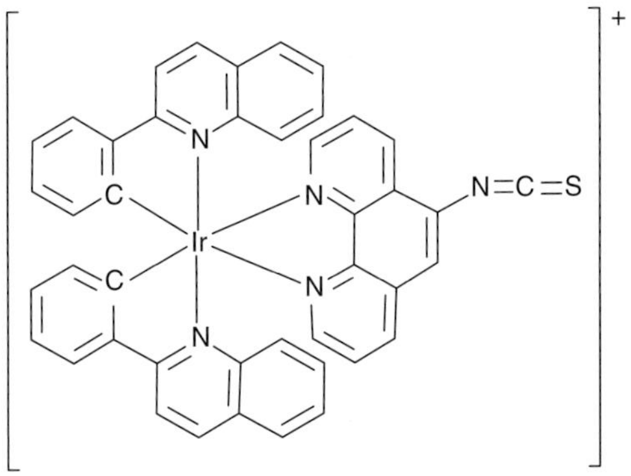

54

Bioorthogonal labeling has emerged as a versatile method to image biomolecules in their native environments [89]. In a typical procedure, a substrate modified with a chemical reporter is incorporated into live cells or organisms, which is then recognized by a bioorthogonal probe that carries the complementary functionality [89]. The iridium(III) complex appended with a dibenzocyclooctyne (DIBO) unit [Ir(ppy-COOH)$_2$(bpy-DIBO)]$^+$ (**55**) was designed as the first phosphorescent bioorthogonal labeling reagent [90]. Since the DIBO unit exhibits high reactivity toward azides, this complex undergoes facile reactions with azide-modified biomolecules including serum albumins and transferrin. As 1,3,4,6-tetra-*O*-acetyl-*N*-azidoacetyl-D-mannosamine (Ac$_4$ManNAz) can be converted by CHO cell via the biosynthetic pathway into *N*-azidoacetyl sialic acids that are localized on the cell surface, the complex has been used to label the cell-surface glycans by bioorthogonal reactions (Fig. 4.18).

55

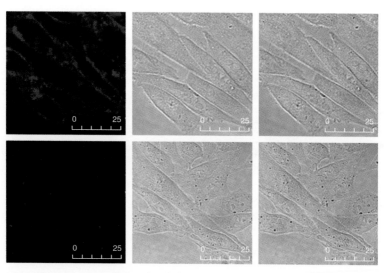

Figure 4.18 *Fluorescence (left), brightfield (middle), and overlaid (right) confocal microscopy images of Ac$_4$ManNAz-treated (top) and -untreated (bottom) CHO cells incubated with complex **55** (30 μM, 37 °C, 5 h). Reprinted with permission from [90] © 2012 Royal Society of Chemistry*

4.5   Conclusion

In this chapter, we have described the basics of photophysical processes involved in luminescent transition metal complexes and lanthanide chelates and fundamental techniques to detect these compounds in an intracellular environment. We have also discussed the cellular uptake properties of luminescent inorganic complexes, design of organelle specific complexes, applications of TPA and upconversion luminescence in cell and organism imaging, and the use of complexes as intracellular sensors and labels. Luminescent transition metal complexes and lanthanide-based chelates display intense and long-lived luminescence with large Stokes' shifts, which allows the detection of intracellular complexes by CLSM, FLIM, and flow cytometry. Many complexes also display TPA and upconversion luminescence, which involves low-energy excitation, allowing deeper penetration, weaker autofluorescence, less photobleaching, and lower phototoxicity. The cellular uptake of these complexes usually depends on their formal charges, lipophilicity, molecular size, and appended biological substrates. By ingenious modification of their coordinating ligands or surrounding chelators, complexes with specific organelle staining properties can be designed. The incorporation of various recognition or reactive functional groups into these complexes will also result in the development of new intracellular sensors and labels for ions and molecules. In conclusion, we anticipate that the rich and advantageous photophysical properties of luminescent transition metal complexes and lanthanide chelates will continue to contribute to the design of cell and organism imaging reagents.

Acknowledgements

We thank the Hong Kong Research Grants Councils (Project Nos. CityU 102212 and CityU 102311) for financial support.

References

1. See for example: A. Periasamy (ed.) *Methods in Cellular Imaging*, American Physiological Society, **2001**; M. Fernandez-Suarez, A. Y. Ting, *Nat. Rev. Mol. Cell Bio.* **2008**, *9*, 929.
2. See for example: I. L. Medintz, H. T. Uyeda, E. R. Goldman, H. Mattoussi, *Nat. Mater.* **2005**, *4*, 435; X. Michalet, F. F. Pinaud, L. A. Bentolila, J. M. Tsay, S. Doose, J. J. Li, G. Sundaresan, A. M. Wu, S. S. Gambhir, S. Weiss, *Science* **2005**, *307*, 538; M. C. Vasudev (ed.) *Quantum Dots as Fluorescent Probes for in Vitro Cellular Imaging: A Multi Labeling Approach*, University of Illinois at Chicago, **2006**.
3. See for example: S. Gross, D. Piwnica-Worms, *Cancer Cell* **2005**, *7*, 5; H. J. Carlson, R. E. Campbell, *Curr. Opin. Biotechnol.* **2009**, *20*, 19; G. Jung (ed.) *Fluorescent Proteins II: Application of Fluorescent Protein Technology*, Springer, **2012**.
4. See for example: (a) V. Balzani, F. Scandola, *Supramolecular Photochemistry*, Ellis Horwood, New York, **1990** (b) K. Kalyanasundaram, *Photochemistry of Polypyridine and Porphyrin Complexes*, Academic Press, San Diego, **1992** (c) D. M. Roundhill,

Photochemistry and Photophysics of Metal Complexes, Plenum Press, New York, **1994** (d) J.-C. G. Bünzli, C. Piguet, *Chem. Soc. Rev.* **2005**, *34*, 1048 (e) J. R. Lakowicz, *Principles of Fluorescence Spectroscopy*, 3rd edn, Springer, **2009** (f) K. M.-C. Wong, V. W.-W. Yam, *Acc. Chem. Res.* **2011**, *44*, 424 (g) V. W.-W. Yam, K. M.-C. Wong, *Chem. Commun.* **2011**, *47*, 11579 (h) X. He, V. W.-W. Yam, *Coord. Chem. Rev.* **2011**, *255*, 2111 (i) K. A. Gschneidner, Jr.,, J.-C. G. Bünzli, V. K. Pecharsky (ed.) *Handbook on the Physics and Chemistry of Rare Earths: Optical Spectroscopy*, Elsevier, **2011**.

5. See for example: (a) J.-C. G. Bünzli, Chem. Lett. **2009**, *38*, 104 (b) J.-C. G. Bünzli, S. V. Eliseeva, *Chem. Soc. Rev.* **2010**, *39*, 189 (c) V. Fernández-Moreira, F. L. Thorp-Greenwood, M. P. Coogan, *Chem. Commun.* **2010**, *46*, 186 (d) K. K.-W. Lo, M.-W. Louie, K. Y. Zhang, *Coord. Chem. Rev.* **2010**, *254*, 2603 (e) Q. Zhao, F. Li, C. Huang, *Chem. Soc. Rev.* **2010**, *39*, 3007 (f) K. K.-W. Lo, K. Y. Zhang, S. P.-Y. Li, *Pure Appl. Chem.* **2011**, *83*, 823 (g) K. K.-W. Lo, K. Y. Zhang, S. P.-Y. Li, *Eur. J. Inorg. Chem.* **2011**, 3551 (h) R. G. Balasingham, M. P. Coogan, F. L. Thorp-Greenwood, *Dalton Trans.* **2011**, *40*, 11663 (i) K. K.-W. Lo, S. P.-Y. Li. K. Y. Zhang, *New J. Chem.* **2011**, *35*, 265 (j) Q. Zhao, C. Huang, F. Li, *Chem. Soc. Rev.* **2011**, *40*, 2508 (k) D. Parker, *Aust. J. Chem.* **2011**, *64*, 239 (l) K. K.-W. Lo, A. W.-T. Choi, W. H.-T. Law, *Dalton Trans.* **2012**, *41*, 6021.

6. See for example: K. Konig, *J. Microsc.* **2000**, *200*, 83; M. D. Cahalan, I. Parker, S. H. Wei, M. J. Miller, **2002**, *2*, 872; S. Yao, K. D. Belfield, *Eur. J. Org. Chem.* **2012**, 3199.

7. S. L. Ashworth, R. M. Sandoval, G. A. Tanner, B. A. Molitoris, *Kidney Int.* **2007**, *72*, 416.

8. G. Stutzmann, *Microsc. Microanal.* **2008**, *14*, 482.

9. J. A. Scherschel, M. Rubart, *Microsc. Microanal.* **2008**, *14*, 492.

10. Y. Imanishi, K. H. Lodowski, Y. Koutalos, Biochemistry **2007**, *46*, 9674.

11. K. Park, *J. Control. Release* **2008**, *132*, 1.

12. Q. Yu, A. A. Heikal, *J. Photochem. Photobiol. B-Biol.* **2009**, *95*, 46.

13. K. Koenig, A. Ehlers, I. Riemann, S. Schenkl, R. Bueckle, M. Kaatz, *Microsc. Res. Tech.* **2007**, *70*, 398.

14. E. S. Kim, H. J. Chun, H. M. Kim, B. R. Cho, *Gastroenterolgy*, **2009**, *136*, A648.

15. N. J. Durr, T. Larson, D. K. Smith, B. A. Korgel, K. Sokolov, A. Ben-Yakar, *Nano Lett.* **2007**, *7*, 941.

16. C. D. Andrade, C. O. Yanez, H.-Y. Ahn, T. Urakami, M. V. Bondar, M. Komatsu, K. D. Belfied, *Bioconjugate Chem.* **2011**, *22*, 2060.

17. K. Schenke-Layland, I. Riemann, O. Damour, U. A. Stock, K. Koenig, *Adv. Drug Deliv. Rev.* **2006**, *58*, 878.

18. See for example: F. Wang, X. Liu, *Chem. Soc. Rev.* **2009**, *38*, 976.

19. J. Zhao, S. Ji, H. Guo, *RSC Adv.* **2011**, *1*, 937.

20. M. Hoppert, *Microscopic Techniques in Biotechnology*, John Wiley & Sons Ltd, **2006**; J. B. Pawley, *Handbook Of Biological Confocal Microscopy*, Springer, **2006**.

21. L.-C. Chen, W. R. Lloyd III,, C.-W. Chang, D. Sud, M.-A. Mycek, *Methods Cell Biol.* **2013**, *114*, 457.

22. M. G. Ormerod, *Flow Cytometry: A Practical Approach*, Oxford University Press, **2000**.

23. F. L. Thorp-Greenwood, V. Fernández-Moreira, C. O. Millet, C. F. Williams, J. Cable, J. B. Court, A. J. Hayes, D. Lloyd, M. P. Coogan, *Chem. Commun.* **2011**, *47*, 3096.

24. S. I. Pascu, P. A. Waghorn, T. D. Conry, B. Lin, H. M. Betts, J. R. Dilworth, R. B. Sim, G. C. Churchill, F. I. Aigbirhio, J. E. Warren, *Dalton Trans.* **2008**, 2107.

25. S. I. Pascu, P. A. Waghorn, B. W. C. Kennedy, R. L. Arrowsmith, S. R. Bayly, J. R. Dilworth, M. Christlieb, R. M. Tyrrell, J. Zhong, R. M. Kowalczyk, D. Collison, P. K. Aley, G. C. Churchill, F. I. Aigbirhio, *Chem. Asian. J.* **2010**, *5*, 506.

26. H. F. VanBrocklin, A. Liu, M. J. Welch, J. P. O'Neil, J. A. Katzenellenbogen, *Steroids* **1994**, *59*, 34.

27. C. A. Puckett, J. K. Barton, *J. Am. Chem. Soc.* **2007**, *129*, 46; C. A. Puckett, J. K. Barton, *Biochemistry* **2008**, *47*, 11711.

28. W. Jiang, Y. Gao, Y. Sun, F. Ding, Y. Xu, Z. Bian, F. Li, J. Bian, C. Huang, *Inorg. Chem.* **2010**, *49*, 3252.

29. K. K.-W. Lo, P.-K. Lee, J. S.-Y. Lau, *Organometallics* **2008**, *27*, 2998.

30. K. Y. Zhang, H.-W. Liu, T. T.-H. Fong, X.-G. Chen, K. K.-W. Lo, *Inorg. Chem.* **2010**, *49*, 5432.

31. N. Viola-Villegas, A. E. Rabideau, J. Cesnavicious, J. Zubieta, R. P. Doyle, *ChemMedChem* **2008**, *3*, 1387.

32. N. Viola-Villegas, A. E. Rabideau, M. Bartholoma, J. Zubieta, R. P. Doyle, *J. Med. Chem.* **2009**, *52*, 5253.

33. O. Warburg, *Science* **1956**, *123*, 309.

34. R. A. Medina, G. I. Owen, *Biol. Res.* **2002**, *35*, 9.

35. H.-W. Liu, K. Y. Zhang, W. H.-T. Law, K. K.-W. Lo, *Organometallics* **2010**, *29*, 3474.

36. M. Gottschaldt, U. S. Schubert, S. Rau, S. Yano, J. G. Vos, T. Kroll, J. Clement, I. Hilger, *ChemBioChem* **2010**, *11*, 649.

37. M.-W. Louie, H.-W. Liu, M. H.-C. Lam, Y.-W. Lam, K. K.-W. Lo, *Chem. Eur. J.* **2011**, *17*, 8304.

38. (a) A. J. Amoroso, M. P. Coogan, J. E. Dunne, V. Fernández-Moreira, J. B. Hess, A. J. Hayes, D. Lloyd, C. Millet, S. J. A. Pope, C. Williams. *Chem. Commun.* **2007**, 3066 (b) V. Fernández-Moreira, F. L. Thorp-Greenwood, A. J. Amoroso, J. Cable, J. B. Court, V. Gray, A. J. Hayes, R. L. Jenkins, B. M. Kariuki, D. Lloyd, C. O. Millet, C. F. Williams, M. P. Coogan, *Org. Biomol. Chem.* **2010**, *8*, 3888.

39. J. Kuil, P. Steunenberg, P. T. K. Chin, J. Oldenburg, K. Jalink, A. H. Velders, F. W. B. van Leeuwen, *ChemBioChem* **2011**, *12*, 1897.

40. J. R. Casey, S. Grinstein, J. Orlowski, *Nat. Rev. Mol. Cell Biol.* **2010**, *11*, 50.

41. A. J. Amoroso, R. J. Arthur, M. P. Coogan, J. B. Court, V. Fernández-Moreira, A. J. Hayes, D. Lloyd, C. Millet, S. J. A. Pope, *New. J. Chem.* **2008**, *32*, 1097.

42. R. W.-Y. Sun, A. L.-F. Chow, X.-H. Li, J. J. Yan, S. S.-Y. Chui, C.-M. Che, *Chem. Sci.* **2011**, *2*, 728.

43. (a) T. Chen, Y. Liu, W.-J. Zheng, J. Liu, Y.-S. Wong, *Inorg. Chem.* **2010**, *49*, 6366 (b) V. Pierroz, T. Joshi, A. Leonidova, C. Mari, J. Schur, I. Ott, L. Spiccia, S. Ferrari, G. Gasser, *J. Am. Chem. Soc.* **2012**, *134*, 20376.

44. J. W. Walton, A. Bourdolle, S. J. Butler, M. Soulie, M. Delbianco, B. K. McMahon, R. Pal, H. Puschmann, J. M. Zwier, L. Lamarque, O. Maury, C. Andraud, D. Parker, *Chem. Commun.* **2013**, *49*, 1600.

45. T. Zhang, X. Zhu, C. C. W. Cheng, W.-M. Kwok, H.-L. Tam, J. Hao, D. W. J. Kwong, W.-K. Wong, K.-L. Wong, *J. Am. Chem. Soc.* **2011**, *133*, 20120.

46. Y. Hai, J.-J. Chen, P. Zhao, H. Lv, Y. Yu, P. Xu, J.-L. Zhang, *Chem. Commun.* **2011**, *47*, 2435.

47. G.-L. Law, K.-L. Wong, C. W.-Y. Man, S.-W. Tsao, W.-T. Wong, *J. Biophoton.* **2009**, *2*, 718.

48. K. K.-W. Lo, M.-W. Louie, K.-S. Sze, J. S.-Y. Lau, *Inorg. Chem.* **2008**, *47*, 602.

49. S. Clède, F. Lambert, C. Sandt, Z. Gueroui, M. Réfrégiers, M.-A. Plamont, P. Dumas, A. Vessières, C. Policar, *Chem. Commun.* **2012**, *48*, 7729.

50. C.-L. Ho, K.-L. Wong, H.-K. Kong, Y.-M. Ho, C. T.-L. Chan, W.-M. Kwok, K. S.-Y. Leung, H.-L. Tam, M. H.-W. Lam, X.-F. Ren, A.-M. Ren, J.-K. Feng, W.-Y. Wong, *Chem. Commun.* **2012**, *48*, 2525.

51. W.-L. Kwong, R. W.-Y. Sun, C.-N. Lok, F.-M. Siu, S.-Y. Wong, K.-H. Low, C.-M. Che, *Chem. Sci.* **2013**, *4*, 747.

52. R. Berezney, K. W. Jeon (ed.) *Nuclear Matrix: Structural and Functional Organization*, Elsevier, **1995**.

53. H. Scheffler, Y. Ya, I. Ott, *Polyhedron* **2010**, *29*, 66.

54. C. P. Bagowski, Y. You, H. Scheffler, D. H. Vlecken, D. J. Schmitz, I. Ott, *Dalton Trans.* **2009**, 10799.

55. B. Önfelt, L. Göstring, P. Lincoln, B. Nordén, A. Önfelt, *Mutagenesis* **2002**, *17*, 317.

56. J. S.-Y. Lau, P.-K. Lee, K. H.-K. Tsang, C. H.-C. Ng, Y.-W. Lam, S.-H. Cheng, K. K.-W. Lo, *Inorg. Chem.* **2009**, *48*, 708.

57. M. R. Gill, J. Garcia-Lara, S. J. Foster, C. Smythe, G. Battaglia, J. A. Thomas, *Nature Chem.* **2009**, *1*, 662.

58. P. E. Nielsen, M. Egholm, R. H. Berg, O. Buchardt, *Science* **1991**, *254*, 1497.

59. E. Ferri, D. Donghi, M. Panigati, G. Precipe, L. D'Alfonso, I. Zanoni, C. Baldoli, S. Maiorana, G. D' Alfonso, E. Lincandro, *Chem. Commun.* **2010**, *46*, 6255.

60. J. Brunner, J. K. Barton, *Biochemistry* **2006**, *45*, 12295.

61. C. A. Puckett, J. K. Barton, *J. Am. Chem. Soc.* **2009**, *131*, 8738.

62. G.-L. Law, C. Man, D. Parker, J. W. Walton, *Chem. Commun.* **2010**, *46*, 2391.

63. N. A. O'Connor, N. Stevens, D. Samaroo, M. R. Solomon, A. A. Martí, J. Dyer, H. Vishwasrao, D. L. Akins, E. R. Kandel, N. J. Turro, *Chem. Commun.* **2009**, 2640.

64. S. W. Botchway, M. Charnley, J. W. Haycock, A. W. Parker, D. L. Rochester, J. A. Weinstein, J. A. G. Williams, *Proc. Natl. Acad. Sci. USA* **2008**, *105*, 16071.

65. A. K. L. Leung, J. S. Andersen, M. Mann, A. I. Lamond, *Biochem. J.* **2003**, *376*, 553.

66. J. Dyckman, J. K. Weltman, *J. Cell. Biol.* **1970**, *45*, 192; R. R. Cowden, S. K. Curtis, *Histochem. J.* **1974**, *6*, 447.

67. K. Y. Zhang, S. P.-Y. Li, N. Zhu, I. W.-S. Or, M. S.-H. Cheung, Y.-W. Lam, K. K.-W. Lo, *Inorg. Chem.* **2010**, *49*, 2530.

68. C.-K. Koo, L. K.-Y. So, K.-L. Wong, Y.-M. Ho, Y.-W. Lam, M. H.-W. Lam, K.-W. Cheah, C C.-W. Cheng, W.-M. Kwok, *Chem. Eur. J.* **2010**, *16*, 3942.

69. J. Yu, D. Parker, R. Pal, R. A. Poole, M. J. Cann, *J. Am. Chem. Soc.* **2006**, *128*, 2294.

70. T. Furuta, S. S.-H. Wang, J. L. Dantzker, T. M. Dore, W. J. Bybee, E. M. Callaway, W. Denk, R. Y. Tsien, *Proc. Natl. Acad. Sci. USA* **1999**, *96*, 1193.

71. A. D'Aléo, A. Bourdolle, S. Brustlein, T. Fauquier, A. Grichine, A. Duperray, P. L. Baldeck, C. Andraud, S. Brasselet, O. Maury, *Angew. Chem. Int. Ed.* **2012**, *51*, 6622.

72. S. W. Botchway, M. Charnley, J. W. Haycock, A. W. Parker, D. L. Rochester, J. A. Weinstein, J. A. G. Williams, *Proc. Natl. Acad. Sci. USA* **2008**, *105*, 16071.

73. J. Zhou, Z. Liu, F. Li, *Chem. Soc. Rev.* **2012**, *41*, 1323.

74. R. A. Jalil, Y. Zhang, *Biomaterials* **2008**, *29*, 4122.

75. S. Cui, D. Yin, Y. Chen, Y. Di, H. Chen, Y. Ma, S. Achilefu, Y. Gu, *ACS Nano* **2013**, *7*, 676.

76. L. Cheng, C. Wang, X. Ma, Q. Wang, Y. Cheng, H. Wang, Y. Li, Z. Liu, *Adv. Funct. Mater.* **2013**, *23*, 272.

77. Q. Liu, T. Yang, W. Feng, F. Li, *J. Am. Chem. Soc.* **2012**, *134*, 5390.

78. Q. Liu, B. Yin, T. Yang, Y. Yang, Z. Shen, P. Yao, F. Li, *J. Am. Chem. Soc.* **2013**, *135*, 5029.

79. S. Zhang, M. Hosaka, T. Yoshihara, K. Negishi, Y. Iida, S. Tobita, T. Takeuchi, *Cancer Res.* **2010**, *70*, 4490.

80. J. Ji, N. Rosenzweig, I. Jones, Z. Rosenzweig, *J. Biomed. Opt.* **2002**, *7*, 404.

81. G.-L. Law, R. Pal, L. O. Palsson, D. Parker, K.-L. Wong, *Chem. Commun.* **2009**, 7321.

82. R. A. Gatenby, R. J. Gillies, E. T. Gawlinski, A. F. Gmitro, B. Kaylor, *Cancer Res.* **2006**, *66*, 5216.

83. Y. Song, D. Salinas, D. W. Nielson, A. S. Verkman, *Am. J. Physiol.: Cell Physiol.* **2006**, *290*, 741.

84. F. L. M. Ricciardolo, B. Gatston, J. Hunt, J. Allergy, *Clin. Immunol.* **2004**, *113*, 610.

85. R. Pal, D. Parker, *Org. Biomol. Chem.* **2008**, *6*, 1020.

86. P.-K. Lee, H.-W. Liu, S.-M. Yiu, M.-W. Louie, K. K.-W. Lo, *Dalton Trans.* **2011**, *40*, 2180.

87. C. Li, M. Yu, Y. Sun, Y. Wu, C. Huang, F. Li, *J. Am. Chem. Soc.* **2011**, *133*, 11231.

88. B. Wang, Y. Liang, H. Dong, T. Tan, B. Zhan, J. Cheng, K. K.-W. Lo, Y. W. Lam, S. H. Cheng, *ChemBioChem* **2012**, *13*, 2729.

89. (a) E. Saxon, C. R. Bertozzi, *Science* **2000**, *287*, 2007 (b) J. A. Prescher, C. R. Bertozzi, *Nat. Chem. Biol.* **2005**, *1*, 13 (c) P. V. Chang, J. A. Prescher, M. J. Hangauer, C. R. Bertozzi, *J. Am. Chem. Soc.* **2007**, *129*, 8400 (b) E. M. Sletten, C. R. Bertozzi, *Angew. Chem. Int. Ed.* **2009**, *48*, 6974.

90. K. K.-W. Lo, B. T.-N. Chan, H.-W. Liu, K. Y. Zhang, S. P.-Y. Li, T. S.-M. Tang, *Chem. Commun.* **2013**, *49*, 4271.

5

Cellular Imaging with Metal Carbonyl Complexes

Luca Quaroni[a] and Fabio Zobi[b]
[a]Swiss Light Source, Paul Scherrer Institute, Switzerland
[b]Department of Chemistry, University of Fribourg, Switzerland

5.1 Introduction

Vibrational spectroscopy in combination with standard group theory methods continues to be among the most commonly used structural techniques in chemical systems. This is particularly the case for metal complexes comprising ligands such as CN^-, CNR^-, CO, and NO, which present unique vibrational signatures in a region of the spectrum amenable to rapid interpretation. Vibrational spectroscopy was revolutionized in the early 1980s by the advent of Fourier-transform instruments, which replaced the old dispersive apparatus and allowed the development of the technique as a reliable analytical tool. In conjunction with this technological leap, standard organometallic compounds with an M–CO functionality began to cross interdisciplinary boundaries and enter the realm of biology, biochemistry, and medicine.

It was the group of Gérard Jaouen who first exploited the unusual vibrational properties of metal carbonyls in biochemical analysis. In the mid-1980s the French scientists developed a special bioanalytical assay, called carbonylmetallo-immunoassay [1] (CMIA), to detect and quantify the presence of clinically relevant molecules [2, 3]. CMIA is a heterogeneous, non-isotopic, competitive-type metallo-immunoassay in which a specific biotracer labeled with metal carbonyl fragments acts as a vibrational probe. Several antiepileptic medications such as carbamazepine, phenobarbital, diphenylhydantoin, and steroid hormones such as cortisol have been labeled in CMIA with different metal carbonyl reporters

Inorganic Chemical Biology: Principles, Techniques and Applications, First Edition. Edited by Gilles Gasser.
© 2014 John Wiley & Sons, Ltd. Published 2014 by John Wiley & Sons, Ltd.

including derivatives of $Cr(CO)_3$, $Co_2(CO)_6$, $CpMn(CO)_3$, and $CpFe(CO)_2$ (Scheme 5.1) [4–8]. Analysis of the characteristic $v(CO)$ stretching vibrations of the metal carbonyl probes allowed quantitative detection of analytes at a concentration limit as low as about 10 nM in complex biological extracts. The assay was originally developed in the area of clinical biology but more recently it has been extended to environmental studies with assays of pesticides such as atrazine and chlortoluron [9, 10].

Since the introduction of CMIA, metal carbonyl complexes have been employed as infrared spectroscopy probes in a large variety of systems, spanning from small peptide, protein, and DNA labeling to read-out sensors in pH changes, π-stacking interactions or alkali metal concentrations [11, 12]. Until recently, however, conventional vibrational spectroscopy was not considered as a method for the visualization and imaging of biological samples. The reason is that in biological samples the background absorbance of water and organic (macro)molecules make large segments of the infrared spectrum opaque and essentially unsuitable for the reliable interpretation of small spectroscopic changes.

However, the same reasons that has hampered the development of vibrational techniques as imaging tools in biochemical sciences, have offered organometallic chemists a unique opportunity. It is well known that when metal carbonyl complexes are exposed to an

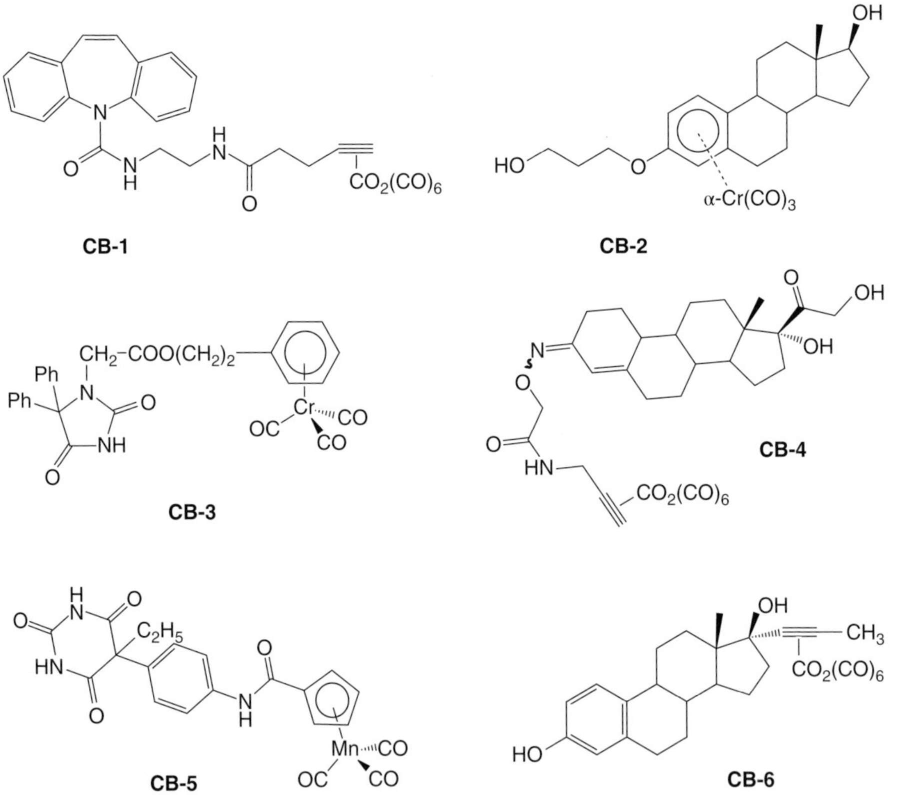

Scheme 5.1 *Structures of selected metal carbonyl labeled biotracers used in CMIA. CB stands for carbonylmetallo-biotracer*

electromagnetic radiation of suitable energy, the bound CO group will resonate with a stretching frequency in the water-transparent region of the vibrational spectrum. This diagnostic region, approximately between 2200 and 1800 cm^{-1}, is virtually void of vibrational signals of organic functional groups. The distinctive features of the metal carbonyl complexes, coupled with technological advances in vibrational microscopy techniques, may thus revolutionize the way in which we are able to visualize cellular structures and dynamic processes in real time.

The inorganic probes, the methodologies and the applications of what may be broadly defined as vibrational cellular imaging are the subjects of this chapter. The general use of metal carbonyl complexes as bioprobe structures will not be addressed here. The field has been reviewed in detail elsewhere and the readers are referred to relevant contributions for a deeper knowledge of the research area [11–13]. In order to guide the reader through the subject matter, this chapter has been organized as follows: in Section 5.2 the fundamental properties of vibrational spectroscopy of metal carbonyl complexes are first described; Section 5.3 provides an overview of vibrational microscopy and cellular imaging; Sections 5.4–5.6 cover in detail the main spectroscopic techniques that make specific use of metal carbonyl bioprobes and which are currently employed in cellular imaging. These are, respectively, infrared (IR), Raman, and one instance of near-field microscopy. Each one of these sections examines first the optical principles and the design of the instruments and then comprehensively summarizes work in the area. Finally, Section 5.7 offers a comparative analysis of the different vibrational techniques described within the context of cellular imaging, highlighting the advantages and limitations associated with each.

5.2   Vibrational Spectroscopy of Metal Carbonyl Complexes

Transition metal complexes tend to contain a large number of atoms, and hence a large number of vibrational degrees of freedom. The interpretation of their vibrational spectra is thus most easily realized for systems containing energetically isolated vibrations. Metal carbonyl complexes are well known for their unique infrared (IR) properties. In general such complexes have at least one diagnostic region of the spectrum that displays relatively few of the $(3N-6)$ normal modes and these will likely not couple very strongly to those vibrations far removed from them. There are four fundamental vibrational modes associated with the M–CO functionality:

1. C–O stretching ν(CO) in the 2150–1750 cm^{-1} region;
2. M–C–O bending δ(MCO) in the 700–500 cm^{-1} region;
3. M–C stretching ν(MC) in the 500–300 cm^{-1} region;
4. C–M–C bending δ(CMC) in the 150–50 cm^{-1} region.

The ν(CO) stretching vibrations are much higher in frequency than all other modes involving the M–C–O unit. Thus, their combination gives rise to normal modes that are nearly pure combinations of the C–O stretching vibrations, with little or no contributions from bending modes. While coupling between different ν(CO) modes occurs, coupling between ν(CO) and any other vibration of the complex can be neglected. In effect, in the interpretation of IR spectra of metal carbonyl complexes one may ignore every vibration except those assigned as ν(CO) stretching modes. The use of vibrational spectroscopy has been so prevalent in metal carbonyl chemistry for just this reason; a molecule such as $Cr(CO)_6$ can be vibrationally analyzed much more simply through consideration of only the six high-energy CO stretching modes rather than the full treatment of all 33 vibrational normal modes.

Gaseous carbon monoxide itself is not a strong IR absorber. The vibrational stretching mode of the diatomic molecule is centered at $2143 \, cm^{-1}$ and gives rise only to weak vibrational signals when exposed to electromagnetic radiations. Yet the vibrational spectra of any transition metal carbonyl complex with terminal CO groups are dominated by very high intensity CO-stretching absorptions. These are most often observed at lower wavenumbers ($2200–1800 \, cm^{-1}$), in a window where water and most organic functional groups do not exhibit vibrational bands (Fig. 5.1). The reasons underlining the shift to lower vibrational frequencies, relative to free CO, are well understood on the basis of the classical synergistic description of the metal–CO bond.

The bonding interaction of carbon monoxide with transition metal ions consists of a concerted electronic delocalization over bonding orbitals of σ and π character. The σ bond originates from an overlap of a symmetric metal orbital with the sp hybrid orbital of the carbon atom, which in free CO formally holds a non-bonded pair of electrons. The π bond ensues from the overlap of an antisymmetric metal orbital of suitable symmetry with the π^* antibonding orbital originating from the CO triple bond. The components of this synergistic interaction are referred to as σ-donation and π-back donation. The non-bonded pair of electrons on free CO is σ-donated to the transition metal. This pair may be then thought to become part of the bonding system of the metal ion, which in turn π-back donates the electronic density to π^* orbital of CO.

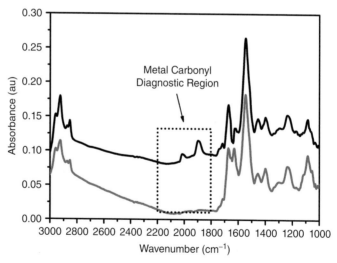

Figure 5.1   *IR spectra recorded in absorbance mode of a single 3T3 fibroblast (bottom, dark gray line) and of the same cell after overnight incubation with the fac-[Re(CO)$_3$(OH$_2$)$_3$]$^+$ complex (black line). Dotted box shows the characteristic metal carbonyl diagnostic region between 2200 and 1800 cm^{-1} (L. Quaroni and F. Zobi unpublished results). The amide I absorption band, between 1600 and 1700 cm^{-1}, is distorted by the elevated contribution from water absorption*

This interaction continues to provide a useful qualitative model to understand and estimate the effects of complexation and the relative shift in the vibrational $v(CO)$ frequencies in carbonyl complexes. Some models have also described these relations quantitatively, allowing the prediction of carbonyl stretching vibrations to a good level of accuracy [14]. The interpretations of the vibrational spectra of the M−CO functionality are generally based on this synergistic electronic delocalization. Thus, arguments are often encountered relating the relative electronic density of the metal ion with the frequency of the $v(CO)$ stretching of these complexes. As the electronic density on the central metal ion increases (i.e., in anionic complexes), the metal tends to exert a more significant π-back donation to the π^* orbital of CO. This effect lowers the CO bond order while concomitantly increasing the M−C bond order. In turn the lower bond order of CO translates into a lower frequency of the $v(CO)$ stretching. The opposite argument holds for cationic or neutral complexes.

The picture, however, is more complicated than the simple argument just presented. Several studies have indicated that polarization effects of the CO bond also play a significant role in the shift of the CO frequencies [15−20]. IR intensities are proportional to the square of the dipole change associated with the vibration. The intensities of the $v(CO)$ stretching vibrations of carbonyl complexes are remarkable in that they can be four to ten times higher than the band of standard organic functional groups. This feature implies a large dipole change when a carbonyl group vibrates. This effect is undoubtedly associated with the concerted σ-donation and π-back donation described previously. The stretching of a CO group bonded to a metal atom leads to a lowering of the energy of the 2π orbital, since the antibonding interaction is weakened and the orbital becomes a better acceptor. At the same time, the overlap between the carbon and the metal orbitals is increased as the M−C distance lessens. There is, however, no linear relationship between $v(CO)$ stretching frequencies and metal electronic density or M−C distances [13, 14, 20, 21]. The vibrational position of the C−O stretching is more closely related to the electronic density on the terminal oxygen atom of the CO group. This observation has been interpreted as a direct measure of the polarization of the C≡O bond, which translates into a lowering of the CO bond order and, consequently, as a measure of a high dipole change and the resonance frequency [21].

Within the context of this chapter, metal carbonyl complexes used as bioprobes, or in bioimaging, contain more than one carbonyl group. Metal complexes with an isolated M−C≡O functionality give rise to a single sharp vibrational band for the $v(CO)$ stretching mode. As the number of CO groups increases in the $[M(CO)_n(L)_{6-n}]^z$ species so does the complexity of the vibrational spectrum (Fig. 5.2). Bis-carbonyl species (i.e., compounds where $n = 2$) generate two distinct signals, a higher-frequency symmetric and a lower-frequency antisymmetric vibrational mode. However, as it will become apparent from the following sections, most metal carbonyl complexes used in imaging of cellular systems comprise species with an *fac*-$[M(CO)_3]^z$ core (i.e., compounds where $n = 3$). The vibrational spectrum of this core shows again a single high-frequency symmetric stretching mode (referred to as the A1 mode, see Fig. 5.2) but two energetically distinct vibrations (E mode). In reality the energy separation of the latter is small. The two transitions are often degenerate appearing as a single broad signal in, for example, the IR spectrum. In general, the position (i.e., the frequency) of these $v(CO)$ stretching modes is influenced by the charge (or oxidation state) of the metal ion, while the vibrational coupling, and hence the separation of the $v(CO)$ bands, is influenced by the geometry of the complex and the spatial arrangement of the CO groups and the ancillary ligands.

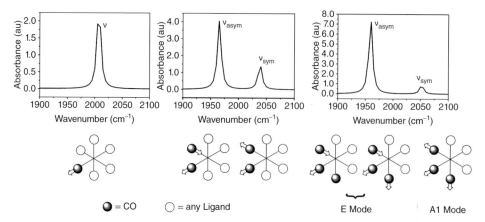

Figure 5.2   *DFT calculated IR spectra of the $v_{sym}(CO)$ and $v_{asym}(CO)$ vibrational stretching modes for carbonyl groups of a generic $[M(CO)_n(OH_2)_{6-n}]$ complex. From left to right n = 1 to 3. The arrows indicate the relative displacement of the COs in the different stretching modes*

5.3   Microscopy and Imaging of Cellular Systems

Microscopy has been for centuries a defining technique for biological and biomedical studies [22], with theories waxing and waning with every wave of improvements in resolution and contrast, the two main properties that define the information content of microscopy techniques, resolution, and contrast.

Resolution is the capability of an optical system to discriminate the images of two points close in space. It has been recognized since the late 19th century that the resolution achievable by a microscope is limited by the diffraction properties of light and is a function of wavelength. In general, the limiting resolution allowed by diffraction is of the order of the wavelength used for image construction, and improves as we move towards the shorter wavelengths of the electromagnetic spectrum. For visible light microscopes this limit is of the order of a few hundred nanometers.

Resolution is not, however, the only quantity that limits our disclosure of the microscopic world. Contrast is as important. Contrast is the ability of an image element to stand out against the background of other adjacent elements. In optical microscopy, contrast in unstained biological samples is generally low. In typical cellular samples, contrast, in the absence of staining, is dominated by optical effects due to discontinuities in the real part of the refractive index of the material, such as refraction, scattering, and interference.

The experimental conditions used for imaging many objects of biological interest, such as cells in an aqueous environment, are such that variations of the refractive index throughout the sample are very small and many objects remain nearly invisible. This limitation has led microscopists to develop contrast-enhancing techniques to improve specimen visibility, image detail, and information content.

In the search for contrast enhancement mechanisms, novel opportunities have been created by implementing vibrational spectroscopy techniques in a microscopy configuration. These techniques rely on the specificity and uniqueness of vibrational spectra of molecular species. As such they provide contrast based on the molecular properties and chemical composition of the sample, often termed chemical contrast. It is precisely within this context that metal carbonyl complexes have been selected as imaging probes based on their unique vibrational features described earlier.

5.3.1 Techniques of Vibrational Microscopy

Numerous techniques are currently in use to obtain vibrational spectra of molecules. From the quantum mechanical point of view they can be classified based on whether they rely on photon absorption or on photon scattering. Conventional infrared absorption spectroscopy, usually simply called IR spectroscopy, as the name implies, is based on the absorption of IR photons and is the prototypical example of the first category. Conventional Raman spectroscopy is based on the inelastic scattering of visible and near-IR photons and is the prototypical example of the second category. Several excellent reviews and books deal with the basic concepts of these two techniques and the reader is directed to these for additional information [23]. Both IR and Raman spectroscopy can be implemented as microscopy techniques, for which the light delivery and collection optics are composed of fast lenses or objectives to allow focusing of the light beam to a small spot [24].

5.4 Infrared Microscopy

Most molecular species, except for homonuclear diatomic molecules, absorb light in the mid-infrared spectral region. As a consequence, most groups of biological molecules have been investigated by IR spectroscopy, including macromolecules, small organic and inorganic molecules, also water, as well as crystalline inclusions. Because of its broad applicability, IR spectroscopy has been extensively used over several decades to study structure and function in biological molecules. IR spectroscopy measurements can be performed using microscopy optics. In this instance the technique is called generally IR microscopy. The terms IR imaging, IR spectromicroscopy (or the equivalent IR microspectroscopy) are used for specific configurations, as detailed later in the chapter. When working in a microscopy configuration, the technique is typically sensitive to mM concentrations of chromophores for a 1 min measurement time, with exact values depending on the extinction coefficients of specific absorption bands [25, 26].

An IR microscope can be thought of as an elaborate sample compartment for an interferometer, containing objectives that act as beam condenser optics. Its overall structure is similar to that of a conventional optical microscope, with a video camera and binoculars for sample inspection, a vertical optical path at the sample and a horizontal x,y stage for sample movement. The objectives focus both the IR beam used for the spectroscopic measurement and the visible light beam used for sample inspection. The optical paths for the two beams are parfocal and parcentric. Therefore they ensure consistent matching of the IR beam location on the sample image and allow selective measurement of specific spatial locations [27]. Fig. 5.3 shows the optical scheme of an IR microscope.

Two basic experimental configurations are used with an IR microscope, often termed a spectromicroscopy (or microspectroscopy) configuration and an imaging configuration.

In a spectromicroscopy measurement an IR microscope is used to condense light into a single small spot and to collect a spectrum of the illuminated portion of the sample. For

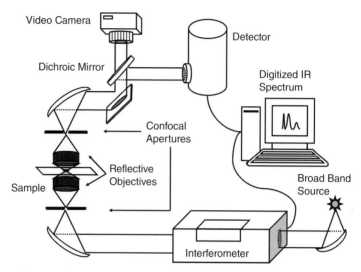

Figure 5.3 *Scheme of an IR microscope. The light coming from the light source (an internal thermal source or an external source, such as a synchrotron) is analyzed by an interferometer. The modulated light is then focused on the sample using reflective optics and a set of confocal apertures to define the measured area. Light transmitted or reflected by the sample is collected and brought to a detector for intensity measurements. The detector can be a single-element detector, for spectromicroscopy measurements, or a multi-element detector, for imaging. A dichroic mirror allows directing visible light to a video camera for sample inspection*

this measurement the microscope is often operated in a confocal geometry. One or more apertures are located in optical planes that are conjugated with the samples plane and are used to limit the size of the incoming beam or the field of view observed by the detector. In most microscopes the optical path can be set so that measurements can be performed in either a transmission or a reflection configuration.

The spatial resolution of an IR microscopy experiment is a critical parameter. Closing the confocal apertures allows measurement of a smaller portion of the sample and provides better spatial resolution, as long as the apertures are larger than the wavelength of the light. As the apertures become smaller than the wavelength, resolution is defined by diffraction effects. Far-field optical elements are used for focusing and the conventional resolution limit defined by Rayleigh, which is generally used for visible microscopy, also applies to IR microscopy (Equation 5.1) [28]:

$$d = 0.61\lambda/NA \qquad (5.1)$$

which is Rayleigh's Criterion for diffraction-limited far-field resolution, where d = minimum distance at which two points can be resolved, λ = wavelength of light, and NA = numerical aperture of the objectives.

The numerical aperture, NA, is a property of the objective design and of the medium through which the light is focused, if it is not air. A typical NA value for commercial objectives is 0.5–0.6, which gives a resolution of approximately λ. For light in the mid-IR region, diffraction-limited resolution is in the range 2.5–25 μm (micrometers). This is of the order of the size of eukaryotic cells, the dimensions of which range from a few to a hundred micrometers.

A set of spectromicroscopy measurements can be used to obtain a 2D distribution of the spectra for each point of the sample by raster scanning it. The resulting spectral data can later be analyzed to provide a map of specific spectral features, such as intensity of a specific absorption band, throughout the sample. The procedure is called IR mapping.

The performance of confocal IR microscopes is poor when they are operated at high spatial resolution, with apertures closed down to a few micrometers. This is due to the elevated loss of light throughput that small apertures impose on the IR beam. Conventional benchtop IR sources are particularly affected because of their limited brightness. As a consequence it is usually difficult to perform IR spectromicroscopy measurements with apertures closed to less than 40–10 µm, below the diffraction limit, the more so when the sample is a complex system such as a cell, with relatively weak absorption bands.

Limitations on throughput can be contained by using a bright synchrotron source of IR radiation in place of the conventional thermal source. The brightness of the source ensures that throughput is little affected until aperture size reaches the diffraction limit, and that some useful throughput is observed also when aperture size is below this limit [29]. Most spectromicroscopy and mapping experiments on small single cells or on subcellular volumes to date have been performed using a synchrotron source, in particular the ones involving living cells in an aqueous medium [26].

During the last 15 years, 2D arrays of detectors for mid-IR radiation have been introduced on the market and gained wide acceptance. These detectors, termed in the IR community focal plane arrays (FPA), allow collecting an extended image of the sample in IR light [27]. When mounted on a microscope coupled to an interferometer, they allow the recording of an IR absorption spectrum for each pixel position. This results in the simultaneous collection of several thousand spectra covering an extended sample. The expression "IR imaging" is currently applied to the collection of an IR image with an FPA detector.

5.4.1 Concentration Measurements with IR Spectroscopy and Spectromicroscopy

For dilute solutions, the concentration C of a chromophore can be calculated from measured IR absorbance using the Lambert–Beer law (Equation 5.2):

$$A = \epsilon \times L \times C \tag{5.2}$$

where in this instance A is the peak absorbance, ϵ is the extinction coefficient at the wavenumber of interest, L is the optical path length, and C is the concentration. In a sandwich-type sample holder this is taken to match the thickness of the spacer used in the sample holder.

Application of Lambert–Beer law is straightforward for macroscopic measurements of uniform samples. However, some cautionary notes must be given about its use in a spectromicroscopy experiment. One issue is that the heterogeneity of the sample must be taken into account in evaluating the accuracy of a quantitative measurement. In the case of the measurement of intracellular chromophores, for example, the gap present between a cell and an optical window must be accounted for in calculating the cellular volume in which a chromophore is distributed. An additional issue is that values of the extinction coefficient ϵ reported in the literature are typically measured using the parallel or only slightly focused beam inside the sample compartment of an interferometer. These values are not exact when using the sharply focused beam of a microscope, for which the average optical path of the rays crossing the sample volume is longer than the nominal thickness of the sample, since the rays come into the sample at an angle. To avoid any differences in optical path, for accurate measurements, particularly when using objectives with a high numerical aperture, the extinction coefficient ϵ must be obtained from standard solutions using the same sample holder as the one used for the sample measurement.

5.4.2 Water Absorption

When live samples are used, it is often necessary to work in the presence of large concentrations of water, intracellular and extracellular. Water is a very strong IR absorber and its presence can obscure several cellular absorption bands. Interference from water is the main limitation to the use of IR spectroscopy on tissue and cell samples. For this reason it is common to remove water from the sample by drying. This approach naturally has the disadvantage that a dried sample is usually no longer alive.

Absorbance spectra are saturated at $1650\,cm^{-1}$ for a water layer thicker than 15 µm, due to absorption from the bending mode of water δ_{H_2O}. The analysis of the amide I band of proteins requires even thinner samples, due to the overlap of this band and δ_{H_2O}, often down to 5–10 µm. Absorption is also saturated in the interval 3200–$3600\,cm^{-1}$ for a layer thicker than 2 µm, because of absorption from the bending modes of water, v_{s,H_2O} and v_{as,H_2O}. These constraints can be a severe limitation when thicker cells or tissue layers are used. A distinct advantage of metal carbonyl complexes is that their characteristic $v(CO)$ modes fall well outside of the water saturated spectral regions, with the result that fairly thick aqueous samples of these complexes can still be studied by IR spectroscopy [26].

5.4.3 Metal Carbonyls as IR Probes for Cellular Imaging

The first example of an organometallic carbonyl tag used for bioimaging in the mid-infrared region was reported by the group of Leong in 2007 [30]. In their work they used water soluble osmium carbonyl clusters **1** and **2** as IR-active probes for the labeling of fatty acids and phosphatidylcholine derivatives. Oral mucosa cells were incubated with **1** and **2** at low micromolar concentrations (32 μM) and it was demonstrated that both compounds could be imaged in the cell by IR spectromicroscopy.

Passive diffusion internalization of clusters **1** and **2** in the cells was demonstrated by washout experiments, concentration dependent studies consistent with lack of adsorption onto the outer surface of the cells, and fractionation analysis of the nucleus, the cytoplasm, the organelles, and the membrane. IR spectromicroscopy also showed no appreciable decomposition or hydrolysis of the clusters suggesting that **1** and **2** are resistant to hydrolysis or metabolism by the cells. Any decomposition affecting the cluster core (i.e., the carbonyl tag) would have been detected as a shift in the pattern of the carbonyl stretching region. Leong and his group further showed that the absorbance of both **1** and **2** across a single mucosa cell had similar intensities and profiles.

The 2D and 3D images of the cellular distribution of the osmium clusters were finally constructed by analysis of the stretching vibrations in the $2010\,cm^{-1}$ region. The resulting distribution associated with **1** is reproduced in Fig. 5.4. IR chemical mapping images (b and c in Fig. 5.4) were closely correlated with the optical image, showing for the first time the potential of IR spectromicroscopy for cellular imaging with metal carbonyl complexes.

Since the original report by Leong, two other studies dealing with the IR cellular imaging with metal carbonyl complexes have appeared. The group of Clotilde Policar in France carried out both investigations. In 2012 synchroton radiation FTIR spectromicroscopy (SR-FTIR-SM) was used for the detection of the rhenium tricarbonyl tamoxifen analogue **3** in MDA-MB-231 breast cancer cells [31]. The hormone independent cancer cells were grown on IR transparent CaF_2 optical disks and then incubated for a period of 1 h with a 10 μM solution of **3**. Subcellular level detection of the carbonyl complex and subsequent single-cell maps could then be generated by plotting the integral under a specific CO stretching vibration band of **3**. In the infrared chemical maps, areas of intense IR-signals at different wavelengths were referred to as *hot spots* and are represented in different colors (Fig. 5.5).

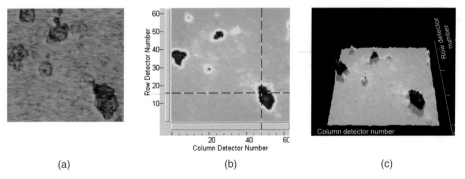

(a)   (b)   (c)

Figure 5.4   (a) Normal optical image of oral mucosa cells, (b) a false-color infrared image, and (c) a 3D display of the image of the oral mucosa cells treated with **1**, taken at $2013\,cm^{-1}$. Adapted with permission from [30] © 2007, American Chemical Society

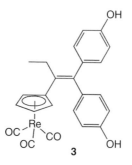

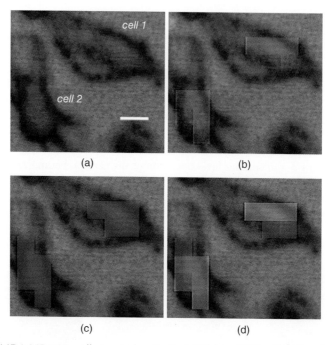

Figure 5.5 *MDA-MB-231 cells treated with **3**. (a) Bright field image (scale bar 10 μm). SR-FTIR-SM mappings: (b) E-band hot spot (red); (c) amide I-band hot spot (blue); (d) E-band hot spot (blue), amide I-band hot spot (red), overlay (magenta). Pixel size: 6×6 μm². Adapted with permission from [31] © 2013, Elsevier*

These workers showed that in all cases mappings of the E-band of **3** and at the amide I-band displayed hot spots in the same location of the cell. This area was associated with a cellular bulge that was topographically interpreted as indication of the nucleus. IR spectra recorded inside the bulge revealed the IR-signatures of both the symmetric A1-band and the E-band of **3**. These were absent in other regions of the same cells providing a strong case for the nuclear localization of the rhenium-tricarbonyl tamoxifen probe. As is clear from Fig. 5.5, the diffraction limited resolution of the IR signal related to the accumulation of **3** in MDA-MB-231 cells is rather coarse if compared with other standard detection techniques such as fluorescence microscopy, magnetic resonance or radio-imaging. However, cellular imaging with metal carbonyl drugs offers the advantage of localizing and imaging the organometallic unit without the need for further derivatizing it with organic chromophores, whose large size is likely to alter the physico-chemical properties of the molecule and hence its cellular distribution.

To partly overcome the limitations described above, the group of Policar has also introduced a rhenium–tricarbonyl probe as a single multimodal agent **4** designed to couple both infrared and luminescent properties for cellular imaging [32]. Probe **4** comprises the *fac*-[Re(CO)$_3$]$^+$ IR-label of C_{3v} local symmetry which, as in the case of **3**, gives rise to two bands of absorption in the IR-transparency window of the cell: an E-band (antisymmetric stretching, doubly degenerate) at about 1920 cm^{-1} and A1-band (symmetric stretching) at about 2020 cm^{-1}. The fluorescent tag was introduced into the coordination sphere of the metal ion as a 4-(2-pyridyl)-1,2,3-triazole ligand. Policar and her group demonstrated that correlative studies can be efficiently carried out with multimodality and showed that infrared and luminescent properties of **4** could be easily recognized from endogenous cell responses.

The approach, known as single core multimodal probe for imaging, (SCoMPI), was successfully employed to localize **4** in the Golgi apparatus of MDA-MB-231 cells (Fig. 5.6). In a first installment MDA-MB-231 cells were observed by wide field fluorescence microscopy with images pointing to a perinuclear localization of **4**. Subsequent SR-FTIR-SM cellular mapping in the mid-IR displayed hot spots of the E- and A1-bands of **4** matching the location of radiative fluorescence emission response (Fig. 5.6). Further results provided evidence for the integrity of the bimodal core of **4** in the intracellular environment underlining the reliability of the bimodal IR and luminescence imaging.

Cellular uptake of carbonyl complexes is typically measured by using concentrations in the range of from 10 μM to 1 mM. Lower concentrations can be used when a longer uptake time is allowed by the experiment. However, the local concentration of carbonyl species accumulated within the cell can be as high as several mM. Caution must be taken to ensure that the viability of the cell is not affected by the relatively high concentration of the carbonyl probe. This control can, of course, be performed as part of the standard toxicity tests for cell exposure to a chemical. In addition, the use of vibrational spectroscopy allows parallel monitoring of toxicity during the very same experiment that monitors metal carbonyl uptake and reactivity. Viability of the treated samples may be checked by comparing IR spectra of the treated and untreated cells. The onset of apoptotic and necrotic processes affects the cellular IR spectrum, leading to changes in the region of

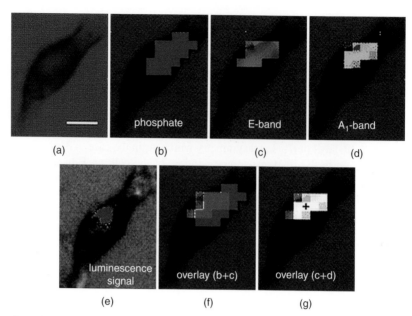

Figure 5.6   *MDA-MB-231 cell incubated with 4. (a) Bright field image (scale bar 10 mm). (b–d) SR-FTIR mappings, hot spots: (b) phosphate antisymmetric stretching (blue), (c) E-band (red), (d) A1-band (cyan). (e) Epifluorescence image, localization of 4 (blue). (f–g) Overlays of SR-FTIR hot spots: (f) overlay (magenta) of (b) (blue) and (c) (red), (g) overlay (white) of (c) (red) and (d) (cyan). Pixel size: 3 × 3 mm². Adapted with permission from [32] © 2012, Royal Society of Chemistry*

carbonyl ester absorption, between 1700 and 1750 cm^{-1}, and the region of amide absorption, between 1600 and 1700 cm^{-1} [33]. Different spectral changes have been reported for different cell lines and tissue types undergoing cell death. Therefore, if changes in these spectral regions are observed during exposure to the CO containing probe, it is advisable to run separate control tests by independently inducing apoptosis and/or necrosis into the cellular sample in the absence of the carbonyl complex and recording the specific IR spectrum. This control will confirm whether the carbonyl imaging agent is actually inducing cell death or whether the observed spectroscopic changes are due to other effects.

5.4.4   *In Vivo* Uptake and Reactivity of Metal Carbonyl Complexes

Most applications of IR microscopy to the study of cellular uptake of metal carbonyl complexes have to date relied on the use of the carbonyl ligands as IR probes. Recent work by our groups is now exploring the use of IR microscopy to study the uptake and biochemistry of carbonyl complexes *in vivo*. Interest in the biochemistry of CO has seen considerable gowth in the last decade, given its physiological role in homeostatis and signalling. Part of this interest has been directed towards pharmacological applications, to take advantage of the cytoprotective properties demonstrated for small concentrations of CO. Pharmacological developments have focused on the design of CO-releasing molecules (CORMs).

Their mechanism of action and their applicability *in vivo* is currently the subject of intense scrutiny (see Chapter 10). To date IR microscopy is the only technique of vibrational imaging that has been used to study the real-time uptake and reactivity of CORMs.

We have shown that the implemention of IR microscopy on samples of living cells provides information about the reactivity of CO containing compounds in the intracellular and the exctracellular environment. The basic approach is to culture cells on an IR transparent optical window, which is also biocompatible. Cells are grown to the desired density [26, 34, 35] and the window is then transferred to a sandwich-type sample holder (Fig. 5.7), with a medium of the desired composition. The CO complex can be added to the medium when enclosing the sample or injected afterwards, depending on the holder design and experimental requirements. Experiments can be performed either as spectromicroscopy measurements or as mapping or imaging experiments.

As an example, Fig. 5.8 shows a spectromicroscopy measurement performed on the bulk DMEM (Dulbecco's Modified Eagle's Medium) of a freshly dissolved sample of *cis*-$[Re(CO)_2(Br)_4]^{2-}$ (**5**, also known as ReCORM-1) [1, 36–38], outside of the cellular environment of a 3T3 fibroblast cell culture. The evolution of the spectra over a period of 30 min shows that the complex is unstable and ligand exchange is occurring. Results are reported as difference spectra from the start of the measurement. The pattern of positive

Figure 5.7   *Sample holder for IR microscopy measurements on live samples. The holder has a sandwich structure, consisting of two IR optical windows enclosing a volume of growth medium containing the cells. A micrometric spacer ring defines the thickness of the sample and prevents cell damage during closing. An outer metal jacket allows temperature control by flowing a thermostatic liquid or by using a thermoelectric unit. Adapted with permission from [26] © 2011 Royal Society of Chemistry*

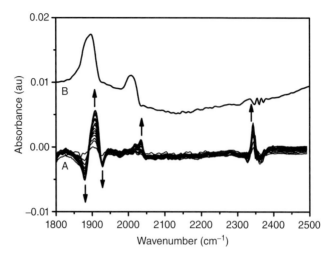

Figure 5.8 *Spectromicroscopy measurements on living single 3T3 fibroblast cells and culture medium. (A) Time resolved difference spectra of the v(CO) region in DMEM medium. The spectra show the hydrolysis of **5** when exposed to an aqueous environment. (B) Accumulated Re–CO complex inside a 3T3 cell after overnight uptake in DMEM medium (L. Quaroni and F. Zobi, unpublished results)*

and negative peaks indicates that aquation of the complex is taking place. This is in qualitative agreement with measurements performed in water solution, although the different kinetics indicate that the composition of the medium has a marked effect on the kinetics of aquation. The band at $2342\,cm^{-1}$ shows a sharp increase over time, in parallel with the hydrolysis process. The band arises from aqueous CO_2, formed in solution by the protonation of HCO_3^- present as a component of the culture medium [39]. The increasing concentration of CO_2 indicates that the acidty of the solution is increasing, in agreement with the proton release step following aquation of **5** [37].

5

Remarkably, time resolved spectromicroscopy measurements on individual cells show that no cellular uptake is taking place over this time scale, either of **5** or of its immediate hydrolysis product. In contrast, spectromicroscopy measurements after overnight uptake in a cellular location show the cellular accumulation of a third compound, the v(CO) bands of which are shown in Fig. 5.8B. The v(CO) band pattern of the accumluated species tentatively indicates the formation of a stable $[Re^I(CO)_3(L)_3]^n$ species (were L = water or other ligands) in which the three carbonyl groups are most probably in a facial arrangment around the central metal ion. Interestingly, this compound is not the active form of **5**, suggesting that, at least in the case of 3T3 cells, **5** cannot express its cytoprotective activity via cellular uptake [38]. The species that is eventually taken up by the cell is a late product of hydrolysis, which is presumably inactive. This observation sets a useful caveat about the interpretation of uptake data in the absence of real-time measurements of chemical events *in vivo*. The information about the identity of the active form and its cellular uptake would be missed by a measurement that describes only the final result of uptake after cell drying. It is also useful to highlight the role that a detailed analysis of the carbonyl spectroscopic properties has in interpreting the processes that occur *in vivo*. The sensitivity of $v(CO)$ bands to other ligands and to the coordination geometry of the metal provides an invaluable probe of reaction dynamics in solution.

These advantages have been recently used by our groups to characterize the intracellular uptake and reactivity of a photosensitive CORM, whose CO releasing properties can be triggered by illumination with visible light [40]. Compound **6** (known as B12-MnCORM-1) is based on the *fac*-$[(CO)_3Mn^I(tacd)]^+$ complex (where tacd = 1,4,8,11-tetraazacyclotetradecane) being linked to the ribose sugar moiety of the vitamin B12. B12-MnCORM-1 takes advantage of the B12 unit as a vector to favor internalization. In recent work we have shown that we can indeed follow the real-time uptake of **6** in 3T3 fibroblast cells over the course of about 1 h [40]. Fig. 5.9 shows that measurements can be performed with a time resolution of 1 min or less, showing the increase of bands at 1931 and 2027 cm^{-1}, at frequencies characteristic for the E- and A1-transitions of B$_{12}$-MnCORM-1. These peaks are positive, indicating that the concentration of **6** is increasing in the measured volume.

Exposure to visible light induces the release of the CO ligands, as indicated by the decrease of absorbance at 1931 and 2027 cm^{-1}, leading to full disappearance of the bands after a few minutes of exposure (see Fig. 5.9).

The subcellular distribution of the complex can be described by mapping the absorption bands of the E- and A1-transitions after a few hours of uptake. Fig. 5.10 shows the distribution of the two bands inside a single intact 3T3 cell. Care must be taken when considering the role of cellular topography in the quantitative interpretation of the maps, as discussed previously. Even with this caveat, we can rule out an exclusively cytoplasmic localization of **6**. A perinuclear distribution, in the nucleus and/or in its proximity, appears to be the most likely interpretation of the maps, in agreement with reports for other metal carbonyl complexes [32].

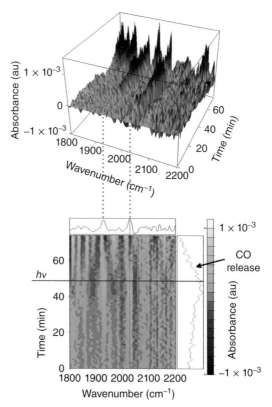

Figure 5.9   *Accumulation and subsequent photolysis of **6** in a single 3T3 fibroblast. Pictures show the time evolution of the IR spectrum of a 3T3 cell incubated with complex **6** in a 2D (bottom) and a 3D (top) representation. The green line in the frame of the 2D spectrum is a "cross-section" along the directions shown in the plot. Marked peaks correspond to the E- and A1-stretching modes of the fac-[Mn^I(CO)$_3$]$^+$ core. Adapted with permission from [40] © 2013 American Chemical Society*

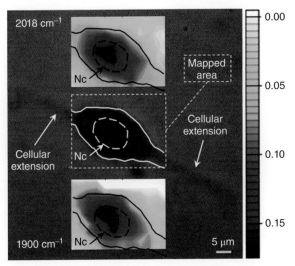

Figure 5.10 *Optical image of a 3T3 fibroblast incubated with* **6**. *Top and bottom inserts show images reconstructed from integrating the intensities of the carbonyl stretching vibrations at frequencies of 2018 and 1900 cm^{-1} respectively. Adapted with permission from [40] © 2013 American Chemical Society*

5.5 Raman Microscopy

Raman spectroscopy has been used extensively for vibrational spectroscopic studies of biomolecules in homogeneous macroscopic samples. The technique is characterized by a wide sensitivity range, depending on the resonance (or lack of) of the excitation laser beam with an electronic transition of the sample. Sensitivity in the range of 10–100 mM can be expected for off-resonance measurements of about 1 min, whereas sensitivity down to 10 μM can be obtained when resonance conditions are satisfied. One feature of the technique is that the weakness of the signal from bulk water allows the use of thick aqueous samples, in the millimeter range, without difficulty [41].

The last 20 years have seen the appearance of microscopy optical configurations for Raman spectromicroscopy and imaging [42]. They consist of a microscope unit, used as a sample compartment, and of a spectrometer used for spectroscopic analysis (Fig. 5.11). The optical path used for sample illumination is matched to the one used for the spectroscopic measurement, ensuring accurate selection of the location for the measurement. In contrast to IR microscopes, Raman microscopes use optical technology similar to that used for fluorescent microscopes, since excitation light is in the visible and near-IR range. In several cases, the microscope unit is itself a standard, off the shelf, optical microscope interfaced with the spectrometer unit that allows analysis of the Raman spectrum. As a consequence, Raman microscopy measurements can be performed under conditions very similar to those for fluorescence measurements and often using the same sample holders. Whereas

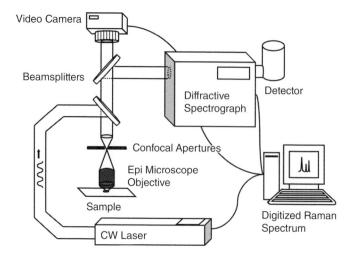

Figure 5.11 *Scheme of a Raman microscope. A laser light source, usually continuous wave (CW) is focused on the sample using visible laser objectives. A confocal aperture can be used to decrease the laser to a diffraction-limited spot. A set of beamsplitters and filters (not shown) allows sample inspection with a video camera and separation on the excitation wavelength from the Raman scattering. The latter is analyzed by a dispersive spectrometer and measured by a spectroscopic grade CCD camera*

several excitation and collection geometries are used in macroscopic Raman experiments, a backscattering setup is the norm in a commercial Raman microscope [42].

Lasers are the conventional light sources for Raman microscopy, providing high brightness and selectivity of the excitation wavelength. The drawback is the high flux density of photons on the sample, which can induce interfering fluorescence emission and sample degradation due to thermal or photochemical effects. These problems are particularly serious when off-resonance measurements are performed, because higher power needs to be used to compensate for the weak scattering. Power levels of several mW are usually necessary for off-resonance measurements, whereas $10-100$ μW are often sufficient for a measurement under resonant conditions.

Raman scattering from water is weak and this ensures that the limitations about sample size and geometry encountered in IR microscopy do not apply to Raman microscopy. Sample thickness is often of the order of a few millimeters, thus ensuring that even the larger eukaryotic cells can be measured [43, 44].

In analogy with what has already been discussed for IR microscopy, Raman microscopy measurements can be performed in a spectromicroscopy or an imaging configuration. In a spectromicroscopy experiment, microscope objectives are used as light condensers, to focus the laser beam into a small spot. Raman backscattered light from this spot is collected by the same objective and sent to the spectrometer for analysis. A confocal aperture can be used to limit the size of the laser spot, down to a diffraction-limited spot [42].

The diffraction limited spatial resolution of a Raman microscopy measurement is also described by Rayleigh's formula (Equation 5.1). In a conventional Raman experiment, the wavelength of the exciting and scattered light is in the UV, visible and near-IR spectral region and the microscope objectives are laser optical objectives, of the same design as the ones used for fluorescence microscopy. As a result, the limiting resolution is in the range achieved by a fluorescence microscope, approximately 100–500 nm.

5.5.1 Concentration Measurements with Raman Spectroscopy and Spectromicroscopy

The intensity of the Raman signal increases with the concentration of the chromophore. However the relationship is complex, given the role played by self-absorption effects and intensity losses due to scattering by other sample components and reflection and refraction by interfaces in the sample, which result in strong matrix effects. The geometry of the sample holder and alignment of the microscope optics also affect the result. Therefore, for quantitative purposes, the general approach is to construct a calibration curve using solutions of known concentration of the analyte in a suitable medium that accounts for the matrix effects of the sample. The same instrument under the same alignment conditions should be used when constructing the calibration curve and performing the measurement.

5.5.2 Metal Carbonyls as Raman Probes for Cellular Imaging

As in the case of IR spectromicroscopy, cellular imaging with Raman active metal carbonyl probes is at its early stages of development and only a few studies have appeared in the literature. In 2010 the groups of Havenith, Meltzer-Nolte, and Schatzschneider reported an investigation on the cellular uptake and intracellular distribution of the photo-inducible cytotoxic CO-releasing molecule *fac*-[Mn(tmp)(CO)$_3$]$^+$ **7** [45].

7

In analogy with compounds **3** and **4**, the local C_{3v} symmetry the *fac*-[Mn(CO)$_3$]$^+$ core gives rise to two absorption bands in **7** but only the antisymmetric E-band could be detected in the Raman spectrum of HT29 cells (human colon cancer) incubated with the complex. Thus optical images of HT29 cells were reconstructed by analysis of the maximum integrated intensity of the CO stretching vibration of the E-band only. As evidenced in Fig. 5.12, Raman images clearly indicated that **7** had become predominantly associated with the nucleus and the nuclear membrane resulting in a distinct circular pattern with further localization in the center of the nucleus. Through the use of Raman spectromicroscopy, the authors were also able to demonstrate that **7** had internalized in the cells by measuring cellular cross-sections along the *x,z* direction. Images reproduced in Fig. 5.12 E–G provided evidence of cellular penetration and nuclear accumulation. This study showed for the

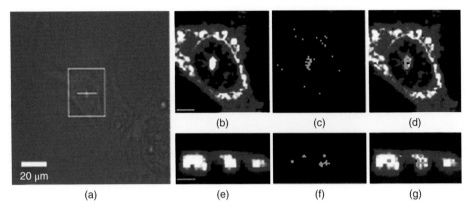

Figure 5.12 *(A) Optical image of an HT29 human colon cancer cell incubated with 7. (B–C) Raman images reconstructed from integrating the intensities of the C–H (2800–3050 cm⁻¹) and C≡O (1945–1965 cm⁻¹) stretching vibrations. (D) Overlaid image of panels (B) and (C). (E–G) Cross-section Raman images along the x,z-direction of the same cell. The scale bar for the Raman images is 6 μm. Adapted with permission from [45] © 2010 Wiley-VCH Verlag GmbH & Co. KGaA, Weinheim*

first time that a stable, water-soluble organometallic carbonyl complex could be used as an intrinsic label-free marker for 3D Raman imaging of living cells. The quality of Raman optical images presented by the German scientists stands out as a remarkable illustration of the potential of this technique.

A second example of metal carbonyl probes used for live-cell imaging was presented in 2012 by the groups of Leong and Olivo [46]. These workers addressed the issue of low Raman scattering cross-section and the related relative high concentration of the metal carbonyl based biotag used for imaging. As a solution to move to low μM concentrations of the metal carbonyl probe (and hence avoiding problems of undesired cytotoxicity of the biotag) they proposed surface-enhanced Raman spectroscopy (SERS) as an approach to signal enhancement of the carbonyl stretching frequencies by coupling the osmium cluster **8** to gold nanoparticles (**NP**, Fig. 5.13). Raman signals of molecules on colloidal gold or silver nanoparticles can be enhanced by several orders of magnitude (typically from 10^6 to 10^{14}) as a result of the strong surface plasmon resonance of the nanostructured surface [47, 48]. This spectroscopic method has been successfully adapted to chemical sensing applications as, for example, in DNA or cellular molecules detection [49–55] and cancer diagnosis [56, 57].

Compound **8-NP** (see Fig. 5.13) was then conjugated to an antibody against epidermal growth factor receptors (anti-EGFR, highly expressed in diverse cancer cells), and incubated with OSCC (epidermoid carcinoma) or SKOV3 (ovarian carcinoma) cells. The cellular distribution of **8-NP*** was then imaged using the SERS-enhanced CO absorption signal at $2030\,\text{cm}^{-1}$, and the observed localization of the carbonyl tag correlated with the bright-field and dark-field microscope images (Fig. 5.14 a–d). Bright scattering spots were clearly correlated with the locations at which the CO vibration signals were detected (Fig. 5.14).

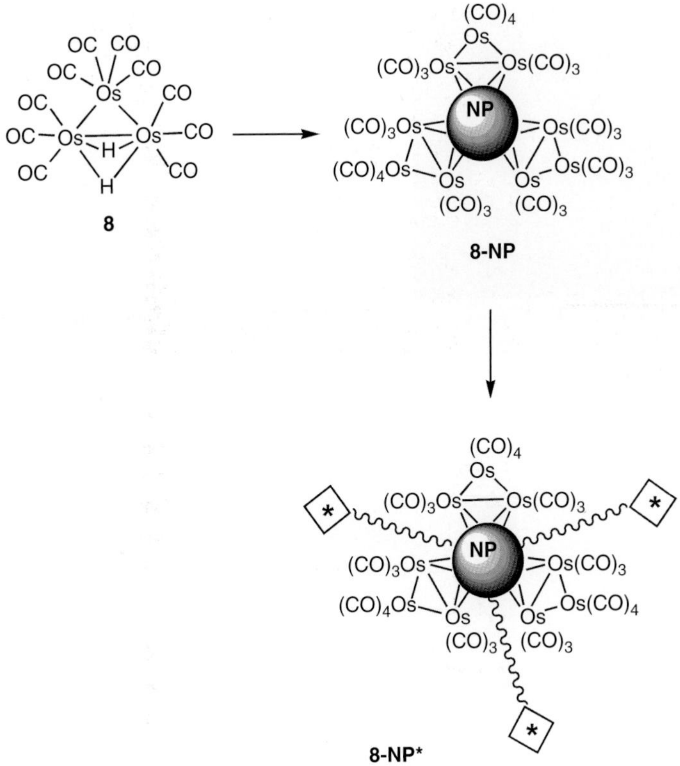

Figure 5.13 *Structure of osmium cluster **8**, and schematic representations of gold nanoparticle conjugate **8-NP** and its anti-EGFR antibody derivative **8-NP***. The anti-EGFR antibody is represented as a "starred box" in **8-NP****

5.6 Near-field Techniques

Both IR and Raman microscopy rely on the use of reflective and refractive optical components to focus the probing beam of light at a distance that is much larger than the wavelength of the light and the size of the source. This focusing condition, defined as the far-field condition, is characterized by Fraunhofer diffraction effects and provides a limiting spatial resolution described by Equation 5.1. This limit can be bypassed by probing a sample with a light source or a detector that are closer and much smaller than the wavelength of light being used. These conditions define a near-field microscopy measurement. Whenever the conditions are satisfied, the limiting spatial resolution is approximately given by the size of the light source or the detector element, depending on the specific configuration in use. A wide range of optical setups for near-field experiments, in the UV, visible and IR spectral regions, have been developed [58–64]. Specific values for the resolution vary with the configuration, but can be as low as $1/400\lambda$ [65]. This resolution is particularly appealing for cellular microscopy with mid-IR radiation since subcellular structures are often in this size range, and too small to be resolved with far-field optics.

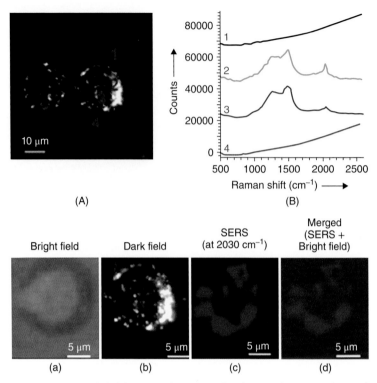

*Figure 5.14   Top: (A) Dark-field image of OSCC cells after incubation with **8-NP***. (B) SERS spectra of OSCC cells at the four different locations indicated. Bottom: Bright-field, dark-field, and SERS mapping images of (a-d) OSCC cells treated with **8-NP***. Adapted with permission from [46] © 2012, Wiley-VCH Verlag GmbH & Co. KGaA, Weinheim*

A variety of optical configurations have been published for near-field spectromicroscopy and imaging in the IR spectral region. To date, only some of these have been used for the study of cellular samples and only one of them has been used to characterize the uptake of a carbonyl complex. Only the latter will be discussed in the following sections.

5.6.1   Concentration Measurements with Near-field Techniques

Most near-field techniques do not easily yield accurate quantitative information about concentration. Although the intensity of the signal generally increases with the concentration of a chromophore, the dependence is complex and very much dependent on the specific optical setup and the topography of the sample. As a consequence, quantification is rarely attempted. However, the optical configurations that are more amenable to quantitative concentration measurements are those based on transmission optics, such as fiber optics and sub-wavelength apertures. In all these cases the construction of calibration curves is necessary, using samples of similar optical properties as the object of the investigation. The topography of the sample often has an effect on the intensity of the signal. Corrections for these effects require separate measurements of sample topography, such as by atomic force microscopy (AFM), and then use of the results obtained to correct the near-field response.

5.6.2   High-resolution Measurement of Intracellular Metal–Carbonyl Accumulation by Photothermal Induced Resonance

In this section we limit ourselves to describing a near-field setup that has been used for the characterization and the imaging of metal carbonyl probes in a cellular system [66]. The setup is based on photothermal detection of IR light absorption (Fig. 5.15) [67, 68]. In this configuration a broadband IR source is used to illuminate the sample. The effect is called photothermal induced resonance (PTIR). Detection is performed by measuring the heat emitted by the sample after absorption of IR light. The wavelength dependence of the emission reproduces the IR absorption pattern; therefore its spectral analysis effectively provides an absorption spectrum of the sample. In a near-field measurement, an IR beam is used to illuminate the whole extension of the sample, while a sub-wavelength thermal probe positioned close to the sample is used for detection. The spatial resolution of the measurement is given by the size of the probe tip, approximately 50 nm. This is sufficient to allow identification of the nucleus and of larger subcellular structures. Scanning the tip over the sample allows the assembly of 2D IR maps of specific bands. The main difficulty of this setup is due to the need for a high power IR source to achieve sufficient sensitivity. In the configuration used by Dazzi *et al.* the light source is a free-electron laser (FEL) [68].

Illumination of the sample with IR light is performed using an ATR prism on which sample cells are deposited or grown (Fig. 5.15). The IR beam is internally reflected at the interface between the prism and the sample, ensuring that light absorption is mediated

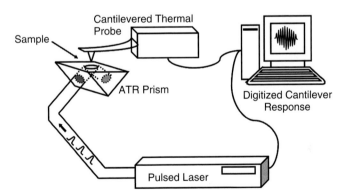

Figure 5.15 *Setup of a near-field photothermal IR microscope. The light from a high-power source, such as a free-electron laser or a benchtop laser, is directed to an ATR prism on which the sample is supported. IR light is totally internally reflected at the prism–sample interface and absorbed by the sample via the mediation of the evanescent field. The absorbed IR light is re-emitted by the sample as heat, which is detected by the proximal cantilevered nanoprobe. Changes in the oscillation of the probe are measured and used to reconstruct the absorption spectrum of the sample*

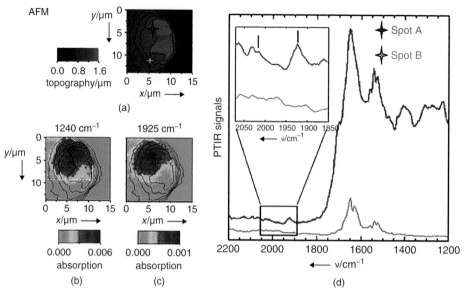

Figure 5.16 *Images of the nucleus of MDA-MB-231 cells treated with **3** and spectromicroscopy. (a) AFM topography. (b, c) PTIR maps with the AFM contours superimposed as black lines recorded at 1240 cm⁻¹ (b) and 1925 cm⁻¹ (c). (d) PTIR spectromicroscopy with an average incident power of 10 mW; spectrum at spot A in blue, spectrum at spot B in green. Inset shows the 1850–2050 cm⁻¹ region; the two ticks indicate the position of the characteristic peaks of **3**. Adapted with permission from [66] © 2011 WILEY-VCH Verlag GmbH & Co. KGaA, Weinheim*

only by the evanescent field extending into the sample. This setup avoids exposing the thermal probe directly to the excitation beam and also allows working in the presence of an aqueous environment. The trade-off for the high resolution of this setup is that relatively high power is needed to obtain a measurable photothermal signal, such that the viability of the sample may be affected. In the example of Fig. 5.16, 10 mW of pulsed IR radiation are used. The same tip that is used as the thermal probe can also function as an AFM tip, allowing measurement of sample topography in the tapping mode with comparable spatial resolution. Comparison of topography and spectromicroscopy data collected at the same spatial resolution is particularly valuable in evaluating the presence of topographic contributions in IR absorption maps.

This specific configuration was successfully used by Policar and coworkers to detect and map the rhenium–tricarbonyl tamoxifen analogue **3** (see Section 5.4.3) in MDA-MB-231 breast cancer cells [66]. The cells were initially located using the AFM methods. This provided the topography of the sample represented as black contours in Fig. 5.16. After incubation with μM solutions of **3**, cells were then mapped using the PTIR setup at several wavelengths. In their contribution these workers provided evidence of the internalization of **3** and showed a prevalent nuclear localization of the carbonyl probe. The nucleus was tentatively located using irradiation at 1240 cm⁻¹ (antisymmetric PO_2^- vibration) and at 1650 cm⁻¹ (amide I band).

5.7 Comparison of Techniques

Table 5.1 provides an overview of the advantages and limitations associated with the different vibrational techniques described within the context of cellular imaging in the preceding sections. Mid-IR microscopy is the most generally applicable of all vibrational microscopy techniques. All heteronuclear polyatomic molecules are detectable. Sensitivity limits vary according to the specific molecule, but are generally around the millimolar region. Its generality means that the technique provides a wealth of information about multiple molecules in a single measurement. This factor is an opportunity as well as a challenge, since complex spectroscopic changes, which are difficult to interpret, are associated with even simple processes. The low power necessary for an IR absorption measurement means that the sample is little perturbed by the measurement, making this a technique of choice for studies on living samples. One of its main limitations is the heavy contribution from water absorption, which severely limits the thickness of the samples, often to just a single cell layer. As a result, the design of the sample holders can be complex, since a sample holder for IR microscopy is a microfluidic device with a variable degree of complexity. Experiment

Table 5.1 *Comparison of vibrational microscopy techniques used for spectromicroscopy and imaging of metal carbonyl complexes in cells and tissue*

Technique	Spatial resolution	Detection limit	Power on sample	Advantages	Limitations
Mid-IR microscopy	2.5–25 μm at the diffraction limit	~1 mM	10–50 μW	General applicability Gentle on sample	High water interference Not selective Low throughput at high spatial resolution
Raman Microscopy	200–500 nm at the diffraction limit	~10 mM	1–10 mW	General applicability Low water Interference	Low sensitivity Harsh on sample (high power) Fluorescence interference
Resonance Raman microscopy	200–500 nm at the diffraction limit	~10 μM	0.1–1 mW	Selective Sensitive Negligible water interference	Applicable to few samples and molecules High fluorescence interference
PTIR	30–50 nm (probe size)	Not defined	10 mW	Very high resolution	Limited availability Harsh on sample (high power)

design and sample preparation protocols also require several modifications from the protocols used for optical microscopy measurements. The other limitation is the relatively low spatial resolution, which creates a challenge for subcellular imaging. The latter implies that mid-IR microscopy is more valuable as a spectromicroscopy technique than as an imaging technique.

Raman microscopy, in contrast, has the same diffraction limited resolution of optical microscopy techniques and is only slightly sensitive to water. These two features imply that the technique can be used with samples and sample holders that are very similar to the ones used for optical microscopy, with little or no adaptation of experimental conditions. Its main drawback is the weakness of the Raman signal, which requires fairly high power, up to several mW, to obtain a useful signal under off-resonance conditions. This requirement can be detrimental to the measurement of sensitive samples and is a constraint to the application of Raman microscopy to living samples. An additional limitation is the heavy interference from fluorescence emission, which happens in the same spectral region as the Raman emission but can be several orders of magnitude stronger, thus covering the Raman signal itself. This is often a problem when measuring biological samples, due to the abundance of natural fluorophores.

These limitations are much reduced when working under resonance conditions with an electronic transition in one of the chromophores. Resonance enhanced signals are several order of magnitude stronger, thus reducing the requirement for high laser power. In some cases they are strong enough to be visible over a fluorescence background. Unfortunately the resonance condition can be achieved only for some specific chromophores, such as porphyrins and carotenoids, thus limiting its applicability. Conditions that allow surface-enhanced Raman spectroscopy measurement also alleviate the limitations of Raman microscopy, providing increased sensitivity and often quenching of fluorescence [47]. However, surface enhancement effects require the proximity of the chromophore to a nanoscopic metal structure. The latter condition can be achieved for some samples but is not always applicable.

Of the vibrational microscopy techniques used to study metal carbonyl compounds, PTIR undeniably provides the highest resolution, allowing measurements on subcellular structures. Its main drawback is the high power required for the measurement, which might affect the viability of the sample after prolonged exposure. An additional issue is the limited accessibility at the present time. Currently a high power free electron laser is necessary to obtain a good thermal signal over a wide spectral range: there is just one facility, at the CLIO FEL, on the Orsay campus of the University Paris-Sud, providing the setup for such a measurement. This situation could change in the near future due to the ongoing introduction of quantum cascade lasers (QCLs) as high power tunable benchtop mid-IR sources.

5.8   Conclusions and Outlook

This chapter describes the state-of-the-art in vibrational spectromicroscopy and imaging of metal carbonyl complexes in cells and tissues. The rich chemistry and unique physic-chemical properties of metal carbonyl species make them exceptional tools for the study cellular systems. As we have emphasized throughout this chapter, it is the characteristic CO stretching frequencies of these species that offer fantastic opportunities in chemical biology. It was thoroughly shown that, even in living cells, when exposed to electromagnetic radiations of suitable energy, metal carbonyl complexes respond with signals in the water-transparent region of the vibrational spectrum, which is clear of the resonances of organic functional groups. Vibrational microscopy techniques, as a way to visualize cellular structures and dynamic processes in real time, are still in their early stages of development. Advances in the field will certainly result from innovative metal carbonyl chemistry coupled with technological progress.

At the time of writing both the techniques involved and their applications to this subject area are seeing a rapid development. This is creating several opportunities that, although not yet explored, are expected to be realized in the next few years or even in the next few months. One such opportunity is the application of other far-field vibrational microscopy techniques to spectromicroscopy and imaging of carbonyl metal complexes in cells and tissue. Several vibrational techniques based on light absorption or scattering have been successfully implemented in a microscopy configuration. Many have already been used for cellular imaging and their use in addressing questions about the uptake and localization of metal carbonyl complexes is just a matter of time. Two promising examples of techniques based on light scattering are CARS (coherent anti-Stokes Raman spectroscopy) [69–71] microscopy and SRS (stimulated Raman scattering) microscopy [72, 73]. Both are multiphoton coherent techniques in which multiple laser beams are matched on the sample to give an emitted scattered signal, which is also a coherent beam. The latter feature ensures high sensitivity, while the use of light wavelengths in the visible and near-infrared allows for correspondingly high diffraction-limited resolution. Both techniques have been successfully used for cellular imaging, and have been able to resolve subcellular structures, such as organelles and lipid membranes, in aqueous samples. As such they are very promising tools for the study of the partition of metal carbonyl complexes within single cells.

Another growing field with promising applications to the study of metal carbonyl complexes is that of near-field vibrational spectroscopy. In addition to the configuration for PTIR described in detail in this chapter, several other setups have been described in the literature, including near-field absorption and near-field scattering configurations [60, 65–71]. A few of these have been applied to cellular imaging and have been shown to achieve subcellular resolution [60, 67]. Their main limitation is the need for fairly high power sources. Although these are easily deployed in the visible spectral region, for example, for TERS (tip enhanced Raman spectroscopy), configurations based on mid-IR light absorption have to rely on a still limited selection of sources that are both broadband and high power. However, current developments in the design and performance of quantum cascade lasers (QCL) are likely to overcome these limitations in the next few years, providing affordable benchtop sources that will greatly expand the applicability of mid-IR spectroscopy.

Acknowledgements

F. Zobi gratefully acknowledges the Swiss National Science Foundation (Grant# PZ00P2_139424) for financial support. L. Quaroni gratefully acknowledges financial support by the Paul Scherrer Institute and Swiss Light Source.

References

1. F. Zobi, L. Kromer, B. Spingler and R. Alberto, Synthesis and reactivity of the 17 e$^-$ complex [Re^{II}(Br)$_4$(CO)$_2$]$^{2-}$: A convenient entry into rhenium(II) chemistry, *Inorg. Chem.*, **48**(18), 8965–8970 (2009).

2. G. Jaouen and A. Vessieres, Transition-metal carbonyl estrogen-receptor assay, *Pure Appl. Chem.*, **57**(12), 1865–1874 (1985).

3. S. Top, G. Jaouen, A. Vessieres, J. P. Abjean, D. Davoust, C. A. Rodger, B. G. Sayer and M. J. Mcglinchey, Chromium tricarbonyl complexes of estradiol derivatives - differentiation of alpha-diastereoisomer and beta-diastereoisomers using one-dimensional and two-dimensional NMR-spectroscopy at 500-Mhz, *Organometallics*, **4**(12), 2143–2150 (1985).

4. A. Varenne, A. Vessieres, P. Brossier and G. Jaouen, Application of the nonradioiso-topic carbonyl metalloimmunoassay (Cmia) to diphenylhydantoin, *Res. Commun. Chem. Pathol. Pharmacol.*, **84**(1), 81–92 (1994).

5. A. Varenne, A. Vessieres, M. Salmain, P. Brossier and G. Jaouen, Production of specific antibodies and development of a nonisotopic immunoassay for carbamazepine by the carbonyl metallo-immunoassay (Cmia) method, *J. Immunol. Methods*, **186**(2), 195–204 (1995).

6. A. Vessieres, K. Kowalski, J. Zakrzewski, A. Stepien, M. Grabowski and G. Jaouen, Synthesis of CpFe(CO)(L) complexes of hydantoin anions (Cp = η^5-C$_5$H$_5$, L = CO, PPh$_3$), and the use of the 5,5-diphenylhydantoin anion complexes as tracers in the non-isotopic immunoassay CMIA of this antiepileptic drug, *Bioconjugate Chem.*, **10**(3), 379–385 (1999).

7. V. Philomin, A. Vessieres and G. Jaouen, New applications of carbonylmetalloim-munoassay (Cmia) - a nonradioisotopic approach to cortisol assay, *J. Immunol. Methods*, **171**(2), 201–210 (1994).

8. M. Salmain, A. Vessieres, P. Brossier, I. S. Butler and G. Jaouen, Carbonylmetal-loimmunoassay (Cmia) a new type of nonradioisotopic immunoassay - principles and application to phenobarbital assay, *J. Immunol. Methods*, **148**(1–2), 65–75 (1992).

9. N. Fischer-Durand, A. Vessieres, J. M. Heldt, F. le Bideau and G. Jaouen, Evaluation of the carbonyl metallo immunoassay (CMIA) for the determination of traces of the herbicide atrazine, *J. Organomet. Chem.*, **668**(1–2), 59–66 (2003).

10. A. Vessieres, N. Fischer-Durand, F. Le Bideau, P. Janvier, J. M. Heldt, S. Ben Rejeb and G. Jaouen, First carbonyl metallo immunoassay in the environmental area: application to the herbicide chlortoluron, *Appl. Organomet. Chem.*, **16**(11), 669–674 (2002).

11. M. Salmain, Labeling of proteins with organometallic complexes: Strategies and applications, in Bioorganometallics: *Biomolecules, Labeling, Medicine*, (ed. G. Jaouen), Wiley-VCH, Weinheim, pp. 181–214 (2006).

12. G. R. Stephenson, Organometallic bioprobes, in *Bioorganometallics: Biomolecules, Labeling, Medicine*, (ed. G. Jaouen), Wiley-VCH, Weinheim, pp. 215–262 (2006).

13. M. Salmain and A. Vessieres, Organometallic complexes as tracers in non-isotopic immunoassay, in *Bioorganometallics: Biomolecules, Labeling, Medicine*, (ed. G. Jaouen), Wiley-VCH, Weinheim, pp. 263–302 (2006).

14. F. Zobi, Parametrization of the contribution of mono- and bidentate ligands on the symmetric C O stretching frequency of fac-[Re(CO)$_3$]$^+$ complexes, *Inorg. Chem.*, **48**(22), 10845–10855 (2009).

15. F. Aubke and C. Wang, Carbon-monoxide as a sigma-donor ligand in coordination chemistry, *Coord. Chem. Rev.*, **137**, 483–524 (1994).

16. G. Frenking and N. Frohlich, The nature of the bonding in transition-metal compounds, *Chem. Rev.*, **100**(2), 717–774 (2000).

17. A. S. Goldman and K. KroghJespersen, Why do cationic carbon monoxide complexes have high C-O stretching force constants and short C-O bonds? Electrostatic effects, not sigma-bonding, *J. Am. Chem. Soc.*, **118**(48), 12159–12166 (1996).

18. A. J. Lupinetti, S. Fau, G. Frenking and S. H. Strauss, Theoretical analysis of the bonding between CO and positively charged atoms, *J. Phys. Chem. A*, **101**(49), 9551–9559 (1997).

19. A. J. Lupinetti, G. Frenking and S. H. Strauss, Nonclassical metal carbonyls: Appropriate definitions with a theoretical justification, *Angew. Chem. Int. Ed.*, **37**(15), 2113–2116 (1998).

20. A. J. Lupinetti, S. H. Strauss and G. Frenking, Nonclassical metal carbonyls, *Prog. Inorg. Chem.*, **49**, 1–112 (2001).

21. F. Zobi, Ligand electronic parameters as a measure of the polarization of the C–O bond in [M(CO)$_x$L$_y$]^n Complexes and of the relative stabilization of [M(CO)$_x$L$_y$]$^{n/n+1}$ species, *Inorg. Chem.*, **49**(22), 10370–10377 (2010).

22. P. Fara, A microscopic reality tale, *Nature*, **459**(7247), 642–644 (2009).

23. C. Matthaeus, B. Bird, M. Miljkovic, T. Chernenko, M. Romeo and M. Diem, Infrared and Raman microscopy in cell biology, in *Biophysical Tools for Biologists, Vol* **2**: *in Vivo Techniques*, (eds J. J. Correia and H. W. Detrich III,), Elsevier, pp. 275-308 (2008).

24. R. Salzer, G. Steiner, H. H. Mantsch, J. Mansfield and E. N. Lewis, Infrared and Raman imaging of biological and biomimetic samples, *Fresenius J. Anal. Chem.*, **366**(6–7), 712–726 (2000).

25. K. L. Goff, L. Quaroni and K. E. Wilson, Measurement of metabolite formation in single living cells of *Chlamydomonas reinhardtii* using synchrotron Fourier-transform infrared spectromicroscopy, *Analyst*, **134**(11), 2216–2219 (2009).

26. L. Quaroni and T. Zlateva, Infrared spectromicroscopy of biochemistry in functional single cells, *Analyst*, **136**(16), 3219–3232 (2011).

27. I. W. Levin and R. Bhargava, Fourier transform infrared vibrational spectroscopic imaging: Integrating microscopy and molecular recognition, *Ann. Rev. Anal. Chem.*, **56**, 429–474 (2005).

28. D. E. Wolf, The optics of microscope image formation, in *Digital Microscopy*, (eds G. Sluder and D. E. Wolf), Elsevier, San Diego, pp. 11–55 (2003).

29. G. L. Carr, Resolution limits for infrared microspectroscopy explored with synchrotron radiation, *Rev. Sci. Instrum.*, **72**(3), 1613–1619 (2001).

30. K. V. Kong, W. Chew, L. H. K. Lim, W. Y. Fan and W. K. Leong, Bioimaging in the mid-infrared using an organometallic carbonyl tag, *Bioconjugate Chem.*, **18**(5), 1370–1374 (2007).

31. S. Clede, F. Lambert, C. Sandt, Z. Gueroui, N. Delsuc, P. Dumas, A. Vessieres and C. Policar, Synchrotron radiation FTIR detection of a metal-carbonyl tamoxifen analog. Correlation with luminescence microscopy to study its subcellular distribution, *Biotechnol. Adv.*, **31**(3), 393–395 (2012).

32. S. Clede, F. Lambert, C. Sandt, Z. Gueroui, M. Refregiers, M. A. Plamont, P. Dumas, A. Vessieres and C. Policar, A rhenium tris-carbonyl derivative as a single core multimodal probe for imaging (SCoMPI) combining infrared and luminescent properties, *Chem. Commun.*, **48**(62), 7729–7731 (2012).

33. H. Y. Holman, M. C. Martin, E. A. Blakely, K. Bjornstad and W. R. McKinney, IR spectroscopic characteristics of cell cycle and cell death probed by synchrotron radiation based Fourier transform IR spectromicroscopy, *Biopolymers*, **57**(6), 329–335 (2000).

34. H. W. Kreuzer, L. Quaroni, D. W. Podlesak, T. Zlateva, N. Bollinger, A. McAllister, M. J. Lott and E. L. Hegg, Detection of metabolic fluxes of O and H atoms into intracellular water in mammalian cells, *PLoS ONE*, **7**(7), doi: 10.1371/journal.pone.0039685 (2012).

35. J. R. Mourant, R. R. Gibson, T. M. Johnson, S. Carpenter, K. W. Short, Y. R. Yamada and J. P. Freyer, Methods for measuring the infrared spectra of biological cells, *Phys. Med. Biol.*, **48**(2), 243–257 (2003).

36. F. Zobi and O. Blacque, Reactivity of 17 e(-) complex $[Re^{II}(Br)_4(CO)_2]^{2-}$ with bridging aromatic ligands. Characterization and CO-releasing properties, *Dalton Trans.*, **40**(18), 4994–5001 (2011).

37. F. Zobi, O. Blacque, R. A. Jacobs, M. C. Schaub and A. Y. Bogdanova, 17 e^- rhenium dicarbonyl CO-releasing molecules on a cobalamin scaffold for biological application, *Dalton Trans.*, **41**(2), 370-378 (2012).

38. F. Zobi, A. Degonda, M. C. Schaub and A. Y. Bogdanova, CO releasing properties and cytoprotective effect of cis-trans- $[Re^{II}(CO)_2Br_2L_2]^n$ complexes, *Inorg. Chem.*, **49**(16), 7313–7322 (2010).

39. M. Carbone, T. Zlateva and L. Quaroni, Monitoring and manipulation of the pH of single cells using infrared spectromicroscopy and a molecular switch, *Biochim. Biophys. Acta*, Apr;1830(4):2989–93 (2013).

40. F. Zobi, L. Quaroni, G. Santoro, T. Zlateva, O. Blacque, B. Sarafimov, M. C. Schaub and A. Y. Bogdanova, Live-fibroblast IR imaging of a cytoprotective photoCORM activated with visible light, *J. Med. Chem.*, **56**(17), 6719–6731 (2013).

41. J. R. Ferraro, K. Nakamoto and C. W. Brown, *Introductory Raman Spectroscopy*, Academic Press (2003).

42. S. Stewart, R. J. Priore, M. P. Nelson and P. J. Treado, Raman Imaging, *Ann. Rev. Anal. Chem.*, **5**, 337–360 (2012).

43. M. M. Mariani, P. J. R. Day and V. Deckert, Applications of modern micro-Raman spectroscopy for cell analyses, *Integr. Biol.*, **2**(2–3), 94–101 (2010).

44. T. Weeks and T. Huser, Raman spectroscopy of living cells, in *Biomedical Applications of Biophysics*, (ed. T. Jue), Humana Press, New York, pp. 185–210 (2010).

45. K. Meister, J. Niesel, U. Schatzschneider, N. Metzler-Nolte, D. A. Schmidt and M. Havenith, Label-free imaging of metal-carbonyl complexes in live cells by Raman microspectroscopy, *Angew. Chem. Int. Ed.*, **49**(19), 3310–3312 (2010).

46. K. V. Kong, Z. Lam, W. D. Goh, W. K. Leong and M. Olivo, Metal carbonyl-gold nanoparticle conjugates for live-cell SERS imaging, *Angew. Chem. Int. Ed.*, **51**(39), 9796–9799 (2012).

47. I. A. Larmour and D. Graham, Surface enhanced optical spectroscopies for bioanalysis, *Analyst*, **136**(19), 3831–3853 (2011).

48. P. L. Stiles, J. A. Dieringer, N. C. Shah and R. R. Van Duyne, Surface-enhanced Raman spectroscopy, *Ann. Rev. Anal. Chem.*, **1**, 601–626 (2008).

49. A. Barhoumi, D. Zhang, F. Tam and N. J. Halas, Surface-enhanced Raman spectroscopy of DNA, *J. Am. Chem. Soc.*, **130**(16), 5523–5529 (2008).

50. S. E. J. Bell and N. M. S. Sirimuthu, Surface-enhanced Raman spectroscopy (SERS) for sub-micromolar detection of DNA/RNA mononucleotides, *J. Am. Chem. Soc.*, **128**(49), 15580–15581 (2006).

51. J. A. Dougan, D. MacRae, D. Graham and K. Faulds, DNA detection using enzymatic signal production and SERS, *Chem. Commun.*, **47**(16), 4649–4651 (2011).

52. K. Kneipp, A. S. Haka, H. Kneipp, K. Badizadegan, N. Yoshizawa, C. Boone, K. E. Shafer-Peltier, J. T. Motz, R. R. Dasari and M. S. Feld, Surface-enhanced Raman spectroscopy in single living cells using gold nanoparticles, *Appl. Spectrosc.*, **56**(2), 150–154 (2002).

53. E. Papadopoulou and S. E. J. Bell, Label-free detection of single-base mismatches in DNA by surface-enhanced Raman spectroscopy, *Angew. Chem. Int. Ed.*, **50**(39), 9058–9061 (2011).

54. E. Papadopoulou and S. E. J. Bell, DNA reorientation on Au nanoparticles: label-free detection of hybridization by surface enhanced Raman spectroscopy, *Chem. Commun.*, **47**(39), 10966–10968 (2011).

55. R. Stevenson, R. J. Stokes, D. MacMillan, D. Armstrong, K. Faulds, R. Wadsworth, S. Kunuthur, C. J. Suckling and D. Graham, *In situ* detection of pterins by SERS, *Analyst*, **134**(8), 1561–1564 (2009).

56. S. Y. Feng, J. Q. Lin, M. Cheng, Y. Z. Li, G. N. Chen, Z. F. Huang, Y. Yu, R. Chen and H. S. Zeng, Gold nanoparticle based surface-enhanced Raman scattering spectroscopy of cancerous and normal nasopharyngeal tissues under near-infrared laser excitation, *Appl. Spectrosc.*, **63**(10), 1089–1094 (2009).

57. D. Graham, R. Stevenson, D. G. Thompson, L. Barrett, C. Dalton and K. Faulds, Combining functionalised nanoparticles and SERS for the detection of DNA relating to disease, *Faraday Discuss.*, **149**, 291–299 (2011).

58. E. Bailo and V. Deckert, Tip-enhanced Raman scattering, *Chem. Soc. Rev.*, **37**(5), 921–930 (2008).

59. A. Hartschuh, Tip-enhanced near-field optical microscopy, *Angew. Chem. Int. Ed.*, **47**(43), 8178–8191 (2008).

60. F. Huth, M. Schnell, J. Wittborn, N. Ocelic and R. Hillenbrand, Infrared-spectroscopic nanoimaging with a thermal source, *Nature Mater.*, **10**(5), 352–356 (2011).

61. L. Novotny, The history of near-field optics, in *Progress in Optics*, *Vol. 50*, (ed. E. Wolf), Elsevier, Amstrdam, pp. 137–184 (2007).

62. B. Pettinger, Tip-enhanced Raman spectroscopy (TERS), *Surface-Enhanced Raman Scattering: Physics and Applications*, **103**, 217–240 (2006).

63. D. W. Pohl, Optics at the nanometre scale, *Philosoph. Trans. Royal Soc. London Ser. Math. Phys. Eng. Sci.*, **362**(1817), 701–717 (2004).

64. D. A. Schmidt, I. Kopf and E. Bruendermann, A matter of scale: from far-field microscopy to near-field nanoscopy, *Laser Photon. Rev.*, **6**(3), 296–332 (2011).
65. J. Houel, E. Homeyer, S. Sauvage, P. Boucaud, A. Dazzi, R. Prazeres and J.-M. Ortega, Midinfrared absorption measured at a lambda/400 resolution with an atomic force microscope, *Opt. Express*, **17**(13), 10887–10894 (2009).
66. C. Policar, J. B. Waern, M. A. Plamont, S. Clede, C. Mayet, R. Prazeres, J.-M. Ortega, A. Vessieres and A. Dazzi, Subcellular IR imaging of a metal-carbonyl moiety using photothermally induced resonance, *Angew. Chem. Int. Ed.*, **50**(4), 860–864 (2011).
67. L. Bozec, A. Hammiche, H. M. Pollock, M. Conroy, J. M. Chalmers, N. J. Everall and L. Turin, Localized photothermal infrared spectroscopy using a proximal probe, *J. Appl. Phys.*, **90**(10), 5159–5165 (2001).
68. A. Dazzi, Photothermal induced resonance. Application to infrared spectromicroscopy, in *Thermal Nanosystems and Nanomaterials*, (ed. C. E. Ascheron), Springer, Heidelberg, pp. 469–503 (2009).
69. C. L. Evans and X. S. Xie, Coherent anti-Stokes Raman scattering microscopy: Chemical imaging for biology and medicine, *Ann. Rev. Anal. Chem.*, **1**, 883–909 (2008).
70. L. G. Rodriguez, S. J. Lockett and G. R. Holtom, Coherent anti-stokes Raman scattering microscopy: A biological review, *Cytometry Part A*, **69A**(8), 779–791 (2006).
71. A. Zumbusch, G. R. Holtom and X. S. Xie, Three-dimensional vibrational imaging by coherent anti-Stokes Raman scattering, *Phys. Rev. Lett.*, **82**(20), 4142–4145 (1999).
72. C. W. Freudiger, W. Min, B. G. Saar, S. Lu, G. R. Holtom, C. He, J. C. Tsai, J. X. Kang and X. S. Xie, Label-free biomedical imaging with high sensitivity by stimulated Raman scattering microscopy, *Science*, **322**(5909), 1857–1861 (2008).
73. C. W. Freudiger and S. Xie, *In vivo* imaging with stimulated Raman scattering microscopy, *Opt. Photonics News*, **22**(12), 27–27 (2011).

6

Probing DNA Using Metal Complexes

Lionel Marcélis,[a] Willem Vanderlinden[b] and Andrée Kirsch-De Mesmaeker[a]
[a]Organic Chemistry and Photochemistry, Université libre de Bruxelles, Belgium
[b]Division of Molecular Imaging and Photonics, Laboratory of Photochemistry and Spectroscopy, Department of Chemistry, KU Leuven, Belgium

6.1 General Introduction

Numerous studies exist on the influence of DNA on the properties of metal complexes. These compounds exhibit the characteristics that are essential to behave as ideal DNA photoprobes, photolabeling agents, photoreagents and even as potential (photo)drugs targeting DNA. The last aspect will, however, not be covered in this chapter: interested readers can refer to various reviews [1-4]. Instead, we will focus on explaining why and how some metal complexes are capable of probing DNA – not only its overall structure or topology but also its local properties and characteristics. In spite of the importance of the Pt complexes, they will not be discussed here because they mainly play a role as anticancer drugs [4]. This chapter will strictly focus on the Ru(II) complexes because for years now they have continued to be studied as DNA (photo)probes. However, it is also necessary to mention that other metal ions such as Cr, Rh, Re and Ir have also been used for such purposes [5-10].

We will not review the whole of the literature in this research field, but we will attempt to explain with specific examples why some complexes in particular have been more successful for probing DNA, not only DNA in a test-tube, but also *in vivo* for cellular biology applications (see for example Fig. 6.1).

Because the role of Ru(II) complexes as DNA probes is mainly an outcome of their light triggered response, the first prerogative is to understand their photophysical properties. Thus, we begin this chapter with a general discussion of the photophysics of Ru(II) complexes, and then more specifically those which we have selected as interesting examples of photoprobes in the later part of the discussion. A comprehensive overview of the photophysics will lead us towards rationalising the interactive behaviour of such complexes with

Inorganic Chemical Biology: Principles, Techniques and Applications, First Edition. Edited by Gilles Gasser.
© 2014 John Wiley & Sons, Ltd. Published 2014 by John Wiley & Sons, Ltd.

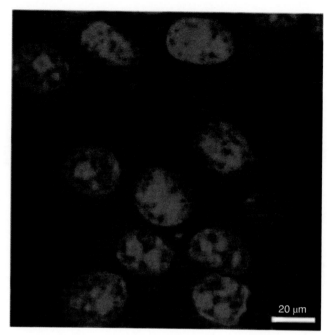

Figure 6.1 *Confocal laser scanning microscopy image of fixed MCF-7 cells incubated with a Ru(II) complex showing nuclear staining. Reproduced with permission from [11] © 2011 Wiley-VCH Verlag GmbH & Co. KGaA, Weinheim*

different DNA types, from local DNA characteristics to large DNA global structures found in living cells. Specific examples of Ru(II) complexes used for photoprobing special DNA properties will then be discussed.

6.2 Photophysics of Ru(II) Complexes

6.2.1 The First Ru(II) Complex Studied in the Literature: $[Ru(bpy)_3]^{2+}$

The first photophysical studies of $[Ru(bpy)_3]^{2+}$ were published in the literature between 1970 and 1980 [12, 13] (Fig. 6.2). Since then this octahedral complex has been a reference model in photophysics for most Ru(II) complexes examined, until now. Therefore its photophysics is briefly described (Fig. 6.3).

$[Ru(bpy)_3]^{2+}$ is orange–red in colour, absorbs in the UV–visible and emits in the visible range of the light spectrum, according to the photophysical scheme depicted in Fig. 6.3. Illumination of this complex populates, after relaxation, the singlet metal to ligand charge transfer (^{1}MLCT) excited state, which crosses over to the triplet MLCT excited state (^{3}MLCT) with 100% efficiency. This process of electron spin inversion, is called inter-system-crossing (ISC) (Fig. 6.2). It is extremely fast (a few hundreds of femtoseconds) so that it can compete with 100% efficiency with the deactivation of the ^{1}MLCT to the ground state; this is due to the Ru heavy atom effect. This ^{3}MLCT state deactivates to

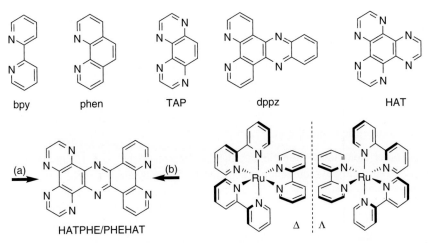

Figure 6.2 *The different polypyridyl ligands discussed in this chapter and an example of a Λ and Δ Ru(II) complex. (a) Chelation site on the HAT part and (b) chelation site on the PHE part of the HATPHE/PHEHAT ligand, respectively. bpy = 2,2'-bipyridine; phen = 1,10-phenanthroline; TAP = 1,4,5,8-tetraazaphenanthrene; dppz = dipyrido[3,2-a:2',3'-c]phenazine; HAT = 1,4,5,8,9,12-hexaazatriphenylene; HATPHE/PHEHAT = 1,10-phenanthrolino[5,6-b]1,4,5,8,9,12-hexaazatriphenylene*

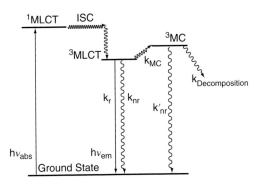

Figure 6.3 *Photophysical scheme for [Ru(bpy)$_3$]$^{2+}$, valid also for other tris-homoleptic Ru(II) complexes*

the ground state with emission of light, whose luminescence lifetime (τ) is controlled by the lifetime of the ^{3}MLCT state, that is, the inverse of the sum of the nonradiative (k_{nr}) and radiative deactivation rate constants (k_r) [($\tau = 1/(k_r + k_{nr})$]. In order to explain the fact that this lifetime decreases when the temperature increases, a third deactivation process, called thermal activation to the metal centred triplet excited state (^{3}MC), has to be added. The ^{3}MLCT state lifetime therefore becomes $\tau = 1/(k_r + k_{nr} + A\exp(-\Delta E/RT))$ (typically $\tau = 620$ ns in water at RT under argon [14]), where ΔE represents the energy difference between the ^{3}MLCT state and the ^{3}MC state and A is a pre-exponential factor. This ^{3}MC state, in addition to its nonradiative deactivation to the ground state, can also give rise to the loss of one coordinating ligand (a bpy), and lead to complex decomposition. Because this higher energy ^{3}MC state does not luminesce, it does not disturb the emission spectrum of the complex which originates exclusively from the ^{3}MLCT state. The luminescent ^{3}MLCT excited state may be regarded as a $[(bpy)_2Ru(III)(bpy^{\bullet-})]^{2+}*$ excited species in which the electron is transferred from the Ru(II) centre to one of the bpy ligands; in simple terms the emitter corresponds to the motif $[Ru(III)(bpy^{\bullet-})]*$.

6.2.2   Homoleptic Complexes

For the other homoleptic complexes such as $[Ru(phen)_3]^{2+}$ and $[Ru(TAP)_3]^{2+}$ [15] (Fig. 6.2), the photophysical scheme depicted in Fig. 6.3 remains valid with of course different associated values for k_{nr}, k_r, and ΔE, which are characteristic, in addition to the absorption and emission spectra, of the spectroscopic properties of each complex. The emitting species in this case are $[Ru(III)(phen^{\bullet-})]*$ and $[Ru(III)(TAP^{\bullet-})]*$. In comparison with $[Ru(bpy)_3]^{2+}$, the main difference for these two complexes is their ΔE values, which for $[Ru(TAP)_3]^{2+}$ in particular, are small enough for the ^{3}MC state to be easily populated at *RT*.

6.2.3   Heteroleptic Complexes

In case of heteroleptic complexes with two phen or two TAP ligands and the third ligand as dppz, HAT, HATPHE or PHEHAT (Fig. 6.2), the photophysical behaviour becomes more complicated. The deviations from the photophysical scheme presented in Fig. 6.3 are discussed below. It is mainly because of the changes of the photophysics modulated by the three ligands and the solvent or DNA micro-environment around the complexes that these heteroleptic complexes are excellent DNA photoprobes. In order to discuss the photophysics, we first have to know, which ^{3}MLCT state has the lowest energy, that is, where the electron excited from the metal centre to the ligand becomes localised, since the emission takes place from the most stabilised ^{3}MLCT state.

6.2.3.1   *[Ru(phen)₂(dppz)]²⁺*

The most stabilised excited state should be the one in which the transferred electron is localised on the ligand with the most stabilised LUMO. Thus with $[Ru(phen)_2(dppz)]^{2+}$ [16, 17], should the emitter correspond to a $[Ru(III)(phen^{\bullet-})]*$ or to a $[Ru(III)(dppz^{\bullet-})]*$ ^{3}MLCT state? Actually, this information can be simply obtained from the electrochemical behaviour of the complex. By characterisation of the first reduction wave of the complex, it can be determined whether the reduction, that is, the addition of the first electron, takes place on the phen or on the dppz ligand. This can be assessed from comparison with the reduction of reference complexes such as $[Ru(phen)_3]^{2+}$ in this case. Proceeding in this way, it has been concluded that the first electron is added to the dppz, which is the most π accepting ligand. Thus the emitting species corresponds to $[Ru(III)(dppz^{\bullet-})]*$ [18]. However for this complex, no emission is detected in water but only in the organic solvents. After numerous debates and publications on the origin of this strange behaviour [16, 17], it has been agreed that the $[Ru(III)(dppz^{\bullet-})]*$ species presents two types of excited states: a dark and a luminescent state, whose energies are very close (Fig. 6.4a). Their relative energy levels depend on the solvent. With water molecules, because of the H-bonding between water and the N atoms of dppz, the dark state is stabilised as compared with the bright one.

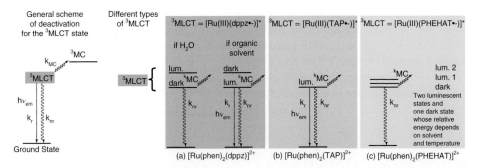

Figure 6.4 *Photophysical scheme for (a) [Ru(phen)$_2$(dppz)]$^{2+}$, (b) [Ru(TAP)$_2$(dppz)]$^{2+}$ and (c) [Ru(phen)$_2$(PHEHAT)]$^{2+}$*

Therefore no luminescence is detected in water. In an organic solvent, the energy level of the dark state is no longer stabilised so the energy of the bright state becomes lower than that of the dark state; therefore the excited complex emits in organic solvents.

6.2.3.2   [Ru(TAP)$_2$(dppz)]$^{2+}$

The case of [Ru(TAP)$_2$(dppz)]$^{2+}$ is quite different [19-21]. This complex emits both in water and in organic solvents. The first reduction wave for this complex corresponds to the addition of the electron on the TAP ligand, but not on the dppz ligand, which indicates that the LUMO is centred on TAP. Thus, for [Ru(TAP)$_2$(dppz)]$^{2+}$, the lowest ^{3}MLCT state corresponds to a [Ru(III)(TAP$^{\bullet-}$)]* emitter to which only one bright state is associated. Therefore a luminescence is observed even in water and is modulated by the solvent (Fig. 6.4b).

6.2.3.3   [Ru(phen)$_2$(HAT)]$^{2+}$

On the basis of previous examples, the case of [Ru(phen)$_2$(HAT)]$^{2+}$ is comparable to that of [Ru(TAP)$_2$(dppz)]$^{2+}$ (Fig. 6.4b) [22]. The HAT ligand is the most π accepting ligand, thus the LUMO is centred on HAT. The [Ru(III)(HAT$^{\bullet-}$)]* emitter has only one bright state, which is responsible for the luminescence in water or organic solvents, as is the case for [Ru(III)(TAP$^{\bullet-}$)]* in the [Ru(TAP)$_2$(dppz)]$^{2+}$ complex. Obviously, excited [Ru(phen)$_2$(HAT)]$^{2+}$ has its own characteristics (k_r, k_{nr}, activation to ^{3}MC).

6.2.3.4   [Ru(phen)$_2$(HATPHE)]$^{2+}$ and [Ru(phen)$_2$(PHEHAT)]$^{2+}$

Replacing the HAT ligand by a PHEHAT/HATPHE ligand (Fig. 6.2) in [Ru(phen)$_2$(HAT)]$^{2+}$ is very interesting from the photophysical point of view. When this ligand is complexed on the HAT side (Fig. 6.2a) to give [Ru(phen)$_2$(HATPHE)]$^{2+}$ [23], its emission behaviour is similar to that of [Ru(phen)$_2$(HAT)]$^{2+}$. In the ^{3}MLCT state, the transferred electron from the Ru(II) centre to the HATPHE ligand is localised on the HAT part of HATPHE, with virtually no influence of the PHE motif. In contrast, when the PHEHAT ligand is complexed on the PHE side, in [Ru(phen)$_2$(PHEHAT)]$^{2+}$ [23] (Fig. 6.2b), the photophysics is amazingly different from that of [Ru(phen)$_2$(HATPHE)]$^{2+}$ and exhibits some similarity with [Ru(phen)$_2$(dppz)]$^{2+}$. Indeed, there is more than one excited state associated with [Ru(III)(PHEHAT$^{\bullet-}$)]* (Fig. 6.4c), the energies of which are very close.

It has been shown [23] that three different excited states contribute to the deactivation of the excited complex: one dark state and two luminescent states (one in which the electron would be transferred on the PHE part and the other in which the electron would be more localised on the HAT part). The relative population of these three states depends on the solvent and temperature. The dark state is populated in water, as for $[Ru(phen)_2(dppz)]^{2+}$, so that no luminescence is detected in aqueous solution; emission from one or the other bright state depends on temperature and is observed in organic solvents such as acetonitrile.

6.2.3.5   *[Ru(TAP)$_2$(PHEHAT)]$^{2+}$*

Again, the case of $[Ru(TAP)_2(PHEHAT)]^{2+}$ is completely different [24] from $[Ru(phen)_2(PHEHAT)]^{2+}$. For the TAP complex, the electrochemistry indicates that the LUMO is centred on the TAP ligand and no longer on the PHEHAT ligand so that the ^{3}MLCT luminophore corresponds to $[Ru(III)(TAP^{\bullet-})]^*$. Although the case is similar to that of $[Ru(TAP)_2(dppz)]^{2+}$, the characteristics of emission (emission spectrum, excited state lifetime and associated rate constants of deactivation) are of course modulated by the third ligand PHEHAT [24]. In spite of several efforts at synthesising the TAP equivalent of $[Ru(phen)_2(HATPHE)]^{2+}$, that is the $[Ru(TAP)_2(HATPHE)]^{2+}$, the latter could not be prepared until now.

6.2.4   Photoinduced Electron Transfer (PET) and Energy Transfer Processes

As explained above, the three ligands modulate the photophysics of the Ru(II) complexes. Moreover, they also influence the electron donor or acceptor properties of the ^{3}MLCT excited state. In the presence of a reducing agent (electron donor) (Fig. 6.5a), an electron can be transferred exergonically from the HOMO of this donor to the half-empty HOMO of the triplet excited complex, thus to its Ru(III) centre. In the presence of an oxidising agent (electron acceptor) (Fig. 6.5b), an electron can be transferred from the half-filled LUMO of the excited complex to the empty LUMO of this acceptor.

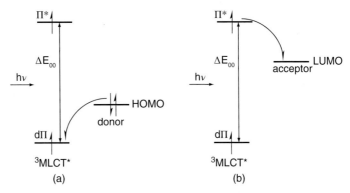

Figure 6.5 *(a) For example, donor = G of DNA and acceptor = [Ru(TAP)$_2$(PHEHAT)]$^{2+*}$; (b) donor = excited complex and the acceptor is benzoquinone, for example*

These photoinduced electron transfer (PET) processes become thermodynamically favourable because the oxidising (Fig. 6.5a) or reducing power (Fig. 6.5b) of the excited complex is enhanced in comparison with its ground state by the energy of the HOMO–LUMO transition (ΔE_{00}) of the complex. When the complex contains at least two π accepting ligands, such as in $[Ru(TAP)_2(PHEHAT)]^{2+}$ (as in the case in Fig. 6.5a), a PET can take place from a guanine (G) base as donor to the excited complex, with formation of an oxidised G ($G^{\bullet+}$) and a reduced complex $[Ru(TAP)(TAP^{\bullet-})(PHEHAT)]^{1+}$ (Equation 6.1). The excited complex represented by $[Ru(III)(TAP^{\bullet-})]^*$ is thus transformed into the reduced species $[Ru(II)(TAP^{\bullet-})]$. Such a PET process is generally accompanied by a proton transfer from $G^{\bullet+}$ to $[Ru(II)(TAP^{\bullet-})]$ [21] (Equation 6.1). These processes can occur with a mononucleotide guanosine monophosphate as well as with a G base inside an oligo- or poly-nucleotide or any other G containing DNA; they cause of course a quenching of the luminescence of the ^{3}MLCT state.

$$[Ru(III)(TAP^{\bullet-})]^* + G(DNA)$$

$$\rightarrow [Ru(II)(TAP^{\bullet-})] + G^{\bullet+}(DNA) \quad \text{or} \quad [Ru(II)(TAPH^{\bullet})] + G(-H)^{\bullet}(DNA) \quad (6.1)$$

$$[Ru(II)(TAP^{\bullet-})] + G^{\bullet+}(DNA) \quad \text{or} \quad [Ru(II)(TAPH^{\bullet})] + G(-H)^{\bullet}(DNA)$$

$$\rightarrow [Ru(II)(TAP)] + G(DNA) \quad (6.2)$$

$$[Ru(II)(TAP^{\bullet-})] + G^{\bullet+}(DNA) \quad \text{or} \quad [Ru(II)(TAPH^{\bullet})] + G(-H)^{\bullet}(DNA)$$

$$\rightarrow\rightarrow\rightarrow \text{irreversible photoadduct} \quad (6.3)$$

$$[Ru(III)(TAP^{\bullet-})]^* + O_2 \rightarrow [Ru(II)(TAP)] + {}^1O_2^* \quad (6.4)$$

In the reaction scheme shown in Equations 6.1 to 6.4, the excited complex is represented by the emitting species.

After the PET process (Equation 6.1), the back electron transfer regenerates the starting material (Equation 6.2). In competition with Equation 6.2, the two radicals that are produced can also react together to give rise, after several steps, to an irreversible adduct of the G base with the complex (Equation 6.3). The structure of such an adduct is presented in Fig. 6.6a. Its production corresponds to an interaction type of the complex with DNA that we call an *irreversible interaction*, as will be discussed further. These adducts can themselves absorb light and in the presence of a second G base, produce a bi-adduct, leading to the addition of two G moieties on the same complex. When these two G bases belong to two DNA strands, as depicted in Fig. 6.6b, this process generates an *irreversible photocrosslinking* between the two strands (see also later) [25].

In addition to the reactions discussed above, the ^{3}MLCT state of the Ru(II) complexes can also photosensitise oxygen by energy transfer (Equation 6.4). Thus the ^{3}MLCT excited states are capable of transferring their electronic energy to the ground state triplet of oxygen, which causes a quenching of luminescence and produces oxygen in its singlet excited state [26]. The latter is a very important reactive transient versus DNA (see later).

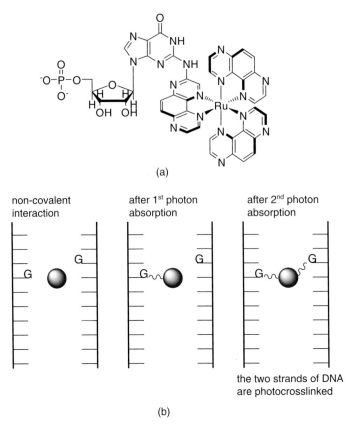

(a)

(b)

Figure 6.6 (a) *Photoadduct of [Ru(TAP)₃]²⁺ with GMP. (b) Cartoon representing the mono-adduct formation with the G of one strand and the bi-adduct formation with the second G of the other strand on the same complex (irreversible photocrosslinking between the two strands). Purple sphere stands for [Ru(TAP)₂(phen)]²⁺ for example*

6.3   State-of-the-art on the Interactions of Mononuclear Ru(II) Complexes with Simple Double-stranded DNA

Based on the main characteristics of the ^{3}MLCT states of Ru(II) complexes described above, in the next paragraphs we will discuss their behaviour in the presence of DNA. However, in view of their potential use as DNA photoprobes, other characteristics of these metallic compounds such as their shapes, charges, conformations and so on, in relation with the type of DNA to be investigated have to be considered. For example, if information on the *local structure* of a simple duplex DNA at low concentration in aqueous solution is needed, usually calf-thymus DNA (CT-DNA) or synthetic oligo- or polynucleotides are used in the presence of the metallic complexes. A short discussion is presented below to summarise the basic analytical methods that can be used and references to important reviews are also provided. Moreover information on the *global structure* or morphology and topology of more complicated DNA is also discussed, for example, on how does a double-stranded

DNA (dsDNA) wind and form higher order structures. In such cases other analytical tools such as scanning force microscopy (SFM) are used (see later). For investigating either the *global* or *local structure* of DNA, the metallic probes can interact *reversibly* or *irreversibly*, depending on their photophysical properties, as described above.

The first studies on the interactions with simple double-stranded DNAs were carried out with CT-DNA due to its commercial availability, and then with synthetic oligonucleotides because of their convenient syntheses with desired lengths and base sequences. In the following we summarise the general conclusions that resulted from these studies.

6.3.1   Studies on Simple Double-stranded DNAs

6.3.1.1   *Ru enantiomers*

Thanks to the separation of the simple mononuclear Ru(II) complexes into their enantiomers Λ and Δ (Fig. 6.2), it was demonstrated that some diastereomeric interactions with DNA are more favourable over others, depending on the chosen partners, that is, the DNA type and the metallic enantiomer. There are many forms of DNA duplexes (A-DNA, B-DNA, Z-DNA, etc.) [27] that have different sizes and shapes of grooves and exhibit different helicities, namely right (the case of A- and B-DNA) and left (the case of Z-DNA). The interactions of these DNAs with Ru(II) enantiomers are discussed in several reviews [28–30].

6.3.1.2   *Interaction geometries*

In addition to the enantiomericity Λ or Δ, another factor which plays an important role in the DNA interaction is the shape of the ligands coordinated to the Ru(II) centre. When one of the three ligands is a large planar ligand such as dppz [28, 31, 32] or PHEHAT [18, 23, 33], it can intercalate between the stacking of the DNA base pairs in contrast to bpy [34], as long as there are intercalation sites available. Generally an intercalated complex occupies two to three base pairs depending on the other two non-intercalated ancillary ligands. An interaction according to an intercalation mode produces a high affinity constant of the metallic complex for DNA. With three small polyazaaromatic ligands, such as phen or TAP, the complex generally adsorbs in the DNA grooves, or some workers have described this interaction geometry as semi-intercalation [35]. Moreover, different intercalation geometries have been detected. For example, in the case of $[Ru(phen)_2(dppz)]^{2+}$, for each enantiomer (Δ or Λ), in the presence of the same oligonucleotide, two types of intercalation can be distinguished [36], namely the symmetrical and canted intercalation (Fig. 6.7, see also later). Thus, a total of four intercalation geometries exist for racemic $[Ru(phen)_2(dppz)]^{2+}$. Furthermore, cooperative and anti-cooperative complex–complex interactions have been evidenced for Λ and $\Delta[Ru(phen/bpy)_2(dppz)]^2$ on interaction with $[poly(dA-dT)]_2$ in its minor groove [37]. The ancillary ligands, bpy or phen, differently influence the cooperative or anti-cooperative interactions.

6.3.1.3   *Techniques and methods for studying the interaction geometries*

The most precise and refined spectroscopy to study the interaction of metal complexes with DNA at the molecular level is X-ray crystallography. However, obtaining suitable

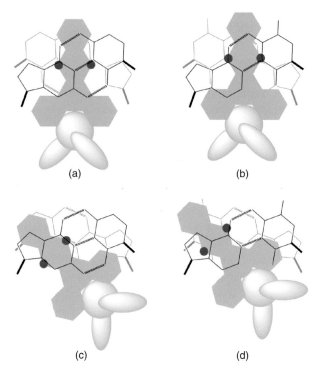

(a)   (b)

(c)   (d)

Figure 6.7 *Cartoons for (a), (b), symmetrical and (c), (d), canted intercalation of Δ-[Ru(L)$_2$(dppz)]$^{2+}$ at (GC)$_2$ (a,c) and (AT)$_2$ (b,d) steps. Reproduced with permission from [36] © 2012 Wiley-VCH Verlag GmbH & Co. KGaA, Weinheim*

crystals is very challenging. Recently published crystal structures of two dppz complexes, Λ-[Ru(phen)$_2$(dppz)]$^{2+}$ [38] and Λ-[Ru(TAP)$_2$(dppz)]$^{2+}$ [39] with oligonucleotides, place the metal complexes in the minor groove; canted and symmetrical intercalations of the dppz ligand (Fig. 6.7) are observed but at different base pairs (G–C or T–A). Moreover, one ancillary ligand of the complex, phen or TAP is also 'semi-intercalated' but in a duplex other than the one where the dppz is intercalated (Fig. 6.8). In other words, one complex interacts with two duplexes in the same crystal.

Although these results are very interesting, they might not be extrapolated to solutions or to living cells as the interaction geometry not only depends on the DNA sequence and structure, but also on the loading level of DNA (or binding density), salt concentrations, temperature and so on. For studying the interaction geometries in solution, a battery of techniques, complementary to each other, has to be used, such as viscosity measurements of the DNA–metallic complex assembly, linear dichroism (LD), circular dichroism (CD), calorimetry, NMR and emission spectroscopy and gel electrophoreses. They are not discussed in this chapter, except for emission in connection with the photophysical properties as developed above.

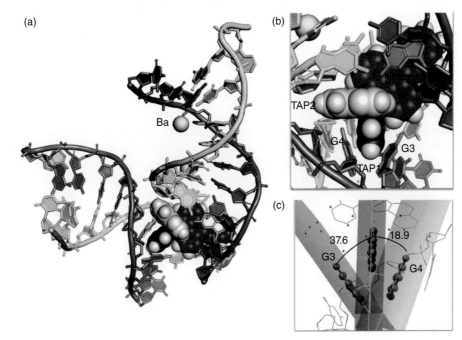

Figure 6.8 *X-ray data structure determination of an intercalating Ru(II)dipyridophenazine complex which kinks DNA by semi-intercalation of a tetraazaphenanthrene ligand. Reproduced with permission from [39] © 2011 National Academy of Sciences, U.S.A.*

6.3.2 Influence of DNA on the Emission Properties

The electronic spectroscopy, mainly luminescence, has played a very important role in studies of binding interactions of the metallic complexes with DNA. Thus the measurements of luminescence intensity under steady-state illumination or the emission lifetimes (i.e. the excited-state lifetimes of the ^{3}MLCT states) under pulsed illumination constitute the tools of choice for investigating the interaction. These studies give access to the binding constant values and different geometries of interaction. The influence of DNA on the emission behaviours of the ^{3}MLCT states can be divided into two categories.

6.3.2.1 *Category 1, increase of luminescence*

In this first category, the emission intensity and lifetimes of the excited states increase by intercalation or semi-intercalation of the complex into the bases stacking. This is generally due to the fact that the nonradiative deactivation rate constant k_{nr} (Fig. 6.4a,c) decreases [40]. Indeed, the DNA scaffold in which the complex is embedded diminishes the vibrational deactivation of the excited state. Moreover this scaffold prevents the quenching of the excited state by energy transfer to oxygen dissolved in the aqueous phase. This is typically the case of $[Ru(phen)_3]^{2+}$ (semi-intercalation), $[Ru(phen)_2(HAT)]^{2+}$ (intercalation of HAT) and $[Ru(phen)_2(HATPHE)]^{2+}$ (intercalation of HATPHE). $[Ru(phen)_2(dppz)]^{2+}$ in which the emitter corresponds to $[Ru(III)(dppz^{\bullet -})]^*$, represents an extremely interesting case. As this excited species does not luminesce in water because of the population of its dark state (Fig. 6.4a), the intercalation of its dppz ligand triggers the luminescence due to the less polar microenvironment of the base pairs and the absence of H-bonding with the water molecules. Therefore the lowest ^{3}MLCT state becomes the bright state (Fig. 6.4a), such that $[Ru(phen)_2(dppz)]^{2+}$ has been called a 'light-switch on' complex in the presence of DNA.

Moreover, if luminescence measurements are performed with an enantiomerically pure complex, even the symmetrical intercalation (Fig. 6.7) can be distinguished from the canted intercalation thanks to the different associated emission lifetimes due to different degrees of protection of the N atoms of dppz from water molecules in the two intercalation modes [36]. Studies of the percentage of contribution of the two emission lifetimes for each enantiomer combined with isothermal titration calorimetry (ITC) data as a function of the binding density, led to the attribution of each emission lifetime. Thus, the long lifetime could be assigned to the canted intercalation and the short one to the symmetrical intercalation [37]. Of note, in spite of its high affinity for DNA, this $[Ru(phen)_2(dppz)]^{2+}$ probe corresponds to a case of *reversible interaction* since it is possible to download the complex from DNA by simple dialysis against water. The other complex, $[Ru(phen)_2(PHEHAT)]^{2+}$, can also behave as a 'light-switch on' with DNA, however in that case, the two bright excited states participate in the total emission.

6.3.2.2   *Category 2, quenching of luminescence*

In contrast to the 'light-switch on' probe of the first category, for the second category of complexes, the emission intensity and lifetimes of the excited states decrease on interaction with DNA [1, 40]. This quenching is attributed to an electron transfer process from a G base to the ^{3}MLCT state of the Ru(II)–TAP compound (Fig. 6.5a and Equation 6.1). For a more detailed discussion, the readers can refer to a review [41]. Moreover, this luminescence inhibition is accompanied by a reaction between the two radicals of the pair (Equation 6.3), which produces an adduct of the metal complex at a G site of DNA (Fig. 6.6a). This is the case for $[Ru(TAP)_3]^{2+}$, $[Ru(TAP)_2(dppz)]^{2+}$ and $[Ru(TAP)_2(PHEHAT)]^{2+}$ because, as explained above, these complexes contain at least two TAP ligands so that their lowest ^{3}MLCT state corresponding to $[Ru(III)(TAP^{\bullet -})]^*$ is very oxidizing (Fig. 6.5a) [1, 40]. This type of complex behaves thus as a 'light-switch off' with DNA and in contrast to the previous case, produces an *irreversible interaction*. Indeed the complex cannot be downloaded from DNA by dialysis and the resulting photoproduct can be observed by gel electrophoresis with oligonucleotides containing G bases. If there is no G base in the sequence, for example with $[poly(dA-dT)]_2$, there is no quenching and the behaviour is similar to that of the first category.

6.4   Structural Diversity of the Genetic Material

In the previous sections dsDNAs such as CT-DNA or ds synthetic oligonucleotides were considered as sole targets for the Ru(II) complex photoprobes and the discussion was focused on the binding interaction and influence on the photophysical properties of the Ru(II) complexes. However, far more elaborate DNA conformations exist in living cells. Thus the genetic material, in addition to its sequence and double-helical nature, also exhibits other characteristics, such as its flexibility, its capacity to wind–unwind around proteins and so on. These local and global properties can be described by different measurable parameters such as mechanical and topological properties, which will be explained first in the following section. These properties can be assessed employing, for example, SFM. After a description of some of these DNA properties, an example of analysis of the DNA superstructures using a Ru(II) complex will be considered.

6.4.1 Mechanical Properties of DNA

In contrast to the DNA used in most *in vitro* experiments, the genetic material present in cells exhibits a far more pronounced structural diversity. DNA in living cells is constantly bent, twisted and coiled by macromolecular machines. Furthermore, complex organisms have evolved a very lengthy genome, as the result of the storage of genetic information in a one-dimensional code. Several levels of organisation and compaction are therefore necessary to make it fit into the tiny volume that is provided in the cell.

The local mechanical properties of DNA are thus of crucial importance. By far the most widely used model for DNA bending is the so-called worm-like chain model (WLC) [42]. DNA is treated as an isotropic, continuously flexible elastic rod. This implies that the energy associated with the bending (ΔG_{bend}, Equation 6.5) of the DNA helix shows a quadratic dependence on the extent of the deformation:

$$\Delta G_{bend} = \frac{1}{2} B \left(\frac{\theta^2}{L} \right) \tag{6.5}$$

where L is the axial length of helix that is bent through an angle θ (in radians) and B is the bending rigidity (in J mol^{-1} bp; length being measured in bp, bp = base pair).

In addition to bending, DNA also exhibits torsional flexibility. This flexibility gives rise to a twisting motion about the DNA helix axis and results in local variations of helical repeat. Again, in most of the cases these fluctuations are modeled using an isotropic elasticity model and the corresponding torsional rigidity is around 440 kJ mol^{-1} bp.

Torsional stiffness does not affect the DNA structure on a global scale in the case of linear DNA chains, but it is an important determinant of the DNA shape in living cells, where free rotation of the DNA strands is impeded in circular bacterial genomes or stable chromatin loops in eukaryotes. It is said that in these cases, the DNA is topologically constrained. In a strict sense, topology can be defined as the study of properties that can only be changed by cutting and rejoining, NOT by mechanical deformations (bending, stretching, twisting, etc.). More specifically, DNA topology deals with the entanglement of DNA and the way Nature copes with it. In the following paragraph, how topological constraints can be exploited to *excite* the DNA will be discussed.

6.4.2 DNA Topology

The impact of DNA topology on cellular processes can be illustrated using the example of DNA transcription (in essence the enzyme-catalysed synthesis of RNA from DNA). For the DNA to be transcribed, the two strands of the double helix need to separate to act as a template. The transcription machinery cannot rotate around the helix because it is too large and/or physically anchored to the surrounding matrix. Therefore torsional stress is generated during the process, both ahead (helix overwinding) of the polymerase as well as in its wake (helix underwinding) [43]. At a certain point, determined by the torsional and bending stiffness, the DNA starts to buckle and coil upon itself. This phenomenon, as a result of over- or underwinding and of topological constraints, is known as DNA supercoiling.

Some concepts of DNA topology and supercoiling are illustrated in Fig. 6.9. Considering a *linear* double-stranded DNA molecule, topological changes can be made by *joining its ends* (depicted in Fig. 6.9 by arrows with full lines). First, considering closing the linear dsDNA into a *circle* by ligation of a single strand of the double-stranded DNA and connecting the 5′ with the 3′ end, free rotation around this newly created bond is still possible, and as a consequence this circular form of DNA mostly adopts an open configuration independently from the geometry of the chain before ring closure. In case both strands of the DNA are closed covalently 5′ to 3′, free rotation of the one strand around the other is no longer possible. Thus, if we take the linear dsDNA at its ends, unwind the double helix a certain number of turns and then close both strands, the dsDNA cannot locally adopt a favourable B-DNA helical structure when it is forced to have an open circular shape. Since intrinsically, dsDNA truly likes to have B-DNA type helix, it starts to buckle and coil in

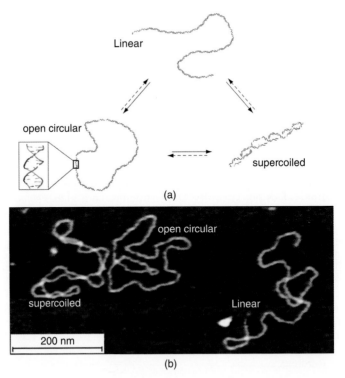

(a)

(b)

Figure 6.9 *(a) Interconversions of topological forms of DNA by cutting (arrows with dashed lines) and joining (arrows with full lines). (b) Scanning force micrograph depicting linear supercoiled and open circular pUC19 plasmid DNA*

3D into a *supercoiled* structure. The number of superhelical crossings (or 'nodes') depends on the number of helix turns that were removed before ring closure. It is of interest that these different topological forms of DNA can also be interconverted by a 'cutting action' as depicted by the arrows with dashed lines in Fig. 6.9.

It is important to see that this supercoiled form of DNA is a *high energy form*. On the one hand this implies that it can counteract the forces and torques imposed by macromolecular machines such as RNA polymerase. Actually, at a certain point, the polymerase will even stall, and it is clear that the cell needs mechanisms to overcome this issue. On the other hand, the cell is opportunistic and exploits the DNA's high energy state. Indeed, essentially all *in vivo* DNA is stored in an underwound (negatively supercoiled) state, which activates the DNA. In a pioneering work, Vologodskii *et al.* [44] stated that for physiological superhelical densities: '*a DNA molecule becomes as it were alive, turning from a stable double helix into a diversely fluctuating entity*'.

It has been shown that several non-traditional sequence-dependent structures are engendered by DNA unwinding: palindromic sequences can result in cruciform-extrusion [45], left-handed Z-DNA might arise in sequences with alternating pyrimidine–purine sequences [46] and intra-molecular triple-helices (H-DNA) are stabilised in mirror-symmetric oligo-purine: oligo-pyrimidine sequences [47]. More recently, Seeman and coworkers have demonstrated (and modeled) the formation of a dumbbell-shaped structure in negatively supercoiled plasmids [48]. The stem of the 'dumbbell' consists of so-called paranemic crossover (PX) DNA, a four-stranded coaxial DNA complex containing a central dyad axis that relates two flanking parallel double helices. PX-DNA provides a means for dsDNA–dsDNA recognition and might be relevant during the initial steps of homologous recombination *in vivo*. The *in vivo* importance (and existence) of cruciforms, Z-DNA, H-DNA and PX-DNA has raised significant debates. In contrast, the biological role of *denaturation bubbles* is relatively undisputed and is generally considered to be of great importance in many nuclear processes [49]. At a sufficiently high magnitude of supercoiling density, these denaturation bubbles can be stable at AT-rich sequences [44, 50]. At moderate negative supercoiling density (homeostatic level), they are metastable at these sequences, stochastically alternating between the double-helix and denatured state [44]. For an 'average' sequence, the underwinding of the DNA is generally set as such that it is in a base-paired form, but requires only the tiniest amount of energy to open up.

Among the different possible DNA structures, the G-quadruplexes DNAs, which are guanine-rich sequences forming G-quartets with eight Hoogsteen-type hydrogen bonds (see later) have also been recognised as playing an important role. Such conformations are mainly present in telomeres, at the ends of eukaryotic chromosomes. Telomeres are repetitive portions of DNA sequences, which protect chromosomes from fusion or deterioration. During each cell replication, a part of this telomeric sequence is lost. Telomeres may thus be regarded as time-regulators for cell divisions, since when they become too short, the replication is stopped.

In conclusion, DNA *in vivo* is generally in a high energy state, and this activates the DNA to adopt specific structures, both at small (local denaturation, Z-DNA, …) as well as at large length scales (supercoiling, …). Moreover, the cell has evolved specific DNA folding patterns for the regulation of cellular processes. How do Ru(II) polypyridyl complexes interact with these relevant forms of DNA, and how can they be properly investigated?

Can these complexes be designed to have certain selectivity for these local features, and can the mechanical and/or topological properties of DNA be irreversibly modified?

6.4.3   SMF Study with [Ru(phen)$_2$(PHEHAT)]$^{2+}$ and [Ru(TAP)$_2$(PHEHAT)]$^{2+}$

Much of our understanding on the interactions between Ru(II) complexes and DNA has come from studies where the effect of DNA on the photophysical properties of the Ru(II)–dye is investigated as outlined above. But how does the Ru(II) complex affect DNA structure? In this context, imaging with scanning force microscopy (SFM) has been shown to be a powerful technique. In contrast to optical microscopy techniques, SFM does not make use of lenses and light. Instead, a very sharp probe-tip is scanned over a surface and records the topographical information of that area. The structure of biomolecules, such as DNA, which are adsorbed from solution onto an ultra-flat surface, can as such be resolved with nanoscale precision. In the following section, we show how SFM can be used to study both the *reversible (in the dark)* as well as the *irreversible (after laser irradiation)* effects of [Ru(TAP)$_2$(PHEHAT)]$^{2+}$ interacting with DNA [51].

From photophysical characterisations (see earlier discussion) it was anticipated that even *in the dark*, [Ru(TAP)$_2$(PHEHAT)]$^{2+}$ interacts strongly with DNA via intercalation of the extended PHEHAT ligand. Intercalation typically increases the DNA curvilinear length in a concentration-dependent manner and therefore SFM constitutes an ideal means to probe this interaction mode in the absence of light. Thus SFM, *a priori* considered as a technique for studying global DNA characteristics, allows also characterising local properties as exemplified in the following discussions.

Based on the SFM topographs of adsorbed linear DNA restriction fragments, contour length distributions can be determined. The DNA fractional occupancies as a function of intercalator concentration are obtained from the contour length distributions under these conditions and binding isotherms can be constructed to allow quantitative fitting to models for ligand binding. A widely used model was developed by McGhee and Von Hippel [52], which might be applied to systems where small ligands bind to homogeneous lattices (DNA is treated as a sequence-neutral entity) in a non-cooperative fashion. This model includes a binding site size n. For n larger than unity, complete saturation of the lattice is more difficult because of the buildup of sites smaller than the binding site size. In addition to the intrinsic association constant K, ligand binding is thus determined by site-exclusion.

The determination of the contour length combined with the distributions has been successfully applied to study the intercalative binding of [Ru(phen)$_2$(PHEHAT)]$^{2+}$ to DNA, yielding an intrinsic association constant ($K = (1.5 \pm 0.2) \times 10^5$ M^{-1}) and binding site size ($n = 2.8 \pm 0.3$) [51], of the same order of magnitude as the value obtained from luminescence measurements [18]. A similar evaluation of the parameters describing the intercalative binding mode of [Ru(TAP)$_2$(PHEHAT)]$^{2+}$ was however not successful, due to an unexpected structural collapse. More specifically, intra- and intermolecular cross-links resulted in the compaction of individual molecules and molecular aggregation, respectively. The cross-linking activity implied that next to the PHEHAT ligand, the TAP ligands are also able to bind DNA.

This brought up the question about how exactly TAP interacts with DNA. At the time of this investigation, the co-crystals of small oligodeoxynucleotides with [Ru(phen)$_2$(dppz)]$^{2+}$ and [Ru(TAP)$_2$(dppz)]$^{2+}$ were also being resolved [38, 39], showing as mentioned before a semi-intercalative binding mode for TAP and phen. However, since the SFM data disclosed no dramatic $[Ru(phen)_2(PHEHAT)]^{2+}$-induced DNA collapse in contrast to the effect by $[Ru(TAP)_2(PHEHAT)]^{2+}$, it was questionable as to whether semi-intercalation could be at the origin of the observed DNA cross-linking. Because in the crystal structures, both phen and TAP induced quite a dramatic bend in the DNA on semi-intercalation, SFM allowed quantitative investigation of semi-intercalation *in vitro*. More specifically, every semi-intercalation event should reduce, via DNA bending, the distance measured between the two ends of a linear DNA fragment (the so-called end-to-end distance). Such an effect could indeed be observed from the titration of linear DNA fragments with the homoleptic $[Ru(TAP)_3]^{2+}$. Nonetheless, since the effect of $[Ru(phen)_3]^{2+}$ and $[Ru(TAP)_3]^{2+}$ on the end-to-end distance of DNA was remarkably similar, semi-intercalation did not explain the dramatic DNA cross-linking observed for $[Ru(TAP)_2(PHEHAT)]^{2+}$ (and not for $[Ru(phen)_2(PHEHAT)]^{2+}$).

Therefore, an alternative binding mode, TAP-mediated hydrogen bonding with DNA, was investigated. Indeed, the additional nitrogen atoms present in TAP as compared with phen could act as hydrogen bond acceptors, whereas hydrogen bond donors could be found in the grooves of the DNA. The possibility of hydrogen bonding was investigated via a competition assay. On incubation of DNA and $[Ru(TAP)_2(PHEHAT)]^{2+}$ with increasing concentrations of urea – a strong hydrogen bond donor – indeed an effective reduction of $[Ru(TAP)_2(PHEHAT)]^{2+}$-mediated DNA cross-linking was observed whereas no effect of urea on the PHEHAT-mediated intercalative binding mode was detected. Thus, it seemed that TAP ligands interact *in vitro* with DNA via a hydrogen binding mode.

In a next step, the impact of the photoreaction sensitised by $[Ru(TAP)_2(PHEHAT)]^{2+}$ on the mesoscale structure of supercoiled plasmid DNA molecules was investigated. Early experiments on the light-reaction between ruthenium-TAP complexes with super-coiled plasmid DNA indicated an efficient formation of nicks in the DNA backbone via photo-oxidation [53, 54]. The dramatic effect of the topochemical change on plasmid DNA conformation (see Fig. 6.9) can be investigated in a quantitative manner from SFM images by counting the number of intrachain DNA crossovers per DNA molecule (we will subsequently refer to this value as the 'nodenumber'). The distribution of nodenumbers in a native, supercoiled plasmid (pUC19 vector) DNA sample mainly exhibits a peak centred at a nodenumber of 7, whereas the nodenumber distribution obtained from a DNA sample treated with a nicking enzyme (which cuts sequence specifically only one of both strands in the double helix), exhibits a peak centred at a nodenumber of 2. In other words, SFM can be used to visualise and quantify the difference between an intact, supercoiled versus a nicked, open circular DNA molecule. Thus, sensitised photocleaving activity can be assessed via the direct SFM imaging of the DNA reaction products that were irradiated for various times in the presence of the $[Ru(TAP)_2(PHEHAT)]^{2+}$ complex. This was confirmed by comparison with $[Ru(phen)_2(PHEHAT)]^{2+}$, a reference photosensitiser, which induces photocleavages in DNA via the light-induced generation of singlet oxygen. On increasing irradiation times of $[Ru(phen)_2(PHEHAT)]^{2+}$, the number of molecules with small nodenumbers (corresponding to nicked plasmids) increased at the expense of the number of molecules with larger nodenumbers (corresponding to intact, supercoiled plasmids).

When $[Ru(TAP)_2(PHEHAT)]^{2+}$ was used as a photosensitiser, unexpectedly, a much more complex behaviour was observed in terms of the evolution of the nodenumber distribution with irradiation time. Instead of finding two well-resolved peaks throughout the node number distributions of the samples at various irradiation times, ever more molecules with intermediate nodenumbers appeared during the first minute (phase I) of sample irradiation. Prolonged irradiation (phase II: 1 min to 1 h) however shifted this fraction of molecules again to higher nodenumbers, indicating yet another product from the photoreaction. The unexpected behaviour was interpreted as resulting from the formation of irreversible covalent photoadducts (cf. Fig. 6.6).

Several models were evaluated to explain the observations, and it turned out that both mechanical as well as topological changes in DNA on photoadduct formation underline the structural changes in circular DNA. More specifically, photoadduct formation on linear DNA fragments was shown to increase the DNA bending rigidity, which can explain the behaviour in phase I: the reduced ability to bend will reduce the nodenumber of supercoiled DNA, as it tries to adopt a more open conformation.

In order to explain the behaviour in phase II, we reasoned that the presence of two TAP ligands might in principle allow for the photocrosslinking of two distant G sites in the supercoiled plasmid DNA, thereby separating the molecule into two topological domains and increasing the average nodenumber in these DNA molecules. This hypothesis was tested by treating $[Ru(TAP)_2(PHEHAT)]^{2+}$-photosensitised plasmids with a restriction endonuclease. The EcoRI enzyme recognises and cuts only a single site in the pUC19 plasmid. Thus, in case irreversible photocrosslinking has indeed occurred, such treatment should only affect the topological state of the domain containing the recognition site. In other words, if the hypothesis is true, one should be able to find linearised molecules with an internal supercoiled domain. This appeared indeed to be the case, proving the formation of *G containing intersegmental photo-cross-links* for the first time (Fig. 6.10).

Finally, the effect of TAP-mediated hydrogen bonding on the outcome of the Ru(II) complex photoreaction was investigated using supercoiled plasmids as a DNA substrate and in the presence of urea. The effect on irreversible photo-adduct formation on DNA structure appeared to be quasi-identical to the case where no urea was added. However, the rate of photocleavage formation appeared to be greatly suppressed in the presence of urea. Thus, somehow the geometry of the TAP-mediated hydrogen bonded state of $[Ru(TAP)_2(PHEHAT)]^{2+}$ with DNA appears to be important for the relative amount observed between photocleavages and photoadducts.

6.5   Unusual Interaction of Dinuclear Ru(II) Complexes with Different DNA Types

So far, we have focussed our discussion on Ru(II) complexes based on a single metal centre. Nevertheless, polynuclear entities can be built with two or more Ru ions. However for such assemblies, the resulting metallo-organic constructions must contain at least one bridging ligand, such as the HATPHE/PHEHAT or HAT (Fig. 6.2), which possesses three chelating sites, or TPAC or μ-11,11'-bidppz (Fig. 6.11). In this section, we will present different cases of dinuclear complexes, which exhibit a specific behaviour in the presence of DNA.

The interest in dinuclear Ru(II) complexes lies in allowing the construction of more elaborate and larger structures, with more complicated shape, as compared with monometallic species. This increasing versatility in size and shape/conformation enables the design of

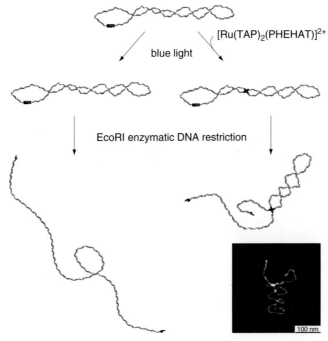

Figure 6.10 *Experimental design to prove the formation of stable intersegmental photo-crosslinks. A supercoiled plasmid DNA containing a single recognition sequence for EcoRI-mediated enzymatic restriction (shown as a red–black rectangle) is irradiated in the absence (left) and presence (right) of [Ru(TAP)₂(PHEHAT)]²⁺. Where [Ru(TAP)₂(PHEHAT)]²⁺ can generate on prolonged light irradiation covalent photocrosslinks (shown as a black cross 'X'), enzymatic linearisation will result in a linear form DNA with internal supercoiled domain, whereas the negative control should never exhibit such a supercoiled region after linearisation. Part of this figure was reproduced with permission from [51] © 2012 American Chemical Society*

complexes which exhibit a specific interaction with classical DNA double helices or constitute specific probes for particular DNA superstructures. The photochemical behaviour of dinuclear complexes in the presence of DNA, is also different as compared with that of their mononuclear analogues. As we will see, with different examples, this is due to both a change in the photophysics and a modification of their interaction with the targets.

It also has to be pointed out that the existence of two Ru(II) centres inside the same complex results in diastereomeric species, since individual metal centres can adopt a Δ or a Λ helicity (Fig. 6.2). The behaviour of each stereoisomer in the presence of DNA can thus be different.

6.5.1 Reversible Interaction of [{(Ru(phen)₂}₂HAT]⁴⁺ with Denatured DNA

The HAT ligand possesses three equivalent chelating sites, allowing the design of mono-, di- or trinuclear complexes. As mentioned before, the monometallic [Ru(phen)₂(HAT)]²⁺ is luminescent either in water or in organic solvent. Like the TAP ligand, the HAT ligand is also highly π-deficient, and therefore the luminophore corresponds to [Ru(III)-HAT•⁻]*.

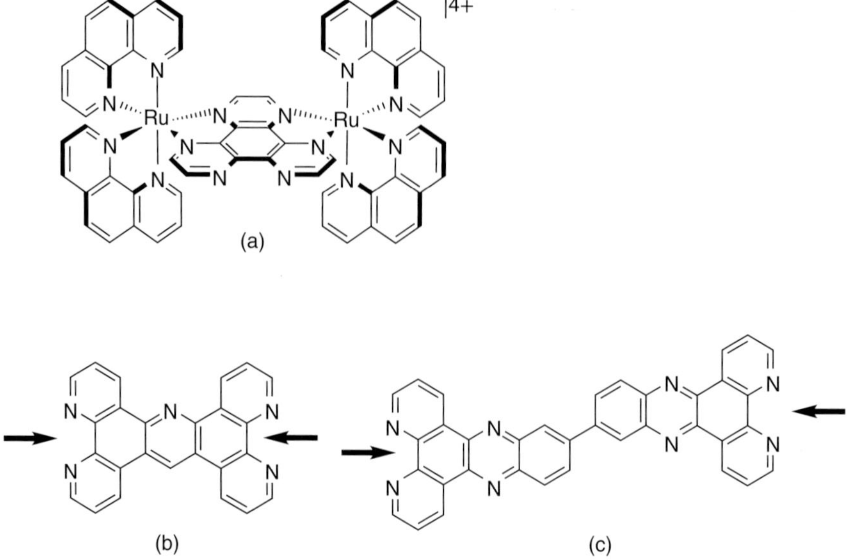

Figure 6.11 Structure of (a) the dinuclear [{(Ru(phen)₂}₂HAT]⁴⁺ complex; (b) the TPAC (tetrapyrido[3,2-a:2',3'-c:3'',2''-h:2''',3'''-j]acridine) and (c) the μ-11,11'-bidppz (11,11'-bi(dipyrido[3,2-a:2',3'-c]phenazinyl)) ligand. The arrows indicate possible sites of coordination

In the presence of DNA, the [Ru(phen)₂(HAT)]²⁺ complex is semi-intercalated via the HAT moiety [55].

For the dinuclear [{(Ru(phen)₂}₂HAT]⁴⁺ (Fig. 6.11a) (the trinuclear complex will not be discussed) [22]), there is no change of luminophore, which continues to be [Ru(III)-HAT•⁻]*; however the photoredox potential is affected by the presence of the second Ru ion. On the basis of thermodynamics, a photoinduced ET (PET) between a guanine base and this dinuclear complex, should be exergonic (cf. Fig. 6.5a). Laser-flash photolysis experiments indeed showed that in the presence of GMP, the [{(Ru(phen)₂}₂HAT]⁴⁺ is photoreduced [56]. As has been explained above, initiated by this PET, a photoadduct could thus be expected. Formation of a photoproduct with GMP was actually detected, but the amount of such a covalent adduct was much lower than that from photoreactions with mononuclear Ru(II) complexes bearing at least two TAP or HAT ligands. It was proposed that this lower yield of photoadduct formation with GMP is due to the presence of aggregates in the ground state between [{(Ru(phen)₂}₂HAT]⁴⁺ and GMP. Indeed, the absorption spectrum of [{(Ru(phen)₂}₂HAT]⁴⁺ is modified in the presence of GMP, which is indicative of an interaction between these species in the ground state. These aggregates, would prevent a favourable organisation of the species in a geometry such that a covalent bond formation could appear between the G base and the bridging HAT ligand after the ET under irradiation.

In the presence of CT-DNA, the emission data show that the dinuclear complex, in contrast to the monometallic analogue [(Ru(phen)₂(HAT)]²⁺ or even [Ru(phen)₃]²⁺, does not

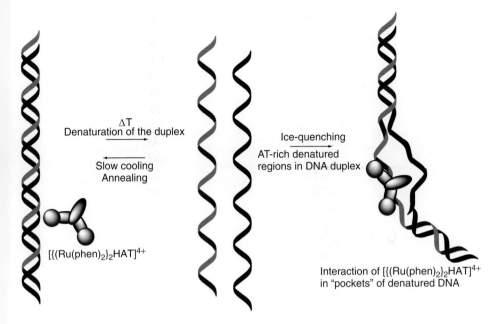

Figure 6.12 *Cartoon for DNA denaturation*

interact with the double helix (i.e. there is no influence of DNA on the emission). However in the presence of denatured DNA, the situation is totally different and an increase in luminescence is observed. Actually, during a denaturation process, initially at high temperature, the two strands of DNA are separated and afterwards, when the two strands are allowed to cool down slowly, the double helix is restored according to the Watson–Crick base pairing (annealing). However if the sample is ice-quenched after heating, denatured DNA where the two strands stick together with some tracks in the double helix but with other regions with defects consisting of non-paired bases forming bulges, for example, is obtained (Fig. 6.12). As the cytosine–guanine base-pairing is more thermodynamically stable than the adenosine–thymine pairing, the denatured sections are more likely to contain A–T than C–G bases.

Thus in the presence of such denatured DNA, the increased luminescence of $[\{(Ru(phen)_2\}_2HAT]^{4+}$ is due to protection of the excited dinuclear complex inside some type of A–T rich pockets or bulges in the denatured DNA, where this dinuclear compound can easily reside. A PET with guanine bases was detected in transient absorption under pulsed laser excitation, but without photoadduct formation [56]. The presence of this PET could be intriguing, since this process should quench the luminescence, in disagreement with the observed increased emission. However, it has to be pointed out that the measured luminescence results from two antagonistic effects: (i) an increase due to protection by A–T rich denatured regions where the complex is mainly hosted, and (ii) a decrease due to a quenching by PET with some G–C regions. As the luminescence is increased with denatured DNA, it indicates that the effect of A–T protection has a greater contribution than the ET quenching by G–C [56].

In conclusion, the difference in behaviour between $[\{(\text{Ru(phen)}_2\}_2\text{HAT}]^{4+}$ and its analogue $[(\text{Ru(phen)}_2(\text{HAT})]^{4+}$ highlights the interest in this dinuclear complex as a denatured DNA photoprobe. The presence of the second ion centre modifies both the photochemistry and the interaction with DNA. A PET process with guanine bases becomes thermodynamically possible because of the presence of the second Ru ion, which increases the π-deficiency on the central bridging HAT ligand. This increasing oxidation power of the excited state favours in this way a PET with G bases. Moreover, although the monometallic complex is quasi intercalated inside the DNA helix, the $[\{(\text{Ru(phen)}_2\}_2\text{HAT}]^{4+}$ does not interact well with a double helix, but strongly interacts with denatured portions of DNA [56]. This specific behaviour makes the $[\{(\text{Ru(phen)}_2\}_2\text{HAT}]^{4+}$ an excellent photoprobe for DNA unpaired portions, such as bulges, hairpins and so on, which play important roles in nuclear processes as mentioned in Section 6.4.2.

Each diastereoisomer of the dinuclear complex has also been isolated, purified and studied [57]. Only slight differences have been observed between the $\Delta\Delta$ and $\Lambda\Lambda$ enantiomers and the *meso* form $\Delta\Lambda$. It turned out that the *meso* form interacts better with denatured DNA than the two enantiomers, but up to now, no explanation has been proposed to rationalise this stereospecificity.

6.5.2   Targeting G-quadruplexes with Photoreactive $[\{\text{Ru(TAP)}_2\}_2\text{TPAC}]^{4+}$

The TPAC ligand (Fig. 6.11b), which was synthesised for the first time by Demeunynck *et al.* [58], based on a central acridine core, has two equivalent chelating sites. Unexpectedly, in contrast to the previous HAT case, the chelation of a second Ru ion to the bridging TPAC ligand, influences neither the photophysical properties nor the redox potentials of the resulting dinuclear complex in comparison with its mononuclear analogue. Therefore, in the $[\{(\text{Ru(TAP)}_2\}_2\text{TPAC}]^{4+}$ complex, the two ruthenium centres behave fairly independently and the luminophore corresponds to the $[\text{Ru(III)-TAP}^{\bullet-}]^*$ sub-units. As a consequence, upon irradiation, this complex should be able to induce an ET with a guanine base followed by a photoadduct formation. On the basis of the previous results with the dinuclear HAT complex, it would also be expected that $[\{(\text{Ru(TAP)}_2\}_2\text{TPAC}]^{4+}$ does not interact well with a classical double stranded DNA. However, it was observed that upon irradiation in the presence of G containing oligonucleotide duplexes (at least with one G base on each strand), not only can a mono-adduct easily form, but after a second photon absorption by the resulting photoadduct, a bi-adduct is also produced. The latter leads to a photocrosslinking between the two DNA strands as shown in the cartoon of Fig. 6.6. The yield of this photocrosslinking is much more important than that for the TAP mononuclear analogues, probably because of the presence of four TAP ligands that enhance the probability of irreversible photoadduct formation with G sites and photocrosslinking [59]. We think that this efficient production of mono-adduct with a first G base would destroy the duplex structure locally. This would allow the photocrosslinking with a G base of the other strand upon absorption of a second photon. This contrasts with the case of the dinuclear HAT complex discussed above. In that case, the bridging HAT ligand is inaccessible to the G bases due to the two Ru(phen)$_2$ moieties (Fig. 6.11a). Therefore, a G adduct can no longer be formed.

The photobridging of two strands via the $[\{(\text{Ru(TAP)}_2\}_2\text{TPAC}]^{4+}$ is not only efficient, but implies two guanine bases that are further away from each other, actually separated

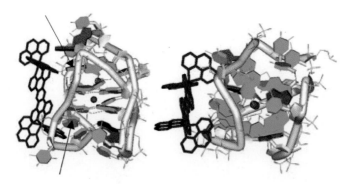

Figure 6.13 *Molecular modeling of the last snapshot of the 2 ns MD run of a G-quadruplex after irreversible photocrosslinking of two G bases by [{(Ru(TAP)$_2$}$_2$TPAC]$^{4+}$. Reproduced with permission from [59] © 2012 Wiley-VCH Verlag GmbH & Co. KGaA, Weinheim*

by four base pairs [25]. Thanks to these properties, the dinuclear [{(Ru(TAP)$_2$}$_2$TPAC]$^{4+}$ is a perfect candidate for targeting G-quadruplexes present in telomeric sequences (see previous discussion) (Fig. 6.13). In cancerous cells, the telomeres lengths are restored after each cell division by the action of the telomerase protein, which enables uncontrolled cell proliferation. Therefore cancerous cells become immortal. G-quadruplexes present in telomeric regions are thus interesting pharmaceutical targets. In this context, the irreversible photobridging obtained between two G bases inside G-quadruplexes by the photoreactive dinuclear TPAC complex (Fig. 6.13), which completely blocks the unfolding, could induce the normal senescence of the cells and stop the cancer proliferation. It turns out that the dinuclear [{(Ru(TAP)$_2$}$_2$TPAC]$^{4+}$ is much more photoreactive towards G-quadruplex structures than any other photoreactive monometallic Ru(II) complex analogue bearing at least two TAP ligands (the photoadduct formation yield is five-times more important in the case of [{(Ru(TAP)$_2$}$_2$TPAC]$^{4+}$ than for the [(Ru(TAP)$_2$TPAC]$^{2+}$ analogue).

6.5.3 Threading Intercalation

As has been pointed out in the previous section, [Ru(phen)$_2$(dppz)]$^{2+}$ is a very powerful probe for DNA, due to its emission, which depends on the intercalation mode of the dppz ligand into the DNA base-pair stacking. Even symmetrical and canted intercalation geometries can be distinguished via the emission lifetime. Nevertheless, [Ru(phen)$_2$(dppz)]$^{2+}$ exhibits a relatively low selectivity towards special structures or sequences of DNA. Therefore, bridging ligands based on a bidppz template such as the μ-11,11′-bidppz [60] and μ-C4(cpdppz) [61] have been developed by Lincoln *et al.* (Fig. 6.11c and 6.14a). As for the bridging TPAC, the two Ru(II) centres chelated to these bidppz ligands do not affect each other. The emission of these dinuclear complexes thus has the same characteristics as those of the mononuclear [Ru(phen)$_2$(dppz)]$^{2+}$ analogue and the luminophore corresponds to [Ru(III)-dppz$^{\bullet-}$]*.

The first studies with the $\Delta\Delta$ and $\Lambda\Lambda$ enantiomers of [μ-(11,11′-bidppz)(phen)$_4$Ru$_2$]$^{4+}$ reported simple binding of this complex inside a groove of DNA [60]. During further

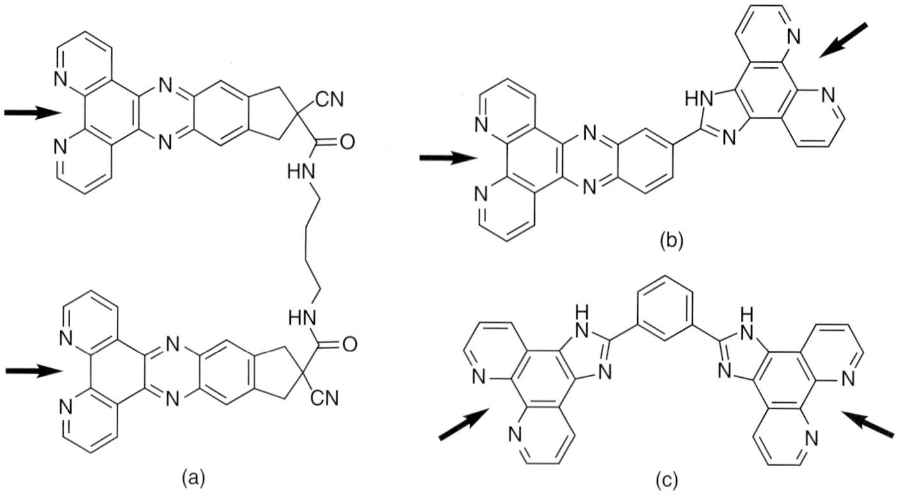

Figure 6.14 *Structure of (a) the µ-C4(cpdppz) (N,N'-bis(12-cyano-12,13-dihydro-11H-cyclopenta[b]dipyrido[3,2-h:2',3'-j]phenazine-12-carbonyl)-1,4-diaminobutane), (b) the µ-dppzip (2-(dipyrido[3,2-a:2',3'-c]phenazin-11-yl)imidazo[4,5-f][1, 10]phenanthroline) and (c) the µ-bipb ligand (1,3-bis(imidazo[4,5-f][1, 10]phenanthrolin-2-yl)benzene). The arrows indicate possible sites of coordination*

research, serendipity played an important role. Analyses of a sample which had been left at room temperature for a week revealed that the complex exhibited a different interaction with DNA than its initial groove binding [62]. It turned out that this geometry rearrangement, which induces an important increase in luminescence intensity, is very slow.

Once this final state is reached, dissociation of the complex from DNA is also very slow in comparison with the initial groove-binding process (several days at 45 °C are needed for SDS-induced dissociation whereas the initial groove binding form dissociates instantaneously). The binding mode proposed for this final state corresponds to a threading intercalation; in other words the intercalation of the bridging bidppz requires that one of the two metal centres threads through the DNA duplex with the phen ancillary ligands of the two metallic centres in opposite grooves of the duplex. Thus, the extremely slow kinetics are due to the fact that one bulky Ru(phen)$_2$ moiety has to pass through the base-pair stacking. This threading intercalation results in an important stability towards dissociation from DNA, in spite of the fact that this interaction is non-covalent. Such a high DNA affinity and slow dissociation kinetics are important in biological applications. Actually, similar threading intercalations are present in nature. For example, the natural antibiotic nogalamycin binds to DNA by threading a bulky sugar moiety through the base-pairs stacking, resulting in a highly kinetically stable interaction, responsible for the cytotoxicity of nogalamycin [63].

The interaction of each diastereoisomer of $[\mu\text{-}(11,11'\text{-bidppz})(phen)_4Ru_2]^{4+}$ has been characterised by emission, LD and CD [64]. Each of the three stereoisomers binds to DNA by threading intercalation, the central bridging biddpz ligand being sandwiched inside the base-pairs in an anti-conformation (Fig. 6.11c) [65]. As the Ru(phen)$_2$ moiety located in the minor groove is more deeply embedded in the DNA duplex, the intercalation is non-symmetrical. By comparison of the meso form with the $\Delta\Delta$ and $\Lambda\Lambda$ enantiomers, it has been concluded that the Λ part is more deeply intercalated in the minor groove, which allows a better fit of the Δ part in the major groove [65].

The base sequence has a major influence on the threading intercalation rate; this process is 65 (for $\Lambda\Lambda$-[μ-(11,11'-bidppz)(phen)$_4$Ru$_2$]$^{4+}$) to 2500 (for $\Lambda\Lambda$-[μ-(11,11'-bidppz)(bpy)$_4$ Ru$_2$]$^{4+}$) times faster in [poly(dAdT)]$_2$ than in CT-DNA [66]. Thus long AT tracks can be targeted by these dinuclear complexes thanks to kinetic recognition. The latter is proposed as being related to duplex breathing dynamics and the enhanced flexibility of AT regions, which favours the threading process. Further studies showed that at least ten consecutive AT base pairs are needed for an efficient threading intercalation [67]. Even if only one base pair has to be melted for one ruthenium centre to thread through the DNA helix [68], the transition state involves a stretch of DNA much larger than the dimensions of the dinuclear complex itself.

It is also worth mentioning that the influence of other parameters on the threading inter-calation process has been examined using other bridging ligands (see Fig. 6.14) (effect of ancillary ligands [64, 69], rigidity of the bridging ligand [70], distance between the two Ru(II) centres and extended aromaticity of the intercalated ligand [71]). It turned out that the kinetic of threading is the slowest for the [μ-(11,11'-bidppz)(phen)$_4$Ru$_2$]$^{4+}$. These studies revealed that the dinuclear complexes with the μ-bipb bridging ligand (Fig. 6.14c) does not intercalate into CT-DNA. However, interestingly, the different stereoisomers of [μ-bipb(phen)$_4$Ru$_2$]$^{4+}$, which are luminescent in water and organic solvent, were used for cell staining, and revealed that the two enantiomers $\Delta\Delta$ and $\Lambda\Lambda$ have different patterns of intra-cellular localisation [72]. These enantiomeric differences in intracellular distribution in fixed cells are due to differential affinities for the cellular components. This effect of chirality on the cellular localisation of dinuclear complexes is unique and promising for the design of efficient photoprobes for sub-cellular structures.

6.6   Conclusions

In this chapter we have tried to show the relations between the photophysics and photo-chemistry of some Ru(II) complexes in water and organic solvents with their properties as DNA probes, more particularly as DNA photoprobes. This is illustrated by a few specific examples chosen in order to highlight this relationship.

In recent years, important milestones have been reached in this research field. The fact that several cocrystals of synthetic oligonucleotides with intercalating Λ/ΔRu(II) complexes have been obtained and analysed [38, 39] represents an important step forward. Moreover, it has even been possible with the Λ or Δ enantiomers of [Ru(phen/bpy)$_2$ (dppz)]$^{2+}$ [37] in the presence of [poly(dA-dT)]$_2$ in solution, to conciliate for one enantiomer, the two excited state lifetimes and the ITC data, with canted and symmetrical intercalations observed from crystallographic data [38, 39]. For the first time, the effects of illumination of a photoreactive TAP intercalating Ru(II) complex have been observed by SFM at the molecular level, on large DNA molecules such as plasmid DNA. Furthermore, in cellular biology the fact that some particular enantiomers or diastereoisomers of dinuclear Ru(II) complexes are able to mark, in particular, some cellular components by their specific luminescence, and this depending on whether the cells are living or dead, is also especially attractive and deserves further research.

Although significant progress has been made in the last few years concerning the geometry of interaction of Ru(II) probes with simple and even more complicated DNA molecules *in vitro* and in live cells, there are still many questions left unanswered. For example, although as mentioned earlier, there are similarities between the results of the studies with complex–DNA solutions and cocrystals of complex–oligonucleotides, there are also differences of behaviour between these complexes cocrystallised with oligonucleotides, as observed by X-ray crystallography and the same complexes interacting with plasmid DNA as observed by SFM. Indeed no H-bonding between the TAP ligand of $[Ru(TAP)_2(PHEHAT)]^{2+}$ and DNA is observed from the X-ray crystallographic analysis, whereas by SFM in the absence of urea and with a certain concentration of DNA, the existence of H-bonding between the TAP ligand and DNA has to be concluded from the occurrence of DNA aggregations (such aggregations are not observed with $[Ru(phen)_2(PHEHAT)]^{2+}$). Moreover, this H-bonding-induced DNA aggregation results under illumination of the Ru(II) complex into a much higher yield of photocleavages as compared with photoadducts (or photocrosslinkings) than in the absence of aggregation (in the presence of urea). This dependence of the Ru(II) complex photochemistry on the DNA superstructure is not yet understood. Maybe the SFM technique with a better lateral resolution or in combination with single molecule optical approaches (such as tip-enhanced Raman spectroscopy) will, in the future, allow those questions to be answered.

It is clear from the discussions here that many of the molecular details of the probe–DNA interaction are still to be explored in a simplified *in vitro* approach. However, understanding the uptake and behaviour of these metallic probes in the complex environment of the cell constitutes a crucial challenge on the road towards their use in living material.

Acknowledgement

L. M. and W. V. thank, respectively, the FNRS (Fonds National de la Recherche Scientifique) and the IWT (Innovation by Science and Technology in Flanders) for a PhD grant. The authors are grateful to the FNRS for the financial support.

References

1. L. Marcélis, C. Moucheron and A. Kirsch-De Mesmaeker, Ru-TAP complexes and DNA. From photo-induced electron transfer to gene photo-silencing in living cells, *Philosophical Transations of the Royal Society A*, **371** n°1995 2012 0131 (DOI 10.1098/rsta.2012.0131) (2013).
2. L. Marcélis, J. Ghesquière, K. Garnir, A. Kirsch-De Mesmaeker and C. Moucheron, Photo-oxidizing RuII complexes and light: Targeting biomolecules via photoadditions, *Coordination Chemistry Reviews*, **256** (15–16), 1569–1582 (2012).
3. U. Schatzschneider, Photoactivated biological activity of transition-metal complexes, *European Journal of Inorganic Chemistry*, (10), 1451–1467 (2010).
4. P.C. Bruijnincx and P.J. Sadler, New trends for metal complexes with anticancer activity, *Current Opinion in Chemical Biology*, **12**(2), 197–206 (2008).

5. J.D. Aguirre, A.M. Angeles-Boza, A. Chouai, C. Turro, J.P. Pellois and K.R. Dunbar, Anticancer activity of heteroleptic diimine complexes of dirhodium: A study of intercalating properties, hydrophobicity and in cellulo activity, *Dalton Transactions*, 10806–10812 (2009).

6. M. Wojdyla, J.A. Smith, S. Vasudevan, S.J. Quinn and J.M. Kelly, Excited state behaviour of substituted dipyridophenazine Cr(III) complexes in the presence of nucleic acids, *Photochemical \& Photobiological Sciences*, **9**, 1196–1202 (2010).

7. J.A. Smith, M.W. George and J.M. Kelly, Transient spectroscopy of dipyridophenazine metal complexes which undergo photo-induced electron transfer with DNA, *Coordination Chemistry Reviews*, **255** (21–22), 2666–2675 (2011).

8. S.J. Burya, A.M. Palmer, J.C. Gallucci and C. Turro, Photoinduced ligand exchange and covalent DNA binding by two new dirhodium bis-amidato complexes, *Inorganic Chemistry*, **51**(21), 11882–11890 (2012).

9. NP. Kane-Maguire, Photochemistry and photophysics of coordination compounds: chromium, *Topics in Current Chemistry*, **280**, 37–67 (2007).

10. K.K.W. Lo and K.Y. Zhang, Iridium(III) complexes as therapeutic and bioimaging reagents for cellular applications, *RSC Advances*, **2**, 12069–12083 (2012).

11. M.R. Gill, H. Derrat, C.G.W. Smythe, G. Battaglia and J.A. Thomas, Ruthenium(II) metallo-intercalators: DNA imaging and cytotoxicity, *ChemBioChem*, **12**(6), 877–880 (2011).

12. B. Durham, J.V. Caspar, J.K. Nagle and T.J. Meyer, Photochemistry of tris(2,2'-bipyridine)ruthenium(2+) ion, *Journal of the American Chemical Society*, **104**(18), 4803–4810 (1982).

13. J.V. Caspar and T.J. Meyer, Photochemistry of tris(2,2'-bipyridine)ruthenium(2+) ion $[Ru(bpy)_3]^{2+}$. Solvent effects, *Journal of the American Chemical Society*, **105**(17), 5583–5590 (1983).

14. A. Juris, V. Balzani, F. Barigelletti, S. Campagna, P. Belser and A. von Zelewsky, Ru(II) polypyridine complexes: photophysics, photochemistry, electrochemistry, and chemiluminescence, *Coordination Chemistry Reviews*, **84**, 85–277 (1988).

15. A. Masschelein, L. Jacquet, A. Kirsch-De Mesmaeker and J. Nasielski, Ruthenium complexes with 1,4,5,8-tetraazaphenanthrene. Unusual photophysical behavior of the tris-homoleptic compound, *Inorganic Chemistry*, **29**(4), 855–860 (1990).

16. E.J.C. Olson, D. Hu, A. Hörmann, A.M. Jonkman, M.R. Arkin, E.DA. Stemp, J.K. Barton and P.F. Barbara, First observation of the key intermediate in the "light-switch" mechanism of $[Ru(phen)_2dppz]^{2+}$, *Journal of the American Chemical Society*, **119**(47), 11458–11467 (1997).

17. J. Olofsson, B. Önfelt and P. Lincoln, Three-state light switch of $[Ru(phen)_2dppz]^{2+}$: Distinct excited-state species with two, one, or no hydrogen bonds from solvent, *Journal of Physical Chemistry A*, **108**(20), 4391–4398 (2004).

18. C. Moucheron, A. Kirsch-De Mesmaeker and S. Choua, Photophysics of $Ru(phen)_2(PHEHAT)^{2+}$: A novel "light switch" for DNA and photo-oxidant for mononucleotides, *Inorganic Chemistry*, **36**(4), 584–592 (1997).

19. I. Ortmans, B. Elias, J.M. Kelly, C. Moucheron and A. Kirsch-De Mesmaeker, $[Ru(TAP)_2(dppz)]^{2+}$: a DNA intercalating complex, which luminesces strongly in water and undergoes photo-induced proton-coupled electron transfer with guanosine-5'-monophosphate, *Dalton Transactions*, 668–676 (2004).

20. C.G. Coates, P. Callaghan, J.J. McGarvey, J.M. Kelly, L. Jacquet and A. Kirsch-De Mesmaeker, Spectroscopic studies of structurally similar DNA-binding ruthenium (II) complexes containing the dipyridophenazine ligand, *Journal of Molecular Structure*, **598**(1), 15–25 (2001).

21. B. Elias, C. Creely, G.W. Doorley, M.M. Feeney, C. Moucheron, A. Kirsch-De Mesmaeker, J. Dyer, D.C. Grills, M.W. George, P. Matousek, A.W. Parker, M. Towrie and J.M. Kelly, Photooxidation of guanine by a ruthenium dipyridophenazine complex intercalated in a double-stranded polynucleotide monitored directly by picosecond visible and infrared transient absorption spectroscopy, *Chemistry – A European Journal*, **14**(1), 369–375 (2008).

22. L. Jacquet and A. Kirsch-De Mesmaeker, Spectroelectrochemical characteristics and photophysics of a series of Ru(II) complexes with 1,4,5,8,9,12-hexaazatriphenylene: effects of polycomplexation, *Journal of the Chemical Society, Faraday Transactions*, **88**, 2471–2480 (1992).

23. A. Boisdenghien, C. Moucheron and A. Kirsch-De Mesmaeker, [Ru(phen)$_2$(PHE HAT)]$^{2+}$ and [Ru(phen)$_2$(HATPHE)]$^{2+}$: Two ruthenium(II) complexes with the same ligands but different photophysics and spectroelectrochemistry, *Inorganic Chemistry*, **44**(21), 7678–7685 (2005).

24. B. Elias, L. Herman, C. Moucheron and A. Kirsch-De Mesmaeker, Dinuclear Ru(II)PHEHAT and -TPAC complexes: Effects of the second Ru(II) center on their spectroelectrochemical properties, *Inorganic Chemistry*, **46**(12), 4979–4988 (2007).

25. L. Ghizdavu, F. Pierard, S. Rickling, S. Aury, M. Surin, D. Beljonne, R. Lazzaroni, P. Murat, E. Defrancq, C. Moucheron and A. Kirsch-De Mesmaeker, Oxidizing Ru(II) complexes as irreversible and specific photo-cross-linking agents of oligonucleotide duplexes, *Inorganic Chemistry*, **48**(23), 10988–10994 (2009).

26. D. Garcìa-Fresnadillo, Y. Georgiadou, G. Orellana, A.M. Braun and E. Oliveros, Singlet-oxygen (1Δg) production by ruthenium(II) complexes containing polyazaheterocyclic ligands in methanol and in water, *Helvetica Chimica Acta*, **79**(4), 1222–1238 (1996).

27. W. Saenger, *Principles of Nucleic Acid Structure*, C.R. Cantor (ed.), Springer-Verlag, New York (1988).

28. K.E. Erkkila, D.T. Odom and J.K. Barton, Recognition and reaction of metallointercalators with DNA, *Chemical Reviews*, **99**(9), 2777–2796 (1999).

29. A.W. McKinley, P. Lincoln and E.M. Tuite, Environmental effects on the photophysics of transition metal complexes with dipyrido[2,3-a:3',2'-c]phenazine (dppz) and related ligands, *Coordination Chemistry Reviews*, **255**(21-22), 2676–2692 (2011).

30. M.R. Gill and J.A. Thomas, Ruthenium(II) polypyridyl complexes and DNA-from structural probes to cellular imaging and therapeutics, *Chemical Society Reviews*, **41**, 3179–3192 (2012).

31. A.E. Friedman, J.C. Chambron, J.P. Sauvage, N.J. Turro and J.K. Barton, A molecular light switch for DNA: Ru(bpy)$_2$(dppz)$^{2+}$, *Journal of the American Chemical Society*, **112**(12), 4960–4962 (1990).

32. C. Hiort, P. Lincoln and B. Norden, DNA binding of DELTA- and LAMBDA-[Ru (phen)$_2$DPPZ]$^{2+}$, *Journal of the American Chemical Society*, **115**(9), 3448–3454 (1993).

33. C. Moucheron and A. Kirsch-De Mesmaeker, New DNA-binding ruthenium(II) complexes as photo-reagents for mononucleotides and DNA, *Journal of Physical Organic Chemistry*, **11**(8-9), 577–583 (1998).
34. J.M. Kelly, A.B. Tossi, D.J. McConnell and C. OhUigin, A study of the interactions of some polypyridylruthenium(II) complexes with DNA using fluorescence spectroscopy, topoisomerisation and thermal denaturation, *Nucleic Acids Research*, **13**(17), 6017–6034 (1985).
35. P. Lincoln and B. Norden, DNA binding geometries of ruthenium(II) complexes with 1,10-phenanthroline and 2,2'-bipyridine ligands studied with linear dichroism spectroscopy. Borderline cases of intercalation, *Journal of Physical Chemistry B*, **102**(47), 9583–9594 (1998).
36. A.W. McKinley, J. Andersson, P. Lincoln and E.M. Tuite, DNA sequence and ancillary ligand modulate the biexponential emission decay of intercalated $[Ru(L)_2dppz]^{2+}$ enantiomers, *Chemistry – A European Journal*, **18**(47), 15142–15150 (2012).
37. J. Andersson, L.H. Fornander, M. Abrahamsson, E. Tuite, P. Nordell and P. Lincoln, Lifetime heterogeneity of DNA-bound dppz complexes originates from distinct intercalation geometries determined by complex–complex interactions, *Inorganic Chemistry*, **52**(2), 1151–1159 (2013).
38. H. Niyazi, J.P. Hall, K. O'Sullivan, G. Winter, T. Sorensen, J.M. Kelly and C.J. Cardin, Crystal structures of Λ-[Ru(phen)$_{(2)}$dppz]$^{(2+)}$ with oligonucleotides containing TA/TA and AT/AT steps show two intercalation modes, *Nature Chemistry*, **4**(8), 621–628 (2012).
39. J.P. Hall, K. O'Sullivan, A. Naseer, J.A. Smith, J.M. Kelly and C.J. Cardin, Structure determination of an intercalating ruthenium dipyridophenazine complex which kinks DNA by semiintercalation of a tetraazaphenanthrene ligand, *Proceedings of the National Academy of Sciences of the United States of America*, **108**, 17610–17614 (2011).
40. B. Elias and A. Kirsch-De Mesmaeker, Photo-reduction of polyazaaromatic Ru(II) complexes by biomolecules and possible applications, *Coordination Chemistry Reviews*, **250**(13–14), 1627–1641 (2006).
41. C. Moucheron, A. Kirsch-De Mesmaeker and J.M. Kelly, Photoreactions of ruthenium (II) and osmium (II) complexes with deoxyribonucleic acid (DNA), *Journal of Photochemistry and Photobiology B: Biology*, **40**(2), 91–106 (1997).
42. O. Kratky and G. Porod, Röntgenuntersuchung gelöster fadenmoleküle, *Recueil des Travaux Chimiques des Pays-Bas*, **68**(12), 1106–1122 (1949).
43. L.F. Liu and J.C. Wang, Supercoiling of the DNA template during transcription, *Proceedings of the National Academy of Sciences of the United States of America*, **84**(20), 7024–7027 (1987).
44. AV. Vologodskii, AV. Lukashin, VV. Anshelevich and MD. Frank-Kamenetskii, Fluctuations in superhelical DNA, *Nucleic Acids Research*, **6**(3), 967–982 (1979).
45. M. Gellert, K. Mizuuchi, M.H. O'Dea, H. Ohmori and J. Tomizawa, DNA gyrase and DNA supercoiling, *Cold Spring Harbor Symposia on Quantitative Biology*, **43**, 35–40 (1979).
46. A. Rich, A. Nordheim and A.H.J. Wang, The chemistry and biology of left-handed Z-DNA, *Annual Review of Biochemistry*, **53**(1), 791–846 (1984).

47. S.M. Mirkin, V.I. Lyamichev, K.N. Drushlyak, V.N. Dobrynin, S.A. Filippov and M.D. Frank-Kamenetskii, DNA H form requires a homopurine-homopyrimidine mirror repeat, *Nature*, **330**(6147), 495–497 (1987).

48. X. Wang, X. Zhang, C. Mao and N.C. Seeman, Double-stranded DNA homology produces a physical signature, *Proceedings of the National Academy of Sciences*, **107**(28), 12547–12552 (2010).

49. S. Dasgupta, D.P. Allison, C.E. Snyder and S. Mitra, Base-unpaired regions in super-coiled replicative form DNA of coliphage M13, *Journal of Biological Chemistry*, **252**(16), 5916–5923 (1977).

50. J.H. Jeon, J. Adamcik, G. Dietler and R. Metzler, Supercoiling induces denaturation bubbles in circular DNA, *Physical Review Letters*, **105**, 208101 (2010).

51. W. Vanderlinden, M. Blunt, C.C. David, C. Moucheron, A. Kirsch-De Mesmaeker and S. De Feyter, Mesoscale DNA structural changes on binding and photoreaction with $Ru[(TAP)_2PHEHAT]^{2+}$, *Journal of the American Chemical Society*, **134**(24), 10214–10221 (2012).

52. J.D. McGhee and P.H. von Hippel, Theoretical aspects of DNA-protein interactions: Co-operative and non-co-operative binding of large ligands to a one-dimensional homogeneous lattice, *Journal of Molecular Biology*, **86**(2), 469–489 (1974).

53. J.M. Kelly, D.J. McConnell, C. OhUigin, A.B. Tossi, A. Kirsch-De Mesmaeker, A. Masschelein and J. Nasielski, Ruthenium polypyridyl complexes; their interaction with DNA and their role as sensitisers for its photocleavage, *Journal of the Chemical Society, Chemical Communications*, 1821–1823 (1987).

54. H. Uji-i, P. Foubert, F.C. De Schryver, S. De Feyter, E. Gicquel, A. Etoc, C. Moucheron and A. Kirsch-De Mesmaeker, $[Ru(TAP)_3]^{2+}$-photosensitized DNA cleavage studied by atomic force microscopy and gel electrophoresis: A comparative study, *Chemistry – A European Journal*, **12**(3), 758–762 (2006).

55. R. Blasius, H. Nierengarten, M. Luhmer, JF. Constant, E. Defrancq, P. Dumy, A. van Dorsselaer, C. Moucheron and A. Kirsch-DeMesmaeker, Photoreaction of $[Ru(hat)_2phen]^{2+}$ with guanosine-5′-monophosphate and DNA: Formation of new types of photoadducts, *Chemistry – A European Journal*, **11**(5), 1507–1517 (2005).

56. O. Van Gijte and A. Kirsch-De Mesmaeker, The dinuclear ruthenium(II) complex $[(Ru(Phen)_2)_2(HAT)]^{4+}$ (HAT = 1,4,5,8,9,12-hexaazatriphenylene), a new photore-agent for nucleobases and photoprobe for denatured DNA, *Journal of the Chemical Society, Dalton Transactions*, 951–956 (1999).

57. A. Brodkorb, A. Kirsch-De Mesmaeker, T.J. Rutherford and F.R. Keene, Stereoselective interactions and photo-electron transfers between mononucleotides or DNA and the stereoisomers of a HAT-bridged dinuclear ruII complex (HAT = 1,4,5,8,9,12-hexaazatriphenylene), *European Journal of Inorganic Chemistry*, 2001(8), 2151–2160 (2001).

58. M. Demeunynck, C. Moucheron and A. Kirsch-De Mesmaeker, Tetrapyrido[3,2-a:2′, 3′-c:3″,2″-h:2‴,3‴-j]acridine (tpac): a new extended polycyclic bis-phenanthroline ligand, *Tetrahedron Letters*, **43**(2), 261–264 (2002).

59. S. Rickling, L. Ghisdavu, F. Pierard, P. Gerbaux, M. Surin, P. Murat, E. Defrancq, C. Moucheron and A. Kirsch-De Mesmaeker, A rigid dinuclear ruthenium(II) complex as an efficient photoactive agent for bridging two guanine bases of a duplex or

quadruplex oligonucleotide, *Chemistry – A European Journal*, **16**(13), 3951–3961 (2010).

60. P. Lincoln and B. Norden, Binuclear ruthenium(II) phenanthroline compounds with extreme binding affinity for DNA, *Chemical Communications*, 2145–2146 (1996).

61. B. Önfelt, P. Lincoln and B. Nordén, Enantioselective DNA threading dynamics by phenazine-linked $[Ru(phen)_2dppz]^{2+}$ Dimers, *Journal of the American Chemical Society*, **123**(16), 3630–3637 (2001).

62. L.M. Wilhelmsson, F. Westerlund, P. Lincoln and B. Norden, DNA-Binding of semi-rigid binuclear ruthenium complex Δ,Δ μ-(11,11'-bidppz)(phen)$_4$Ru$_2$]$^{4+}$: Extremely slow intercalation kinetics, *Journal of the American Chemical Society*, **124**(41), 12092–12093 (2002).

63. Y.C. Liaw, Y.G. Gao, H. Robinson, G.A. Van der Marel, J.H. Van Boom and A.H.J. Wang, Antitumor drug nogalamycin binds DNA in both grooves simultaneously: molecular structure of nogalamycin-DNA complex, *Biochemistry*, **28**(26), 9913–9918 (1989).

64. F. Westerlund, P. Nordell, J. Blechinger, T.M. Santos, B. Nordén and P. Lincoln, Complex DNA binding kinetics resolved by combined circular dichroism and luminescence analysis, *Journal of Physical Chemistry B*, **112**(21), 6688–6694 (2008).

65. L.M. Wilhelmsson, E.K. Esborner, F. Westerlund, B. Norden and P. Lincoln, Meso stereoisomer as a probe of enantioselective threading intercalation of semirigid ruthenium complex $[\mu$-(11,11'-bidppz)(phen)$_4$Ru$_2$]$^{4+}$, *Journal of Physical Chemistry B*, **107**(42), 11784–11793 (2003).

66. P. Nordell, F. Westerlund, L.M. Wilhelmsson, B. Nordén and P. Lincoln, Kinetic recognition of AT-rich DNA by ruthenium complexes, *Angewandte Chemie International Edition*, **46**(13), 2203–2206 (2007).

67. P. Nordell, F. Westerlund, A. Reymer, A.H. El-Sagheer, T. Brown, B. Nordèn and P. Lincoln, DNA polymorphism as an origin of adenine-thymine tract length-dependent threading intercalation rate, *Journal of the American Chemical Society*, **130**(44), 14651–14658 (2008).

68. T. Paramanathan, F. Westerlund, M.J. McCauley, I. Rouzina, P. Lincoln and M.C. Williams, Mechanically manipulating the DNA threading intercalation rate, *Journal of the American Chemical Society*, **130**(12), 3752–3753 (2008).

69. F. Westerlund, P. Nordell, B. Nordén and P. Lincoln, Kinetic characterization of an extremely slow DNA binding equilibrium, *Journal of Physical Chemistry B*, **111**(30), 9132–9137 (2007).

70. F. Westerlund, M.P. Eng, M.U. Winters and P. Lincoln, Binding geometry and photophysical properties of DNA-threading binuclear ruthenium complexes, *Journal of Physical Chemistry B*, **111**(1), 310–317 (2007).

71. J. Andersson, M. Li and P. Lincoln, AT-specific DNA binding of binuclear ruthenium complexes at the border of threading intercalation, *Chemistry – A European Journal*, **16**(36), 11037–11046 (2010).

72. F. Svensson, J. Andersson, H. Åmand and P. Lincoln, Effects of chirality on the intracellular localization of binuclear ruthenium(II) polypyridyl complexes, *Journal of Biological Inorganic Chemistry*, **17**, 565–571 (2012).

7

Visualization of Proteins and Cells Using Dithiol-reactive Metal Complexes

Danielle Park, Ivan Ho Shon, Minh Hua, Vivien M. Chen and Philip J. Hogg
Lowy Cancer Research Centre and Prince of Wales Clinical School,
University of New South Wales, Australia

7.1 The Chemistry of As(III) and Sb(III)

Trivalent arsenicals and antimonials form high-affinity ring structures with closely spaced dithiols (Fig. 7.1). For example, the crystal structure of the *p*-tolylarsenoxide and 2,3-dimercaptopropanol complex shows a five-membered ring in which both sulfur atoms are complexed to arsenic [1]. The cyclic structure is much more stable than the linear structures formed between As(III) and monothiols because of entropic considerations [2]. The effective local concentration of the second thiol and singly bound As(III) is very high, which drives the reaction (Fig. 7.1). The spacing of the thiol groups influences the affinity of their interactions with As(III). The optimal spacing of cysteine sulfur atoms for reaction with As(III) is 3–4 Å [1, 3].

The affinity of As(III) for differently spaced sulfur atoms has been measured using the organoarsenical, 4-(*N*-(*S*-glutathionylacetyl)amino)phenylarsonous acid (GSAO) [4]. GSAO is a conjugate of phenylarsonous acid and the tripeptide, glutathione (Fig. 7.2A). GSAO forms high-affinity complexes with the small dithiols, dimercaptopropanol, 6,8-thioctic acid and dithiothreitol (Table 7.1), but not with monothiols. The affinity correlates with the number of atoms in the ring structure with the As(III). For example, the apparent dissociation constant for dimercaptopropanol, which forms a five-membered ring with the As(III), increases from 130 to 420 nM for dithiothreitol, which forms a seven-membered ring with the As(III). GSAO also complexes with peptide and protein dithiols (Table 7.1). The peptides bind GSAO with a dissociation constant of ~1 μM, while the arsenical binds the active site dithiol of the protein reductant, thioredoxin [5],

Inorganic Chemical Biology: Principles, Techniques and Applications, First Edition. Edited by Gilles Gasser.
© 2014 John Wiley & Sons, Ltd. Published 2014 by John Wiley & Sons, Ltd.

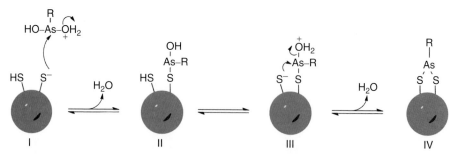

Figure 7.1 *Mechanism of cross-linking of two closely spaced cysteine sulfur atoms by As(III). (I) Owing to buffer action, a small percentage of RAs(OH)$_2$ will be protonated at the hydroxyl oxygen, which is easily displaced as water by attack of a thiolate anion on arsenic. (II and III) Action of the buffer can allow an effective proton transfer between the remaining thiol and hydroxyl, whereby an intramolecular displacement of another molecule of water will proceed through attack of the remaining thiolate anion on As. The co-location of the second thiol with the singly bound As means that the effective local concentration of these reactants is enormous, which drives the reaction. (IV) The result is the cross-linked trivalent arsenic derivative of the protein*

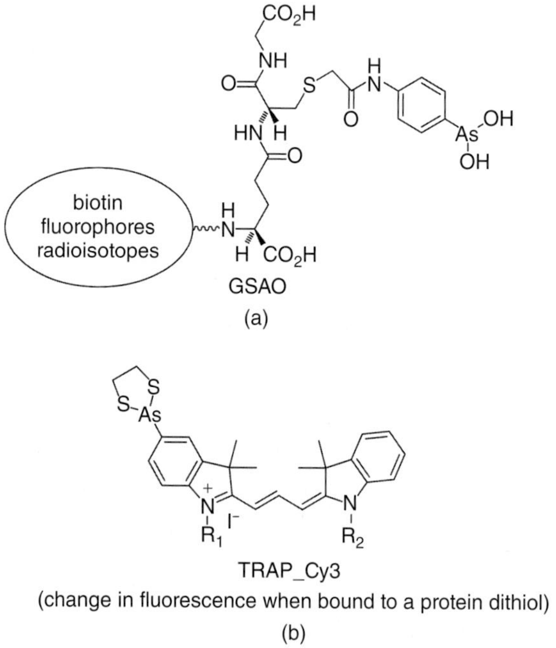

GSAO

(a)

TRAP_Cy3
(change in fluorescence when bound to a protein dithiol)

(b)

Figure 7.2 *Labeling of native proteins with organoarsenicals. A, GSAO; different reporter groups are linked through the amine of the γ-glutamyl residue of the tripeptide. B, TRAP-Cy3; the fluorescence properties of the Cy3 fluorophore change when the compound is bound to protein dithiols*

Table 7.1 Dissociation constants for binding of a trivalent organoarsenical to synthetic and protein dithiols. Table adapted from reference [4]

Dithiol	Ring size[a]	Dissociation constant (nM)
2,3-Dimercapto-1-propanol	5	130
6,8-Thioctic acid	6	200
Dithiothreitol	7	420
TrpCysGlyProCysLys[b]	15	1,420
TrpCysGlyHisCysLys[c]	15	870
Thioredoxin	15	370

[a]Number of atoms in the ring structure with the As(III) of GSAO.
[b]TrpCysGlyProCysLys corresponds to the active-site sequence of thioredoxin [7].
[c]TrpCysGlyHisCysLys corresponds to the active-site sequence of protein disulfide isomerase [8].

with a dissociation constant of 370 nM. There are 15 atoms in the ring structures of the peptides and thioredoxin with the As(III) of GSAO. The higher affinity for thioredoxin is probably because the active site cysteine thiols are closer and possibly more favourably ionised in the protein than the peptides, which is brought about by the secondary structure of thioredoxin [6].

7.2 Cysteine Dithiols in Protein Function

Most of the closely spaced cysteine dithiols in nature are functional, that is involved in the action of the protein. The functional cysteine pairs are the active site dithiols of oxidoreductases or the reduced forms of allosteric disulfide bonds.

Protein disulfide bonds, or cystine residues, are the links between the sulfur atoms of two cysteine amino acids; 15 662 disulfide bonds have been defined in 3758 human proteins to date. About half of the disulfide bonds (8183) are in a third of the proteins (1204) that are secreted or function in the endoplasmic reticulum (ER), Golgi and endosome. A similar number (7097) are in a third or so of the proteins that reside in the plasma or organelle membranes (1989). Disulfide bonds are formed in proteins as they mature in the cell [9], which occurs in the ER, Golgi complex, post-Golgi complex vesicles and mitochondrial inter-membrane space in eukaryotic cells [10] and in the periplasmic space in bacteria [11].

Disulfide bonds perform either a structural or a functional role and there are two types of functional disulfide: the catalytic and allosteric bonds [12]. The correct pairing of cysteines to cystines in maturing proteins is assisted by the catalytic disulfides of the oxidoreductases, such as protein disulfide isomerase (PDI) [8] and thioredoxin [5]. These proteins have a reactive dithiol/disulfide in their active site that undergoes cycles of oxidation/reduction with disulfides or dithiols in the protein substrate [13]. The allosteric disulfides are bonds that control mature protein function by being reductively cleaved in a regulated manner [14–16]. This is the third type of posttranslational modification [12, 16]. The allosteric disulfides are cleaved by oxidoreductases or by thiol–disulfide exchange.

About 30 mammalian, plant, bacterial or viral proteins have been found to contain allosteric disulfides to date [12]. Their cleavage has been reported to change ligand binding [17, 18], substrate hydrolysis [19], proteolysis [20–22], or oligomer formation [23, 24] of the protein. Some of the human proteins are involved in immune response (CD4, interleukin receptor subunit gamma), thrombosis (tissue factor, β2-glycoprotein I, factor XI, von Willebrand factor), blood pressure regulation (angiotensinogen), inflammation (transglutaminase 2, C-reactive protein) and cancer (MICA, lymphangiogenic growth factors, C-terminal SRC kinase).

Trivalent arsenicals have been used to study functional dithiols since the 1980s. By cross-linking the dithiol they usually inhibit protein function. Both individual proteins and cellular pathways have been probed with As(III), most often with the small organoarsenical, phenylarsonous acid. The proteins studied include thioredoxin [25], lecithin–cholesterol acyltransferase [26], tyrosyl phosphatases [27] and lipoamide in the pyruvate dehydrogenase complex [28]. The cellular pathways studied include metabolic regulation and cell stress [29], hexose transport [30] and ubiquitin-dependent protein degradation [31]. Trivalent arsenical affinity chromatography has also been employed to purify dithiol-containing proteins [32] (see also Chapter 1 dedicated, in part, to the purification of biomolecules using metal complexes).

Conversely, small synthetic dithiols are used as antidotes for As(III) poisoning. For example, 2,3-dimercaptopropanol (also known as British anti-Lewisite or BAL) was produced during World War II to combat the weaponised organoarsenical, 2-chloroethenyl dichloroarsine (also known as Lewisite) [2].

7.3   Visualization of Dithiols in Isolated Proteins with As(III)

The visualization of functional dithiols in isolated proteins has been achieved using a biotin-tagged conjugate of the tripeptide trivalent arsenical, GSAO (Fig. 7.2A) [4]. Incorporation of the biotin-tagged GSAO into proteins is assessed by blotting the labelled proteins with streptavidin-peroxidase. For example, binding of GSAO–biotin to the allosteric dithiol of Tissue Factor (TF) is shown in Fig. 7.3 [33].

7.4   Visualization of Dithiols on the Mammalian Cell Surface with As(III)

GSAO–biotin has also been used to visualize mammalian cell surface proteins containing closely spaced cysteine dithiols, as this conjugate is effectively membrane-impermeable. Up to 12 distinct proteins were labelled with GSAO–biotin on the surface of vascular endothelial and human fibrosarcoma cells and the pattern of labelled proteins differed between the two cell types. It is likely that some, perhaps many, cell surface proteins containing closely spaced thiols were not identified because of their low abundance.

The oxidoreductase, PDI, was one of the proteins detected on both cell surfaces. PDI is involved in redox control of protein dithiols/disulfides on the surface of fibroblasts [35], lymphocytes [36, 37] and platelets [38]. For instance, the thiol content of 11 proteins on the fibrosarcoma cell surface was increased with overexpression of PDI, while the thiol content of three of the 11 proteins was decreased with underexpression of PDI [35].

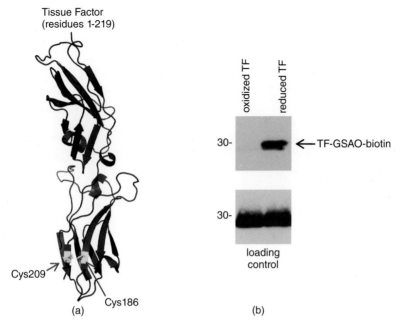

Figure 7.3 *Labeling of the reduced allosteric disulfide bond in Tissue Factor with an organoarsenical. A, Ribbon structure of the extracellular part of human TF. The unpaired Cys186 and Cys209 residues are shown as sticks (yellow). The reduced protein is represented by deleting the disulfide bond in the oxidised structure (PDB ID 2HFT [34]). B, Labelling of the unpaired Cys186 and Cys209 thiols with GSAO–biotin. The Cys186–Cys209 disulfide bond of TF was reduced with dithiothreitol and the protein incubated GSAO–biotin. A sample was resolved on SDS-PAGE and incorporation of GSAO–biotin measured by blotting with strepta-vidin-peroxidase. The bottom panel shows equivalent loading of TF by Western blot and the position of the 30 kDa size marker is indicated. Adapted with permission from [33] © 2011 Biochemical Society*

7.5   Visualization of Dithiols in Intracellular Proteins with As(III)

A membrane-permeable conjugate of As(III) with the cyanine dye, Cy3, has been made to image proximal dithiols in living cells [39] (Fig. 7.2B). The polarization of the Cy3 dye changes when the compound is bound to cysteine dithiols. It was used to probe adaptive microbial responses to increases in oxygen and light levels, where dithiol oxidation is one component of the response. The photosynthetic microbe, *Synechococcus*, has been studied, but this compound should also be useful for monitoring protein dithiol oxidation in mammalian and in other cells.

7.6   Visualization of Tetracysteine-tagged Recombinant Proteins in Cells with As(III)

Trivalent bisorganoarsenicals have been used to image proteins engineered to contain two pairs of adjacent cysteine residues [40, 41]. The biarsenical reagents, known as

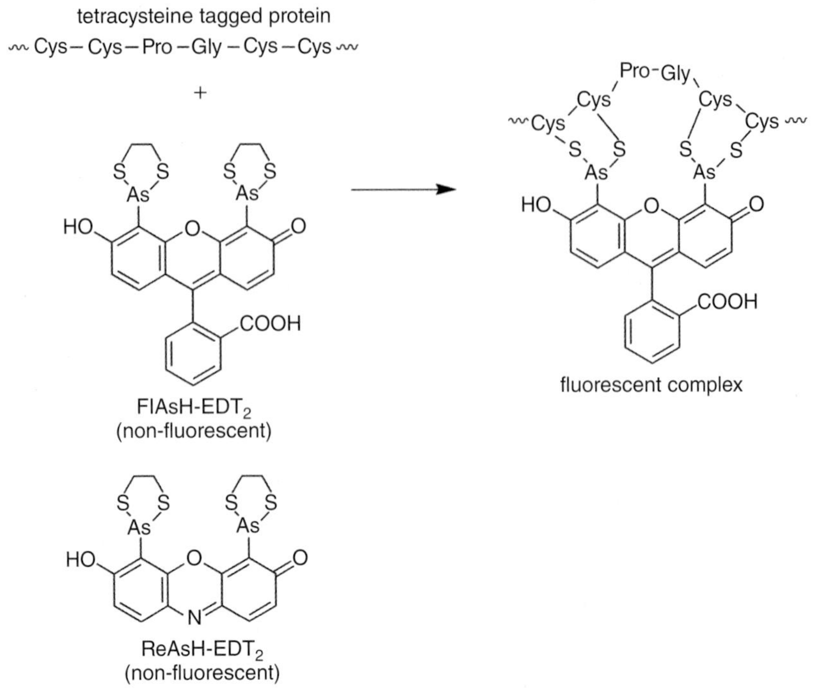

Figure 7.4 *Labelling of engineered proteins with a biarsenical. The biarsenicals, FlAsH-EDT$_2$ and ReAsH-EDT$_2$, become fluorescent when they bind to recombinant proteins containing a tetracysteine tag*

FlAsH-EDT$_2$ and ReAsH-EDT$_2$, become fluorescent when they bind to recombinant proteins containing the tetracysteine tag, Cys-Cys-Pro-Gly-Cys-Cys (Fig. 7.4). This tag forms a short α-helix that positions the cysteine thiols so that they react with the biarsenical and has been placed at protein termini, within flexible loops, or incorporated into secondary structure elements [41]. FlAsH-EDT, a derivative of fluorescein, fluoresces in the green and ReAsH-EDT, a derivative of resorufin, fluoresces in the red. Photostable FRET-competent FlAsH probes have also been developed [42].

This technology has been used to track the subcellular localization of expressed proteins and for visualizing live cells [40, 41, 43]. It has also been used for protein purification, protein–protein, protein stability and protein aggregation studies [41, 44].

7.7 Visualization of Cell Death in the Mouse with Optically Labelled As(III)

7.7.1 Cell Death in Health and Disease

Cell death plays an integral role in human physiology. In the gut, for instance, intestinal epithelial cells have a turnover of about five days, whilst immune cells such as neutrophils have a lifespan measured only in hours [45, 46]. Cell turnover is tightly regulated and usually occurs without collateral damage to the body, however imbalance of this process can occur in certain pathologies. Ischemic injuries such as myocardial infarction and stroke result in excessive cell death due to oxygen deprivation. Cell death is also a feature of neurodegenerative diseases such as Alzheimer's.

In situations where cell death is inhibited or outweighed by cell proliferation such as cancer, death may be induced by therapy. The aim of most chemotherapeutic and radiotherapeutic agents is to kill tumour cells and thus eliminate the tumour mass [47, 48]. To determine if the therapy is effective, tumour size is usually assessed by computed tomography after 2–3 months of treatment. This result is used in decisions about further therapy. The management of this disease would be improved by technologies that rapidly assess tumour cell proliferation and death in response to treatment [49]. Such a technology would also enhance the pace of new pharmaceutical development [49, 50]. By measuring the efficacy of new cytotoxic drugs and possibly the toxicity of these drugs in other tissues, 'go/no go' decisions will be easier to make.

There has been a focus on the development of imaging agents, including an organoarsenical, that can visualize cell death in different tissues. Before these are discussed, the most common types of cell death will be outlined. Apoptosis is a programmed form of cell death essential to the elimination of superfluous or damaged cells in a range of physiological scenarios [51]. Unlike necrosis, apoptosis involves the controlled packaging of cellular constituents into membrane-enclosed vesicles called apoptotic bodies and removal by phagocytes without eliciting a full immune response. One of the earliest events characterizing apoptosis is a reduction in cell size due to an efflux of water, followed by membrane blebbing. As the apoptotic process proceeds, phospholipid asymmetry is lost in the plasma membrane, the mitochondrial membrane potential is lost, nuclear chromatin is condensed and the DNA fragmented. Various apoptotic pathways have been characterized, including the mitochondrial mediated 'intrinsic' pathway and a death receptor mediated 'extrinsic pathway,' however, common to each of these is the end-point activation of a group of cysteine proteases called caspases. Caspases, so named for their cleavage of substrates after an aspartate residue, can be classified as either initiator or executioner caspases. Whilst the role of the former (caspases 8, 9 and 10) is to activate other caspases by cleavage, the latter (caspases 3, 6 and 7) are responsible for the degradation of cellular constituents.

Necrosis represents a cell's response to gross physical or chemical injury, including mechanical stress, osmotic shock, freeze thawing and heat. This form of cell death is characterized by immediate and irreparable damage to the cytoplasmic membrane, resulting in an influx of water and cellular oedema. Consecutive rupture of external and internal membranes leads to the release of harmful lysosomal and cytoplasmic constituents, which in turn elicit an inflammatory response [52]. Unlike apoptosis, which is likened to programmed cell suicide, necrosis is akin to accidental cell death or 'murder.'

Necrosis and apoptosis have long been viewed as opposing arms of a classic dichotomy; however it now appears that somewhat of a continuum exists between the two, with the emergence of secondary necrosis following apoptosis [53]. Secondary necrosis describes the terminal stages of apoptotic cell death when phagocytic clearance is impeded. Typically, apoptotic cells are scavenged early in the apoptotic programme by surrounding macrophages and neutrophils. In the absence of macrophages, neighbouring epithelial, endothelial or dendritic cells may also perform this role. Sequestration of apoptotic cells can occur prior to DNA fragmentation and perhaps even before significant morphological changes are observed. This is certainly the case in highly regulated physiological scenarios such as embryogenesis, however in a number of pathologies scavenger cells are often insufficient in number and/or functionally incapacitated. This may be the case in tumours and damaged tissues where the relative paucity of macrophages to dying cells results in prolonged residency of apoptotic cells. In this scenario, the usual morphological and biochemical changes of apoptosis are observed until the completion of the apoptotic programme. In the absence of phagocytic clearance, however, the cell transitions into a secondary necrosis whereby the plasma membrane integrity is lost and the cell eventually ruptures [54].

The anucleate blood cell, the platelet, also undergoes necrosis. The haemostatic system is tightly regulated and designed for the rapid formation of a stable adhesive clot, limited to the area of vascular damage. Platelets are key players in this intricately complex system and have a multitude of physiological roles, which include activation, granule secretion and procoagulant functions. Platelets exhibit heterogeneity in response to agonist stimulation in respect to morphology and surface characteristics and this is influenced by environmental and intrinsic factors [55]. Not all platelets are activated to the same extent. Loosely bound discoid platelets with low activation states are observed on the outer layer of the growing thrombus, whereas platelets in the thrombus core show higher activation states with sustained calcium spikes and tight adhesive contacts [56, 57]. The procoagulant platelets in the thrombus core have features indicative of necrotic cell death [58]. An excess of necrotic platelets may lead to the development of occlusive thrombi seen in coronary artery disease and stroke.

7.7.2   Cell Death Imaging Agents

Of the cell death imaging probes, those that detect phosphatidylserine elaborated on the surface of apoptotic cells have received the most attention [59–67]. In healthy cells, the ATP-dependent enzymes aminophospholipid translocase and floppase work in concert to maintain lipid bilayer asymmetry, whereby cationic phospholipids such as phosphatidylcholine and sphingomyelin are pumped to the outer leaflet and phosphatidylserine is restricted to the inner leaflet [68]. In the early stages of apoptosis, translocase and floppase are inactivated in a calcium dependent manner, whilst scramblase is simultaneously activated, leading to re-distribution of phosphatidylethanolamine and phosphatidylserine across the membrane. The main purpose of phosphatidylserine externalization in the apoptotic programme appears to be as a recognition signal for surrounding macrophages, facilitating controlled disposal of apoptotic cells without inciting an inflammatory response [69]. These phosphatidylserine ligands, though, can also bind this phospholipid elaborated on viable cells, such as activated platelets, macrophages, endothelial cells and aging erythrocytes. Cells under stress, such as hypoxia, will also transiently expose phosphatidylserine.

Other cell death imaging probes target an activated caspase and a cytoplasmic antigen. Caspase-3 is a central effector caspase in the demolition and clearance of apoptotic cells [70] and a radio-labelled small molecule inhibitor of activated caspase-3/7, an isatin-5 sulfonamide known as ICMT-11, has been used to image treatment-related lymphoma cell death in mice [71]. High hepatic uptake of ICMT-11 precludes imaging of the abdominal region, however, and there is possible cross-reactivity with cathepsins. The RNA metabolism protein, La antigen, translocates from the nucleus to the cytoplasm during apoptosis [72] and is recognized by antibodies when the integrity of the plasma membrane is compromised during late apoptosis [73, 74]. A radio-labelled anti-La antibody has been developed that shows promise for imaging of tumour cell death mediated by DNA alkylating agents [73, 74].

7.7.3   Visualization of Cell Death in Mouse Tumours, Brain and Thrombi with Optically Labelled As(III)

Optically labelled versions of the tripeptide As(III), GSAO, have been employed to image mammalian cell death in culture and in the living mouse [75–77]. Tagged GSAO accumulates in the cytosol of apoptotic and necrotic cells co-incident with loss of plasma membrane integrity, and is retained in the cytoplasm by reacting predominantly with heat shock protein-90 (Hsp90) [75]. Hsp90 is an abundant cytoplasmic chaperone that plays a fundamental role in cellular homeostasis and tumorigenesis. The Hsp90 chaperone machinery protects a number of mutated and overexpressed oncoproteins from misfolding and degradation, and is considered an important factor in oncogene addiction and tumour cell survival [78, 79]. The As(III) atom of GSAO cross-links the Cys597, Cys598 dithiol of Hsp90, forming a stable dithioarsinite (Fig. 7.5A).

Fluorescently tagged GSAO specifically labels apoptotic and necrotic cells in culture and a biotin tagged GSAO labels cells of the same morphology in murine tumours [75–77]. A conjugate of GSAO with the near-infrared fluorophore, Alex Fluor 750 (GSAO-AF750), has been used to non-invasively image cyclophosphamide-induced tumour cell death in orthotopic human mammary tumours in mice [76]. The Alex Fluor 750 allows for maximum tissue penetration of the fluorescent signal and minimises the complications of tissue auto-fluorescence [81]. Importantly, the GSAO-AF750 does not accumulate in healthy organs or tissues in the mouse and the unbound compound is cleared from the circulation via the kidneys in about 3 h. The favourable biodistribution of GSAO-AF750, the nature of its tumour cell target and its capacity to non-invasively detect tumour cell death should see application of this compound in studies of the efficacy of existing and new chemotherapeutics.

GSAO-AF750 has also been used to label apoptotic and necrotic cell death in traumatic brain injury, which is a major public health issue [77]. Anatomical imaging is most often used to assess traumatic brain injuries and there is a need for imaging modalities that provide cellular information. GSAO-AF750 is a very effective imager of cell death in brain lesions in mice. An optimal signal-to-background ratio was observed as early as 3 h after injection of GSAO-AF750 and the signal intensity positively correlated with both lesion size and probe concentration (Fig. 7.5B).

The abundance of Hsp90 within the cytosol (1–2% of total cellular protein) bodes well for the suitability of GSAO conjugates for imaging cell death. The effectiveness of an imaging agent is determined in part by how much of the agent accumulates in a given volume. A high concentration of imaging agent at the target results in better limits of detection and resolution. The abundance of Hsp90 in the cytosol allows for high levels of GSAO conjugates in apoptotic/necrotic cells and therefore superior detection and resolution of cell death.

GSAO–fluorophore also labels necrotic platelets at sites of vascular injury in the circulation. Necrotic platelets can be generated *in vitro* by activation with both thrombin and collagen, and only these platelets label with GSAO–fluorophore. Whereas the main protein

human Hsp90α
residues 293–732

non-invasive imaging of brain
cell death with GSAO-AF750

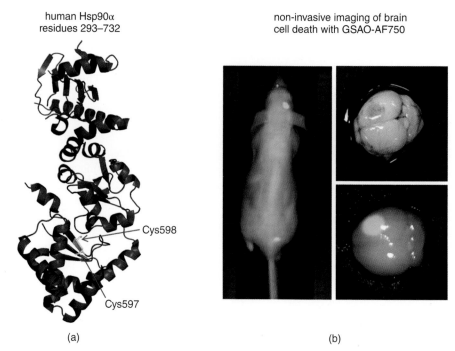

(a)   (b)

Figure 7.5 *Non-invasive imaging of brain cell death with an optically labelled organoarsenical. A, Ribbon structure of residues 293–732 of human Hsp90α. The unpaired Cys597 and Cys598 residues are shown as sticks (yellow). The structure is that of PDB ID 3Q6M [80]. B, In vivo and ex vivo imaging of brain lesion cell death in mice with GSAO-AF750. A 60 s brain cryolesion was induced in the front part of the right parietal lobe, followed by tail vein injection of 1 mg kg⁻¹ GSAO-AF750. Whole body fluorescence imaging was performed 24 h after probe injection. The brain of the mouse was excised for ex vivo imaging. A bright field image showing red discolouration of the lesion site is shown at the top, while the fluorescence image of the same brain is shown at the bottom. Adapted by permission from Macmillan Publishers Ltd [77] © 2013*

dithiol target for GSAO in tumour cells is Hsp90 (Fig. 7.5A), a main target in necrotic platelets appears to be hexokinase I. This enzyme and its ATP co-substrate catalyses the conversion of glucose into glucose-6-phosphate, which is the first step in glycolysis. It appears that the As(III) of GSAO cross-links Cys237 and Cys256 of hexokinase I. From the crystal structure of the enzyme [82], the sulfur atoms of these cysteine thiols are 3.6 Å apart, which is close enough to react with As(III).

There is a significant overlap in cellular markers of activation, apoptosis and necrosis in the anucleate platelet. GSAO–fluorophore is a useful tool for differentiating these populations when used in combination with other activation markers. GSAO–fluorophore labels platelet aggregates in collagen-dependent models of thrombosis in murine arterioles, where platelets are activated by both collagen and thrombin (Fig. 7.6). There is very little labelling of platelets in collagen-independent models of thrombosis. This organoarsenical is being used to probe the biology of necrotic platelets and their contribution to thrombus formation in the living body.

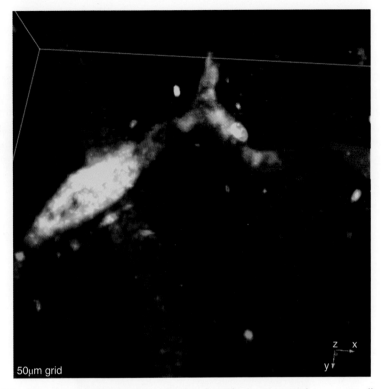

50µm grid

z x

y

Figure 7.6   *Visualization of necrotic platelet thrombi in vivo with an optically labelled organoarsenical. 3D reconstruction of intravital fluorescence microscopy images of a thrombus triggered by FeCl₃ damage to a mouse cremaster muscle arteriole. Shown is co-localisation of GSAO-Oregon Green (green) with platelets (red, DyLight488 conjugated anti-CD42 antibody) in the thrombus. Merge of the GSAO-Oregon Green and platelet signal appears yellow*

7.8   Visualization of Cell Death in Mouse Tumours with Radio-labelled As(III)

The preceding section has highlighted the utility of an optically labelled organoarsenical for the detection of dying and dead cells in culture as well as in mice. The ability to image dying and dead cells *in vivo* also has the potential to be an extremely useful research and clinical tool in man, for example in the assessment of treatment response to cytotoxic chemotherapy and radiotherapy in oncology. The optical reporting groups are not suitable for human use at this stage. Although molecular imaging in humans may be achieved by radioisotope methods and magnetic resonance imaging (MRI), in clinical practice MRI remains largely an anatomic imaging modality and is much less sensitive than radioisotope methods for detection of molecular signals [83]. Therefore, radioisotope techniques have been the focus for potential translation of organoarsenical imaging into man.

GSAO has been radio-labelled using two bifunctional metal chelators, diethylen-etriaminepentaacetic acid (DTPA) and 4,7,10-tetra-azacyclododecanetetraacetic acid (DOTA), which both conjugate to GSAO via the amine group of the γ-glutamyl residue

(Fig. 7.2A) and are able to chelate trivalent metal ions (such as indium and gallium). Initial *in vitro* studies demonstrated significant uptake of indium-111 (^{111}In) DTPA GSAO into apoptotic cancer cells in culture and Lewis lung carcinoma tumour xenografts in C57BL/6 mice [75]. Of the normal organs, renal uptake was highest as it is the route of excretion of ^{111}In DTPA GSAO. Using the same murine model, ^{111}In DTPA GSAO was compared with technetium-99m (^{99m}Tc)-Annexin V (another agent currently under investigation for imaging of cell death). There was similar tumour and renal uptake of both agents, however there was higher liver uptake of ^{99m}Tc Annexin V, consistent with its route of clearance [75]. Biodistribution studies with ^{111}In DTPA GSAO found that there was relatively high and persistent uptake within the tumour that peaked at 2.5 h post-injection

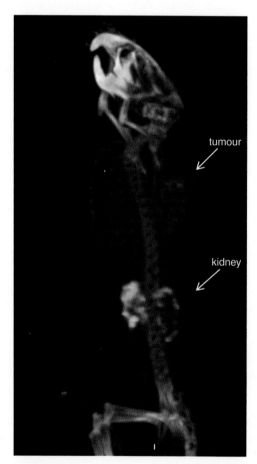

Figure 7.7 *Non-invasive imaging of tumour cell death with a radio-labelled organoarsenical. Micro SPECT/CT acquired 2 h after intravenous administration of ^{67}Ga DOTA GSAO in an immuno-compromised mouse bearing a subcutaneous human prostate tumour xenograft (arrowed). Note the punctuate uptake within the tumour and low uptake within normal organs except the kidneys*

with a slow decline thereafter. Excretion of the tracer was relatively rapid with only 24% of the compound retained at 2.6 h. Gallium-67 (^{67}Ga) DOTA GSAO has also been studied and found to have more favourable biodistribution characteristics than ^{111}In DTPA GSAO. ^{67}Ga DOTA GSAO demonstrated a slightly higher peak tumour uptake with less short-term retention in the kidneys, liver and spleen (Fig. 7.7).

Preliminary studies have also been undertaken to assess if there is an increase in ^{67}Ga DOTA GSAO uptake within tumours in untreated versus treated mice bearing subcutaneous human prostate cancer xenografts. These studies have failed to demonstrate an absolute difference in ^{67}Ga DOTA GSAO uptake between untreated and treated cohorts. Although the reasons for this remains to be elucidated, one potential explanation is that the absolute uptake may be less useful than the incremental change following therapy, and as such frequent sequential imaging prior to and following therapy may be important. Both ^{67}Ga and ^{111}In have long physical half-lives that make frequent sequential imaging difficult or impossible. Gallium-68 (^{68}Ga) is a positron-emitting radioisotope, which can be substituted for ^{67}Ga. Its short half-life of 68 min will allow frequent sequential imaging, and positron emission tomography also has a higher sensitivity and resolution and is quantitative. Work with ^{68}Ga DOTA GSAO is on going and future studies will focus on assessing the incremental changes in cell death in response to therapy with ^{68}Ga DOTA GSAO.

7.9 Summary and Perspectives

Organometalloids, in particular conjugates of As(III) with various reporter groups, have proven very useful for visualizing proteins in isolation, in cultured cells and in tissues in the living mouse body that contain appropriately spaced cysteine dithiols. This is because As(III) reacts selectively and tightly with closely spaced sulfur atoms in biological milieu. As(III) conjugates have been used to image recombinant proteins tagged with cysteine dithiols and functional dithiols that are the reduced forms of catalytic or allosteric disulfide bonds. The identification and study of allosteric disulfides/dithiols has really only just begun. The number involved in mammalian, plant, bacterial and viral protein function is unknown but it is likely to be many. As(III) has and will prove useful for the identification, study and visualization of these functional motifs in proteins.

References

1. Adams E, Jeter D, Cordes AW, Kolis JW. Chemistry of organometalloid complexes with potential antidotes - Structure of an organoarsenic(Iii) dithiolate ring. Inorg Chem. 1990;29(8):1500–1503.
2. Stocken LA, Thompson RH. British anti-Lewisite: 2. Dithiol compounds as antidotes for arsenic. Biochem J. 1946;40(4):535–548.
3. Bhattacharjee H, Rosen BP. Spatial proximity of Cys113, Cys172, and Cys422 in the metalloactivation domain of the ArsA ATPase. J Biol Chem. 1996;271(40):24465–24470

4. Donoghue N, Yam PT, Jiang XM, Hogg PJ. Presence of closely spaced protein thiols on the surface of mammalian cells. Protein Sci. 2000;9(12):2436–2445.
5. Holmgren A, Lu J. Thioredoxin and thioredoxin reductase: current research with special reference to human disease. Biochem Biophys Res Commun. 2010;396(1): 120–124.
6. Weichsel A, Gasdaska JR, Powis G, Montfort WR. Crystal structures of reduced, oxidized, and mutated human thioredoxins: evidence for a regulatory homodimer. Structure. 1996;4(6):735–751.
7. Holmgren A. Thioredoxin and glutaredoxin systems. J Biol Chem. 1989;264(24): 13963–13966.
8. Gilbert HF. Protein disulfide isomerase and assisted protein folding. J Biol Chem. 1997;272(47):29399–29402.
9. Depuydt M, Messens J, Collet JF. How proteins form disulfide bonds. Antioxid Redox Signaling. 2011;15(1):49–66.
10. Braakman I, Bulleid NJ. Protein folding and modification in the mammalian endoplasmic reticulum. Annu Rev Biochem. 2011;80:71–99.
11. Nakamoto H, Bardwell JC. Catalysis of disulfide bond formation and isomerization in the Escherichia coli periplasm. Biochim Biophys Acta. 2004;1694(1–3):111–119.
12. Cook KM, Hogg PJ. Posttranslational control of protein function by disulfide bond cleavage. Antioxid Redox Signal. 2013; 18:1987–2015.
13. Berndt C, Lillig CH, Holmgren A. Thioredoxins and glutaredoxins as facilitators of protein folding. Biochim Biophys Acta. 2008;1783(4):641–650.
14. Hogg PJ. Disulfide bonds as switches for protein function. Trends Biochem Sci. 2003; 28(4):210–214.
15. Schmidt B, Ho L, Hogg PJ. Allosteric disulfide bonds. Biochemistry. 2006;45(24): 7429–7433
16. Hogg PJ. Targeting allosteric disulphides in cancer. Nat Rev Cancer. 2013;13:425–431.
17. Metcalfe C, Cresswell P, Barclay AN. Interleukin-2 signalling is modulated by a labile disulfide bond in the CD132 chain of its receptor. Open Biol. 2012;2(1):110036.
18. Wang MY, Ji SR, Bai CJ, El Kebir D, Li HY, Shi JM, et al. A redox switch in C-reactive protein modulates activation of endothelial cells. FASEB J. 2011;25(9):3186–3196.
19. Jin X, Stamnaes J, Klock C, DiRaimondo TR, Sollid LM, Khosla C. Activation of extracellular transglutaminase 2 by thioredoxin. J Biol Chem. 2011;286(43):37866–37873.
20. Lay AJ, Jiang XM, Kisker O, Flynn E, Underwood A, Condron R, et al. Phosphoglycerate kinase acts in tumour angiogenesis as a disulphide reductase. Nature. 2000;408(6814):869–873.
21. Kaiser BK, Yim D, Chow IT, Gonzalez S, Dai Z, Mann HH, et al. Disulphide-isomerase-enabled shedding of tumour-associated NKG2D ligands. Nature. 2007;447(7143): 482–486.
22. Zhou A, Carrell RW, Murphy MP, Wei Z, Yan Y, Stanley PL, et al. A redox switch in angiotensinogen modulates angiotensin release. Nature. 2010;468(7320): 108–111.
23. Maekawa A, Schmidt B, Fazekas de St Groth B, Sanejouand YH, Hogg PJ. Evidence for a domain-swapped CD4 dimer as the coreceptor for binding to class II MHC. J Immunol. 2006;176(11):6873–6878.

24. Ganderton T, Wong JW, Schroeder C, Hogg PJ. Lateral self-association of VWF involves the Cys2431-Cys2453 disulfide/dithiol in the C2 domain. Blood. 2011; 118(19):5312–5318.
25. Brown SB, Turner RJ, Roche RS, Stevenson KJ. Conformational analysis of thioredoxin using organoarsenical reagents as probes. A time-resolved fluorescence anisotropy and size exclusion chromatography study. Biochem Cell Biol. 1989;67(1): 25–33.
26. Jauhiainen M, Stevenson KJ, Dolphin PJ. Human plasma lecithin-cholesterol acyltransferase. The vicinal nature of cysteine 31 and cysteine 184 in the catalytic site. J Biol Chem. 1988;263(14):6525–6533.
27. Zhang ZY, Davis JP, Van Etten RL. Covalent modification and active site-directed inactivation of a low molecular weight phosphotyrosyl protein phosphatase. Biochemistry. 1992;31(6):1701–1711.
28. Stevenson KJ, Hale G, Perham RN. Inhibition of pyruvate dehydrogenase multienzyme complex from Escherichia coli with mono- and bifunctional arsenoxides. Biochemistry. 1978;17(11):2189–2192.
29. Gitler C, Mogyoros M, Kalef E. Labeling of protein vicinal dithiols: role of protein-S2 to protein-(SH)2 conversion in metabolic regulation and oxidative stress. Methods Enzymol. 1994;233:403–415.
30. Frost SC, Lane MD. Evidence for the involvement of vicinal sulfhydryl groups in insulin-activated hexose transport by 3T3-L1 adipocytes. J Biol Chem. 1985;260(5): 2646–2652.
31. Klemperer NS, Pickart CM. Arsenite inhibits two steps in the ubiquitin-dependent proteolytic pathway. J Biol Chem. 1989;264(32):19245–19252.
32. Kalef E, Gitler C. Purification of vicinal dithiol-containing proteins by arsenical-based affinity chromatography. Methods Enzymol. 1994;233:395–403.
33. Liang HP, Brophy TM, Hogg PJ. Redox properties of the tissue factor Cys186-Cys209 disulfide bond. Biochem J. 2011;437(3):455–460.
34. Muller YA, Ultsch MH, de Vos AM. The crystal structure of the extracellular domain of human tissue factor refined to 1.7 Å resolution. J Mol Biol. 1996;256(1):144–159.
35. Jiang XM, Fitzgerald M, Grant CM, Hogg PJ. Redox control of exofacial protein thiols/disulfides by protein disulfide isomerase. J Biol Chem. 1999;274(4):2416–2423.
36. Lawrence DA, Song R, Weber P. Surface thiols of human lymphocytes and their changes after *in vitro* and *in vivo* activation. J Leukoc Biol. 1996;60(5):611–618.
37. Tager M, Kroning H, Thiel U, Ansorge S. Membrane-bound proteindisulfide isomerase (PDI) is involved in regulation of surface expression of thiols and drug sensitivity of B-CLL cells. Exp Hematol. 1997;25(7):601–607.
38. Burgess JK, Hotchkiss KA, Suter C, Dudman NP, Szollosi J, Chesterman CN, et al. Physical proximity and functional association of glycoprotein 1balpha and protein-disulfide isomerase on the platelet plasma membrane. J Biol Chem. 2000; 275(13):9758–97566.
39. Fu N, Su D, Cort JR, Chen B, Xiong Y, Qian WJ, et al. Synthesis and application of an environmentally insensitive Cy3-based arsenical fluorescent probe to identify adaptive microbial responses involving proximal dithiol oxidation. J Am Chem Soc. 2013;135(9):3567–3575.

40. Griffin BA, Adams SR, Tsien RY. Specific covalent labeling of recombinant protein molecules inside live cells. Science. 1998;281(5374):269–272.

41. Pomorski A, Krezel A. Exploration of biarsenical chemistry--challenges in protein research. Chembiochem. 2011;12(8):1152–1167.

42. Spagnuolo CC, Vermeij RJ, Jares-Erijman EA. Improved photostable FRET-competent biarsenical-tetracysteine probes based on fluorinated fluoresceins. J Am Chem Soc. 2006;128(37):12040–12041.

43. Estevez JM, Somerville C. FlAsH-based live-cell fluorescent imaging of synthetic peptides expressed in Arabidopsis and tobacco. Biotechniques. 2006;41(5):569–570, 72-4.

44. Thorn KS, Naber N, Matuska M, Vale RD, Cooke R. A novel method of affinity-purifying proteins using a bis-arsenical fluorescein. Protein Sci. 2000;9(2):213–217.

45. Marshman E, Booth C, Potten CS. The intestinal epithelial stem cell. Bioessays. 2002;24(1):91–98.

46. Summers C, Rankin SM, Condliffe AM, Singh N, Peters AM, Chilvers ER. Neutrophil kinetics in health and disease. Trends Immunol. 2010;31(8):318–324.

47. Rupnow BA, Knox SJ. The role of radiation-induced apoptosis as a determinant of tumor responses to radiation therapy. Apoptosis. 1999;4(2):115–143.

48. Thompson CB. Apoptosis in the pathogenesis and treatment of disease. Science. 1995;267(5203):1456–1462.

49. Weber WA, Czernin J, Phelps ME, Herschman HR. Technology insight: novel imaging of molecular targets is an emerging area crucial to the development of targeted drugs. Nat Clin Pract Oncol. 2008;5(1):44–54.

50. Weissleder R, Pittet MJ. Imaging in the era of molecular oncology. Nature. 2008;452(7187):580–589.

51. Taylor RC, Cullen SP, Martin SJ. Apoptosis: controlled demolition at the cellular level. Nat Rev Mol Cell Biol. 2008;9(3):231–241.

52. Al-Rubeai M, Fussenegger M. Apoptosis. Boston: Kluwer Academic Publishers; 2004.

53. Silva MT, do Vale A, dos Santos NM. Secondary necrosis in multicellular animals: an outcome of apoptosis with pathogenic implications. Apoptosis. 2008;13(4):463–482.

54. Elliott MR, Ravichandran KS. Clearance of apoptotic cells: implications in health and disease. J Cell Biol. 2010;189(7):1059–1070.

55. Munnix IC, Kuijpers MJ, Auger J, Thomassen CM, Panizzi P, van Zandvoort MA, et al. Segregation of platelet aggregatory and procoagulant microdomains in thrombus formation: regulation by transient integrin activation. Arterioscler Thromb Vasc Biol. 2007;27(11):2484–2490.

56. Munnix IC, Cosemans JM, Auger JM, Heemskerk JW. Platelet response heterogeneity in thrombus formation. Thromb Haemost. 2009;102(6):1149–1156.

57. Nesbitt WS, Westein E, Tovar-Lopez FJ, Tolouei E, Mitchell A, Fu J, et al. A shear gradient-dependent platelet aggregation mechanism drives thrombus formation. Nat Med. 2009;15(6):665–673.

58. Jackson SP, Schoenwaelder SM. Procoagulant platelets: are they necrotic? Blood. 2010;116(12):2011–2018.

59. Schellenberger EA, Bogdanov A, Jr., Petrovsky A, Ntziachristos V, Weissleder R, Josephson L. Optical imaging of apoptosis as a biomarker of tumor response to chemotherapy. Neoplasia. 2003;5(3):187–192.

60. Dechsupa S, Kothan S, Vergote J, Leger G, Martineau A, Berangeo S, *et al.* Quercetin, Siamois 1 and Siamois 2 induce apoptosis in human breast cancer MDA-mB-435 cells xenograft *in vivo*. Cancer Biol Ther. 2007;6(1):56–61.

61. Beekman CA, Buckle T, van Leeuwen AC, Valdes Olmos RA, Verheij M, Rottenberg S, *et al.* Questioning the value of (99m)Tc-HYNIC-annexin V based response monitoring after docetaxel treatment in a mouse model for hereditary breast cancer. Appl Radiat Isot. 2011;69(4):656–662.

62. Lederle W, Arns S, Rix A, Gremse F, Doleschel D, Schmaljohann J, *et al.* Failure of annexin-based apoptosis imaging in the assessment of antiangiogenic therapy effects. EJNMMI Res. 2011;1(1):26.

63. Wang F, Fang W, Zhao M, Wang Z, Ji S, Li Y, *et al.* Imaging paclitaxel (chemotherapy)-induced tumor apoptosis with 99mTc C2A, a domain of synaptotagmin I: a preliminary study. Nucl Med Biol. 2008;35(3):359–364.

64. Zhao M, Beauregard DA, Loizou L, Davletov B, Brindle KM. Non-invasive detection of apoptosis using magnetic resonance imaging and a targeted contrast agent. Nat Med. 2001;7(11):1241–1244.

65. Smith BA, Akers WJ, Leevy WM, Lampkins AJ, Xiao S, Wolter W, *et al.* Optical imaging of mammary and prostate tumors in living animals using a synthetic near infrared zinc(II)-dipicolylamine probe for anionic cell surfaces. J Am Chem Soc. 2010;132(1): 67–69.

66. Smith BA, Xiao S, Wolter W, Wheeler J, Suckow MA, Smith BD. *In vivo* targeting of cell death using a synthetic fluorescent molecular probe. Apoptosis. 2011;16(7): 722–731.

67. Xiong C, Brewer K, Song S, Zhang R, Lu W, Wen X, *et al.* Peptide-based imaging agents targeting phosphatidylserine for the detection of apoptosis. J Med Chem. 2011;54(6):1825–1835.

68. Fadeel B. Plasma membrane alterations during apoptosis: role in corpse clearance. Antioxid Redox Signaling. 2004;6(2):269–275.

69. Wang RF. Progress in imaging agents of cell apoptosis. Anti-Cancer Agents Med Chem. 2009;9(9):996–1002.

70. Porter AG, Janicke RU. Emerging roles of caspase-3 in apoptosis. Cell Death Differ. 1999;6(2):99–104.

71. Nguyen QD, Smith G, Glaser M, Perumal M, Arstad E, Aboagye EO. Positron emission tomography imaging of drug-induced tumor apoptosis with a caspase-3/7 specific [18F]-labeled isatin sulfonamide. Proc Natl Acad Sci USA. 2009;106(38): 16375–16380.

72. Ayukawa K, Taniguchi S, Masumoto J, Hashimoto S, Sarvotham H, Hara A, *et al.* La autoantigen is cleaved in the COOH terminus and loses the nuclear localization signal during apoptosis. J Biol Chem. 2000;275(44):34465–34470.

73. Al-Ejeh F, Darby JM, Pensa K, Diener KR, Hayball JD, Brown MP. *In vivo* targeting of dead tumor cells in a murine tumor model using a monoclonal antibody specific for the La autoantigen. Clin Cancer Res. 2007;13(18 Pt 2):5519s–5527s.

74. Al-Ejeh F, Darby JM, Tsopelas C, Smyth D, Manavis J, Brown MP. APOMAB, a La-specific monoclonal antibody, detects the apoptotic tumor response to life-prolonging and DNA-damaging chemotherapy. PLoS One. 2009;4(2):e4558.

75. Park D, Don AS, Massamiri T, Karwa A, Warner B, Macdonald J, *et al*. Noninvasive imaging of cell death using an hsp90 ligand. J Am Chem Soc. 2011;133(9): 2832–2835.

76. Park D, Xie B-W, Van Beek ER, Blankevoort V, Que I, Löwik CWGM, *et al*. Optical imaging of treatment-related tumour cell death using a heat shock protein-90 alkylator. Mol Pharm. 2013;10:3882–3891.

77. Xie BW, Park D, Van Beek ER, Blankevoort V, Orabi Y, Que I, *et al*. Optical imaging of cell death in traumatic brain injury using a heat shock protein-90 alkylator. Cell Death Dis. 2013;4:e473.

78. Trepel J, Mollapour M, Giaccone G, Neckers L. Targeting the dynamic HSP90 complex in cancer. Nat Rev Cancer. 2010;10(8):537–549.

79. Wandinger SK, Richter K, Buchner J. The Hsp90 chaperone machinery. J Biol Chem. 2008;283(27):18473–18477.

80. Lee CC, Lin TW, Ko TP, Wang AH. The hexameric structures of human heat shock protein 90. PLoS One. 2011;6(5):e19961.

81. Xie BW, Mol IM, Keereweer S, van Beek ER, Que I, Snoeks TJ, *et al*. Dual-wavelength imaging of tumor progression by activatable and targeting near-infrared fluorescent probes in a bioluminescent breast cancer model. PLoS One. 2012;7(2):e31875.

82. Aleshin AE, Kirby C, Liu X, Bourenkov GP, Bartunik HD, Fromm HJ, *et al*. Crystal structures of mutant monomeric hexokinase I reveal multiple ADP binding sites and conformational changes relevant to allosteric regulation. J Mol Biol. 2000;296(4): 1001–1015.

83. Levin CS. Primer on molecular imaging technology. Eur J Nucl Med Mol Imaging. 2005;32 Suppl 2:S325–S345.

8

Detection of Metal Ions, Anions and Small Molecules Using Metal Complexes

Qin Wang and Katherine J. Franz
Department of Chemistry, Duke University, USA

8.1 How Do We See What is in a Cell?

Biology is a complex network of molecules that interact, organize, and communicate. We cannot readily visualize most of these species, and yet we are interested in tracking their presence within cells, tissues, or biological fluids in order to provide a deeper understanding of their physiological significance. There has been considerable effort, therefore, to develop chemical sensors that can "see" specific analytes in biological systems.

A chemical sensor reports the presence of an analyte by converting a recognition event (usually, but not always, a binding event) into a spectroscopic signal. The analytes of interest include a broad swath of molecule types, including anions, reactive small molecules, proteins, nucleic acids, metal ions, or any other biological or foreign molecule. The spectroscopic signals can be colorimetric, luminescent, magnetic, electrochemical, or radioactive. Currently used sensors largely rely on organic dyes, fluorescent proteins, nanoparticles, and metal complexes for responsive signal transduction. In this chapter, we focus exclusively on the use of metal complexes as design features for chemical sensors. Subsets of this category include radiometals and electroactive metals, which are not covered further here. Instead, we direct interested readers to a recent comprehensive review of the coordination chemistry of radiometals and their application as probes in PET and SPECT imaging [1], as well as reviews on the use of electroactive transition metal receptors for electrochemical sensing applications [2, 3].

Inorganic Chemical Biology: Principles, Techniques and Applications, First Edition. Edited by Gilles Gasser.
© 2014 John Wiley & Sons, Ltd. Published 2014 by John Wiley & Sons, Ltd.

8.1.1   Why Metal Complexes as Sensors?

A metal complex comprises a central metal ion surrounded by supporting ligands. The variety of choices for these components affords metal complexes with a wide diversity of structural, electronic, and coordination features that are amenable to designing probes for biological applications [4]. The Lewis acid characteristics of metal ions with high electron affinity make them good receptors for donor Lewis bases, which include many of the analytes we are interested in sensing. Furthermore, metal complexes can undergo metal–ligand exchange reactions with various analytes, allowing detection processes through a displacement mechanism. These ligand exchange reactions may be tuned for both thermodynamic and kinetic features by altering the type of central metal or the supporting ligands. Additionally, the partially filled d orbitals of the transition metals (or f orbitals for lanthanides) impart desirable electronic and magnetic properties to metal complexes, making them favorable chromophores, luminophores, or paramagnetic centers as sensor building blocks. Based on these features, metal coordination can therefore be employed in many aspects of sensor design.

8.1.2   Design Strategies for Sensors Built with Metal Complexes

The two functional units of a sensor include a recognition site for interaction with the analyte of interest, and a responsive group for signaling the binding event between the sensor and the analyte. The responsive group can be a chromophore, luminophore (fluorophore/phosphor), or a paramagnetic chelate that exhibits different optical or magnetic features in the absence and presence of the analyte.

There are three general strategies for constructing a sensor based on a metal complex (Fig. 1). The most common approach is to link the recognition site and the responsive group via a covalent tether. In this tether approach, the metal complex can function as either the responsive group (1a) or the recognition site (1b). Tethered sensors that detect other metal ions, for example, usually link a chelator (metal ion receptor) to a photoluminescent metal complex. The resulting sensors therefore belong to category 1a, where metal complexes function as responsive units. In contrast, the inherent Lewis acidity of metal ions can be exploited to accept electrons from anions or neutral molecules. In such cases, metal complexes serve as recognition sites for these Lewis basic analytes, following the general formula of 1b.

An alternative strategy is a displacement approach, where the analyte interacts with the metal complex by displacing either the central metal ion (2a), or one of the original ligands (2b) to elicit a distinct spectroscopic change.

In both the tether and displacement strategies, the analyte–sensor interaction and the change in spectroscopic feature are in principle reversible. A dosimeter approach, however, involves a chemical reaction between the sensor and the analyte of interest, which is usually irreversible but highly selective. The analyte can either act as a reactant (3a) or a catalyst (3b) to induce modification of the chemodosimeter with a concomitant change in optical characteristics.

In the following sections, we expand on these general design strategies with representative examples of metal complexes used to probe biologically important metal ions, anions, and small molecules. The examples are by no means exhaustive, but were selected either because they nicely illustrate a design principle or because they have been used to study a biological process.

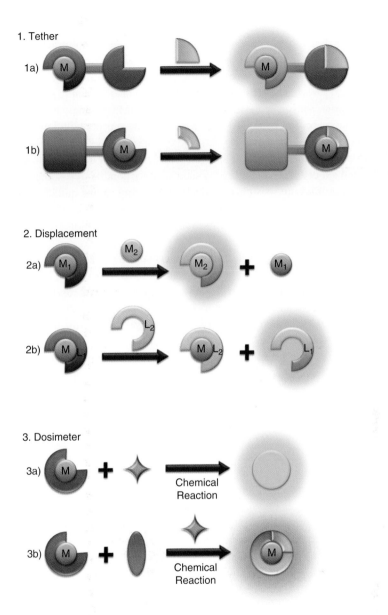

Figure 8.1 *Three main strategies for designing sensors built with metal complexes. In the tether approach, the signal transduction unit is covalently linked to the recognition unit, and a metal complex can be used for either function (1a and 1b, respectively). The displacement approach involves either the replacement of one metal for another (2a), or the replacement of a ligand at the metal center (2b) in ways that the final product transduces a signal. Dosimeters operate by irreversible chemical reaction, where the analyte of interest reacts directly with a metal complex (either at the metal center or with one of its supporting ligands) to form a new signal-transducing product (3a). Alternatively, the analyte can catalyze the reaction of two species to form a new signal-transducing product (3b)*

8.1.3   General Criteria of Metal-based Sensors for Bioimaging

To be applicable for studying biological systems, sensors should minimally perturb the system and be non-toxic at their working concentration range. More information about how the sensor might otherwise perturb a biological system is obviously preferable, but not always known *a priori*. The localization of a sensor is also critical [5]. Depending on different applications, sensors with varying lipophilicity can target analytes in the extracellular or intracellular space. For example, most water-soluble sensors with low lipophilicity are confined to the extracellular environment, whereas lipophilic sensors may cross lipid membranes for detection of intracellular species. Other molecular design features can localize a probe within subcellular compartments such as mitochondria, lysosomes or nucleus [6]. If localization profiles are well characterized, this feature can be quite powerful for probing specific organelles. If localization is not well characterized, however, it may impart a sensing bias due to preferential concentration or (un)intended biological interactions that can be misinterpreted.

High selectivity and appropriate affinity for the analytes of interest are desirable features for a sensor. Given the complexity of biological systems, recognition between the sensor and the analyte can be challenging in the presence of various competing species. Selectivity can be achieved by modulating key factors of recognition units, such as coordinating atom, coordination number, geometry, or cavity size. Principles of coordination chemistry, such as the hard–soft acid base (HSAB) principle and Irving–Williams series provide general guidelines for designing selective recognition units. Affinity, often expressed as the equilibrium dissociation constant (K_d) for the analyte, should be compatible with the median concentration of the analyte in the sample. Depending on the local concentration of the analyte, the ideal K_d value may range from millimolar to nanomolar or even lower.

An ideal sensor should exhibit a readily detectable response to the analyte. For example, colorimetric sensors with color changes that can be visualized by the naked eye are most suitable for *in vitro* rapid detection of analytes at suitable concentrations. For luminescent probes, a "turn-on" detection with an analyte-induced fluorescence enhancement, rather than fluorescence quenching ("turn-off"), is favored. Similarly, a significant increase in relaxivity that contributes to a higher contrast is desirable for MRI imaging. While "on" and "off" sensors report on the presence of a species, ratiometric probes are more amenable to quantitative reporting of the concentration of the analyte and less prone to artifacts. A change in absorption or emission upon analyte recognition is monitored at two wavelengths and the ratio of their signal intensity depends on the concentration of the analyte.

8.2   Metal Complexes for Detection of Metal Ions

Metal ions play fundamental roles in a wide range of biological processes (see also Chapter 9 on the photo-release of metal ions in living cells). Among the essential transition metals in the human body, iron (4–5 g), zinc (2–3 g), and copper (250 mg) are the three most abundant elements [7]. While adequate levels of these metal ions are essential for growth and development, disruption of metal homeostasis is associated with pathological conditions including cardiovascular diseases, cancer, and neurodegenerative disorders [8–10].

Some heavy metal ions, such as mercury and lead, are categorized as hazardous substances due to their potent toxicity. The concentration limits of these metal ions in drinking water are therefore strictly defined by the World Health Organization [11]. Traditional quantitation of these metal ions mainly relies on expensive analytical instruments with tedious sample preparation procedures [12]. Hence, development of small molecule sensors that offer rapid detection, immediate signal feedback and *in vivo* tracking of metal ions represents an attractive direction for heavy metal sensing and quantification.

Monitoring local concentrations as well as global distributions of both essential and toxic metals is highly desirable. Optical techniques based on luminescent or colorimetric sensors represent the main strategy for visualizing metal ions in biological fluids or cells [13]. Most optical sensors for metal ions rely on metal coordination-induced alteration in emission intensity, wavelength or lifetime of chromophores including traditional organic dyes, fluorescent proteins, and emissive metal complexes. Our focus in this section is specifically metal complexes as chromophores. Their advantages over organic dyes include a large Stokes shift, long emission lifetimes, enhanced sensitivity, and high photostability (see also Chapter 4 for the use of metal complexes for cell and organism imaging).

8.2.1 Tethered Sensors for Detecting Metal Ions

The general framework for this class of sensor is a metal ion receptor (recognition unit) covalently connected to an emissive metal complex (responsive unit), as depicted schematically in Fig. 8.1, 1a. Upon coordination of the sensed metal ion to the apo receptor, the charge or energy interaction between the metalated receptor and the responsive metal complex results in an alteration of the latter's photophysical properties. Most luminescent sensors are based on luminescent lanthanide complexes (Sm(III), Eu(III), Tb(III), Dy(III), Yb(III)) or transition metal complexes with d^6, d^8, and d^{10} electron configurations (Ru(II), Ir(III), Pt(II), and Cu(I)) [14, 15].

8.2.1.1   *Luminescent Ir(III) complexes for sensing Zn^{2+} and Cu^{2+}*

Cyclometalated iridium(III) complexes have low-lying triplet excited states and microsecond lifetimes, making them promising luminophores in construction of sensors for various analytes, including metal ions [16]. Among these luminophores, the cationic $[Ir(N^\wedge C\text{-ppy})_2(N^\wedge N)]^+$ complexes (where $N^\wedge C$-ppy stands for 2-phenylpyridine and $N^\wedge N$ represents a diimine ligand, such as bipyridine or phenanthroline) are most frequently employed, owing to their synthetic convenience and facile functionalization of diimine ligands to tune emissive states. The $N^\wedge N$ diimine ligand can thus be derivatized with a metal chelation site to afford a luminescent sensor for other metal ions.

A series of Zn^{2+}-responsive sensors have been developed by conjugating DPA (2,2′-dipicolyamine) as a metal ion receptor onto heteroleptic Ir(III) luminophores (Fig. 8.2). The Ir(III) complex **1** with two chelation arms conjugated onto a bipyridine ligand displays a selective Zn^{2+}-induced ratiometric modulation of emission wavelength and lifetime in organic solutions [17]. Lo and coworkers developed a series of luminescent cyclometalated Ir(III) complexes containing a single DPA moiety connected to a 1,10-phenanthroline ligand, one example of which is shown as **2** in Fig. 8.2 [18]. Upon binding one equivalent of Zn^{2+}, these compounds exhibit 1.2−5.4-fold enhancement of

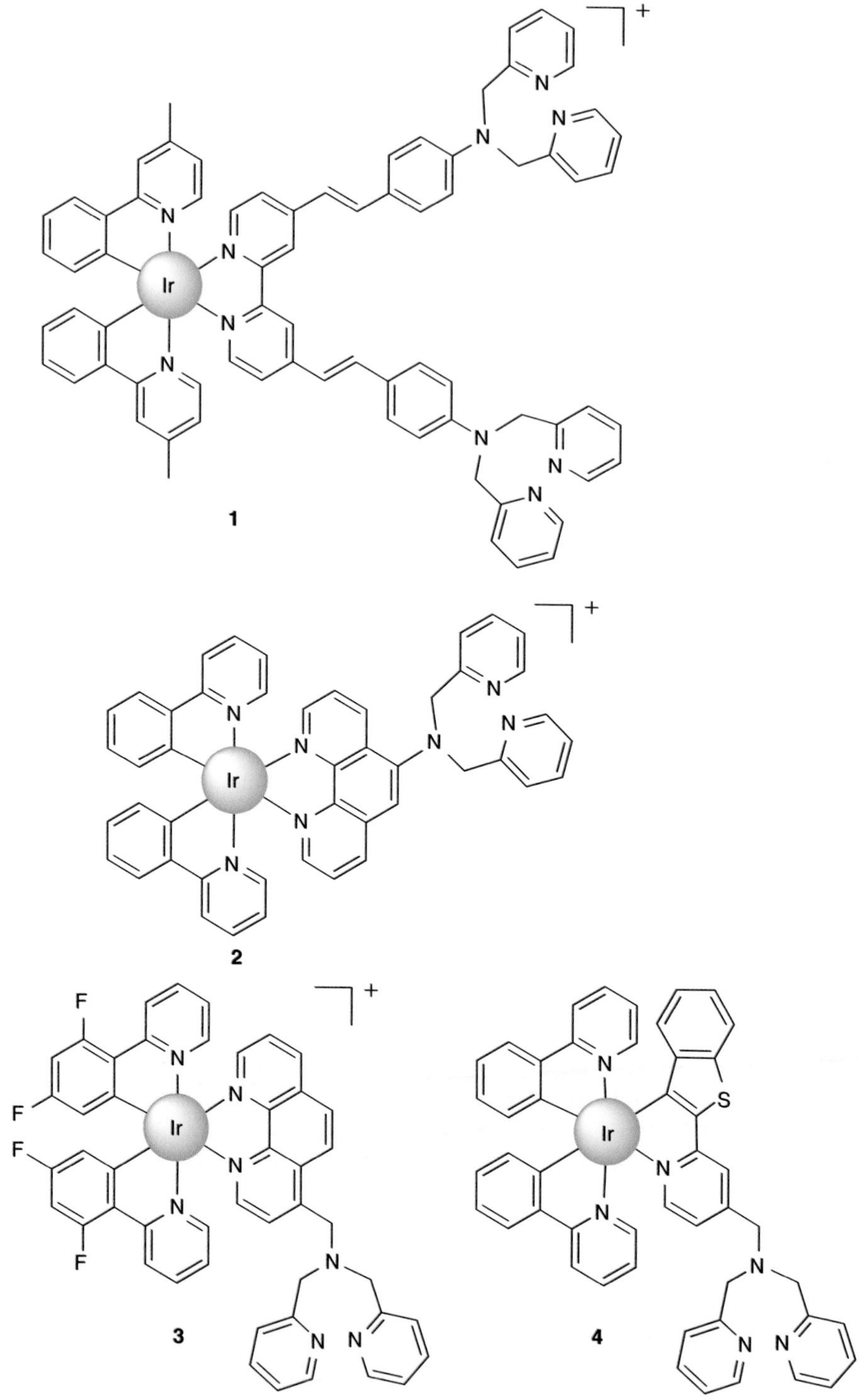

Figure 8.2   Cyclometalated Ir(III) complexes for sensing Zn^{2+} and Cu^{2+}

emission compared with their apo complexes, with K_d values on the order of 10^{-5} M. While the strong emission of these Ir(III) probes is retained in the intracellular environment, they induce moderate cytotoxicity with micromolar IC_{50} values. Later, Nam and Lippard and coworkers reported the DPA-modified Ir(III) complex **3** with two blue-phosphorescent (difluorophenyl)pyridine ligands and one yellow-phosphorescent phenanthroline ligand [19]. Sensor **3** thus exhibits a dual emission in blue (461 nm) and yellow (528 nm) regions in its apo form. Complexation with Zn^{2+} leads to a 12-fold selective turn-on of the yellow phosphorescence. The ratiometric sensing behavior of **3** has also been applied in live cell imaging of intracellular Zn^{2+} [19]. Further addition of a competitive chelator TPEN in cells significantly reduced the yellow phosphorescence signal, demonstrating promising reversibility.

It is important to note that the DPA receptor in these sensors is not selective for Zn^{2+} over other metal ions from a thermodynamic perspective. In fact, these compounds bind more strongly to $Fe^{2+}, Co^{3+}, Ni^{2+}$, and Cu^{2+}, but because these paramagnetic metal cations quench the phosphorescence, the "turn-on" response itself provides the discrimination for Zn^{2+}. By cleverly tweaking the ratiometric platform by appending DPA onto a benzothienyl pyridine (btp) instead of phenanthroline or bipyridine, Nam, Lippard and coworkers created **4** as a sensor for Cu^{2+} [20]. The apo complex **4** exhibits dual phosphorescence with green emission from the ppy ligands and red from the btp ligand, while addition of Cu^{2+} results in preferential quenching of the red phosphorescence and thus an approximately 4-fold increase in intensity ratio of green emission to red emission (I_{ppy}/I_{btp}) *in vitro*. The promising reversibility and selectivity of **4** for Cu^{2+} enables quantification of intracellular Cu^{2+} in fixed HeLa cells through the generation of phosphorescence intensity ratio images of green and red channels (Fig. 8.3e).

Although the cytotoxicity of Ir(III) complexes limits their biological application for live cell studies, these present achievements have provided a valuable starting point for further development of metal complex-based luminescent sensors for imaging of intracellular mobile metal ions.

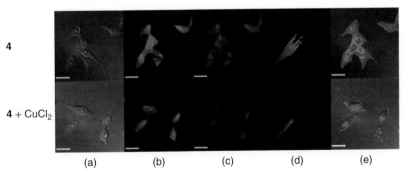

Figure 8.3 *Images of fixed HeLa cells that were first pre-incubated with 500 μM CuCl$_2$ (bottom) or not (top), then incubated with **4** before imaging. a, Contrast images. b, Phosphorescence acquired through a green channel. c, Phosphorescence acquired through a red channel. d, Co-localization scatter plot of green and red channels. e, Phosphorescence intensity ratio images of green and red channels. Reproduced with permission from [20] © 2011 American Chemical Society*

8.2.2 Displacement Sensors for Detecting Metal Ions

A displacement approach for sensing metal ions with a metal complex-based sensor involves two possible mechanisms: central metal displacement or ligand displacement, as shown schematically in Fig. 8.1, 2a and 2b, respectively.

8.2.2.1 Metal displacement: displacing Zn^{2+} to detect Cu^{2+}

Paramagnetic metal centers usually quench fluorescence, making turn-on sensors for these metals a challenging prospect. In contrast, filled-shell d^{10} ions such as Zn^{2+} are not inherent quenchers. Replacing one metal for another therefore represents one sensing strategy. The following example illustrates this general principle, which could be more broadly applied to other systems, and stands as a reminder of how complex equilibria could affect sensor outputs. Wei and coworkers reported a fluorescent zinc complex **5-Zn** (Fig. 8.4) with multiple benzene and pyridine rings as a potential "turn-off" sensor for Cu^{2+} in organic solution [21]. The apo ligand **5** itself exhibits relatively weak fluorescence emission at 375 nm. Complexation with Zn^{2+} in a 1:1 stoichiometry hinders free rotation around the C–C bonds between two aromatic rings, resulting in a 6.4-fold enhancement of fluorescent intensity. In contrast, subsequent addition of Cu^{2+} to the Zn(II) complex significantly quenches the fluorescent signal, due to the partially filled d shell of Cu^{2+}. While the apo ligand **5** possesses selectivity exclusively for Cu^{2+} and Zn^{2+} over other metal cations, its affinity for Cu^{2+} is higher than that for Zn^{2+}, allowing selective central metal displacement of Zn^{2+} with Cu^{2+} and fluorescent "turn-off" detection of Cu^{2+}.

8.2.2.2 Ligand displacement: exchanging a ligand to detect a metal

In a different twist of a displacement strategy, Gunnlaugsson and coworkers took advantage of clever ligand-exchange reactions to develop a sensor for Fe^{2+} in aqueous buffered solution from a ternary Eu^{3+} complex **6-BPS** (Fig. 8.5). The water soluble 4,7-diphenyl-1,10-phenanthroline-disulfonate (BPS) acts as an antenna to sensitize Eu emission [22]. Addition of Fe^{2+} to the emitting **6-BPS** system displaces the BPS ligand, resulting in formation of the Fe-BPS$_3$ complex and complex **6**, which is poorly luminescent in the absence of its BPS antenna. BPS itself is a known ligand for colorimetric sensing of Fe^{2+} [23–25]. Merging BPS onto an Eu(III) complex sensitizes its luminescent emission, which significantly enhances the detection limit of Fe^{2+} to ~10 pM. Additionally, the alkyl thiol group of **6-BPS** facilitates its incorporation onto gold nanoparticles and further benefits evaluation of their potential application as Fe^{2+} displacement sensors in biological media.

8.2.3 MRI Contrast Agents for Detecting Metal Ions

Despite the considerable development of optical sensors for *in vitro* and *in cellulo* metal imaging, optical techniques have intrinsic limitations, one being the limited depth of light penetration. Magnetic resonance imaging (MRI) does not have this limitation and thus allows three-dimensional visualization of internal structures of living specimens in a non-invasive procedure without the use of ionizing radiation [26]. The trade-off is relatively low resolution and sensitivity compared with fluorescence. The utility of MRI therefore

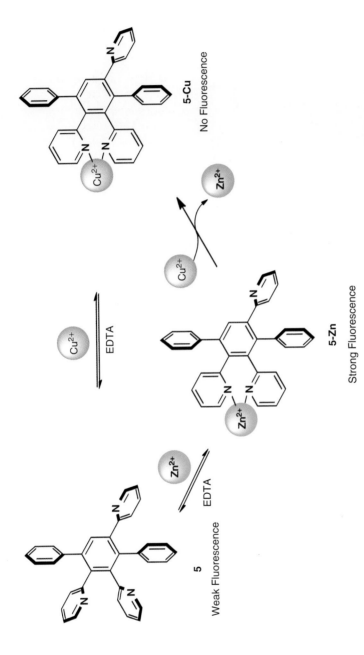

Figure 8.4 *Zn²⁺ complex for fluorescent "turn-off" detection of Cu²⁺ through central metal displacement [21]*

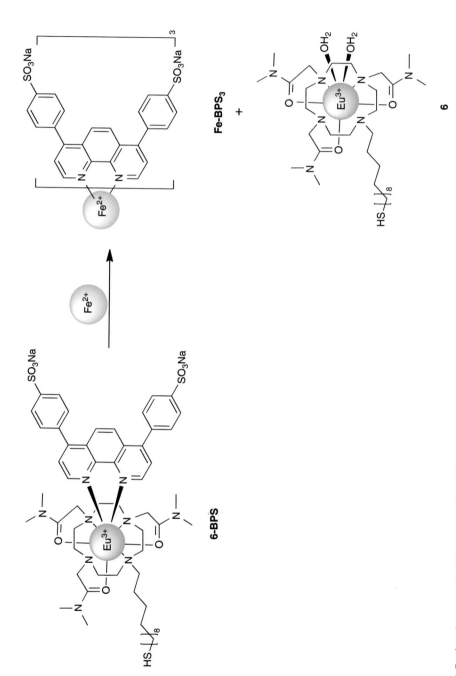

Figure 8.5 *Luminescent Eu(III) complex for Fe^{2+} detection through ligand displacement. Fe^{2+} competes with the Eu macrocyclic complex for binding BPS, thereby preventing the sensitized emission of* **6** *[22]*

rests on imaging the global response from multicellular structures as opposed to imaging intracellular substructures or species.

Water is the most abundant molecule in biological tissues. MRI images are derived from nuclear magnetic resonance of water protons, in which the signal intensity is proportional to the relaxation rates of the nuclear spins. Since most MR samples are heterogeneous, the intrinsic contrast between organs or tissues may be adequately differentiated by various water concentrations and local environments. In 50% of clinical MRI scans [27, 28], however, contrast agents are used to further improve signal resolution and sensitivity. These contrast agents contain paramagnetic metal centers with open coordination sites for water molecules to access the inner sphere, in order to enhance paramagnetic relaxation of water protons. Gd^{3+}-based complexes with seven unpaired electrons ($S = 7/2$) are most commonly used owing to their high magnetic moments and long electron spin relaxation time (10^{-9} s). A few contrast agents were constructed based on other paramagnetic ions including high-spin Mn^{2+}, Mn^{3+}, Fe^{3+}, and Cu^{2+}. The efficacy of the contrast agents to accelerate water proton relaxation, defined as relaxivity (r_i), is influenced by the properties of the contrast agents, including the number of water molecules directly bound to the metal ion (hydration number, q), the mean residence lifetime of bound water molecules (τ_m), and the rotational correlation time (τ_R) [27, 29].

Current Gd^{3+}-based contrast agents used in clinical medicine are nonspecific and typically restricted to the extracellular space, which limits their use to depicting anatomical structures. Many physiological processes, however, occur intracellularly by specific biochemical events. The desire to visualize biological processes by MRI has sparked a trend in the development of next generation MRI contrast agents that can be biologically activated by a specific event. By optimizing the three parameters of relaxivity (q, τ_m, and τ_R), these "smart" contrast agents can selectively undergo relaxivity change in response to a specific biomarker or analyte of interest, for example, metal ions.

Manipulation of the hydration number q is the most common strategy for development of metal ion-activated MRI contrast agents. According to relaxation theory, enhanced water accessibility to the paramagnetic metal center yields higher relaxivity. As illustrated in Fig. 8.6, the Gd^{3+}-based contrast agents contain a paramagnetic Gd^{3+} chelate core and at least one metal ion receptor. In the apo form of the Gd^{3+} complex, an arm of the metal receptor binds the Gd^{3+} center and blocks water access. In the presence of the sensed metal ion, this metal receptor dissociates from the Gd^{3+} center in favor of binding the sensed metal. This intramolecular ligand displacement reaction opens an additional binding site for one water molecule to reach the inner sphere of Gd^{3+}, changing q from 0 to a higher level ($q = 1$) upon metalation and providing enhanced signal contrast.

8.2.3.1 Calcium sensing

Meade and coworkers pioneered the strategy of ligand displacement in designing Gd-DOPTA, a Ca^{2+} sensor that represents the first metal-responsive MRI contrast agent (Fig. 8.7) [30, 31]. It contains two macrocyclic Gd–DOTA cores connected by a modified version of the Ca^{2+} receptor, BAPTA. In the absence of the Ca^{2+}, the carboxylate arms of the BAPTA metal binding site coordinate to the Gd^{3+} centers on both sides, as depicted in **7**. Upon interaction with Ca^{2+}, the carboxylic oxygen donors switch from Gd^{3+} to bind with Ca^{2+}, as depicted in **8**. This conformational change increases the accessibility of water

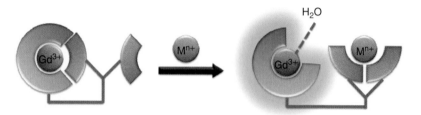

Figure 8.6　*Manipulation of hydration number to modulate relaxivity of metal ion-responsive MRI contrast agents*

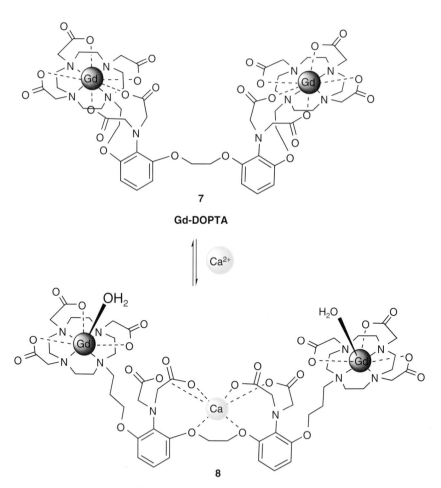

Figure 8.7　*Representation of Gd–DOPTA reagent coordinating with Ca^{2+} to increase coordinated water molecules in the first coordination shell of Gd^{3+} [30, 31]*

to the Gd^{3+} inner sphere, with an approximately 80% relaxivity increase. Complex **7** binds with Ca^{2+} in the low micromolar range, corresponding to intracellular Ca^{2+} concentration levels. However, the shortcomings of cell impermeability and relatively low relaxivity render it more reasonable to target extracellular Ca^{2+}, since sensing extracellular Ca^{2+} at concentrations in the millimolar range is more compatible with the relatively high doses of contrast agent **7** needed for acquisition of adequate resolution.

The same displacement approach has been used to design a series of Gd^{3+} contrast agents for extracellular Ca^{2+} sensing [32–34]. Most of them are bimetallic complexes involving two macrocyclic DOTA-derived Gd^{3+} chelate cores, bridged with a spacer containing a Ca^{2+} receptor. Unlike the high-affinity BAPTA unit used in complex **7**, the central metal receptors in these complexes are designed to have a lower affinity for Ca^{2+}, adapted to detection of extracellular Ca^{2+} with minimal interference from Mg^{2+} or Zn^{2+} [35].

A problem for these Gd–DOTA based contrast agents is a loss of relaxivity in physiological fluids compared with that in water. It is hypothesized that water molecules in the metalated complex are displaced from Gd^{3+} by anions in the biological buffers. However, at least one example that has only one Gd–DOTA core produces a 36% and a 25% Ca^{2+}-dependent relaxivity enhancement in simulated cerebrospinal fluid and extracellular matrix, respectively, which could be sufficient for detection of Ca^{2+} in the brain [35].

8.2.3.2   *Copper sensing*

Copper is an essential element for life, serving as a catalytic cofactor in enzymes such as cytochrome c oxidase, superoxide dismutase, and tyrosinase [9]. The catalytic activity of copper derives from redox cycling between cuprous (Cu^+) and cupric (Cu^{2+}) oxidation states. Owing to its redox activity, misregulation of copper homeostasis is implicated in many serious diseases, including neurodegenerative pathologies as well as Wilson's disease or Menkes syndrome, characterized by an inability to appropriately distribute copper to cells and tissues.

The extracellular copper levels vary in different compartments of the body, with a concentration of $10-25$ μM in the blood serum, 30 μM in the synaptic cleft, and $0.5-2.5$ μM in the cerebrospinal fluid. Intracellular copper levels within neurons can reach $2-3$ orders of magnitude higher concentrations and the +1 oxidation state of copper dominates intracellularly [13].

Que and Chang introduced a series of copper-responsive MRI contrast agents for $Cu^{1+/2+}$ based on a Gd–DOTA core incorporated with a pendant receptor for copper (Fig. 8.8). However, the first-generation compound **9** containing an iminodiacetate moiety for Cu^{2+} lacks selectivity over Zn^{2+} [36]. To overcome this problem, later-generation sensors were developed by changing the copper receptor from a carboxylate-based moiety to a softer thioether-based moiety [37]. Introduction of N and S donors in **10–13** results in high selectivity for Cu^+ over other competing ions including Cu^{2+} and Zn^{2+} with turn-on responses of relaxivity of up to 360%, while **14** containing a combination of N, S and O donors in the binding site responds equally to Cu^+ and Cu^{2+}. In particular, compound **10** exhibits the most promising *in vitro* behavior, including high selectivity and affinity for Cu^+ ($K_d =$ 0.26 pM), as well as remarkable relaxivity enhancement (360%).

Nevertheless, like other typical Gd^{3+}-based contrast agents, sensor **10** is largely confined to the extracellular environment due to cell impermeability. In order to target the *in*

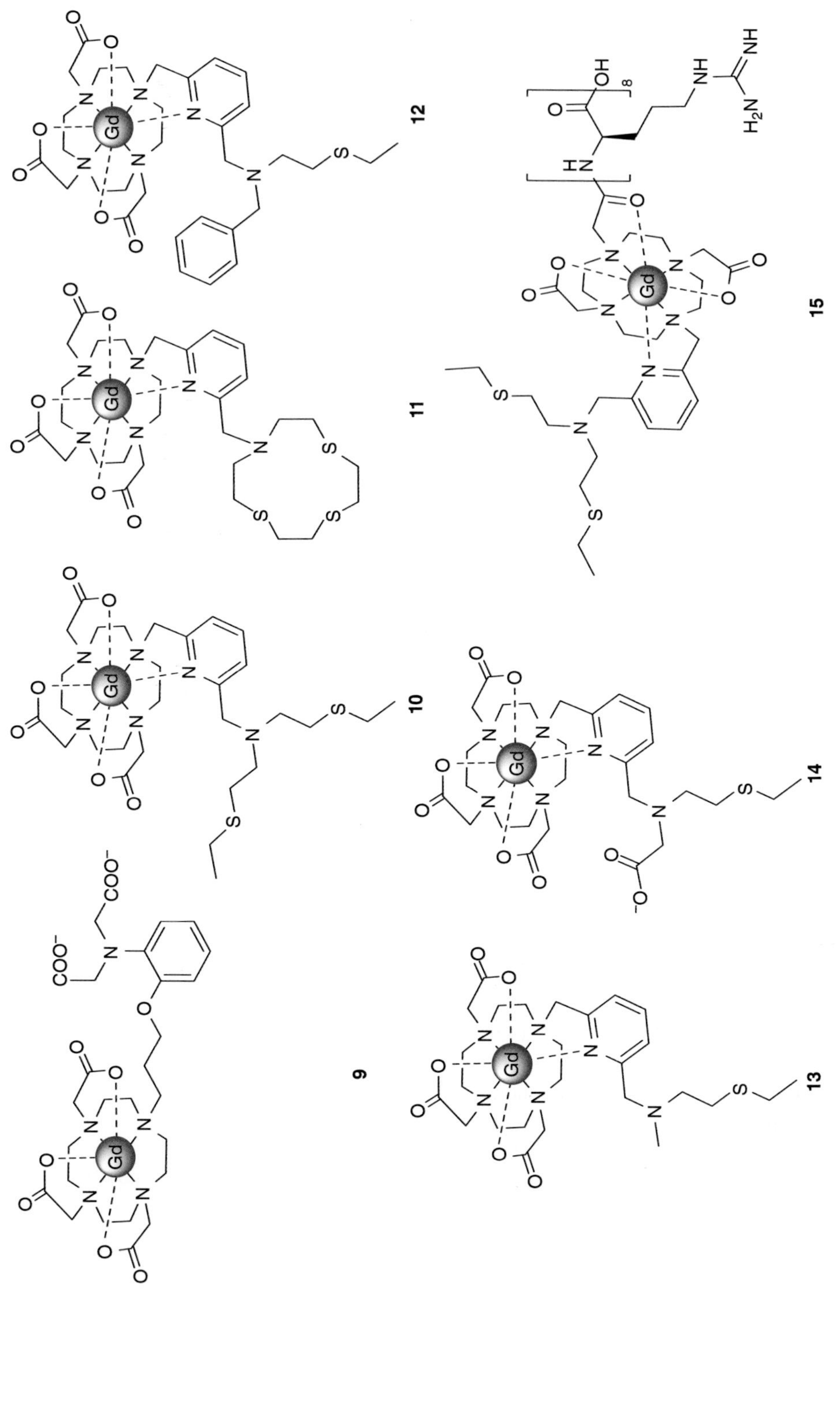

Figure 8.8   Copper-responsive MRI contrast agents based on Gd(III) complexes

cellulo Cu^+, it was modified with an octaarginine (Arg_8) tail to afford **15** for improved cellular uptake and retention. Derivative **15** accumulates in live cells at concentrations 9-fold greater than the parent sensor **10**, and was able to discriminate copper levels between a wildtype and a Menkes disease model cell line with a mutant copper efflux protein [38]. This encouraging progress may shed light on the *in vivo* application of **15** as a copper sensor to report on biological regulation of exchangeable copper pools [38].

It is also worth noting that **15** exhibits cytotoxicity with an IC_{50} value of 295 μM in HEK293 cells after a 24 h incubation, compared with >2000 μM for the extracellular parent compound **10** [38]. This reduced cell viability is estimated to originate from the introduction of the octaarginine group and is consistent with other observations of octaarginine-bearing compounds [39].

8.2.3.3 Zinc sensing

Most biological zinc is bound to proteins and appears as an essential structural element, for example in zinc finger transcription factors, or as a catalytic cofactor, as in several hydrolase enzymes. The remaining mobile zinc exists in certain organs, such as the brain, retina, pancreas, and prostate. The total cellular zinc concentration, including the bound and unbound forms, is in the range of a few hundred micromolar, whereas the unbound mobile zinc is present in the picomolar range intracellularly [40]. In the extracellular space, the average concentration of free zinc ion has been estimated to rise to 8 μM in human plasma [41]. Since Zn^{2+} is also an important signaling ion, the highest concentration of zinc in the body can be found in vesicles of certain types of glutamatergic neuronal, prostatic, and pancreatic cells, reaching as high as 300 μM [42].

Owing to the spectroscopically silent d^{10} electronic configuration, visualization of biological zinc largely relies on fluorescent-based sensors that can selectively and reversibly detect zinc in cells or tissue slices, with all the limitations of optical techniques. With regard to the MRI contrast agents designed for Zn^{2+} sensing, one strategy reported by Nagano and coworkers is based on the incorporation of a central Gd–DTPA core with flanking Zn^{2+}-receptors, either DPA or a combination of carboxylate and pyridyl side chains (Fig. 8.9, complexes **16** and **17**). The main drawback of this system is that the Zn^{2+} coordination geometry hinders the access of water molecule to the Gd^{3+} inner sphere, leading to a decrease in relaxivity [43, 44].

An alternative approach is to modify the general template of Gd^{3+}-based probes for Ca^{2+} sensing by decreasing the number of carboxylic acid arms to lower the affinity for "harder" Ca^{2+} while retaining adequate affinity for Zn^{2+}. Meade and coworkers developed Zn^{2+} sensors based on the Gd–DOTA core with an iminodiacetate moiety, or one pyridine and one acetate arm for Zn^{2+} binding (Fig. 8.9, complexes **18** and **19**). Both compounds have over 100% increase in relaxivity in response to Zn^{2+}, while compound **18** also responds to Cu^{2+} [45, 46]. This lack of selectivity is consistent with the observation for the Cu^{2+}-responsive MRI agent **9** (Fig. 8.7), probably due to the structural similarity of these two sensors.

A novel Gd–DOTA-based Zn^{2+} sensor **20** has been reported by Sherry and coworkers, which bridges two DPA moieties for Zn^{2+} binding with a macrocyclic Gd^{3+} core [47]. Its apo form Gd **20** exhibits a modest relaxivity increase in the presence of 2 equivalents of Zn^{2+}, while a substantially higher relaxivity enhancement (165%) is observed when the metalated Gd **20**–$2Zn^{2+}$ complex is bound to human serum albumin (HSA). Although **20**

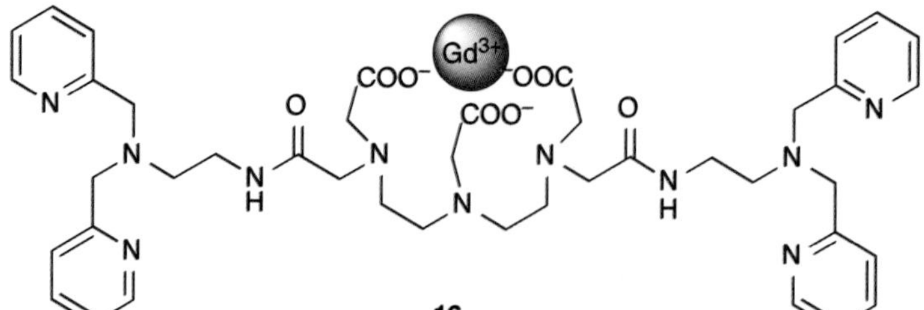

Figure 8.9   *Responsive MRI contrast agents for Zn²⁺ sensing*

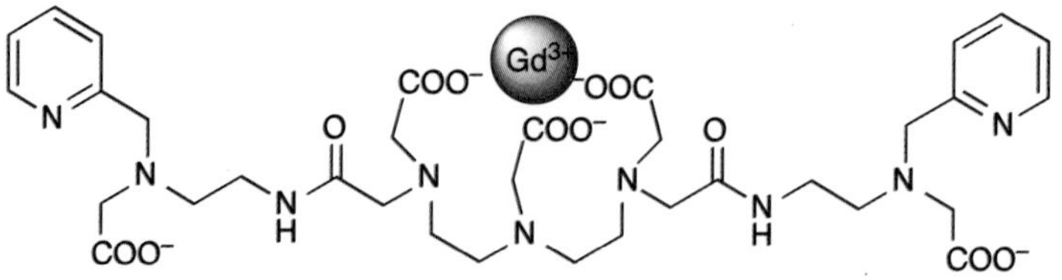

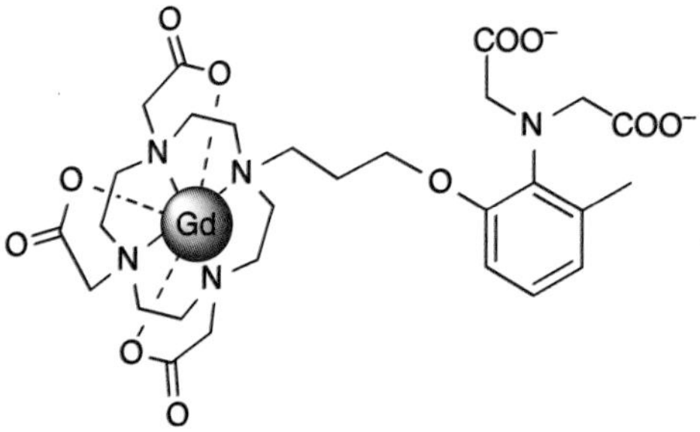

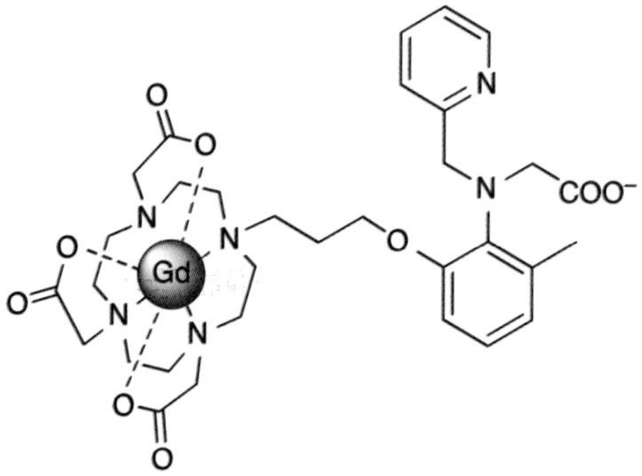

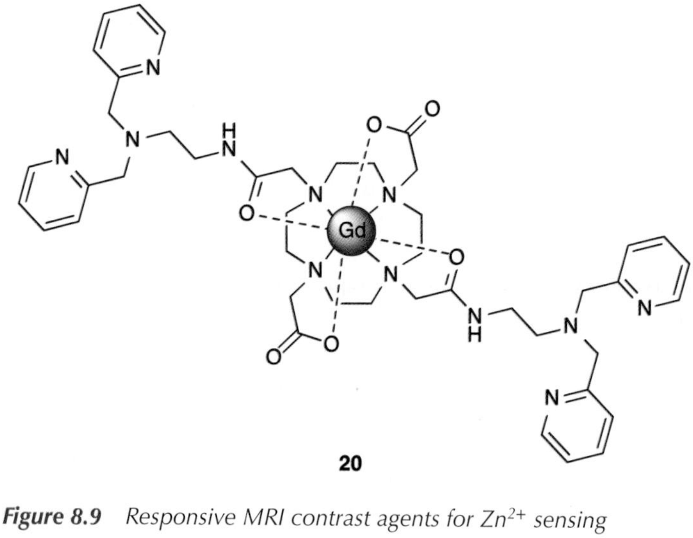

also responds to Cu^{2+}, this competition is hypothesized not to interfere with *in vivo* detection of Zn^{2+}, as Cu^{2+} is typically found at much lower concentrations. Indeed, it has been tested in mice to visualize the release of Zn^{2+} from pancreatic β-cells to extracellular space upon glucose-stimulated insulin secretion [48]. Sensor **20** is confined to the extracellular space as anticipated. The released Zn^{2+} forms a 2:1 complex with the sensor extracellularly, which eventually exhibits a relaxivity increase upon complexation with HSA. More importantly, cellular status of both cell expansion and loss of β-cell function, which signals the progression of diabetes, can be identified *in vivo* from the contrast images [48]. This Zn^{2+}-responsive MRI sensor potentially provides a noninvasive tool for *in vivo* monitoring of β-cell function during development of diabetes.

A porphyrin scaffold **21** for dual-function fluorescence–MRI sensing of Zn^{2+} has been established by Lippard and coworkers (Fig. 8.10) [49]. The metal-free porphyrin **21** is fluorescent while the Mn^{3+}–porphyrin complex **22** is a promising MRI contrast agent, which allows dual imaging by either optical detection or magnetic resonance. Complex **22** was shown to distribute in different regions of the brain in a Zn^{2+}-dependent manner for *in vivo* imaging [50]. This Mn^{3+}–porphyrin platform is the first reported cell-permeable MRI sensor for a metal ion. With all these desirable features, this example may represent a novel direction for development of responsive MRI contrast agent without Gd^{3+}.

Despite the significant progress in developing responsive MRI agents for metal ions, only a few have advanced to *in cellulo* or *in vivo* imaging. The current pattern of designing these "smart" MRI agents largely relies on derivatization of a nonspecific metal-based MRI contrast agent with a selective metal receptor. A major challenge remains in delivering these MRI reagents to sites of interest other than the extracellular environment. Several cell delivery vehicles could be employed for improving cellular uptake, including encapsulation into liposomes [51] and conjugation to macromolecular transporters such as peptides [52, 53], dendrimers [54], dextrans [55], TiO_2 [56], or gold nanoparticles [57]. However, as cellular accumulation and retention increase, the intracellular versions modified with these delivery vehicles may exhibit increased cytotoxicity compared with the parent compound. Moreover, the safety issue of Gd^{3+} being released from the complex is more profound in the intracellular environment, and thus higher-affinity Gd^{3+} chelates or cell-permeable alternatives based on other paramagnetic metal centers are desirable.

8.2.4   Chemodosimeters for Metal Ions

If the recognition of an analyte by a sensor is based on an irreversible chemical reaction, then the sensor is defined as a chemodosimeter. Dosimetric sensing offers a highly selective and sensitive detection method with a rapid response for metal ions. At least two general scenarios can be envisioned for metal complex based chemodosimeters for sensing metal ions. In the first, the sensed metal ion reacts directly with the dosimeter in a way that chemically modifies the sensor and changes its photophysical properties, either turning off or on a signal (Fig. 8.1, 3a). In the second scenario, the sensed metal ion catalyzes a chemical reaction between two species to form a new product with enhanced photophysical properties (Fig. 8.1, 3b).

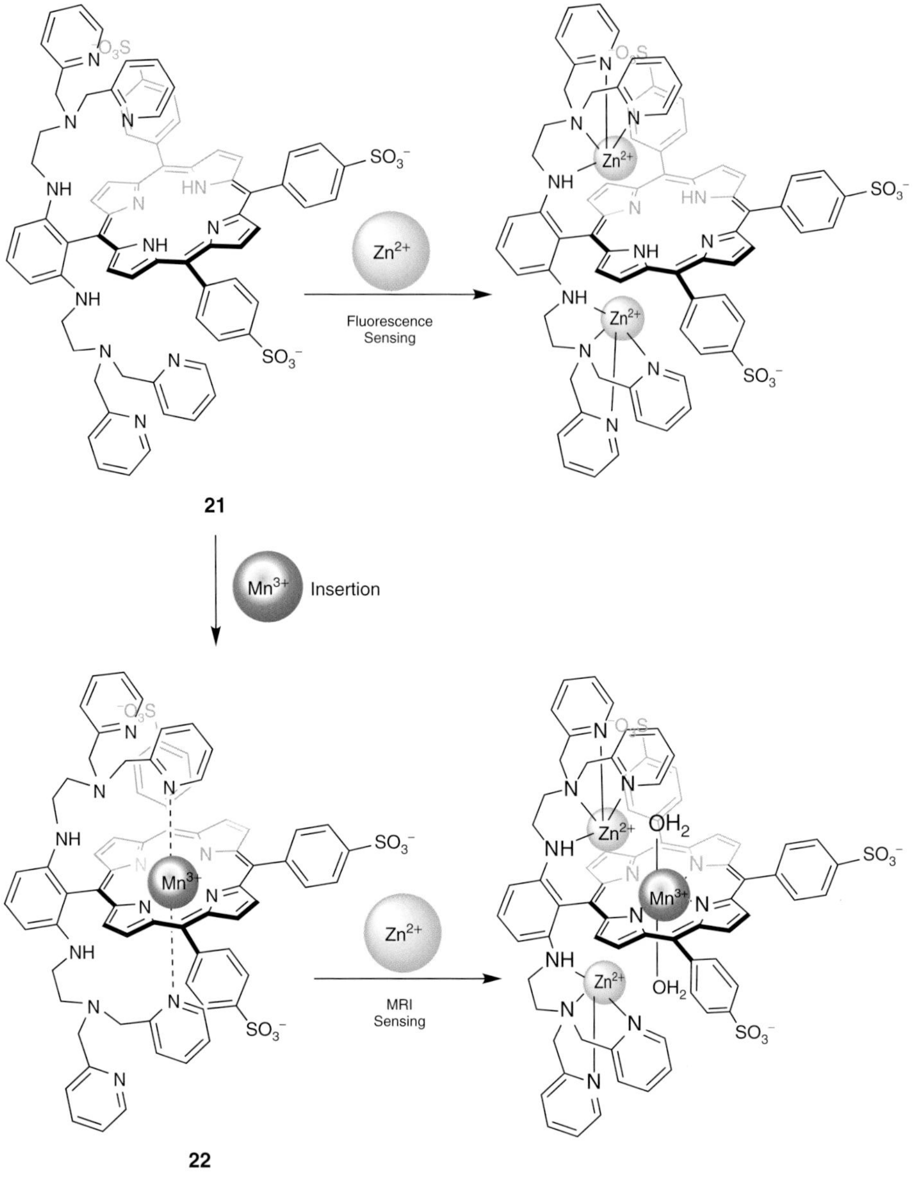

Figure 8.10   *Representation of an Mn^{3+}–porphyrin dual sensing platform for detection of Zn^{2+}. Adapted with permission from [49] © 2007 National Academy of Sciences, U.S.A*

8.2.4.1   *Phosphorescent chemodosimeters for Hg^{2+}*

Several phosphorescent chemodosimeters for Hg^{2+} based on Ir(III) complexes have been developed (Fig. 8.11). For chemodosimeters **23**, **24**, and **25**, the underlying sensing principle involves recognition of Hg^{2+} by the S atoms embedded in the Ir complex, causing

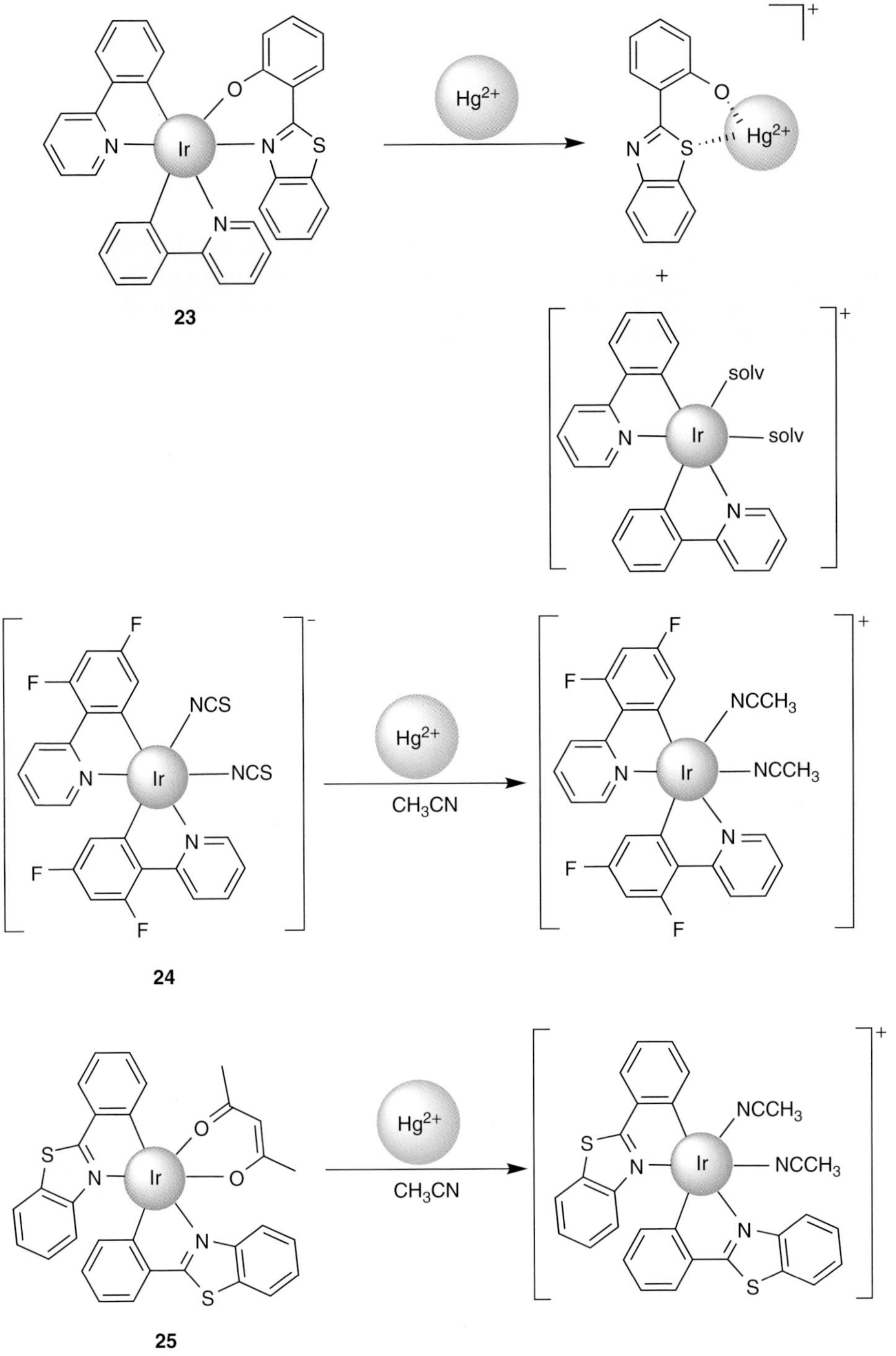

Figure 8.11   *Phosphorescent Ir(III) dosimeters (far left) react with Hg²⁺ and provide a turn-off signal (far right, top two reactions) or a change in emission wavelength that provides a ratiometric response (far right, bottom)*

Hg^{2+}-induced ligand displacement or decomposition with a phosphorescent "turn-off" or ratiometric change [58–60]. The ratiometric behavior of **25** has been further applied in live cells for tracking intracellular Hg^{2+}. This example represents the first neutral Ir complex that exhibits adequate cell membrane permeability and ratiometric sensing capacity for Hg^{2+}[60].

8.2.4.2　*Luminescent Ru(II) dosimeters for Cu^{2+}*

A "turn-on" luminescent chemodosimeter, **26**, for sensing Cu^{2+} has been reported by Gopidas and coworkers, which takes advantage of the ability of Cu^{2+} to oxidize aromatic sulfur in organic solution (Fig. 8.12) [61]. The analyte Cu^{2+} serves as an oxidant and reacts with the phenothiazine (Ptz) moiety in the dosimeter. In the absence of Cu^{2+}, excitation of the $[Ru(bpy)_3]^{2+}$ causes electron transfer from the Ptz moiety to the excited $[Ru(bpy)_3]^{2+}$ luminophore, quenching its luminescence. Addition of excess Cu^{2+} in acetonitrile readily oxidizes the Ptz reactive moiety to give a stable 5-oxide entity, which prevents the electron donation process and restores the emission signal at 620 nm. Since only Cu^{2+} is capable of such oxidation reactions, this example provides a highly selective approach for "turn-on" detection of micromolar concentration of Cu^{2+}. It should be noted that part of the driving force for this reaction comes from the stabilization of the resulting Cu^+ by acetonitrile. Applying such a strategy to a biological problem will therefore require an assessment of the redox potential of the mobile copper to be sensed.

8.2.4.3　*Dosimetry based on Cu^+-catalyzed click chemistry*

An illustrative example of analyte-catalyzed dosimetric sensing approach is an Eu^{3+}-based "turn-on" luminescent sensor for Cu^+ developed by Viguier and Hulme (Fig. 8.13) [62]. The Eu^{3+} complex **27** itself is non-emissive, however, the Cu^+-triggered reaction of **27** with dansyl-containing azide **28** gives the sensitized emissive complex **29**. This sensing process is based on the highly efficient Cu^+-catalyzed click reaction, where in the alkynyl moiety of the Eu^{3+} complex **27** specifically interacts with the azide tail of **28** through Cu^+-catalyzed Huisgen 1,3-dipolar cycloaddition. Since the click reaction is bioorthogonal and exclusively catalyzed by Cu^+ species, this combination of alkynyl-luminophore and azide-antenna provides a general pattern for designing selective Cu^+ sensors in biological contexts.

8.3　Metal Complexes for Detection of Anions and Neutral Molecules

A metal complex, with a positively charged Lewis acidic metal center framed by one or more ligands, can exhibit unique metal-related reactivity including metal–ligand substitutions and metal-mediated redox reactions. These attractive characteristics afford an important and complementary strategy to traditional organic fluorescent probes for detecting biologically important anions and neutral molecules with inherent Lewis basicity. In this section, we highlight a selection of metal-based sensor scaffolds that exemplify the three main design approaches outlined in Fig.8.1. At this stage, we also encourage interested readers to look at Chapter 10, which describes not the detection but the release of bioactive molecules using metal complexes.

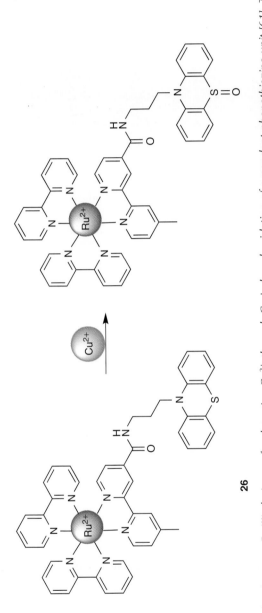

26

Figure 8.12 *Luminescent Ru(II) dosimeter for detecting Cu^{2+} through Cu-induced oxidation of a pendant phenothiazine unit [61]. The phenothiazine quenches emission from the Ru complex on the left, whereas the oxidized product on the right provides an emission signal*

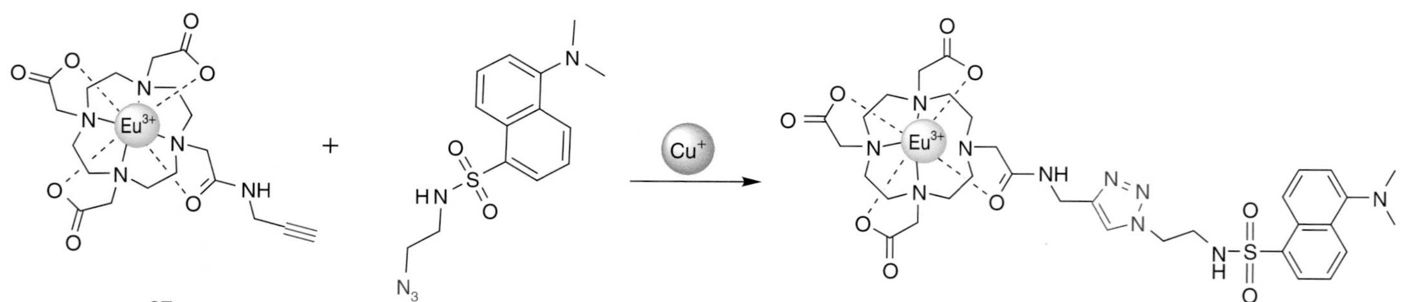

27 28 29

Figure 8.13 *Luminescent Eu(III) dosimeter for detecting Cu^+ through Cu^+-catalyzed click reaction [62]. The Eu complex* ***27*** *is not emissive, whereas the product* ***29*** *allows for sensitized Eu emission*

8.3.1   Tethered Approach: Metal Complex as Recognition Unit

In the tethered sensors described previously in Section 8.2.1, the metal complex is the responsive unit. In sensors designed to detect anions or small molecules, this paradigm is reversed, and instead the metal complex acts as the recognition unit for analytes that coordinate directly to the metal center. The surrounding ligands of the complex provide additional molecular recognition for analyte specificity. Common fluorophores such as rhodamine, coumarin or fluorescein are often incorporated as responsive groups for signaling analyte–sensor recognition (Fig. 8.1, 1b). Alternatively, the metal complex can be both the recognition site *and* the reporting site, as is the case with luminescent lanthanide complexes [5, 6].

8.3.1.1   Detecting phospho species by binding to Zn(II) complexes

Phospho species including inorganic phosphate and pyrophosphate (PPi), nucleoside pyrophosphates such as ATP, as well as phosphoproteins and phospholipids, are biologically important targets that have attracted particular attention. For example, ATP is known as both a universal energy source and an extracellular signaling messenger, whereas PPi is the product of ATP hydrolysis in cells. Protein phosphorylation status is part of a ubiquitous regulatory machinery of cellular signal transduction.

Employing a metal complex as the binding site for phospho species takes advantage of the strong host–guest binding affinity between the central metal ion and phospho analytes to enable detection in aqueous systems. Among the metal complexes explored as phospho receptors, Zn^{2+}–DPA complexes and their analogues have been studied extensively. Fig. 8.14 represents the general scaffold of tethered sensors for detection of phospho species. Typically, two Zn^{2+}–DPA arms are incorporated into a fluorophore through a spacer that determines the distance between the two phospho receptors. The binuclear scaffold of these receptors is critical to their recognition function, since mononuclear versions exhibit little affinity for phospho species. Selectivity of these sensors towards one particular phospho species over others remains a challenging issue. For PPi sensors, for example, discrimination of PPi from phosphate can be achieved by tuning the length of the spacer. However, distinguishing between PPi and ATP is more challenging and requires careful design of the corresponding receptors based on their differences in total anionic charge densities [63].

In the absence of analyte, coordination of DPA with the second Zn^{2+} is not favored due to electrostatic repulsion between two Zn^{2+} centers, which allows electron transfer from DPA amine to the fluorophore and concomitant fluorescence quenching. Synergic coordination of phospho species with both receptors effectively reduces the electrostatic repulsion owing to the negative charge of phospho anions, thereby favoring bimetallic Zn^{2+} chelation with both DPA moieties and suppression of the quenching mechanism [14].

Yoon and coworkers developed a "turn-on" fluorescent probe **30** for PPi by integrating two Zn^{2+}–DPA moieties into fluorescein, wherein the conjugated aromatic rings of fluorescein also provide the spacer that favors selective recognition of PPi (Fig. 8.15) [64]. Binding with PPi in buffered aqueous solution induces a slight red shift of emission and a 1.5-fold fluorescence enhancement derived from synergic chelation of the two receptors.

The dinuclear Zn^{2+}–DPA platforms have further been exploited for selective detection of anionic phospholipids in biological contexts. The Smith group reported a Zn^{2+}–DPA-based complex **31** with affinity for anionic phosphatidylserine (PS) enriched in apoptotic cell

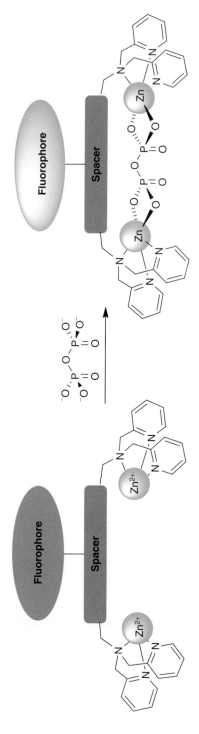

Figure 8.14 *General scaffold of tethered sensors for detecting phospho species*

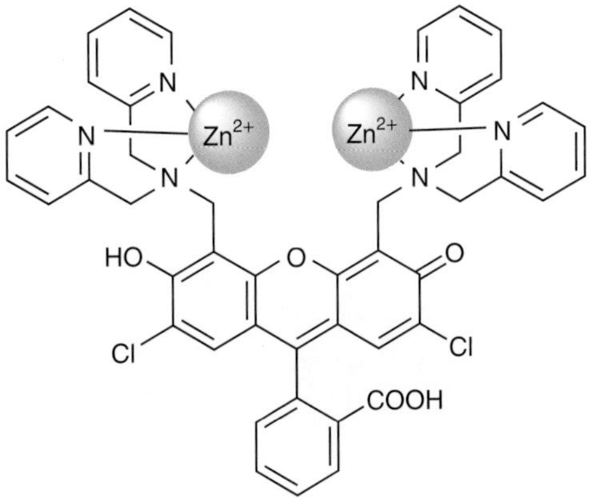

30

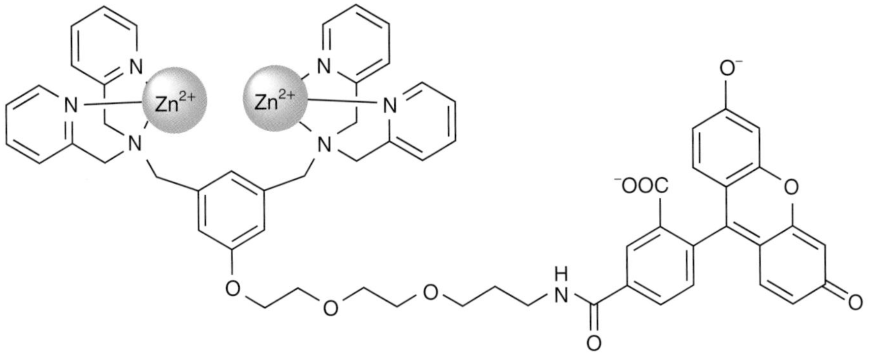

31

Figure 8.15 *Examples of tethered fluorescent sensors based on Zn^{2+}–DPA for recognition of phospho species*

membranes (Fig. 8.15) [65]. PS is a type of phospholipid that normally resides on the inner cytosolic side of cell membranes. At the early stage of programmed cell death (apoptosis), PS is no longer restricted to the inner side, but rather exposed on the outer leaflet of membranes, which renders it an excellent biomarker of apoptosis. Because **31** cannot permeate cell membranes, its extracellular fluorescence intensity reflects translocation of PS from the inner to the outer leaflet, which is proportional to the extent of cell death. An effective assay based on probe **31** and its biotin or quantum dot versions has thus been established for *in cellulo* differentiation of healthy versus apoptotic cells, which circumvents some limitations of the sophisticated protein-based annexin V method [65, 66].

8.3.1.2   *Measuring bicarbonate by lanthanide complexes*

The Parker group has pioneered the development of luminescent lanthanide complexes as cellular probes responsive to a host of biologically important species, including pH, metal

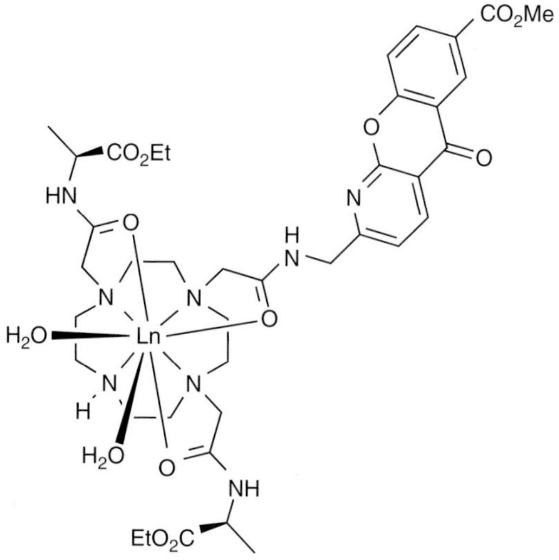

Figure 8.16 *Luminescent Eu(III) and Tb(III) complexes that provide ratiometric analyses of bicarbonate [67, 68]*

ions, and oxyanions such as phosphate, lactate, citrate, and bicarbonate [5, 6]. The general architecture of these probes is a central luminescent lanthanide, typically Tb(III) or Eu(III), contained within a macrocyclic chelate ring with pendant arms that help form an appropriate binding site for the analyte and that link to an aromatic antenna for sensitized luminescence emission (Fig. 8.16). The reversible displacement of coordinated water molecules by the sensed anion not only increases the emission intensity, but also distinctively modulates the spectral form and circularly polarized emission from these lanthanide complexes, which facilitates ratiometric analyses. A recent advance from this group includes a pair of Tb/Eu complexes that can measure bicarbonate concentrations rapidly in human serum as well as directly within mitochondria of living cells [67, 68]. A key design feature of **32** is the intentional choice of the amide-substituted azanthone sensitizer, which helps to promote uptake of the probe by macropinocytosis and subsequent localization preferentially to mitochondria.

8.3.2   Displacement Approach: Metal Complex as Quencher

Metal centers with partially filled orbitals normally function as fluorescence quenchers through electron- or energy-transfer quenching pathways. As a result, metal–ligand substitutions offer a versatile approach to alter fluorescence outputs. The general mechanism takes advantage of the metal–analyte recognition event to displace a metal-coordinated fluorophore and unquench its fluorescence through ligand substitutions. Two different strategies may be employed: analyte-induced displacement of the ligand/fluorophore from the metal center, or analyte-induced metal removal.

8.3.2.1 Sensing nucleoside polyphosphates by intramolecular fluorophore displacement

The strategy based on a binuclear Zn^{2+}–DPA system incorporated into a chromophore for intramolecular ligand displacement has enabled detection of phospho derivatives in aqueous solutions. In particular, Hamachi and coworkers introduced a xanthene-bridged bis(Zn^{2+}–DPA) complex **33** for fluorescent imaging of intracellular ATP [69, 70]. In the absence of ATP, the carbonyl oxygen of the xanthone fluorophore coordinates with the bimetallic Zn^{2+} moieties, which effectively quenches fluorescence. Upon binding of a nucleoside polyphosphate substrate such as ATP to the Zn^{2+} sites, displacement of the pendant fluorophore from the metal coordination sphere results in recovery of the conjugation in the xanthene ring and a "turn-on" emission *in vitro* (Fig. 8.17) [69]. Acetylation of both hydroxyl groups affords a cell-permeable version capable of conversion into sensor **33** by intracellular esterases. The fluorescence intensity of **33** in live Jurkat cells exhibits an ATP-dependent pattern that is consistent with the images obtained with quinacrine, a known probe of ATP stores. However, while this sensor demonstrates adequate selectivity towards polyphosphates over mono- or diphosphate anions, it cannot sufficiently discriminate among polyphosphate anions, including other nucleoside triphosphates and inositol triphosphate [69].

8.3.2.2 Sensing NO at a metal by fluorophore displacement

Nitric oxide (NO) is implicated in a wide range of biological processes and it has also been identified as an important gaseous messenger in signal transduction pathways, which sparked interest in the development of detection methods for sensing NO in real-time and *in vivo*.

The Lippard group reported a series of metal-based platforms for NO sensing through the general mechanism that NO selectively displaces ligand/fluorophores from the quenching metal centers with a concomitant fluorescence turn-on. Their initial sensor **34** is composed of a Co^{2+} center anchored by aminotroponiminate ligands attached to dansyl fluorophores that are quenched due to their proximity to the paramagnetic Co^{2+} center [71]. Recognition with NO induces dissociation of one fluorescent ligand upon formation of a metal–dinitrosyl adduct, producing significant enhancement of fluorescence emission (Fig. 8.18).

A ligand displacement strategy has also been applied to other platforms for NO detection based on iron–cyclam, ruthenium–porphyrin, and dirhodium–tetracarboxylate complexes [72–74]. While these examples are mechanistically related, the ruthenium–porphyrin system **35** involves NO-induced release of free ligand/fluorophore Ds-im from the original metal complex, rather than ligand displacement in an intramolecular fashion observed in the cobalt case (Fig. 8.18). In addition to release of the fluorescent axial ligand, the reaction of **35** with NO also involves release of CO and formation of NO_2^-, which ends up coordinated to the Ru product.

However, application of these displacement strategies are largely confined to organic solvents with high concentrations of NO gas bubbled into a cuvette, since competitive H_2O coordination in aqueous solution may interrupt the NO-induced ligand substitution.

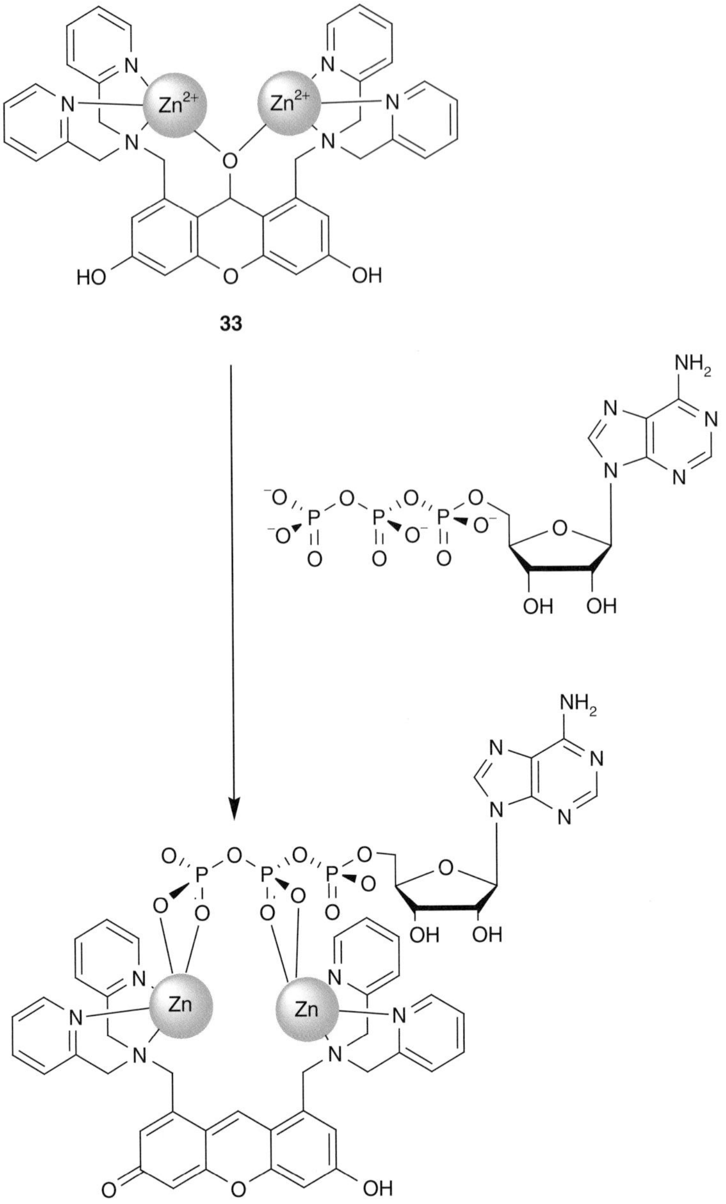

Figure 8.17 *Schematic representation of xanthene-bridged bis(Zn^{2+}–DPA) complex for detecting ATP via a displacement approach [69]. Prior to ATP binding, the fluorescence of **33** is quenched, whereas the ATP-bound structure on the right is fluorescent*

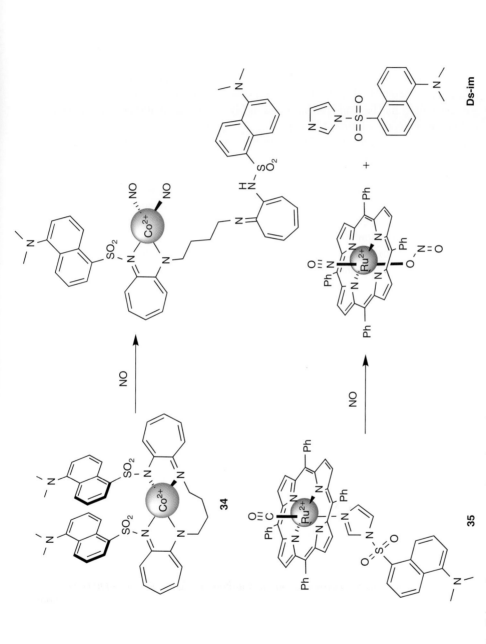

Figure 8.18 *Schematic representation of metal complexes for NO sensing through ligand displacement [71, 72]. The fluorophores on the left are quenched by the paramagnetic metal centers, whereas the products on the right show increased fluorescence*

8.3.2.3 *Sensing cyanide by metal displacement*

The widespread industrial use of cyanide and its extreme toxicity have necessitated the development of cyanide sensing methods with low detection limits. Attributed to the high binding affinity of cyanide anion with copper ion, most turn-on optical sensors for cyanide typically consist of a chromophore appended with ligands that are metalated with Cu^{2+} quenching centers. Coordination of cyanide to the Cu^{2+} centers induces removal of para-magnetic metal quenchers, causing fluorescence return and subsequent formation of stable $[Cu(CN)_x]^{n-}$ products (Fig. 8.19). Representative scaffolds of cyanide sensors that rely on a demetalation mechanism are illustrated in Fig. 8.20, which allows turn-on detection of cyanide in aqueous solutions [75–78].

Notably, luminescent sensor **38** containing a Cu^{2+}–DPA moiety incorporated into an Ir^{3+}–polypyridyl luminophore is capable of cyanide detection in aqueous environments at concentrations lower than the standard safety limits, and in live cells pre-exposed to a 0.2 ppm aqueous cyanide solution [77]. Yoon and coworkers introduced sensor **39** for *in vivo* imaging using nematode *C. elegans* as the testing organism. Different levels of cyanide exposure, with concomitant removal of copper from **39**, results in varying degrees of fluorescein emission recovery [78, 79].

An alternative to the Cu-based fluorimetric cyanide sensors relies on cyanide binding to the Co center of cobalamin to give a deep violet-colored complex [80]. A solid-phase extraction method has been developed that takes advantage of this intense color change to detect cyanide in blood samples [81].

8.3.3 Dosimeter Approach

The overall strategy for designing chemodosimeters relies on exploiting selective analyte-induced reaction chemistry for its detection. This reaction-based sensing chemistry represents an attractive direction for developing novel chemoselective probes for tracking biologically important signaling and stress molecules. Metal-based dosime-ters for anions and small molecules can be classified based on their involvement in the dosimetric reaction. In the first case, the metal center mediates redox or organometallic reactions directly with the analyte, which leads to a new signal response from the reporter. This strategy has been applied in the development of a number of NO, CO, and H_2O_2 chemodosimeters. In the second case, the metal center is not directly involved in the

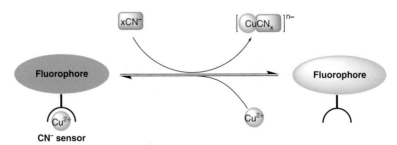

Figure 8.19 *Schematic representation of Cu(II) complexes for detection of cyanide anion by cyanide-induced demetalation*

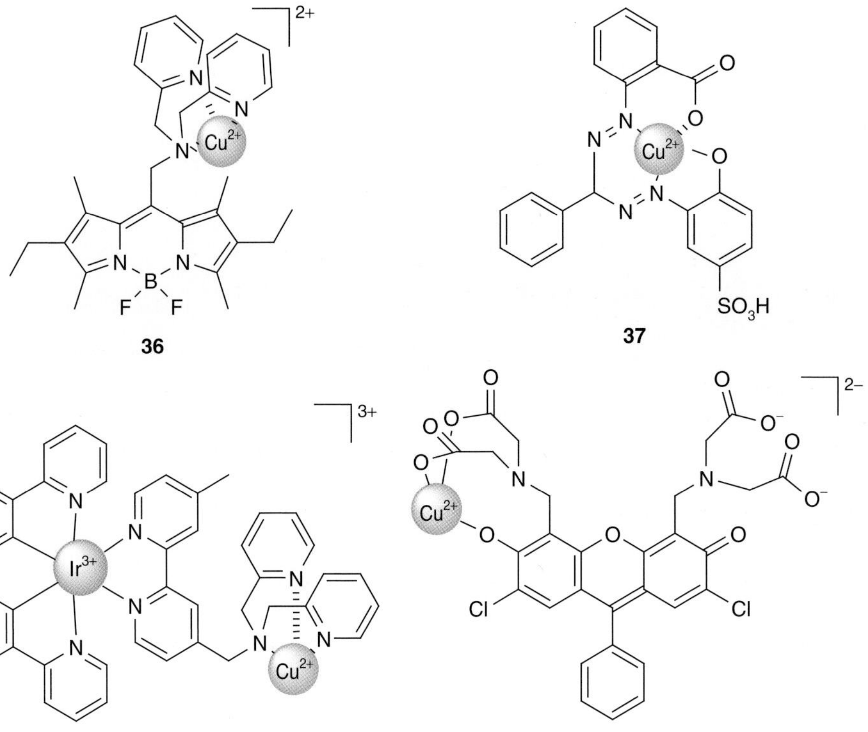

Figure 8.20   *Cu(II) complexes as cyanide sensors*

reaction. Instead, the analyte reacts directly with an organic component of the sensor, which then changes the signal from the metal-based reporter unit. Representative examples include chemodosimeters for detection of reactive oxygen species (ROS).

8.3.3.1   Metal-mediated reaction strategy

Detecting NO by copper-mediated reductive nitrosylation.   The main limitations of the aforementioned displacement sensors for NO include water incompatibility and cross-reactivity with O_2 or other species. In order to circumvent these drawbacks, Lippard and coworkers developed Cu^{2+}-based complex **40** that enables NO detection both in aqueous solution and in live cells [82]. It consists of a Cu^{2+}-bound aminoquinoline conjugated with a fluorescein that is quenched by the paramagnetic Cu^{2+} center. In the presence of NO, the Cu^{2+}-mediated reductive nitrosylation at the secondary amine releases the metal quencher, with a concomitant enhancement of fluorescein emission (Fig. 8.21). This reaction-based sensing process demonstrates high selectivity towards NO over a variety of RNS and ROS species. Based on these desirable features, complex **40** and its later-generation analogues have been used as a tool to study the role of a bacterial NO synthase through imaging of endogenous NO production in biological models ranging from bacteria, macrophages to olfactory bulb brain slices [82–84].

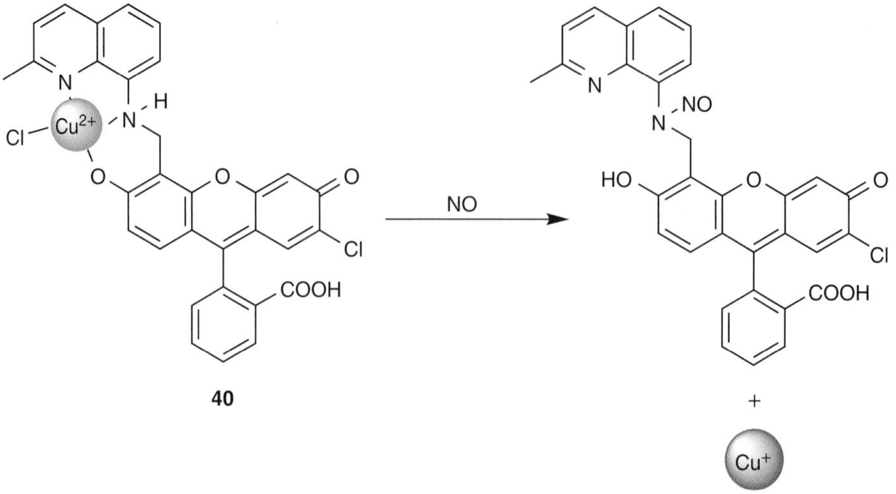

Figure 8.21   *The Cu(II) complex* **40** *is weakly emissive, but reaction with NO induces reductive nitrosylation of the ligand, release of Cu+, and restoration of fluorescence intensity [82]*

Detecting CO by palladium-mediated carbonylation.   While carbon monoxide (CO) is best known for its toxicity, it is also a gasotransmitter, similar to NO, that plays important roles in regulating response to both physical and chemical stresses (see also Chapter 10). The real-time tracking of CO at a cellular level remains an unsolved challenge. Unlike NO, CO is not redox-reactive. It does, however, have a rich history in organometallic chemistry. Therefore, a novel chemodosimeter **41** based on metal-mediated carbonylation chemistry has been developed by Chang and coworkers [85]. Conjugation of a cyclopalladated moiety to a boron dipyrromethene difluoride (BODIPY) fluorophore is anticipated to quench the emission, while reaction with CO results in Pd-mediated carbonylation, release of Pd(0) and concomitant fluorescence recovery (Fig. 8.22). The cellular fate of the spent Pd(0) has

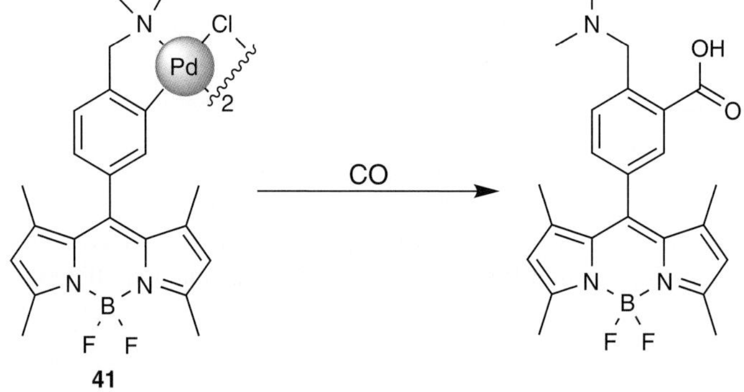

Figure 8.22   *Cyclopalladated dosimeter for sensing CO through Pd-mediated carbonylation [85]. The complex on the left has diminished fluorescence intensity compared to the product on the right*

not yet been addressed. The robust turn-on response, the cell permeability, and low toxicity of **41** make it attractive for imaging CO both in aqueous buffered solutions and in live cells with high selectivity over a range of reactive species.

Detecting H_2O_2 by iron-mediated oxidative cleavage. In addition to reductive and organometallic reactions, metal-mediated oxidative mechanisms can also be applied in chemoselective imaging. Two elegant examples are selective H_2O_2 dosimeters based on iron-peroxide redox reactivity, where metal ions are reaction sites for H_2O_2 (Fig. 8.23). Hitomi *et al.* recently developed probe **42** containing an Fe^{3+}–polypyridine conjugated to a non-emissive 3,7-dihydroxyphenoxazine amide [86], whereas the Nam group directly adopted the scaffold of the classic zinc sensor ZP1 and further employed its bis-Fe^{2+} metalated version as an H_2O_2 sensor **43** [87]. Addition of H_2O_2 results in iron-mediated oxidation, cleavage, and release of the corresponding emissive fluorescent dyes resorufin and bis-carboxy fluorescein, respectively. **43** is reported to track changes in H_2O_2 levels localized to lysosomes in live cells [87]. The signal transduction mechanism was speculated to involve a reactive iron-peroxo or oxo center capable of mediating N-dealkylation chemistry. The fate of the released iron and whether it continues to propogate reactive oxidation chemistry has not yet been addressed.

8.3.3.2 Organic Reaction Strategy

Detecting ROS by profluorophore oxidation. The high reactivity of ROS has been widely utilized to design chemoselective reaction-based sensors. This general concept is the "profluorophore" strategy that involves modification of a fluorophore with a redox reactive handle that effectively quenches its emission. Selective redox reaction of the handle with ROS causes cleavage of the handle or addition of the analytes to further reveal the fluorophore with a subsequent recovery of fluorescence. Within the scope of metal-based dosimeters, metal ions anchored by suitable ligands function as the reporter units, rather than recognition sites, in the following examples.

Representative examples of metal-based ROS sensors using a "profluorophore" strategy are illustrated in Fig. 8.24. Chang and coworkers introduced a Tb(III)-based boronate-protected probe **44** [88]. In the absence of H_2O_2, the boronate-capped aromatic ring is a poor lanthanide sensitizer, while H_2O_2-triggered oxidative cleavage reaction produces an electron-rich aniline antenna that enables energy transfer to the lanthanide center with a turn-on luminescent response. Importantly, **44** is capable of *in cellulo* monitoring of endogenous H_2O_2 production using oxidatively stressed macrophages as a cell model.

Sensor **45** is another example that uses a change in the sensitizing antenna to turn on lanthanide luminescence. As an antenna precursor, trimesate is a weak chelator and poor sensitizer for Tb(III). However, reaction with hydroxyl radicals (•OH) converts it into an adequate bidentate chelator that functions as a nearby antenna and efficient sensitizer for Tb(III) luminescence, resulting in an 11-fold enhancement of emission intensity [89].

Sensor **46** also undergoes a ligand-centered oxidative reaction that increases its luminescent emission, but in a slightly different platform. In this case, reaction with ClO^- relieves the luminescence quench caused by rapid isomerization of the C=N−OH derivative on the bipyridyl ligand of cyclometalated Ir(III) complex **46** [90]. All three dosimeters in Fig. 8.24 demonstrate high selectivity for the corresponding analyte over other reactive species.

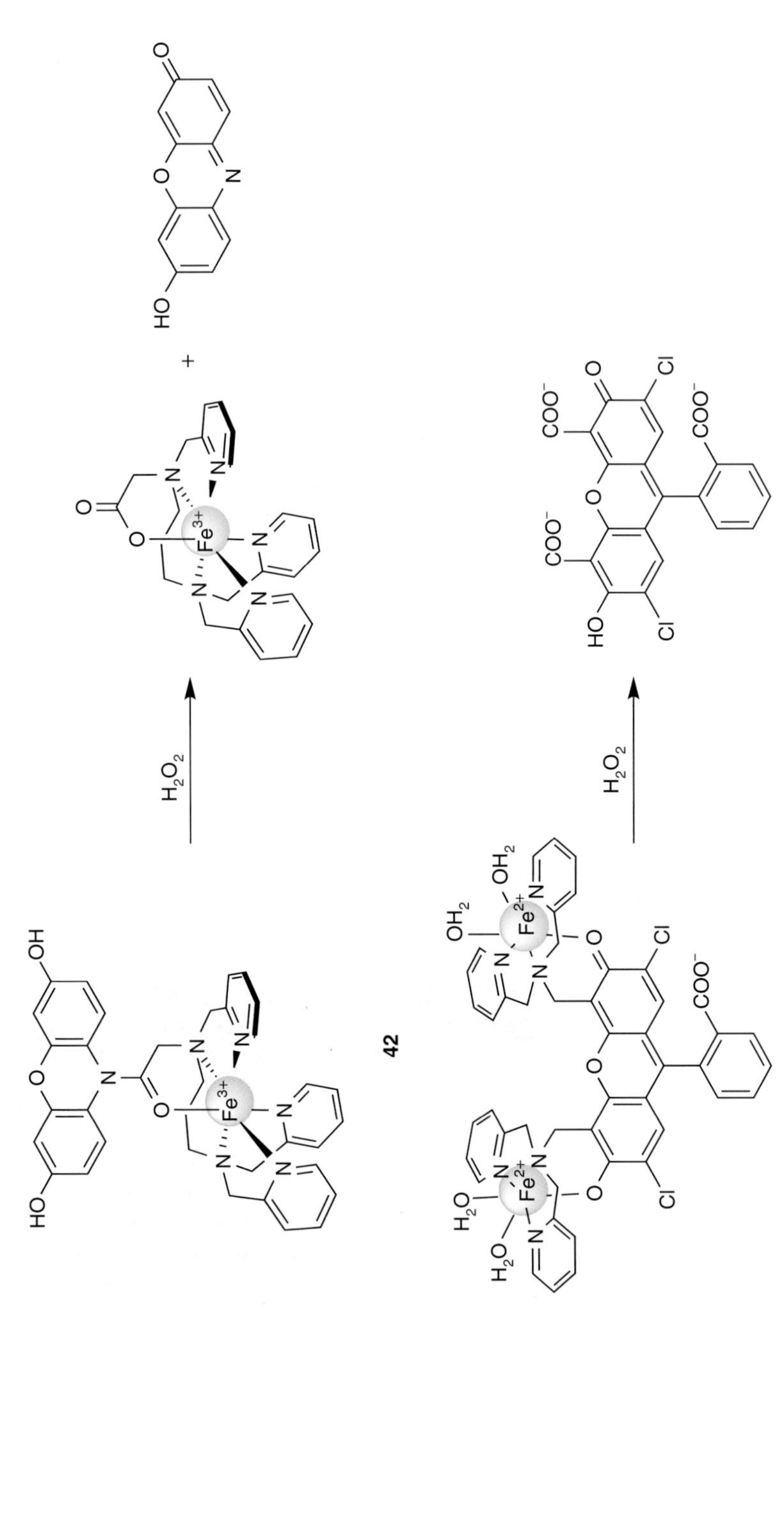

Figure 8.23 *Iron complexes as dosimeters for detecting H_2O_2 through Fe-mediated oxidative ligand cleavage to unquench and release a fluorophore [86, 87]*

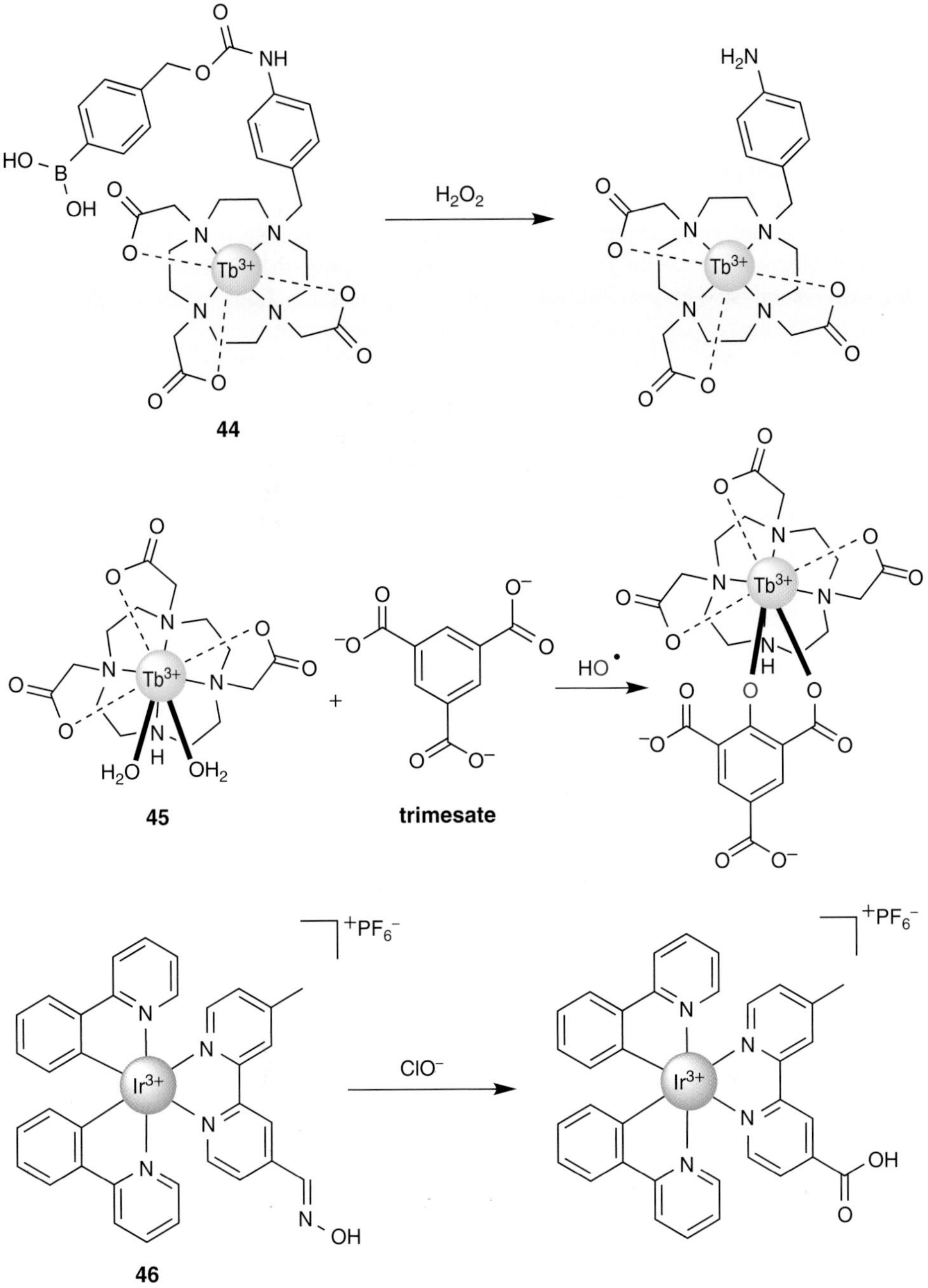

Figure 8.24 *Metal-based "profluorophores" for detection of various reactive species. The complexes on the right are more emissive than those on the left*

8.4 Conclusions

Compared with the large volume of probes and sensors built from traditional organic dyes as signal-transducing elements, successful examples of metal-based reagents are relatively limited. This disparity is evident by scanning through catalogs of commercial fluorescent probes, for example. However, the examples presented in this chapter attest to the vast scope available for creative sensor design based on various aspects of metal coordination chemistry. The design principles elucidated in the previous sections provide general frameworks in which the metal can be used as the recognition site, the reporter site, the reactive site, or the catalyst. The possibilities of optimizing these functions by tuning electronic, steric, and geometric parameters of metal complexes are limitless.

Creating these probes requires a grasp of inorganic chemistry, fundamental organic transformation, organometallic reactions, and an understanding of the photophysical underpinnings of the signaling mechanism. Utilizing these probes effectively to learn about biology requires an understanding of cell and molecular biology and a deep appreciation for the complexity of living systems. Ongoing and future work in this area therefore draws on all of these disciplines for further advancement. Challenges for the future include a better understanding of the uptake and subcellular partitioning of these probes, a better handle on knowing the intrinsic capabilities of a particular probe to avoid misinterpreting its response, creating multimodal systems that combine multiple imaging techniques, and moving beyond "one analyte, one probe" into a more systems approach. These challenges are indeed forefront challenges for bioprobes in general and are not unique to metal-based probes. Given their unique properties, however, metal complexes may provide unique solutions.

Acknowledgements

We thank the National Institutes of Health (GM084176) and the National Science Foundation (CHE-1152054) for supporting our laboratory's work on topics related to this review.

Abbreviations

ATP	adenosine triphosphate
BAPTA	1,2-bis(2-aminophenoxy)ethane- *N,N,N',N'*-tetraacetic acid
BODIPY	boron dipyrromethene
BPS	4,7-diphenyl-1,10-phenanthroline-disulfonate
bpy	2,2'-bipyridine
btp	2-(2'-benzothienyl)pyridine
DOTA	1,4,7,10-tetraazacyclododecane-1,4,7,10-tetraacetic acid
DPA	2,2'-dipicolyamine
Ds-im	dansyl imidazole
DTPA	2-[bis[2-[bis(carboxymethyl)amino]ethyl]amino]acetic acid
HSA	human serum albumin

MRI	magnetic resonance imaging
PET	positron emission tomography
PPi	pyrophosphate
ppy	2-phenylpyridine
PS	phosphatidylserine
Ptz	phenothiazine
RNS	reactive nitrogen species
ROS	reactive oxygen species
SPECT	single photon emission computed tomography
ZP1	4',5'-bis[bis(2-pyridylmethyl)aminomethyl]-2',7'-dichlorofluorescein

References

1. Wadas, T.J., Wong, E.H., Weisman, G.R., and Anderson, C.J. (2010) Coordinating radiometals of copper, gallium, indium, yttrium, and zirconium for PET and SPECT imaging of disease. Chem Rev. 110: 2858–2902.
2. Beer, P.D. and Cadman, J. (2000) Electrochemical and optical sensing of anions by transition metal based receptors. Coord. Chem. Rev. 205: 131–155.
3. Tucker, J.H. and Collinson, S.R. (2002) Recent developments in the redox-switched binding of organic compounds. Chem. Soc. Rev. 31: 147–156.
4. Haas, K.L. and Franz, K.J. (2009) Application of metal coordination chemistry to explore and manipulate cell biology. Chem. Rev. 109: 4921–4960.
5. Montgomery, C.P., Murray, B.S., New, E.J., Pal, R., and Parker, D. (2009) Cell-penetrating metal complex optical probes: Targeted and responsive systems based on lanthanide luminescence. Acc. Chem. Res. 42: 925–937.
6. New, E.J., Parker, D., Smith, D.G., and Walton, J.W. (2010) Development of responsive lanthanide probes for cellular applications. Curr. Opin. Chem. Biol. 14: 238–246.
7. Jones, C.J. and Thornback, J., (2007) Medicinal Applications of Coordination Chemistry. Cambridge: Royal Society of Chemistry.
8. Andrews, N.C. and Schmidt, P.J. (2007) Iron homeostasis. Annu Rev Physiol. 69: 69–85.
9. Kim, B.-E., Nevitt, T., and Thiele, D.J. (2008) Mechanisms for copper acquisition, distribution and regulation. Nat. Chem. Biol. 4: 176–185.
10. Crichton, R.R., Dexter, D.T., and Ward, R.J. (2008) Metal based neurodegenerative diseases–From molecular mechanisms to therapeutic strategies. Coord. Chem. Rev. 252: 1189–1199.
11. World Health Organization, (2004) Guidelines for Drinking-water Quality. 3rd edn. Geneva: World Health Organization.
12. Nolan, E.M. and Lippard, S.J. (2008) Tools and tactics for the optical detection of mercuric ion. Chem Rev. 108: 3443–3480.
13. Que, E.L., Domaille, D.W., and Chang, C.J. (2008) Metals in neurobiology: Probing their chemistry and biology with molecular imaging. Chem. Rev. 108: 1517–1549.
14. Liu, Z., He, W., and Guo, Z. (2013) Metal coordination in photoluminescent sensing. Chem. Soc. Rev. 42: 1568–600.

15. Parker, D. (2011) Critical design factors for optical imaging with metal coordination complexes. Aust. J. Chem. 64: 239–243.

16. Dixon, I.M., Collin, J.P., Sauvage, J.P., Flamigni, L., Encinas, S., and Barigelletti, F. (2000) A family of luminescent coordination compounds: iridium(III) polyimine complexes. Chem. Soc. Rev. 29: 385–391.

17. Araya, J.C., Gajardo, J., Moya, S.A., Aguirre, P., Toupet, L., Williams, J.A.G., Escadeillas, M., Le Bozec, H., and Guerchais, V. (2010) Modulating the luminescence of an iridium(III) complex incorporating a di(2-picolyl)anilino-appended bipyridine ligand with Zn^{2+} cations. New J. Chem. 34: 21–24.

18. Lee, P.K., Law, W.H.T., Liu, H.W., and Lo, K.K.W. (2011) Luminescent cyclometalated iridium(III) polypyridine di-2-picolylamine complexes: Synthesis, photophysics, electrochemistry, cation binding, cellular Internalization, and cytotoxic activity. Inorg. Chem. 50: 8570–8579.

19. You, Y., Lee, S., Kim, T., Ohkubo, K., Chae, W.S., Fukuzumi, S., Jhon, G.J., Nam, W., and Lippard, S.J. (2011) Phosphorescent sensor for biological mobile zinc. J. Am. Chem. Soc. 133: 18328–18342.

20. You, Y., Han, Y., Lee, Y.M., Park, S.Y., Nam, W., and Lippard, S.J. (2011) Phosphorescent sensor for robust quantification of copper(II) ion. J. Am. Chem. Soc. 133: 11488–11491.

21. Li, Z.X., Zhang, L.F., Wang, L.N., Guo, Y.K., Cai, L.H., Yu, M.M., and Wei, L.H. (2011) Highly sensitive and selective fluorescent sensor for Zn^{2+}/Cu^{2+} and new approach for sensing Cu^{2+} by central metal displacement. Chem. Commun. 47: 5798–5800.

22. Kotova, O., Comby, S., and Gunnlaugsson, T. (2011) Sensing of biologically relevant d-metal ions using a Eu(III)-cyclen based luminescent displacement assay in aqueous pH 7.4 buffered solution. Chem. Commun. 47: 6810–6812.

23. Pehkonen, S. (1995) Determination of the oxidation-states of iron in natural-waters - a review. Analyst. 120: 2655–2663.

24. Hase, U. and Yoshimura, K. (1992) Determination of trace amounts of iron in highly purified water by ion-exchanger phase absorptiometry combined with flow-analysis. Analyst. 117: 1501–1506.

25. Achterberg, E.P., Holland, T.W., Bowie, A.R., Fauzi, R., Mantoura, C., and Worsfold, P.J. (2001) Determination of iron in seawater. Anal. Chim. Acta. 442: 1–14.

26. Tsien, R.Y. (2003) Imagining imaging's future. Nat. Rev. Mol. Cell Biol.: Ss16–Ss21.

27. Caravan, P., Ellison, J.J., McMurry, T.J., and Lauffer, R.B. (1999) Gadolinium(III) chelates as MRI contrast agents: Structure, dynamics, and applications. Chem Rev. 99: 2293–2352.

28. Werner, E.J., Datta, A., Jocher, C.J., and Raymond, K.N. (2008) High-relaxivity MRI contrast agents: Where coordination chemistry meets medical imaging. Angew. Chem. Int. Edit. 47: 8568–8580.

29. Bonnet, C.S. and Toth, E. (2010) MRI probes for sensing biologically relevant metal ions. Future Med. Chem. 2: 367–84.

30. Li, W.H., Parigi, G., Fragai, M., Luchinat, C., and Meade, T.J. (2002) Mechanistic studies of a calcium-dependent MRI contrast agent. Inorg. Chem. 41: 4018–4024.

31. Li, W.H., Fraser, S.E., and Meade, T.J. (1999) A calcium-sensitive magnetic resonance imaging contrast agent. J. Am. Chem. Soc. 121: 1413–1414.

32. Dhingra, K., Fouskova, P., Angelovski, G., Maier, M.E., Logothetis, N.K., and Toth, E. (2008) Towards extracellular Ca^{2+} sensing by MRI: synthesis and calcium-dependent H-1 and O-17 relaxation studies of two novel bismacrocyclic Gd^{3+} complexes. J. Biol. Inorg. Chem. 13: 35–46.

33. Mishra, A., Fouskova, P., Angelovski, G., Balogh, E., Mishra, A.K., Logothetis, N.K., and Toth, E. (2008) Facile synthesis and relaxation properties of novel bispolyaza-macrocyclic Gd^{3+} complexes: An attempt towards calcium-sensitive MRI contrast agents. Inorg. Chem. 47: 1370–1381.

34. Angelovski, G., Fouskova, P., Mamedov, I., Canals, S., Toth, E., and Logothetis, N.K. (2008) Smart magnetic resonance imaging agents that sense extracellular calcium fluctuations. Chembiochem. 9: 1729–1734.

35. Dhingra, K., Maier, M.E., Beyerlein, M., Angelovski, G., and Logothetis, N.K. (2008) Synthesis and characterization of a smart contrast agent sensitive to calcium. Chem. Commun.: 3444–3446.

36. Que, E.L. and Chang, C.J. (2006) A smart magnetic resonance contrast agent for selective copper sensing. J. Am. Chem. Soc. 128: 15942–15943.

37. Que, E.L., Gianolio, E., Baker, S.L., Wong, A.P., Aime, S., and Chang, C.J. (2009) Copper-responsive magnetic resonance imaging contrast agents. J. Am. Chem. Soc. 131: 8527–8536.

38. Que, E.L., New, E.J., and Chang, C.J. (2012) A cell-permeable gadolinium contrast agent for magnetic resonance imaging of copper in a Menkes disease model. Chem. Sci. 3: 1829–1834.

39. Jones, S.W., Christison, R., Bundell, K., Voyce, C.J., Brockbank, S.M.V., Newham, P., and Lindsay, M.A. (2005) Characterisation of cell-penetrating peptide-mediated peptide delivery. Br. J. Pharmacol. 145: 1093–1102.

40. Maret, W. and Li, Y. (2009) Coordination dynamics of zinc in proteins. Chem. Rev. 109: 4682–4707.

41. Zalewski, P., Truong-Tran, A., Lincoln, S., Ward, D., Shankar, A., Coyle, P., Jayaram, L., Copley, A., Grosser, D., Murgia, C., Lang, C., and Ruffin, R. (2006) Use of a zinc fluorophore to measure labile pools of zinc in body fluids and cell-conditioned media. Biotechniques. 40: 509–520.

42. Frederickson, C.J., Koh, J.Y., and Bush, A.I. (2005) The neurobiology of zinc in health and disease. Nat. Rev. Neurosci. 6: 449–462.

43. Hanaoka, K., Kikuchi, K., Urano, Y., and Nagano, T. (2001) Selective sensing of zinc ions with a novel magnetic resonance imaging contrast agent. J. Chem. Soc., Perkin Trans. 2: 1840–1843.

44. Hanaoka, K., Kikuchi, K., Urano, Y., Narazaki, M., Yokawa, T., Sakamoto, S., Yamaguchi, K., and Nagano, T. (2002) Design and synthesis of a novel magnetic resonance imaging contrast agent for selective sensing of zinc ion. Chem. Biol. 9: 1027–1032.

45. Major, J.L., Parigi, G., Luchinat, C., and Meade, T.J. (2007) The synthesis and *in vitro* testing of a zinc-activated MRI contrast agent. Proc. Natl. Acad. Sci. USA. 104: 13881–13886.

46. Major, J.L., Boiteau, R.M., and Meade, T.J. (2008) Mechanisms of Zn(II)-activated magnetic resonance imaging agents. Inorg. Chem. 47: 10788–10795.

47. Esqueda, A.C., Lopez, J.A., Andreu-De-Riquer, G., Alvarado-Monzon, J.C., Rat-nakar, J., Lubag, A.J.M., Sherry, A.D., and De Leon-Rodriguez, L.M. (2009) A new gadolinium-based MRI zinc sensor. J. Am. Chem. Soc. 131: 11387–11391.

48. Lubag, A.J.M., De Leon-Rodriguez, L.M., Burgess, S.C., and Sherry, A.D. (2011) Noninvasive MRI of beta-cell function using a Zn^{2+}-responsive contrast agent. Proc. Natl. Acad. Sci. USA. 108: 18400–18405.

49. Zhang, X.A., Lovejoy, K.S., Jasanoff, A., and Lippard, S.J. (2007) Water-soluble porphyrins as a dual-function molecular imaging platform for MRI and fluorescence zinc sensing. Proc. Natl. Acad. Sci. USA. 104: 10780–10785.

50. Lee, T., Zhang, X.A., Dhar, S., Faas, H., Lippard, S.J., and Jasanoff, A. (2010) *In vivo* imaging with a cell-permeable porphyrin-based MRI contrast agent. Chem. Biol. 17: 665–673.

51. Kabalka, G.W., Davis, M.A., Moss, T.H., Buonocore, E., Hubner, K., Holmberg, E., Maruyama, K., and Huang, L. (1991) Gadolinium-labeled liposomes containing various amphiphilic Gd-Dtpa derivatives - targeted MRI contrast enhancement agents for the liver. Magn. Reson. Med. 19: 406–415.

52. Allen, M.J. and Meade, T.J. (2003) Synthesis and visualization of a membrane-permeable MRI contrast agent. J. Biol. Inorg. Chem. 8: 746–750.

53. Bhorade, R., Weissleder, R., Nakakoshi, T., Moore, A., and Tung, C.H. (2000) Macrocyclic chelators with paramagnetic cations are internalized into mammalian cells via a HIV-tat derived membrane translocation peptide. Bioconjugate Chem. 11: 301–305.

54. Wiener, E.C., Konda, S., Shadron, A., Brechbiel, M., and Gansow, O. (1997) Targeting dendrimer-chelates to tumors and tumor cells expressing the high-affinity folate receptor. Invest. Radiol. 32: 748–54.

55. Casali, C., Janier, M., Canet, E., Obadia, J.F., Benderbous, S., Corot, C., and Revel, D. (1998) Evaluation of Gd-DOTA-labeled dextran polymer as an intravascular MR contrast agent for myocardial perfusion. Acad. Radiol. 5: S214–S218.

56. Endres, P.J., Paunesku, T., Vogt, S., Meade, T.J., and Woloschak, G.E. (2007) DNA-TiO_2 nanoconjugates labeled with magnetic resonance contrast agents. J. Am. Chem. Soc. 129: 15760–15761.

57. Song, Y., Xu, X., MacRenaris, K.W., Zhang, X.Q., Mirkin, C.A., and Meade, T.J. (2009) Multimodal gadolinium-enriched DNA-gold nanoparticle conjugates for cellular imaging. Angew. Chem. Int. Ed. 48: 9143–9147.

58. Liu, Y., Li, M.Y., Zhao, Q., Wu, H.Z., Huang, K.W., and Li, F.Y. (2011) Phosphorescent iridium(III) complex with an N boolean and O ligand as a Hg^{2+}-selective chemodosimeter and logic gate. Inorg. Chem. 50: 5969–5977.

59. Yang, H., Qian, J.J., Li, L.T., Zhou, Z.G., Li, D.R., Wu, H.X., and Yang, S.P. (2010) A selective phosphorescent chemodosimeter for mercury ion. Inorg. Chim. Acta. 363: 1755–1759.

60. Wu, Y.Q., Jing, H., Dong, Z.S., Zhao, Q., Wu, H.Z., and Li, F.Y. (2011) Ratiometric phosphorescence imaging of Hg(II) in living cells based on a neutral iridium(III) complex. Inorg. Chem. 50: 7412–7420.

61. Ajayakumar, G., Sreenath, K., and Gopidas, K.R. (2009) Phenothiazine attached Ru(bpy)$_{(3)}^{(2+)}$ derivative as highly selective "turn-ON" luminescence chemodosimeter for Cu^{2+}. Dalton Trans.: 1180–1186.

62. Viguier, R.F.H. and Hulme, A.N. (2006) A sensitized europium complex generated by micromolar concentrations of copper(I): Toward the detection of copper(I) in biology. J. Am. Chem. Soc. 128: 11370–11371.

63. Kim, S.K., Lee, D.H., Hong, J.I., and Yoon, J. (2009) Chemosensors for pyrophosphate. Acc. Chem. Res. 42: 23–31.

64. Jang, Y.J., Jun, E.J., Lee, Y.J., Kim, Y.S., Kim, J.S., and Yoon, J. (2005) Highly effective fluorescent and colorimetric sensors for pyrophosphate over H_2PO^{4-} in 100% aqueous solution. J. Org. Chem. 70: 9603–9606.

65. Hanshaw, R.G., Lakshmi, C., Lambert, T.N., Johnson, J.R., and Smith, B.D. (2005) Fluorescent detection of apoptotic cells by using zinc coordination complexes with a selective affinity for membrane surfaces enriched with phosphatidylserine. Chembiochem. 6: 2214–2220.

66. van Engeland, M., Nieland, L.J.W., Ramaekers, F.C.S., Schutte, B., and Reutelingsperger, C.P.M. (1998) Annexin V-affinity assay: A review on an apoptosis detection system based on phosphatidylserine exposure. Cytometry. 31: 1–9.

67. Smith, D.G., Law, G.-l., Murray, B.S., Pal, R., Parker, D., and Wong, K.-L. (2011) Evidence for the optical signalling of changes in bicarbonate concentration within the mitochondrial region of living cells. Chem. Commun. 47: 7347–7349.

68. Smith, D.G., Pal, R., and Parker, D. (2012) Measuring equilibrium bicarbonate concentrations directly in cellular mitochondria and in human serum using europium/terbium emission intensity ratios. Chem. Eur. J. 18: 11604–11613.

69. Ojida, A., Takashima, I., Kohira, T., Nonaka, H., and Hamachi, I. (2008) Turn-on fluorescence sensing of nucleoside polyphosphates using a xanthene-based Zn(II) complex chemosensor. J. Am. Chem. Soc. 130: 12095–12101.

70. Ojida, A., Nonaka, H., Miyahara, Y., Tamaru, S.I., Sada, K., and Hamachi, I. (2006) Bis(Dpa-Zn-II) appended xanthone: Excitation ratiometric chemosensor for phosphate anions. Angew. Chem. Int. Ed. 45: 5518–5521.

71. Franz, K.J., Singh, N., and Lippard, S.J. (2000) Metal-based NO sensing by selective ligand dissociation. Angew. Chem. Int. Ed. 39: 2120–2122.

72. Lim, M.H. and Lippard, S.J. (2004) Fluorescence-based nitric oxide detection by ruthenium porphyrin fluorophore complexes. Inorg. Chem. 43: 6366–6370.

73. Katayama, Y., Takahashi, S., and Maeda, M. (1998) Design, synthesis and characterization of a novel fluorescent probe for nitric oxide (nitrogen monoxide). Anal. Chim. Acta. 365: 159–167.

74. Hilderbrand, S.A., Lim, M.H., and Lippard, S.J. (2004) Dirhodium tetracarboxylate scaffolds as reversible fluorescence-based nitric oxide sensors. J. Am. Chem. Soc. 126: 4972–4978.

75. Guliyev, R., Buyukcakir, O., Sozmen, F., and Bozdemir, O.A. (2009) Cyanide sensing via metal ion removal from a fluorogenic BODIPY complex. Tetrahedron Lett. 50: 5139–5141.

76. Lou, X.D., Zhang, L.Y., Qin, J.G., and Li, Z. (2008) An alternative approach to develop a highly sensitive and selective chemosensor for the colorimetric sensing of cyanide in water. Chem. Commun.: 5848–5850.

77. Reddy, G.U., Das, P., Saha, S., Baidya, M., Ghosh, S.K., and Das, A. (2013) A CN-specific turn-on phosphorescent probe with probable application for enzymatic assay and as an imaging reagent. Chem. Commun. 49: 255–257.

78. Chung, S.Y., Nam, S.W., Lim, J., Park, S., and Yoon, J. (2009) A highly selective cyanide sensing in water via fluorescence change and its application to *in vivo* imaging. Chem. Commun.: 2866–2868.

79. Chen, X., Nam, S.-W., Kim, G.-H., Song, N., Jeong, Y., Shin, I., Kim, S.K., Kim, J., Park, S., and Yoon, J. (2010) A near-infrared fluorescent sensor for detection of cyanide in aqueous solution and its application for bioimaging. Chem. Commun. 46: 8953–8955.

80. Zelder, F.H. (2008) Specific colorimetric detection of cyanide triggered by a conformational switch in vitamin B12. Inorg. Chem. 47: 1264–1266.

81. Mannel-Croise, C. and Zelder, F. (2012) Rapid visual detection of blood cyanide. Anal. Meth. 4: 2632–2634.

82. Lim, M.H., Xu, D., and Lippard, S.J. (2006) Visualization of nitric oxide in living cells by a copper-based fluorescent probe. Nat. Chem. Biol. 2: 375–380.

83. McQuade, L.E., Ma, J., Lowe, G., Ghatpande, A., Gelperin, A., and Lippard, S.J. (2010) Visualization of nitric oxide production in the mouse main olfactory bulb by a cell-trappable copper(II) fluorescent probe. Proc. Natl. Acad. Sci. USA. 107: 8525–8530.

84. Pluth, M.D., Chan, M.R., McQuade, L.E., and Lippard, S.J. (2011) Seminaphtho fluorescein-based fluorescent probes for imaging nitric oxide in Live cells. Inorg. Chem. 50: 9385–9392.

85. Michel, B.W., Lippert, A.R., and Chang, C.J. (2012) A reaction-based fluorescent probe for selective imaging of carbon monoxide in living cells using a palladium-mediated carbonylation. J. Am. Chem. Soc. 134: 15668–15671.

86. Hitomi, Y., Takeyasu, T., Funabiki, T., and Kodera, M. (2011) Detection of enzymatically generated hydrogen peroxide by metal-based fluorescent probe. Anal. Chem. 83: 9213–9216.

87. Song, D., Lim, J.M., Cho, S., Park, S.J., Cho, J., Kang, D., Rhee, S.G., You, Y., and Nam, W. (2012) A fluorescence turn-on H_2O_2 probe exhibits lysosome-localized fluorescence signals. Chem. Commun. 48: 5449–5451.

88. Lippert, A.R., Gschneidtner, T., and Chang, C.J. (2010) Lanthanide-based luminescent probes for selective time-gated detection of hydrogen peroxide in water and in living cells. Chem. Commun. 46: 7510–7512.

89. Page, S.E., Wilke, K.T., and Pierre, V.C. (2010) Sensitive and selective time-gated luminescence detection of hydroxyl radical in water. Chem. Commun. 46: 2423–2425.

90. Zhao, N., Wu, Y.H., Wang, R.M., Shi, L.X., and Chen, Z.N. (2011) An iridium(III) complex of oximated 2,2'-bipyridine as a sensitive phosphorescent sensor for hypochlorite. Analyst. 136: 2277–2282.

9

Photo-release of Metal Ions in Living Cells

Celina Gwizdala[a] and Shawn C. Burdette[b]
[a]Department of Chemistry, University of Connecticut, USA
[b]Department of Chemistry and Biochemistry, Worcester Polytechnic Institute, USA

9.1 Introduction to Photochemical Tools Including Photocaged Complexes

A greater emphasis on controlling processes instead of relying on observational research is the hallmark of modern experimental biology. Combining observation and manipulation can help establish a more detailed understanding of complex processes. Synthetic mediators including fluorescent sensors and actuators have become increasingly popular for monitoring and attenuating biological processes, respectively [1–3]. Since even miniscule changes in concentrations of molecules and ions can have profound effects on homeostasis, signal transduction, and metabolism, synthetic mediators are often chosen to probe these phenomena. Development and application of such tools requires significant expertise in several other diverse areas including synthetic chemistry and cell biology. The multi-disciplinary nature of these types of investigations requires collaboration between scientists with different backgrounds. The end users must convey the requirements and needed improvements to tool developers, and synthetic chemists must be cognizant to pursue high-priority targets.

Metal ions trigger, modulate, and terminate numerous biological processes in all living organisms [4]. The study of metal ion homeostasis, transport, and regulation has a profound impact on understanding basic physiology. Fluorescent sensors for metal ions have been developed extensively over the last three decades (Fig. 9.1) [5]. Seminal work by multiple international research groups has facilitated a sophisticated understanding of sensor photochemistry as well as the specific requirements for applying these tools to biological

Inorganic Chemical Biology: Principles, Techniques and Applications, First Edition. Edited by Gilles Gasser.
© 2014 John Wiley & Sons, Ltd. Published 2014 by John Wiley & Sons, Ltd.

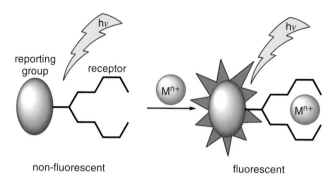

Figure 9.1   *Graphical representation of the signal transduction mechanism in an off–on fluorescent metal ion sensor. In the absence of analyte, light excitation of the reporting group – typically an organic fluorophore – results in no signal. Metal ion binding to the receptor interrupts a quenching pathway so excitation of the reporting group results in a fluorescence output*

problems. Imaging studies with fluorescent sensors have provided important information about basal concentrations levels, fluctuations and mobility, and pathway disruptions of metal ions. While small molecule sensor development has become relatively standardized, the development of metal ion actuators for biological applications has received less attention.

The role of Ca^{2+} as a biological signaling agent began receiving increased attention in the 1970s, but the development of new photochemical tools ushered in a new era of bioinorganic chemistry. The research group led by Roger Tsien played a key role in the development of small molecule fluorescent sensors for Ca^{2+} [6]. Using a Ca^{2+}-selective EGTA (ethylene glycol-bis(2-aminoethylether)-N,N,N',N'-tetraacetic acid) chelator and different organic fluorophores, the Tsien group took advantage of several signal transduction mechanisms to develop a family of probes engineered to interrogate Ca^{2+} signals in a variety of systems (Fig. 9.2). These Ca^{2+} probes as well as the sensors developed in the ensuing years for other metal ions, allowed biological events to be visualized that would otherwise only be detectible by indirect techniques.

Sensors are powerful tools for observation, but lack the ability to manipulate biological events. Achieving a complete understanding of metal ion homeostasis requires additional chemical tools with the ability to precisely control the delivery of metal ions. The simultaneous development of photocaged Ca^{2+} complexes by Tsien and others provided complementary tools for both monitoring and controlling Ca^{2+} concentrations ($[Ca^{2+}]$). Photocaged complexes are small molecule metal ion chelators that undergo a light-initiated change in binding affinity (Fig. 9.3). To date, photocaged complexes have been almost exclusively based on the photochemistry of nitrobenzyl groups, which were developed originally as protecting groups in organic synthesis [10, 11] and to control the delivery of bioactive small molecules like amino acids [12].

To be compatible with biological experimentation, photocaged complexes must satisfy several requirements [1, 2, 13, 14]. From a chemical perspective, an ideal photocaged complex should undergo an efficient photolysis reaction that results in formation of photoproduct(s) with significantly reduced affinity for the metal ion compared with the parent chelator. The photocaged complex must effectively isolate the metal ion from

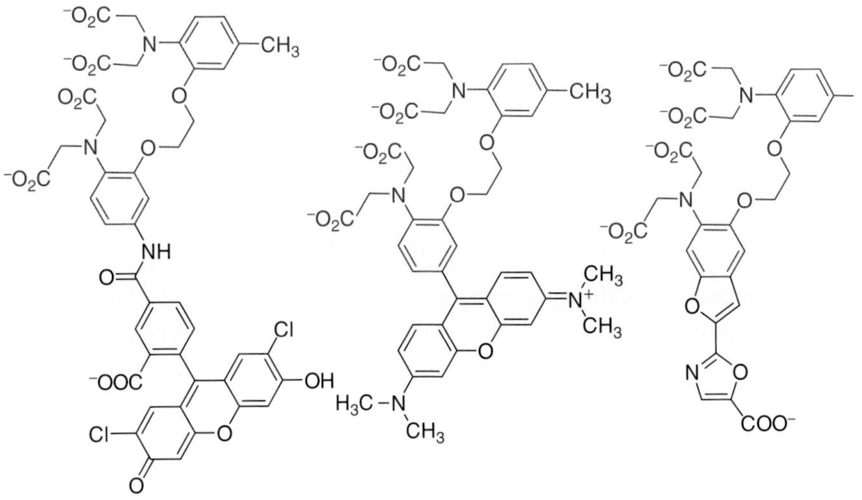

CG-1   Rhod-1   Fura-2

Figure 9.2   *Representative fluorescent sensors using the Ca²⁺-selective BAPTA receptor (magenta). CG-1 [7] and Rhod-1 [8] are off–on sensors while Fura-2 is a ratiometric probe, where the excitation wavelength changes upon Ca²⁺ binding [9]*

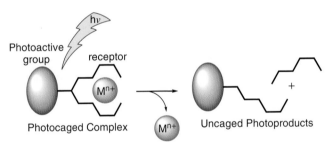

Figure 9.3   *Graphical representation of the metal ion releasing mechanism in a photocaged complex. In the absence of light, the receptor effectively sequesters the metal ion from other chelators. Upon exposure to light, a chemical reaction changes the metal-binding properties of the receptor, which releases the metal ion*

ligand exchange with proteins and other biological ligands, while the photoproducts must bind more weakly than the targeted cellular receptors. In the context of photocages, an efficient reaction proceeds with a high quantum yield and fast kinetics. In addition to a fast organic reaction, the kinetics of the metal ion release from the chelator following photolysis should be rapid enough to simulate biological events. The photocage should be stable in cytosol in its complexed and apo form, and the photocage as well as the corresponding photoproducts should have limited toxicity.

Photocaged complexes have many similarities to fluorescent sensors, but also have key differences with respect to photophysics and metal ion binding properties. The uncaging process requires radiation of sufficient energy to break chemical bonds, so the ability to employ longer wavelength irradiation is inherently limited using standard excitation techniques. Unlike sensors, photocaged complexes need not possess absolute metal ion binding selectivity when applied in metallated form if ligand exchange reactions with endogenous metal ions are slow.

9.2   Calcium Biochemistry and Photocaged Complexes

9.2.1   Strategies for Designing Photocaged Complexes for Ca^{2+}

A significant proportion of the early work with metal ion signaling focused on intracellular Ca^{2+} concentration ($[Ca]_i$) fluctuations. Photocaged complexes provided a convenient methodology to modulate $[Ca]_i$ semi-quantitatively with temporal and spatial control, replacing less precise techniques [15–17]. Controlling $[Ca]_i$ by microinjection of calcium salts or increasing cell membrane Ca^{2+} permeability lacks such specificity or reproducibility. In addition, photocaged complexes allow rapid changes in intercellular $[Ca^{2+}]$ by a predetermined amount and maintain the elevated $[Ca^{2+}]_i$ for a sufficient amount of time to measure physiological or biochemical process [18].

The Ca^{2+} photocaged complexes developed by the Tsien group utilized the nitrobenzyl photoprotective group incorporated into a benzhydrol-derived ligand [13, 19]. The benzhydrol functionality is integrated into a BAPTA chelator (1,2-bis(*o*-aminophenoxy) ethane-*N,N,N′,N′*-tetracidic acid), which is based on the Ca^{2+}-selective EGTA ligand, but shows greater pH insensitivity owing to the replacement of the aliphatic amines with aniline groups. Photolysis converts the substituent para to the Ca^{2+}-binding aniline into an electron withdrawing carbonyl group, which decreases the relative metal ion affinity of the ligand. The reduction of electron density on the Ca^{2+}-bound nitrogen atom stems from the creation of a resonance interaction between the aniline lone pair and the carbonyl oxygen (Fig. 9.4) [20, 21]. These photocaged complexes make Ca^{2+} biologically unavailable prior to photolysis and are collectively referred to as *Nitr* photocages (Fig. 9.5).

The first generation of the Nitr compounds represented a major advance in Ca^{2+} delivery, but also were limited with respect to biological applications [22]. While the photolysis of the chelators proceeds within milliseconds (Table 9.1; Nitr-5 = 0.27 ms, Nitr-7 = 1.8 ms), the quantum yields are relatively low (5%), and the Ca^{2+} affinity difference between the parent chelator and the photoproducts changes only about 40-fold. In an effort to increase the magnitude of change in Ca^{2+} binding affinity, the number of bound nitrogen atoms affected by photolysis was doubled by preparing Nitr-8, which contains a nitrobenzyl group on each of the aromatic rings of BAPTA [13]. While the improvement in affinity changes was significant, approximately 3000-fold, the quantum yields remained low (2.6%) and the kinetics of metal ion release were still slow. The most recent addition to the Nitr family of Ca^{2+} photocaged complexes, Nitr-T, mimics the design of Nitr-8. Nitr-T incorporates both the BAPTA receptor and two nitrobenzyl groups [23]. Photolysis first results in the formation of an asymmetric mononitrosobenzophenone intermediate that undergoes the second photoreaction to yield the final dinitrosobenzophenone photoproduct. The photoproducts were observed by both UV–Vis and ^{1}H-NMR spectroscopy. The Ca^{2+} affinity of Nitr-T changes over 3000-fold from 0.52 μM to 1.6 mM following photolysis but the quantum yield is still relatively low (12%) because of the photochemistry of nitrobenzyl photocaging groups.

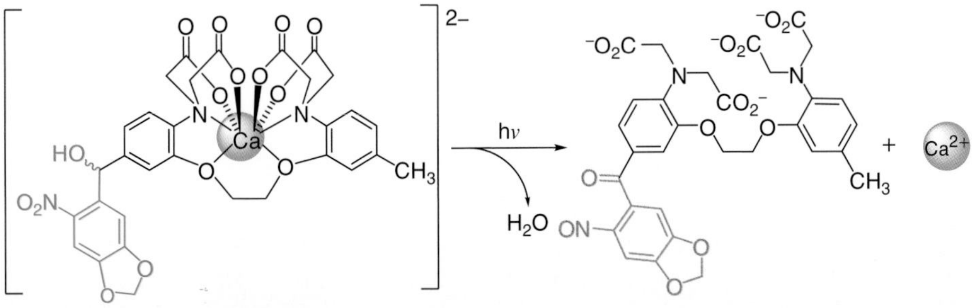

$[Ca(Nitr-5)]^{2+}$

Figure 9.4 *Mechanism of Ca^{2+} release from the photocaged complex of Nitr-5. Upon photolysis, the nitrobenzhydrol group (green) of Nitr-5 is converted into a benzophenone photoproduct (dark yellow). Resonance delocalization of the aniline lone pair onto the carbonyl reduces the Ca^{2+}-binding affinity of the BAPTA receptor, which increases the "free" metal ion concentration by shifting the binding equilibrium*

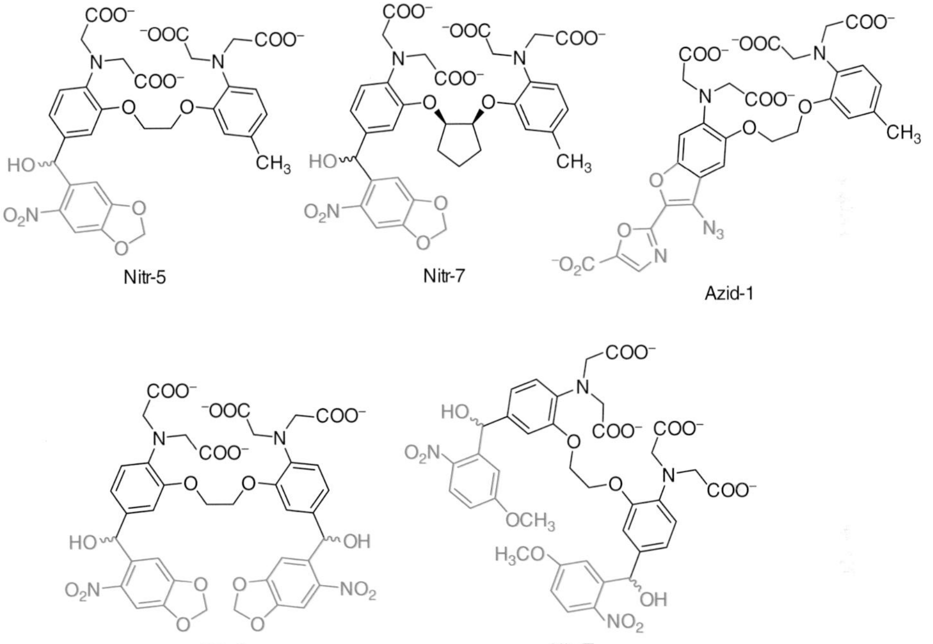

Nitr-5

Nitr-7

Azid-1

Nitr-8

Nitr-T

Figure 9.5 *Photocaged complexes for Ca^{2+} based on attenuation of aniline–Ca^{2+} interactions. The Nitr photocaged complexes share the same release mechanism illustrated in Fig. 9.4. Nitr-8 and Nitr-T contain two aniline groups conjugated to nitrobenhydrol groups. Although the photochemistry is different, the metal-release mechanism in Azid-1 is analogous to the Nitr complexes*

Table 9.1 *Properties of photocaged complexes for Ca²⁺*

| Photocage | Photocage K_d | | Photoproduct K_d | | | | Rate of photolysis (s^{-1}) | Rate of Ca²⁺ release (s^{-1}) |
	Ca²⁺	Mg²⁺	Ca²⁺	$\Delta K_d{}^a$	Φ_{Apo}	Φ_{Ca}		
Nitr-5	145 nM	8.5 mM	6.3 μM	40	0.012	0.035	2.5×10^3	ND
Nitr-7	54 nM	5.4 mM	3.0 μM	50	0.011	0.042	2.5×10^3	ND
Nitr-T	520 nM	ND	1.6 mM	>3000	0.056	0.12	ND^b	ND
Azid-1	230 nM	8.0 mM	120 μM	500	0.9	1.0	ND	ND
DM-Nitrophen	7.0 nM	2.5 μM	3.1 mM, 89 μM	>400 000	0.18	0.18	8×10^4	3.8×10^4
NP-EGTA	80 nM^c	9.0 mM	1.0 mM^c	>10 000	0.2	0.23	5×10^5	6.8×10^4
DMNPE-4	48 nM	10 mM	2.0 mM	>40 000	0.09	0.09	3.3×10^4	4.8×10^4
NBF-EGTA	14 nM^d	15 mM	1.0 mM^d	>70 000	0.7	0.7	ND	2.0×10^4

$^a \Delta K_d = (K_d$ photoproduct)/(K_d photocage). ΔK_d provides a measure of the magnitude of binding constant change for comparing different photocaged complexes.
^{b} Not determined.
^{c} pH 7.2.
^{d} pH 7.5.

Improving the quantum yield in Nitr-like compounds has proven impossible because of the identical uncaging photochemistry of the nitrobenzyl groups. To overcome these limitations, Azid-1 with a structure inspired by Ca²⁺ fluorescent sensor fura-2, was developed [24]. Azid-1 incorporates a BAPTA Ca²⁺ receptor and an azido substituted benzofuran group that functions as the photocaging group [25]. Photolysis results in the formation of an electron withdrawing benzofurane-3-one with a quantum yield of nearly 100% (Fig. 9.6). The enhanced photo-sensitivity, rapid photolysis rate, and over 500-fold change in Ca²⁺ affinity are ideal photocage properties but a complicated synthesis has prevented further use and commercial availability.

An alternative strategy for photocaging Ca²⁺ was developed by Tsien's contemporaries Kaplan and Graham, which involves reduction in metal denticity as the Ca²⁺ releasing strategy [26, 27]. In these Ca²⁺ photocaged complexes, the photolysis of the nitrobenzyl

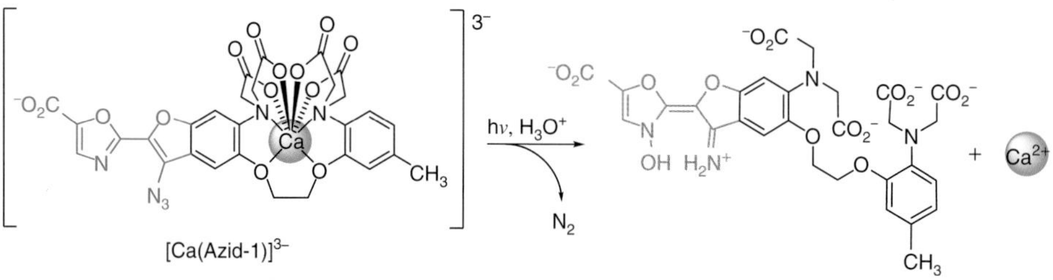

Figure 9.6 *Mechanism of Ca²⁺ release from the photocaged complex of Azid-1. The photolysis of the azide group (green) introduces an electron withdrawing imine (dark yellow) in the position para to a Ca²⁺-bound aniline ligand. The reduced electron density on the aniline nitrogen atom shifts the metal-binding equilibrium toward "free" Ca²⁺*

group results in the scission of the backbone of the chelator, which leads to reduction in the chelate effect. As exemplified by DM–Nitrophen, directly linking the nitrobenzyl group to an amine of an EDTA (ethylenediamine-*N,N,N',N'*-tetraacetic acid) receptor breaks the chelator into two fragments upon exposure to light (Fig. 9.7). Compared with the Nitr photocages, DM–Nitrophen exhibits a markedly higher quantum yield of photolysis (18%) and a large difference in affinity between the photocaged complex (7 nM) and the photoproduct (Table 9.1; 4 mM) [27, 28]. The approximately 50 000-fold change of the Ca^{2+} affinity is partially offset by the Ca^{2+} buffering capacity of the photoproducts and the high affinity of the photocaged complex for Mg^{2+} ($K_d = 1.7 \mu M$) [29, 30].

To circumvent the selectivity issue, the second-generation photocage, NP–EGTA utilized an EGTA receptor (Fig. 9.8). Photolysis leads to rapid carbon–nitrogen bond cleavage ($\tau = 2 \mu s$) and the affinity decreases from 80 nM for the photocaged complex to 1 mM for the photoproducts, which improves Ca^{2+} buffering capabilities [31–33]. While the quantum yield of the NP–EGTA (23%) exceeds those of Nitr compounds, the efficiency of photolysis is seriously limited by the compound's low absorbance (975 $M^{-1} cm^{-1}$). Replacing the simple nitrobenzyl group with the analogous dimethoxy derivative provides DMNPE-4 and doubles the photolysis efficiency by significantly increasing absorbance (5100 $M^{-1} cm^{-1}$). Despite the improved optical properties, the rate of photolysis decreases ($\tau = 30 \mu s$) [34].

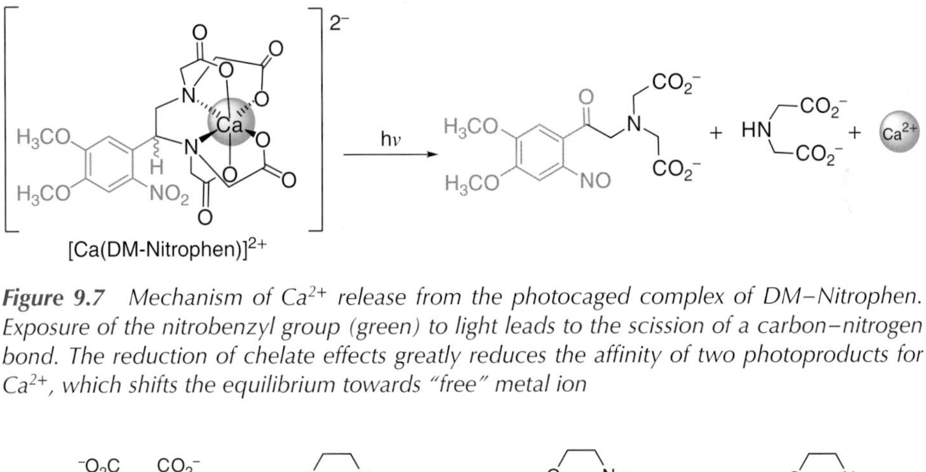

Figure 9.7 *Mechanism of Ca^{2+} release from the photocaged complex of DM–Nitrophen. Exposure of the nitrobenzyl group (green) to light leads to the scission of a carbon–nitrogen bond. The reduction of chelate effects greatly reduces the affinity of two photoproducts for Ca^{2+}, which shifts the equilibrium towards "free" metal ion*

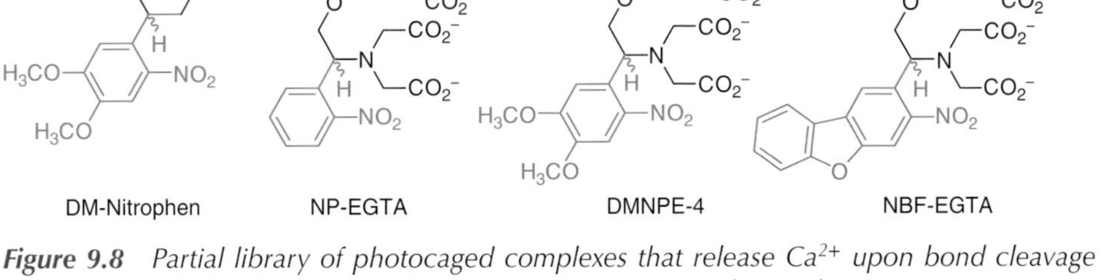

DM-Nitrophen	NP-EGTA	DMNPE-4	NBF-EGTA

Figure 9.8 *Partial library of photocaged complexes that release Ca^{2+} upon bond cleavage illustrated for DM–Nitrophen in Fig. 9.7. While DM–Nitrophen utilizes an EDTA receptor, the remaining photocaged complexes incorporate an EGTA chelator that has an enhanced selectivity for Ca^{2+} over Mg^{2+}*

The most recent photocage developed by the Ellis-Davies research team utilizes EGTA chelator with a novel chromophore of nitrodibenzofuran (NBF). While nitrobenzyl photocages are notoriously resistant to multi-photon excitation, NBF allows for two-photon excitation and makes NBF–EGTA the first photocaged complex without the need for near-UV excitation. NBF–EGTA binds Ca^{2+} with a K_d of 100 nM, which decreases to about 1 mM upon photolysis. In addition to retaining favorable metal ion binding properties, the photocaged complex also has a high quantum yield (70%) and extinction coefficient (18 400 M^{-1} cm^{-1}) [35].

9.2.2   Biological Applications of Photocaged Ca^{2+} Complexes

Steady increases in $[Ca]_i$ from constant light exposure and incremental increases from partial photocaged complex photolysis have been achieved with Nitr-5 and Azid-1. The constant re-establishment of equilibration between high affinity photocages and low affinity photoproducts occurs at a rate that is much faster than the rate of photolysis. The $[Ca]_i$ rises continuously as determined by the total concentration of the low and high affinity species present [20, 36]. The concentration of photoproducts, released analyte, and unreacted photocaged complex can be calculated using the total concentration of the photocage, the intensity of the light source and the quantum yield of the photoreaction [37, 38].

Predicting the amount of Ca^{2+} released becomes more complicated in large cells because light intensity decreases going through the cytosol [39]. In Nitr-5, light absorbance by photoproducts (apo, 24 000 M^{-1} cm^{-1}; $[Ca(Nitr-5)]^{2-}$, 10 000 M^{-1} cm^{-1}) also leads to inefficient photolysis due to screening of excitation light. When photoproduct light absorbance is significantly lower in Azid-1, the situation is reversed. Circumventing these complications when applying Nitr photocaged complexes to biological problems requires correcting for the spatial distribution of light intensity (Beer's law) and diffusion [40]. The spatial and temporal changes in $[Ca]_i$ have been confirmed with Ca^{2+} sensors [41]. Quantifying changes in $[Ca]_i$ for experiments utilizing NP–EGTA follow analogous procedures used with Nitr compounds. Modeling the behavior of DM–Nitrophen remains difficult due to competitive binding with Mg^{2+} and competing endogenous buffers on total concentrations of Ca^{2+} [28, 42]. Using DM–Nitrophen requires experimentation under Mg^{2+} depleted conditions.

Ca^{2+} triggers numerous important biological processes (Fig. 9.9). Since concentration of Ca^{2+} in extracellular fluids (approximately 10^{-3} M) is 10^4 times higher than the resting concentration of the intercellular Ca^{2+} (approximately 10^{-7} M), cells utilize Ca^{2+} as a second messenger [44]. The movement of Ca^{2+} from intracellular stores as well as in and out of the cell via plasma-membrane channels is promoted by various extracellular stimuli [45, 46]. The ability to quantitatively manipulate intracellular Ca^{2+} concentration became an extremely important technique for identifying these stimuli.

Early application of Ca^{2+} photocaged complexes involved studying Ca^{2+}-dependent ion channels. Photocaged complexes allowed the linear dependence of K^+ and nonspecific cation currents in *Aplysia* neurons to be correlated with $[Ca^{2+}]_i$ [41]. Photolysis of Nitr-2 in the neurons under two-electrode voltage clamp conditions lead to induction of a large K^+ channel current despite the low quantum yield (1%) and slow photolysis (5 s^{-1}). Similar experiments explored the role of Ca^{2+}-activated K^+ currents in gastric smooth muscle and the M-current (I_M, muscarine-blocked K^+ current) in sympathetic neurons [47, 48].

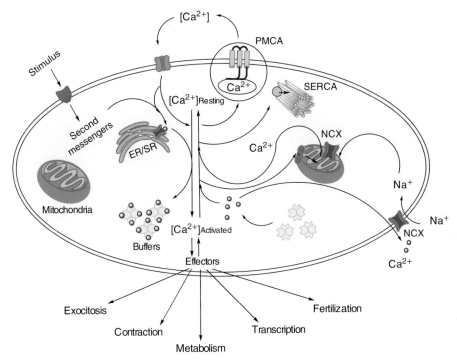

Figure 9.9 *Calcium homeostasis and signaling. Triggered by external stimuli and/or internal messengers, extracellular Ca^{2+} enters the cell and triggers release of Ca^{2+} stored in organelles of endoplasmic or sarcoplasmic reticulum (ER/SR). Released free calcium ions then trigger various cellular processes while any excess is buffered by Ca^{2+} binding proteins. Removal of Ca^{2+} that leaves the effectors or buffers from cytoplasm is accomplished by exchangers such as Na^+/Ca^{2+} exchanger (NCX) and plasma-membrane Ca^{2+}–ATPase (PMCA). Intracellular Ca^{2+} is also pumped into ER via sercoplasmic reticulum Ca^{2+}–ATPase (SERCA) and mitochondria through uniporter. Adapted by permission from Macmillan Publishers Ltd [43] © 2003*

The mechanism of I_M modulation is still not fully understood mainly due to involvement of a large number of receptors [49, 50]. The role of $[Ca^{2+}]_i$ influx in receptor activation and resultant M-current suppression was controversial due to initially conflicting research results [51]. Experiments using Nitr-5 uncovered several types of M-current modulation by $[Ca^{2+}]_i$ [48]. Modest increases in $[Ca^{2+}]_i$ (approximately 100 nM) generated using Nitr-5 were found to enhance the I_M, while large increases in $[Ca^{2+}]_i$ from prolonged photolysis inhibited I_M. I_M suppression can be reduced and subsequently restored with repeated photolysis of Nitr-5 during continues exposure to muscarine.

In another study of M-current modulation, Nitr-5 and Fura-2 were used to simultaneously release and detect Ca^{2+} [48]. The resting $[Ca^{2+}]_i$ between consecutive illumination episodes was measured with Fura-2 during M-current inhibition. However, excessive photolysis periods cause probe photobleaching, which reduced $[Ca^{2+}]_i$ measurement accuracy. Nitr and DM–Nitrophen photocaged complexes were also used in an analogous manner to study Ca^{2+} currents in dorsal root ganglion neurons [52], atrial and ventricular cells [53] as well as the regulation Mg^{2+}-nucleotide L-type Ca^{2+} currents in cardiac cells [54, 55] and to initiate muscle contraction [56, 57].

Ca^{2+} photocaged complexes have elucidated the roles of Ca^{2+} as a neurotransmitter and in hormone release thereby advancing the understanding of basic synaptic and endocrine physiology [33]. Postsynaptic currents that precede and follow Ca^{2+} photorelease in presynaptic neurons cultured in Ca^{2+}-free media reveal that membrane potential has no direct effect on transmitter release [58]. These findings were disputed [59], but eventually were confirmed using squid giant synapses and crayfish neuromuscular junctions [37, 42]. Mimicking naturally occurring $[Ca^{2+}]_i$ transients at Ca^{2+} channels with DM-Nitrophen during an action potential initiated transmitter release. The persistence of the release, which lasts about 15 ms, correlates with Fura-2 $[Ca^{2+}]_i$ measurements. Secretory phases induced by increased $[Ca^{2+}]_i$ from DM−Nitrophen in bovine chromaffin cells and rat melanotrophs demonstrate that Ca^{2+} was responsible for exocytosis and mobilization of vesicles pools in endocrine cells [60].

Comprehensive understanding of the synaptic transmission, facilitation, and depression has been established through rigorous investigation of Ca^{2+}-dependent vesicle mobilization [61]. Experiments using photocaged Ca^{2+} complexes have revealed kinetic aspects and cooperativity in Ca^{2+} binding to the secretory trigger, as well as the release kinetics dependence on local $[Ca^{2+}]_i$ changes at the giant synapse of Held calyx [62], cochlear hair cells [63], and photoreceptors [64]. These and other studies provide a roadmap that can be applied to additional metal ions such as Zn^{2+} that have emerged recently as potential signaling agents.

9.3   Zinc Biochemistry and Photocaged Complexes

9.3.1   Biochemical Targets for Photocaged Zn^{2+} Complexes

Despite a relatively low abundance in tissue of between 10 and $100 \, \mu g \, g^{-1}$, Zn^{2+} is a vital element for human biology because of its versatility. Estimates suggest that 10% of the human genome encodes Zn^{2+}-binding motifs rich in histidine and cysteine residues [65, 66], which accounts for at least 3000 zinc proteins. In comparison, genome sequence information indicates that there are roughly 150 copper proteins [67] and 250 nonheme iron proteins [68]. Proteins utilize Zn^{2+} for both structural and catalytic purposes, but elucidation of the mechanisms through which zinc reaches these proteins only began recently.

Quantitative control and distribution of Zn^{2+} in the cell and the body is carried out by two families of Zn^{2+} transporters (importers and exporters) [69], regulatory zinc sensors, metallothioneins [70], and proteins involved in Zn^{2+}-dependent gene transcription (Fig. 9.10) [71]. While Zn^{2+} may enter/exit cells by different mechanisms, influx/efflux by diffusion is unfavorable. So far, 14 ZIP proteins that shuttle the ions into the cytosol and 10 ZnT proteins that transport zinc out of the cytosol have been characterized [72–74], which indicates these transporters are the major players in homeostatic regulation of cellular Zn^{2+}. Both the detailed mechanism of Zn^{2+} transport as well as the fate of the transported Zn^{2+} have clinical and physiological relevance [75].

Metallothioneins (MT) are metal ion-binding proteins that can accommodate up to seven Zn^{2+} ions in a tetrathiolate coordination environment within two cysteine rich clusters [76].

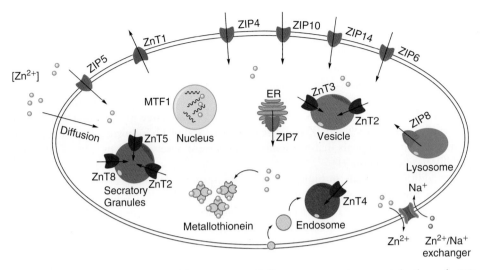

Figure 9.10 *Zinc trafficking and homeostasis. Zn^{2+} enters cytoplasm mainly through ZIP transporters and to a smaller extent by diffusion and Na^+/Zn^{2+} exchangers. Once in the cell Zn^{2+} is buffered by metallothioneins (MT) or is transferred into organelles through ZnT transporters. Otherwise excess zinc ions are expelled out of the cell via Zn^{2+}/Na^+ exchanger and ZnT1*

The possible functions of MTs have been debated extensively because the protein does not fit neatly into any specific class Zn^{2+} proteins. Despite sharing similar coordination environments, affinities for Zn^{2+} ions differ by four orders of magnitude between the seven different binding sites with affinities for the MT-2 haloform ranging from nanomolar to picomolar [77]. Affinity variation under physiological conditions renders MT capable of being either a Zn^{2+} donor or acceptor depending on the extent of cysteine oxidation.

Since redox-inactive Zn^{2+} only binds reduced thiol/thiolate donors in MTs, cellular redox signals can be transduced into Zn^{2+} signals, where $[Zn^{2+}]_i$ increases under oxidizing conditions and decreases under reducing conditions [78]. MT nitrosylation by $[Ca^{2+}]_i$-initiated release of nitric oxide from nitric oxide synthase releases Zn^{2+} from MT increasing $[Zn^{2+}]_i$ [79]. This simple signal transduction highlights the delicate balance between physiological and pathological levels of Zn^{2+} since excessive MT oxidation can release cytotoxic levels of Zn^{2+} [80, 81]. Zn^{2+} homeostasis appears to be maintained by different MTs, which are regulated and trafficked in and out of cytoplasm as well as in the mitochondria and nucleus. The driving force and possible role of Zn^{2+} in MT trafficking is still unclear.

For decades, functions as a regulator of cellular signal transduction were hypothesized for Zn^{2+} but evidence to support this claim emerged much later [82]. Investigations into the possible involvement of Zn^{2+} in intracellular signal transduction can be facilitated by Zn^{2+} photocaged complexes in a manner analogous to the studies with Ca^{2+}. Conclusive proof of the Zn^{2+} signaling pathway would require molecular targets for $[Zn^{2+}]_i$ fluctuations to be identified. While increases in $[Zn^{2+}]_i$ can be generated by MT oxidation [83], stimulation of protein kinase C (PKC) [84] and transport from intracellular compartments [85] or from the extracellular environment could also create Zn^{2+} signals [86]. Photocaged complexes could mimic these Zn^{2+} releasing processes, by providing spatial and temporal control over $[Zn^{2+}]_i$ changes that could be monitored by orthogonal analytical techniques.

The various types of Zn^{2+} signals needed to mediate biological processes require different changes in total Zn^{2+} influx as well as the release time. While Ca^{2+} signals are immediate, Zn^{2+} signals can last for days. Fast zinc signals, lasting less than a minute, were observed in cardiomyocytes [87], monocytes [88], fibroblasts [84], and from intracellular stores [89]. Slower zinc signals, designated as Zn^{2+} waves due to the initial delay, were observed in mast cells where $[Zn^{2+}]_i$ rises over a period of 1 h following the initial release [85]. Even slower signals that last hours or days were observed during cell proliferation and differentiation [90, 91].

The additional signals in biological processes provide evidence for the broad importance of Zn^{2+}. Zn^{2+} can modulate phosphorylation signals by inhibition of protein tyrosine phosphatases (PTP) possibly by binding an active thiolate site [92]. It influences glucose transport by phosphorylation of the insulin receptor, which affects lipids and glucose metabolism [93]. Sequestering Zn^{2+} with the heavy metal ion chelator TPEN (N,N,N',N'-tetra(2-pyridyl)-1,2-ethanediamine) suppresses phosphorylation and the downstream pathways in insulin signaling while increasing $[Zn^{2+}]_i$ mediated by ZIP7 inhibited phosphatases [90, 94]. In a similar manner, Zn^{2+} effects epidermal growth factor (EGF) signaling [95], as well as pathways involving mitogen-activated protein kinase (MAPK) [96] and protein kinase C (PKC) [97].

Intercellular Zn^{2+} signal transduction requires high $[Zn^{2+}]$ to be released. Since normal $[Zn^{2+}]_i$ must be low to maintain homeostasis, large amounts of free Zn^{2+} must be mobilized from vesicular stores [98]. Analysis of isolated vesicles by X-ray absorption fine structure (XAFS) indicate that the pools of low molecular weight Zn^{2+} available for ligand exchange reactions and chelation inside is bound to sulfur, nitrogen, and oxygen containing ligands [99]. Concentration measurements of chelatable Zn^{2+} with fluorescent sensors indicated that most eukaryotic cells contain low nanomolar to high picomolar $[Zn^{2+}]_i$ [87, 100, 101]. Accumulation of Zn^{2+} in vesicles requires active transport that is mediated by ZnT-2 [102]. Vesicular zinc can be found in nerve terminals, somatotrophic cells of the pituitary gland, zymogene granules in pancreatic acinar cells, β-cells of the islets of Langerhans, cells of intestinal crypt, prostate cells, epididymal ducts, and osteoblasts [103]. Function of the released, extracellular Zn^{2+} is still under investigation.

9.3.2   Strategies for Designing Photocaged Complexes for Zn^{2+}

The emerging interest in studying Zn^{2+} homeostasis and signaling encouraged our group to initiate the development of Zn^{2+} photocaged complexes. The initial design strategy adopts the concepts pioneered for Ca^{2+} photocaged complexes by integrating nitrobenzyl groups into Zn^{2+}-binding receptors. The first class of photocages, inspired by Ellis-Davies, DM-Nitrophen, have been designated *ZinCleav* compounds to denote the metal ion of interest (*zinc*) and the release strategy (bond *cleavage*) [104, 105]. Similar to DM-Nitrophen, ZinCleav photocages bind tightly before photolysis, but the scission of the receptor backbone decreases Zn^{2+} affinity by reducing chelate effects (Fig. 9.11).

ZinCleav-1 utilizes the tetradentate EBAP (ethylene-bis-α,α'-(2-aminomethyl)-pyridine) receptor and binds Zn^{2+} with moderately high affinity ($K_d = 0.23$ pM, Table 9.2) [105]. Upon irradiation with 350 nm light, ZinCleav-1 splits into two weakly binding fragments

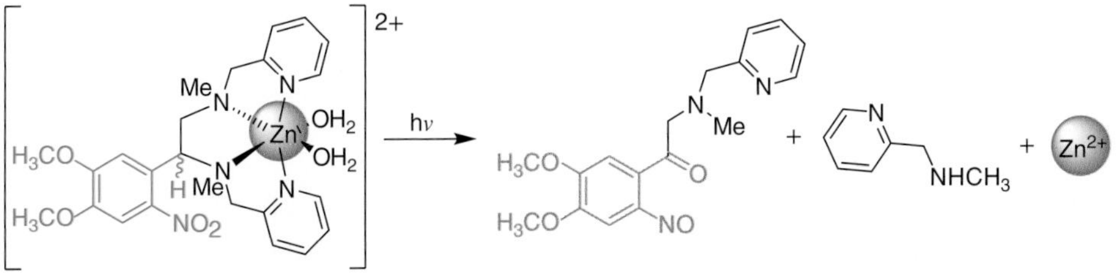

[Zn(ZinCleav-1)(OH₂)₂]²⁺

Figure 9.11 *Mechanism of Zn²⁺ release from the photocaged complex of ZinCleav-1. Exposure of the nitrobenzyl group (green) to light leads to the scission of a carbon–nitrogen bond similarly to DM–Nitrophen, as shown in Fig. 9.7. Reduction of chelate effects greatly reduces the affinity of two photoproducts for Zn²⁺, which shifts the equilibrium toward "free" metal ion*

Table 9.2 *Properties of photocaged complexes for Zn²⁺*

Photocage	Photocage	Photoproduct	ΔK_d	Φ_{Apo}	Φ_{Zn}
	K_d for Zn²⁺ complexes[a]				
ZinCast-1	14 μM[b]/61 μM[c]	5.6 mM[b]/5.6 mM[c]	400[b]/92[c]	0.007[c]	0.022[d]
ZinCast-2	0.25 mM[d]/8.6 μM[e]	NA[d]/48 μM[e]	NA[d]/6[e]	0.004[c]	0.018[d]
ZinCast-3	3.1 mM[d]/33 μM[e]	NA[d]/3.8mM[e]	NA[d]/115[e]	0.002[c]	0.012[e]
ZinCast-4	11 μM[c]/12 μM[d]	56 μM[c]/11 μM[d]	4[b]/1[d]	0.006[c]	NA
ZinCast-5	14 μM[c]/15 μM[d]	NA[c]/NA[d]	NA[c]/NA[d]	0.010[c]	NA
MonoCast	14 μM[d]/12 μM[e]	0.19 mM[d]/14 μM[e]	14[d]/1[e]	ND	ND
DiCast-1	20 μM[d]/12 μM[e]	3.8 mM[d]/27 μM[e]	190[d]/2[e]	ND	ND
DiCast-2	8.7 μM[d]/10 μM[e]	4.2 mM[d]/3.3 μM[e]	480[d]/330[e]	ND	ND
DiCast-3	7.8 μM/5.0 μM[d]	ND	NA[f]	ND	ND
ZinCleav-1	0.23 pM	~40 mM	~10¹¹	0.022	0.017
ZinCleav-2	0.9 fM	0.158 μM	~10⁸	0.047	0.023

[a]The binding constants were determined in aqueous buffer solution (50 mM HEPES, 100 mM KCl, and pH = 7.0) unless otherwise indicated.
[b]20% DMSO–aqueous buffer (50 mM HEPES, 100 mM KCl, and pH = 7.0).
[c]20% EtOH–aqueous buffer (50 mM HEPES, 100 mM KCl, and pH = 7.0).
[d]EtOH.
[e]CH₃CN.
[f]Not available.

containing 2-(aminomethyl)pyridine ligands ($K_d \approx 40$ mM). The large 10^8-fold decrease in Zn²⁺ affinity plus moderate quantum yield of [Zn(ZinCleav-1)]²⁺ (1.7%) make ZinCleav-1 attractive for biological studies. However, many cells maintain basal Zn²⁺ concentrations at levels where addition of [Zn(ZinCleav-1)]²⁺ would perturb homeostasis prior to photolysis. ZinCleav-1 is most applicable for cells where free Zn²⁺ concentrations are maintained above 2 nM.

Intracellular Zn²⁺ has been estimated to reach mM levels implying an abundance of Zn²⁺ in the cellular environment but the total concentrations of free or loosely chelated zinc could be as low as 2 nM or one zinc ion per cell in certain organisms and types of cells [106].

This six order of magnitude difference in concentration suggests that $[Zn^{2+}]_i$ is highly regulated. In order to expand the capacity of ZinCleav photocaged complexes to answer biological questions, work on the second-generation compound focused on increasing the affinity of the Zn^{2+} receptor. Substitution of tetradentate EBAP with a hexadentate TPEN ligand provides ZinCleav-2 with an increased binding affinity for Zn^{2+} ($K_d = 0.9$ fM), which reduces ligand exchange between proteins and $[Zn(ZinCleav-2)]^{2+}$ near the limit possible with small molecule chelators (Fig. 9.12) [104]. Photolysis of ZinCleav-2 leads to the formation of two photoproducts each with a dipicolylaniline (DPA) ligand that binds Zn^{2+} with affinity ($K_d = 158$ nM) which exceeds that of the ZinCleav-1 photoproducts. Increased complex stabilities for ZinCleav-2 and its photoproducts will lower the resting concentration of $[Zn^{2+}]_i$ before and after photolysis compared with ZinCleav-1. Since intracellular concentration of Zn^{2+} can vary between different types of cells, the resting concentration of Zn^{2+} had been determined with the HySS computer program for six hypothetical scenarios, where physiological levels of $[Zn^{2+}]$ vary between 1 fM and 790 pM. Increased buffering capacity of ZinCleav-2 allows for sustainment of low resting Zn^{2+} concentrations while permitting significant enough release of free Zn^{2+} following photolysis. Lower stability for the $[Zn(ZinCleav-1)]^{2+}$ requires an excess amount of apo-ZinCleav-1 be added before photolysis to maintain sub-nM physiological levels of $[Zn^{2+}]_i$.

The ZinCast photocaged complexes, which are alternatives to ZinCleav complexes, combine the nitrobenzhydrol group utilized in Nitr compounds with Zn^{2+} binding ligands (Fig. 9.13) [107–109]. Like ZinCleav photocages, the ZinCast nomenclature evokes both the metal ion (*zinc*) as well as the release mechanism (*casting* off). Exposure of ZinCast to light converts the benzhydrol to a benzophenone that reduces Zn^{2+} affinity due to the partial delocalization of the anilino lone pair onto the carbonyl oxygen (Fig. 9.14). In contrast to ZinCleav compounds, the nitrosobenzophenone photoproducts (ZinUnc) are often stable enough to be isolated for analysis, allowing for greater accuracy in affinity change calculations. The first photocaged complex in the series, ZinCast-1, incorporates the tridentate DPA receptor and binds Zn^{2+} with $K_d = 14$ μM. The 400-fold decrease in Zn^{2+} affinity following irradiation is eight times greater than that measured for similar Ca^{2+}

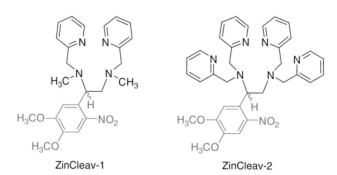

ZinCleav-1   ZinCleav-2

Figure 9.12 *ZinCleav-1 and ZinCleav-2 are the first two members of a family of photocaged complexes for Zn^{2+} that utilize the release mechanism illustrated in Fig. 9.11. By virtue of using a TPEN ligand, ZinCleav-2 possesses fM affinity for Zn^{2+}, which prevents metal ion removal by protein ligands prior to photolysis*

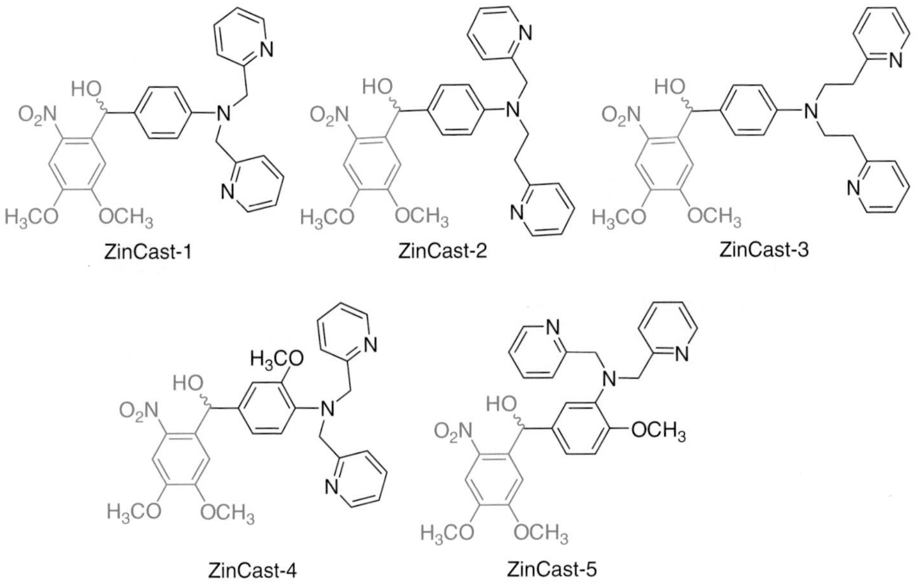

Figure 9.13 *The complete family of ZinCast photocages that utilize the metal releasing mechanism illustrated in Fig. 9.14. Modulating the ligand structure changes the affinity of the photocaged complex for Zn^{2+}*

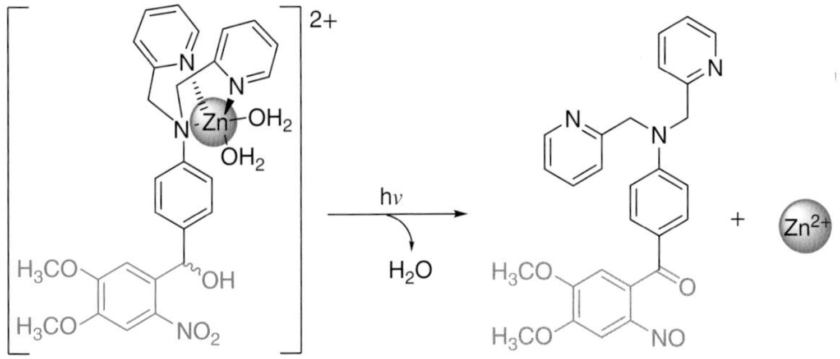

Figure 9.14 *Mechanism of Zn^{2+} release from the photocaged complex of ZinCast-1. Upon photolysis, the nitrobenzhydrol group (green) of ZinCast-1 is converted into a benzophenone photoproduct (dark yellow) similar to the process illustrated for Nitr-5 in Fig. 9.4. Resonance delocalization of the aniline lone pair onto the carbonyl reduces the Zn^{2+}-binding affinity of the DPA receptor, which increases the "free" metal ion concentration by shifting the binding equilibrium*

photocaged complexes [108]. However this magnitude of change is strongly influenced by solvent [109].

Comparative studies with the two ZinCast-1 analogues, namely ZinCast-2 and ZinCast-3, revealed systematic trends in metal binding properties as a function of chelate ring size (Table 9.2) [109]. Expansion of the alkyl linker between the aniline nitrogen atom and the pyridine moiety from a methylene to an ethylene leads to formation of 5,5-, 5,6-, and

6,6- chelate rings for ZinCast-1, ZinCast-2, and ZinCast-3, respectively. The three ZinCast photocaged complexes appear to experience $N_{aniline}-Zn$ bond lengthening as chelate size increases. Shortening of the $Zn-N_{aniline}$ bond in $[Zn(ZinCast-3)]^{2+}$ results in the largest change in Zn^{2+} affinity after photolysis for the series. The stronger $Zn-N_{aniline}$ interaction appears to increase the quantum yield for $[Zn(ZinCast-3)]^{2+}$ photolysis in comparison with the other ZinCast complexes. Unfortunately weak binding constant for ZinCast-3 and low water solubility render this compound unfit for any biological application.

Inclusion of multiple nitrobenzhydrol groups into a ZinCast-like construct has a similar impact on changes in metal binding affinity upon uncaging. The *DiCast* series of compounds adopt the *Cast* nomenclature with the prefix *Di-* to denote the presence of two nitrobenzyl groups (Fig. 9.15) [107]. To increase Zn^{2+} offloading capabilities, while preserving the μM affinity of ZinCast-1, a DPA chelator was expanded by additional aniline to form the four nitrogen donor receptor in DiCast-1 that was then functionalized with two nitrobenzhydrol groups. To quantify the effect of the second chromophore on affinity changes following photolysis, the related photocage MonoCast with only one nitrobenzyl group was prepared. Photolysis of $[Zn(DiCast-1)]^{2+}$ decreases Zn^{2+} affinity from 20 μM to 3.8 mM as measured in ethanol. The second nitrobenzyl group significantly increases Zn^{2+} release in DiCast-1, which exhibits a 190-fold decrease in Zn^{2+} affinity upon photolysis. While decrease in affinity after photolysis for DiCast-1 is considerably larger than that for all related Cast photocages functionalized with one nitrobenzyl group under the same

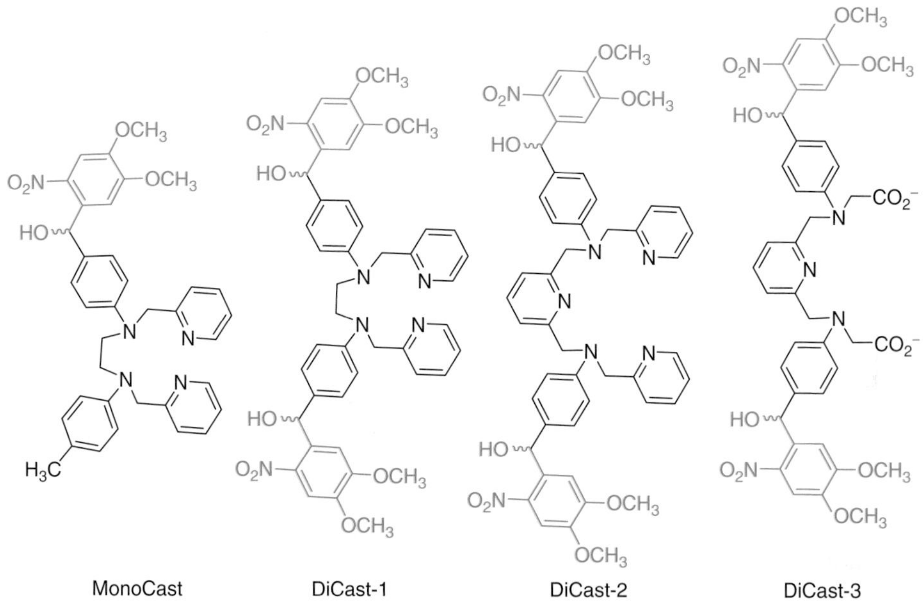

MonoCast   DiCast-1   DiCast-2   DiCast-3

Figure 9.15 *DiCast photocaged complexes incorporating ligand design concepts from the ZinCast series with the multiple nitrobenzyl groups found in Nitr-8 and Nitr-T (Fig. 9.5). Similar to the Ca^{2+} analogues, these photocaged complexes exhibit larger changes in binding affinity on photolysis*

conditions, the changes in Nitr-8 and Nitr-T significantly exceed the ones determined for DiCast-1 [13, 23], and all the experiments were performed in non-aqueous solvent.

To increase DiCast complex stability, a novel receptor with a bridging pyridine that brings together two aniline photocaging ligands was designed. The chelator also incorporates two additional pendant ligands to provide the tripyridyl photocages DiCast-2 and DiCast-3 that contains two carboxylate groups. DiCast-2 binds Zn^{2+} with a $K_d = 8.7$ μM in ethanol. Unlike other Cast photocaged complexes where increasing the Zn^{2+} affinity reduces affinity changes upon uncaging, the change in metal affinity for DiCast-2 is 2.5 times larger than that of DiCast-1. In DiCast-3, the charged carboxylates enhance the water solubility enough to conduct all the experiments in aqueous solution. DiCast-3 possesses a K_d of 7.8 μM, but photoproduct instability makes quantifying affinity changes upon photolysis impossible.

While *in vitro* proof-of-concept has been established for Zn^{2+} photocaged complexes, the corresponding experiments in cells and tissues have yet to be described. The scope of future investigations is currently limited in some respects by the properties of the photocaged complexes reported to date. The binding affinity ZinCast photocages (μM) is weaker than receptors and binding sites within typical endogenous proteins such as metallothioneins, so the application of these tools would be very limited since Zn^{2+} homeostasis could be perturbed prior to photolysis. The data available suggest that designing a ZinCast photocaged complex that binds Zn^{2+} strongly enough before and releases a significant concentration after photolysis will be difficult.

In contrast to ZinCast photocaged complexes, ZinCleav-1 and ZinCleav-2 are suitable for biological investigations. The high affinity and large change in binding constant upon photolysis are both very desirable properties. While these photocaged complexes will prove useful, several design challenges remain. The Zn^{2+} complex of both compounds carries a charge, so delivering the photocages across cellular membranes will require more invasive techniques such as microinjection to conduct intracellular studies. In addition, the quantum yields are lower than desired. Future research directions include both the application of existing photocaged complexes as well as refining the design to circumvent the issues related to biological investigations.

9.4 Photocaged Complexes for Other Metal Ions

9.4.1 Photocaged Complexes for Copper

Cu^{2+} and Cu^+ participate in various biological processes in a protein-bound form. Copper redox activity drives cofactor function in cytochrome c oxidase (CcO) [110], superoxide dismutase (SOD) [111], ceruloplasmin [112], dopamine β monooxygenase [113], and other enzymes that catalyze oxidation–reduction reactions [114, 115]. The interconversion between cuprous and cupric forms is utilized to facilitate electron transfer reactions, but uncontrolled redox process can also contribute to oxidative damage through the formation of reactive oxygen species (ROS) [116]. To prevent cellular damage, cysteine-rich metallothionein and phytochelatin sequester and export excess copper to maintain strict homeostasis [117]. Oxidative stress and elevated copper levels in the brain have a correlation with neurodegenerative disorders, so neural copper homeostasis has received increasing attention [118].

Like Zn^{2+}, copper distributions vary in the brain with the highest concentrations reaching mM levels in the part of brainstem involved in responses to stress and panic and the region of midbrain implicated in dopamine production [119, 120]. Transport and regulation of neural and hepatic copper is carried out by ATPases (ATP7A and ATP7B), and genetic mutations that lead to malfunction of these proteins are associated with inherited neural copper deficiency or Wilson disease [121]. While transmembrane transport proteins are involved in cellular uptake of copper (Ctr) in yeast and mammals [122], prion protein (PrP) [123] and amyloid precursor protein [124] are copper uptake/efflux proteins found only in brain. Micromolar levels of intracellular copper trigger rapid endocytosis and degradation of Ctr [125], while millimolar concentrations prompt endocytosis of PrP [123].

Copper chaperones conduct intracellular Cu^+ transport to cellular targets. Three mammalian chaperons have been identified: Atox1 that loads Cu^+ into Golgi ATPases [126], CCS that delivers Cu^+ to SOD [127], and Cox17 that transports Cu^+ to mitochondrial CcO [128]. The remarkable similarity in mammalian and yeast copper metabolism has helped elucidate mechanisms of human copper homeostasis since yeast can be readily manipulated genetically [129]. However, there are still many unanswered questions regarding function of "free" copper in higher organism neurophysiology [130] and neuropathology [131].

Metallochaperones and other regulatory proteins traffic Cu^+ and maintain homeostasis. Protein facilitated copper homeostasis is particularly important since Cu^+ disproportionates in water and readily reacts with O_2. While Cu^{2+} exists transiently during electron transfer and O_2 reaction in copper proteins, the reducing environment of the cell renders Cu^+ the most prevalent oxidation state. During breakdowns in homeostasis, Cu^{2+} can potentially participate in oxidative stress by catalyzing the formation of hydroxyl radicals from hydrogen peroxide. A Cu^{2+} photocage would allow such processes to be simulated and also provides a tool to generate hydroxide radicals inside cells in a controlled manner. ZinCleav and ZinCast bind Cu^{2+} more tightly than Zn^{2+}, and Cu^{2+} readily displaces Zn^{2+} from the complexes; however, photocaged complexes also have been designed specifically for Cu^{2+}.

The first generation photocage, H_2cage, incorporates a tetradentate bis(pyridylamide) chelator that fragments at a $C–N_{amide}$ bond site to form two bidentate photoproducts (Fig. 9.16) [132]. The Cucage binds Cu^{2+} tightly (16 pM) at pH 7.4 with the loss of two protons to afford the neutral bisamide complex Cucage (Table 9.3). The high quantum

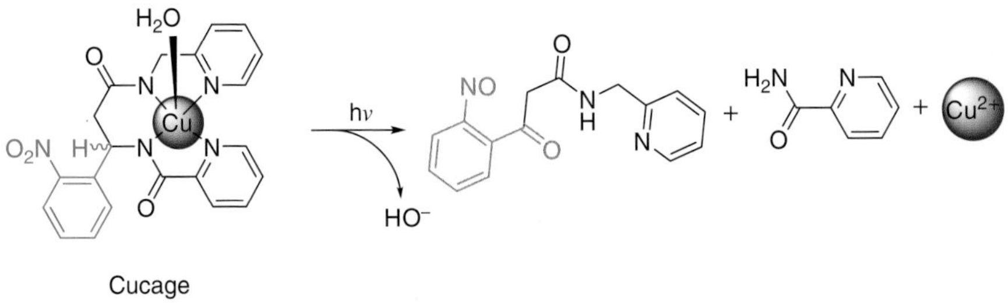

Cucage

Figure 9.16 *Mechanism of Cu^{2+} release from the photocaged complex Cucage (Cu^{2+} complex of H_2cage). The nomenclature formulation in Cu^{2+} photocaged complexes reflects the deprotonation of the amide nitrogen atoms upon metal ion binding. Exposure of the nitrobenzyl group (green) to light leads to the scission of a carbon–nitrogen bond similarly to DM–Nitrophen and ZinCleav-1. Loss of Cu^{2+} leads to the protonation of the amide nitrogen atoms in addition to the metal ion release*

Table 9.3   *Properties of photocaged complexes for Cu²⁺*

Photocage	K_d for Cu²⁺ complexes[a]		ΔK_d	Φ_{Apo}	Φ_{Cu}
	Photocage	Photoproduct			
H₂cage	16 pM	21 nM	1300	0.73	0.32
Imcage	NA	14 nm	NA	NA	NA
Amcage	200 pM	22 nM	110	NA	NA
3arm-3	NA	13 nM	NA	NA	NA
3arm-1	NA	0.92 nM	NA	NA	NA
3Gcage	0.18 fM	NA	NA	0.66	0.43
ZinCleav-1	~0.05 fM	0.23 pM	~4600	0.022	0.017[b]

[a]50 mM HEPES buffered to either pH 7 or 7.4.
[b]Φ of Zn²⁺ complex.

yield (32%) compared with structurally similar photocaged complexes is attributed to the amide group. ZinCleav and ZinCast photolysis efficiency is significantly reduced by resonance interactions between electron lone pairs on nitrogen and the *aci*-nitro intermediate formed upon excitation of the nitrobenzyl chromophore (Fig. 9.17) [104]. In Cucage, the amide carbonyl provides an alternative resonance interaction for the nitrogen lone pair, which probably accounts for the increased quantum yield. While delocalization of the nitrogen lone pair onto the amide carbonyl improves the efficiency of photolysis, it also leads to reduced Cu²⁺ affinity when compared with ZinCleav-1. However, a positive artifact of the amide/amidate ligand is the low affinity of the photoproducts for metal ions, meaning there is essentially no buffering of the released copper. While the oxidation state of the released copper was not determined, Cu⁺ is the most likely product of the photoreaction, which then undergoes quick oxidation to Cu²⁺.

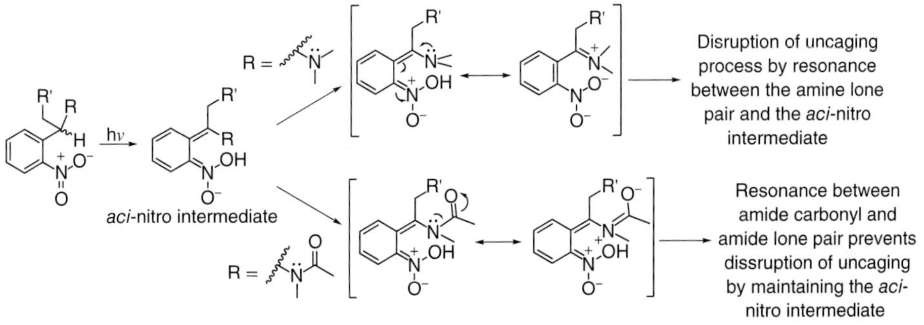

Figure 9.17   *Mechanism of uncaging in ZinCleav- and DM–Nitrophen-like photocaged complexes (top) and H₂cage/Cucage-like species (bottom). The uncaging of nitrobenzyl photocages requires the initial formation of an aci-nitro intermediate that undergoes subsequent reaction steps. The presence of a lone pair of electrons that can participate in a resonance interaction with the aci-nitro intermediate disrupts the uncaging by generating an intermediate with different properties. When the lone pair participates in resonance with an adjacent acceptor group like the carbonyl group in H₂cage, the photocage exhibits photochemical behavior typical of nitrobenzyl compounds*

To increase Cu^{2+} affinity of the photocages, a series of second-generation compounds was prepared [133]. Replacement of a pyridine ligand with an imidazole group (Imcage) or amine donor (Amcage) alleviates steric restraints and allows for the formation of a square planar complex with increased stability (Fig. 9.18). Replacing the pyridine donor with an aliphatic amine in Amcage or the expansion of the 5-membered ring into a 6-membered one in Imcage makes the corresponding Cu^{2+} complexes less stable. The addition of an extra pyridyl group in 3arm-1 and 3arm-3 also failed to increase the affinity for copper.

ZinCleav-1 binds Cu^{2+} more tightly than CuCage (Table 9.3). Based on the results from synthetic and metal binding studies, the preparation of 3Gcage in which one of the amide groups of H_2Cage was replaced by a secondary amine to enhance the binding affinity was undertaken [133]. The substitution increases Cu^{2+} affinity by nearly five orders of magnitude ($K_d = 0.18$ fM) while the quantum yield remains nearly constant (43%). While both ligands function efficiently as Cu^{2+} photocaged complexes, an intrinsic redox activity of copper triggers production of hydroxyl radicals following photolysis with UV light [134]. The cytotoxic activity of Cu3G, the Cu^{2+} complex of 3Gcage, in cancer cells was investigated by releasing Cu^{2+} with UV light. Photoinduced Cu^{2+} release in HeLa cells triggers hydroxyl radical production by a Fenton-like reaction and leads to reduction in the viability of the cells with an IC_{50} value of ~75 μM. The combination of the Cu delivery through Cu3G photolysis and a nontoxic level of H_2O_2 initiated nonapoptotic cell death and further lowered the IC_{50} to ~30 μM.

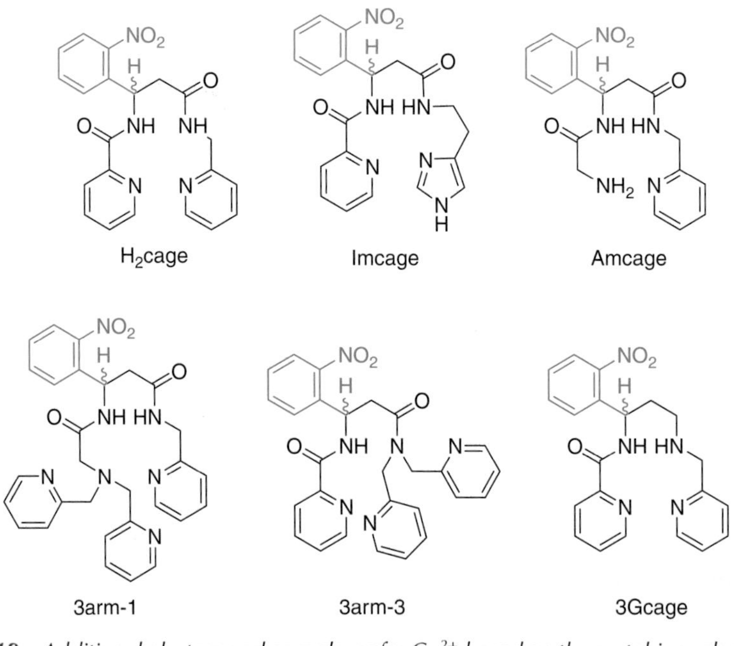

Figure 9.18 *Additional photocaged complexes for Cu^{2+} based on the metal-ion release mechanism illustrated for Cucage in Fig. 9.16. Like the nomenclature for H_2cage/Cucage, the nomenclature reflects whether the ligand is in the apo or metallated form. 3Gcage is designated Cu3G when complexed and Amcage becomes CuAm*

In addition to transport and delivery, Cu^+ has also been implicated in signaling [135, 136]. Designing a photocage for Cu^+ presents both synthetic and receptor design challenges. CuproCleav-1 was prepared by a pathway derived from the synthesis of the ZinCleav photocages (Fig. 9.19) [137]. The receptor consists of two bis-(2-ethylthioethyl)amine groups. The thioether groups provide both selective binding as well as Cu^+ stabilization in water. CuproCleav-1 binds Cu^+ with a K_d of 54 pM, interacts very weakly with Zn^{2+} and Cu^{2+}, and has photophysical properties similar to the ZinCleav compounds. Despite representing a significant advance in photochemical tools for Cu^+, the hydrophobic thioether ligands compromise the water solubility of the photocage in addition to providing the necessary metal binding properties. The biological importance of Cu^+ warrants future efforts to design more bio-compatible CuproCleav photocaged complexes.

9.4.2   Photocaged Complexes for Iron

Like copper, iron has two common oxidation states in biology, and similar functions in electron transfer and dioxygen chemistry. Iron is implicated in many critical biological processes such as respiration [138], DNA synthesis [139], energy generation [140], photosynthesis [141], and nitrogen fixation [142]. At physiological pH, the solubility of Fe^{3+} in water is very low (approximately 10^{-17} M) and while Fe^{2+} solubility is significantly higher (approximately 10^{-1} M), Fe^{2+} readily oxidizes to Fe^{3+} forming insoluble oxides and hydroxides. Although iron solubility under physiological conditions presents a transport problem, Fe^{2+}/Fe^{3+} redox cycling can be deleterious to cells and tissues [143, 144]. For these reasons, cells have developed an intricate system of proteins that permit transport, storage, and homeostasis of iron [145]. Ferric iron is reduced to Fe^{2+}, which is transported across the enterocyte membrane by the divalent metal transporter 1 (DMT1) [146]; however, knowledge of intracellular processes is limited. Iron enters the circulatory system through ferroportin transport, and after oxidation, transferrin (Tf) scavenges Fe^{3+} [147]. Tf binds Fe^{3+} very strongly (10^{-22} M) and prevents reaction with hydrogen peroxide [148].

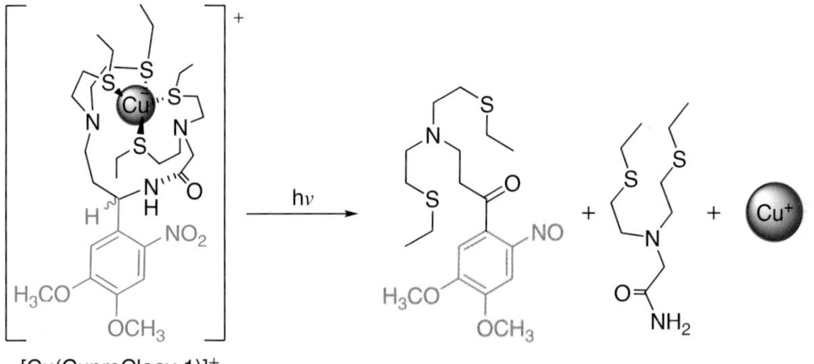

[Cu(CuproCleav-1)]$^+$

Figure 9.19 *Mechanism of Cu^+ release from the photocaged complex of CuproCleav-1. Exposure of the nitrobenzyl group (green) to light leads to the scission of a carbon–nitrogen bond similar to Cucage, ZinCleav-1, and DM–Nitrophen. The reduction of chelate effects greatly reduces the affinity of two photoproducts for Cu^+, which shifts the equilibrium toward "free" metal ion*

Cellular uptake of iron proceeds through endocytosis when diferric Tf binds to its cellular receptor (TfR1) [149]. Subsequently iron released from Tf is reduced to Fe^{2+} and internalized by DMT1 into endosomes.

While the majority of iron in cells is are tightly complexed in proteins, a small amount of iron remains in the free state [150, 151]. Weakly bound intracellular Fe^{2+} in the labile iron pool (LIP) consists of low molecular weight iron complexes and high molecular weight intermediates [152]. Weakly bound iron could readily participate in oxidative damage by ROS generation, so other methods of intracellular iron transport have been considered [153, 154]. To prevent oxidative damage, cells tightly control iron uptake, metabolism, and storage. Cellular iron metabolism is controlled by the iron responsive proteins IRP1 and IRP2, which respond to fluctuations in labile iron. IRPs regulate expression of proteins involved in iron homeostasis through translational control [151]. Systemic iron levels are coordinated by hepcidin, a peptide that triggers ferroportin degradation and blocks the iron release from enterocytes [155]. Prolonged disturbance in iron absorption and trafficking can lead to pathological conditions and death [156]. Many aspects of iron metabolism and homeostasis are still not completely understood, particularly with respect to the LIP [157]. Both fluorescent sensor and photocaged complex development are in the early stages [158, 159].

Environmental iron is widely available in the form of water insoluble iron(III) oxide. Acquisition of this biologically essential metal by many aquatic microorganisms is facilitated by secretion of low molecular weight Fe^{3+} chelators called siderophores [160]. Siderophores utilize one of three types of chelators to coordinate environmental Fe^{3+}: catechols, hydroxamic acids, and α-hydroxy-carboxylic acids. While catecholate siderophores are photoreactive even in their apo form [161], α-hydroxy-carboxylate siderophores become photosensitive only upon Fe^{3+} complex formation [162]. Photolysis of ferric complexes of α-hydroxy-carboxylic acid siderophores with UV light induces ligand oxidation and release of CO_2, which is complemented by reduction of Fe^{3+} to Fe^{2+} [163, 164]. Studies of citric acid siderophores called aerobactins lead to the conclusion that photooxidized aerobactin remains in the complexed form with Fe^{3+} due to comparable stability constants for the aerobactin and its photoproduct [164]. While the Fe^{2+} is effectively released by the photolyzed aerobactin, quick re-oxidation of released iron leads to formation of the Fe^{3+} complex with a newly formed photoproduct. Ferric complexes of β-hydroxyaspartate siderophores known as aquachelins also undergo UV photolysis with oxidation of the ligand and reduction of Fe(III), but reaction is believed to be radical-based [162]. An Fe^{3+} complex is formed by coordination of two hydroxamate groups and one β-hydroxyaspartate residue found at the peptidic group of the fatty acid tails of aquachelins. Determination of the binding constants for aquachelin and its photoproduct revealed that Fe^{3+} affinity decreases from $10^{12.2}$ M^{-1} to $10^{11.5}$ M^{-1} following partial photolytic ligand breakdown. While the fate of the photoreleased Fe^{2+} is not known, it is believed that some of the Fe^{2+} is available for biological reuptake while the rest becomes quickly re-oxidized and scavenged by strong Fe^{3+} ligands.

Inspired by the photochemistry of siderophores and a need for iron releasing photocaged complexes, Baldwin and coworkers prepared a suit of five α-hydroxyacid acid containing chelates designed to bind Fe^{3+} and photorelease Fe^{2+} [165]. Each of the ligands contributes a bidentate salicylidene moiety (Sal) and a bidentate α-hydroxyacid acid (AHA) moiety that functions also as a light sensitive unit (Fig. 9.20). Fe^{3+} complexes were characterized as

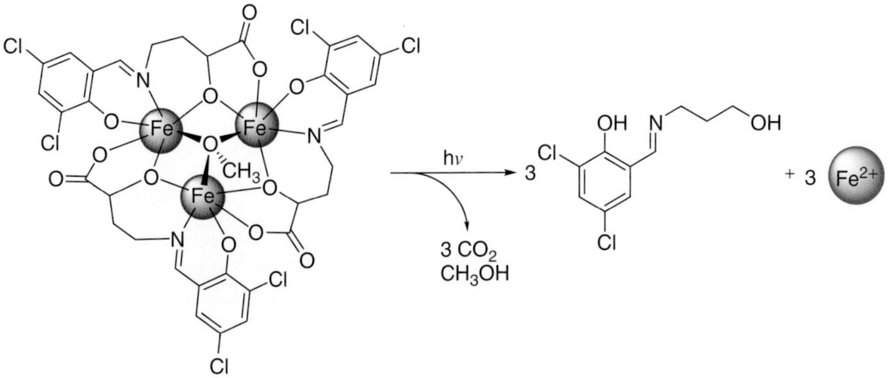

Figure 9.20 *Mechanism of Fe^{3+} release from the photocaged complex [Fe_3(3,5-diCl-Sal-AHA)$_3$(μ_3-OCH_3)]$^+$ where 3,5-diCl-Sal-AHA is 4-((3,5-dichloro-2-hydroxybenzylidene) amino)-2-hydroxybutanoic acid. Light excitation of a phenolate-Fe^{3+} ligand to metal charge transfer band leads to the decarboxylation of the α-hydroxy acid and photoreduction to form Fe^{2+}*

trinuclear clusters with the alkoxo oxygen from the AHA moiety bridging the individual iron ions. Analysis of electrospray ionization mass spectra of all five complexes revealed that the trimeric structures predominate in methanolic solutions. Stability constants for four of the complexes had been determined with cyclic voltammetry and their values place Sal–AHA chelates between those found for hydroxamate containing siderophores (log K < 32) and those of the catecholate containing siderophores (log K > 43). Rates of Fe^{2+} production were observed to be effected by the electron withdrawing/donating groups on the phenolate moiety, where an increase in rate of photolysis was observed in chelates that demonstrated reduced electron donor ability on the phenolate group.

An attempt at development of the Fe^{3+} photocage that utilizes the *o*-nitrobenzyl group and the *N*-phenyl-1-oxa-4,10-dithia-7-azacyclododecane (AT$_2$12C4) macrocycle was inspired by a selective Fe^{3+} fluorescent sensor [166]. FerriCast binds Fe^{3+} ions in organic solvents in 1:1 and 1:2 stoichiometry and while conversion of the photocage with UV light to its photoproduct leads to reduction in Fe^{3+} affinity, introduction of oxygen containing solvents to assays triggers rapid decomplexation of metalated FerriCast (Fig. 9.21) [159].

9.4.3   Photocaged Complexes for Other Metal Ions

While photocaged complexes for other metal ions may be useful, there are relatively few available. CrownCast-1, which utilizes the same metal releasing mechanism as Nitr and ZinCast compounds [167], functions as a group II cation receptor with modest selectivity and affinity in a non-aqueous solution. Additional CrownCast photocages can selectively bind Hg^{2+} and Pb^{2+}, but the metal ion affinity changes upon photolysis are negligible [168]. Similar metal binding behavior has been observed in ArgenCast complexes that utilize a 3,6,12,15-tetrathia-9-azaheptadecane receptor for Ag^+ in mixed solvent [169]. None of the compounds developed to date are suitable for biological applications, which presents an opportunity for future investigations.

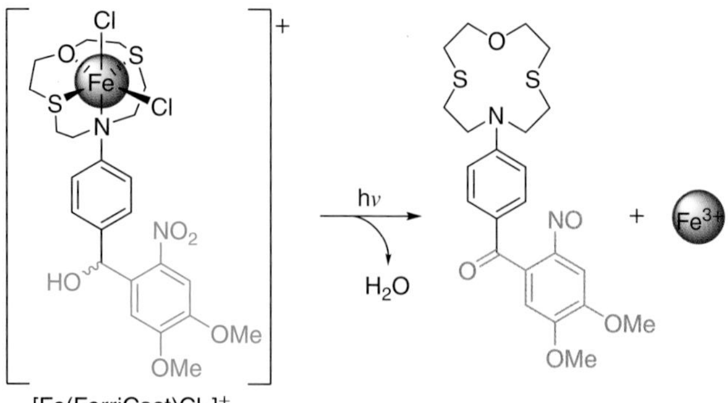

Figure 9.21   *Mechanism of Fe^{3+} release from the photocaged complex of FerriCast-1. Upon photolysis, the nitrobenzhydrol group (green) of FerriCast-1 is converted into a benzophenone photoproduct (dark yellow) similar to the process illustrated for Nitr-5 and ZinCast-1. Despite working in a non-aqueous solvent, the receptor cannot stabilize Fe^{3+} complexes in the presence of oxygen donor groups*

9.5   Conclusions

In addition to necessary signals, disruption in the acquisition and trafficking of metal ions may contribute to neurodegenerative diseases. While significant progress has been made in the fields of molecular and cellular biology, the role of metal ions in biological and pathological processes are beginning to be defined. Despite the evident importance of conducting research in metallobiology, investigations are limited by the available tools. Fluorescent sensors have been invaluable for studying the functions of Ca^{2+}, Zn^{2+}, Cu^+, and many other metal ions. The corresponding photochemical tools for delivering Ca^{2+} into cells and tissues have seen widespread use in uncovering and deciphering signaling pathways, and provide an important roadmap for developing the methodologies necessary to investigate other metal ions. As evidenced by the early work with Zn^{2+}, $Cu^{+/2+}$, and Fe^{3+}, new functions for these metal ions may soon be uncovered, and previous hypotheses can be tested.

Acknowledgment

This work was supported by the NSF grant CHE-0955361 and Worcester Polytechnic Institute.

References

1. C. Gwizdala and S. C. Burdette, Following the Ca^{2+} roadmap to photocaged complexes for Zn^{2+} and beyond, *Curr. Op. Chem. Biol.*, 137–142 (2012).
2. H. W. Mbatia and S. C. Burdette, Photochemical tools for studying metal ion signaling and homeostasis, *Biochemistry*, **51**, 7212–7224 (2012).

3. A. P. Pelliccioli and J. Wirz, Photoremovable protecting groups: reaction mechanisms and applications, *Photochem. Photobiol. Sci.*, **1**(7), 441–458 (2002).
4. E. J. Martinez-Finley, S. Chakraborty, S. J. Fretham and M. Aschner, Cellular transport and homeostasis of essential and nonessential metals, *Metallomics*, **4**(7), 593–605 (2012).
5. D. W. Domaille, E. L. Que and C. J. Chang, Synthetic fluorescent sensors for studying the cell biology of metals, *Nat. Chem. Biol.*, **4**(3), 168–175 (2008).
6. R. Y. Tsien, Fluorescent probes of cell signaling, *Ann. Rev. Neurosci.*, **12**, 227–253 (1989).
7. M. Oheim, M. Naraghi, T. H. Muller and E. Neher, Two dye two wavelength excitation calcium imaging: results from bovine adrenal chromaffin cells, *Cell Calcium*, **24**(1), 71–84 (1998).
8. A. Minta, J. Kao and R. Y. Tsien, Fluorescent indicators for cytosolic calcium based on rhodamine and fluorescein chromophores, *J. Biol. Chem.*, **264**(14), 8171–8178 (1989).
9. M. Poenie, J. Alderton, R. Tsien and R. Steinhardt, Changes of free calcium levels with stages of the cell division cycle, *Nature*, **315**(6015), 147–149 (1985).
10. T. W. Greene, P. G. Wuts and J. Wiley, *Protective Groups in Organic Synthesis*. John Wiley & Sons, Inc.: New York, 1999; Vol. 168.
11. V. R. Pillai, Photoremovable protecting groups in organic synthesis, *Synthesis*, **1980**(01), 1–26 (2002).
12. H. Yu, J. Li, D. Wu, Z. Qiu and Y. Zhang, Chemistry and biological applications of photo-labile organic molecules, *Chem. Soc. Rev.*, **39**(2), 464–473 (2010).
13. S. R. Adams and R. Y. Tsien, Controlling cell chemistry with caged compounds, *Ann. Rev. Physiol.*, **55**(1), 755–784 (1993).
14. G. C. R. Ellis-Davies, Caged compounds: photorelease technology for control of cellular chemistry and physiology, *Nat. Methods*, **4**(8), 619–628 (2007).
15. T. D. Carter, R. G. Bogle and T. Bjaaland, Spiking of intracellular calcium ion concentration in single cultured pig aortic endothelial cells stimulated with ATP or bradykinin, *Biochem. J.*, **278**(Pt 3), 697 (1991).
16. L. Heilbrunn, The action of calcium on muscle protoplasm, *Physiol. Zool.*, **13**(1), 88–94 (1940).
17. D. Thuringer and R. Sauvé, A patch-clamp study of the Ca^{2+} mobilization from internal stores in bovine aortic endothelial cells. II. Effects of thapsigargin on the cellular Ca^{2+} homeostasis, *J. Membr. Biol.*, **130**(2), 139–148 (1992).
18. G. C. Faas and I. Mody, Measuring the kinetics of calcium binding proteins with flash photolysis, *Biochem. Biophys. Acta, Gen. Subjects*, **1820**(8), 1195–1204 (2012).
19. B. Herman and J. J. Lemasters, *Optical Microscopy: Emerging Methods and Applications*. Academic Press: 1993.
20. S. Adams, J. P. Kao, G. Grynkiewicz, A. Minta and R. Tsien, Biologically useful chelators that release Ca^{2+} upon illumination, *J. Am. Chem. Soc.*, **110**(10), 3212–3220 (1988).
21. R. Y. Tsien and R. S. Zucker, Control of cytoplasmic calcium with photolabile tetracarboxylate 2-nitrobenzhydrol chelators, *Biophys. J.*, **50**(5), 843–853 (1986).

22. R. Y. Tsien, New calcium indicators and buffers with high selectivity against magnesium and protons: design, synthesis, and properties of prototype structures, *Biochemistry*, **19**(11), 2396–2404 (1980).

23. J. Cui, R. A. Gropeanu, D. R. Stevens, J. Rettig and A. del Campo, New photolabile BAPTA-based Ca^{2+} cages with improved photorelease, *J. Am. Chem. Soc.*, **134**(18), 7733–7740 (2012).

24. G. Grynkiewicz, M. Poenie and R. Y. Tsien, A new generation of Ca^{2+} indicators with greatly improved fluorescence properties, *J. Biol. Chem.*, **260**(6), 3440–3450 (1985).

25. S. R. Adams, V. Lev-Ram and R. Tsien, A new caged Ca^{2+}, azid-1, is far more photosensitive than nitrobenzyl-based chelators, *Chem. Biol.*, **4**(11), 867–78 (1997).

26. G. Ellis-Davies and J. Kaplan, A new class of photolabile chelators for the rapid release of divalent cations: generation of caged calcium and caged magnesium, *J. Org. Chem.*, **53**(9), 1966–1969 (1988).

27. J. H. Kaplan and G. Ellis-Davies, Photolabile chelators for the rapid photorelease of divalent cations, *Proc. Nat. Acad. Sci. USA*, **85**(17), 6571–6575 (1988).

28. E. Neher and R. Zucker, Multiple calcium-dependent processes related to secretion in bovine chromaffin cells, *Neuron*, **10**(1), 21–30 (1993).

29. R. K. Ayer Jr, and R. S. Zucker, Magnesium binding to DM-nitrophen and its effect on the photorelease of calcium, *Biophys. J.*, **77**(6), 3384–3393 (1999).

30. G. C. Faas, K. Karacs, J. L. Vergara and I. Mody, Kinetic properties of DM-nitrophen binding to calcium and magnesium, *Biophys. J.*, **88**(6), 4421–4433 (2005).

31. F. DelPrincipe, M. Egger, G. Ellis-Davies and E. Niggli, Two-photon and UV-laser flash photolysis of the Ca^{2+} cage, dimethoxynitrophenyl-EGTA-4, *Cell Calcium*, **25**(1), 85–91 (1999).

32. G. Ellis-Davies and J. H. Kaplan, Nitrophenyl-EGTA, a photolabile chelator that selectively binds Ca^{2+} with high affinity and releases it rapidly upon photolysis, *Proc. Nat. Acad. Sci. USA*, **91**(1), 187–191 (1994).

33. G. C. Ellis-Davies, Development and application of caged calcium, *Methods Enzymol.*, **360**, 226–238 (2003).

34. G. Ellis-Davies and R. J. Barsotti, Tuning caged calcium: photolabile analogues of EGTA with improved optical and chelation properties, *Cell Calcium*, **39**(1), 75–83 (2006).

35. A. Momotake, N. Lindegger, E. Niggli, R. J. Barsotti and G. C. Ellis-Davies, The nitrodibenzofuran chromophore: a new caging group for ultra-efficient photolysis in living cells, *Nat. Methods*, **3**(1), 35–40 (2005).

36. C. C. Ashley, I. Mulligan and T. J. Lea, Ca^{2+} and activation mechanisms in skeletal muscle, *Q. Rev. Biophys.*, **24**, 1–73 (1991).

37. L. Lando and R. S. Zucker, Ca^{2+} cooperativity in neurosecretion measured using photolabile Ca^{2+} chelators, *J. Neurophysiol.*, **72**(2), 825–830 (1994).

38. T. Lea and C. Ashley, Ca^{2+} release from the sarcoplasmic reticulum of barnacle myofibrillar bundles initiated by photolysis of caged Ca^{2+}, *J. Physiol.*, **427**(1), 435–453 (1990).

39. K. König, Multiphoton microscopy in life sciences, *J. Microsc.*, **200**(2), 83–104 (2000).

40. R. Zucker, Photorelease techniques for raising or lowering intracellular Ca^{2+}, *Methods Cell Biol.*, **40**, 31–63 (1994).

41. L. Lando and R. S. Zucker, "Caged calcium" *in* Aplysia pacemaker neurons. Characterization of calcium-activated potassium and nonspecific cation currents, *J. Gen. Physiol.*, **93**(6), 1017–1060 (1989).

42. K. Delaney and R. Zucker, Calcium released by photolysis of DM-nitrophen stimulates transmitter release at squid giant synapse, *J. Physiol.*, **426**(1), 473–498 (1990).

43. M. J. Berridge, M. D. Bootman and H. L. Roderick, Calcium signalling: dynamics, homeostasis and remodelling, *Nat. Rev. Mol. Cell. Biol.*, **4**(7), 517–529 (2003).

44. D. E. Clapham, Calcium signaling, *Cell*, **131**(6), 1047–1058 (2007).

45. T. Balla, Regulation of Ca^{2+} entry by inositol lipids in mammalian cells by multiple mechanisms, *Cell Calcium*, **45**(6), 527–534 (2009).

46. T. Gunter, K. K. Gunter, S.-S. Sheu and C. Gavin, Mitochondrial calcium transport: physiological and pathological relevance, *Am. J. Physiol. Cell Physiol.*, **267**(2), C313–C339 (1994).

47. A. Carl, N. McHale, N. Publicover and K. Sanders, Participation of Ca^{2+}-activated K^+ channels in electrical activity of canine gastric smooth muscle, *J. Physiol.*, **429**(1), 205–221 (1990).

48. N. Marrion, R. Zucker, S. Marsh and P. Adams, Modulation of M-current by intracellular Ca^{2+}, *Neuron*, **6**, 533–545 (1991).

49. P. Adams and D. Brown, Luteinizing hormone-releasing factor and muscarinic agonists act on the same voltage-sensitive K+-current in bullfrog sympathetic neurones, *Br. J. Pharmacol.*, **68**(3), 353–355 (1980).

50. P. R. Adams, D. Brown and A. Constanti, Pharmacological inhibition of the M-current, *J. Physiol.*, **332**(1), 223–262 (1982).

51. N. V. Marrion, Control of M-current, *Ann. Rev. Physiol.*, **59**(1), 483–504 (1997).

52. M. Morad, N. W. Davies, J. H. Kaplan and H. D. Lux, Inactivation and block of calcium channels by photo-released Ca^{2+} in dorsal root ganglion neurons, *Science*, **241**(4867), 842–844 (1988).

53. P. Charnet, S. Richard, A. Gurney, H. Ouadid, F. Tiaho and J. Nargeot, Modulation of Ca currents in isolated frog atrial cells studied with photosensitive probes. Regulation by cAMP and Ca^{2+}: A common pathway?, *J. Mol. Cell. Cardiol.*, **23**(3), 343–356 (1991).

54. S. Bates and A. Gurney, Ca^{2+}-dependent block and potentiation of L-type calcium current in guinea-pig ventricular myocytes, *J. Physiol.*, **466**(1), 345–365 (1993).

55. B. O'Rourke, P. H. Backx and E. Marban, Phosphorylation-independent modulation of L-type calcium channels by magnesium-nucleotide complexes, *Science*, **257**(5067), 245–248 (1992).

56. S. Gyorke and M. Fill, Ryanodine receptor adaptation: control mechanism of Ca^{2+}-induced Ca^{2+} release in heart, *Science*, **260**(5109), 807–809 (1993).

57. M. Nabauer and M. Morad, Ca^{2+}-induced Ca^{2+} release as examined by photolysis of caged Ca^{2+} in single ventricular myocytes, *Am. J. Physiol. Cell Physiol.*, **258**(1), C189–C193 (1990).

58. R. S. Zucker and P. G. Haydon, Membrane potential has no direct role in evoking neurotransmitter release, *Nature*, **335**(6188), 360–362 (1988).

59. B. Hochner, H. Parnas and I. Parnas, Membrane depolarization evokes neurotransmitter release in the absence of calcium entry, *Nature*, **342**, 433–435 (1989).
60. C. Heinemann, R. H. Chow, E. Neher and R. S. Zucker, Kinetics of the secretory response in bovine chromaffin cells following flash photolysis of caged Ca^{2+}, *Biophys. J.*, **67**(6), 2546–2557 (1994).
61. B. Pan and R. S. Zucker, A general model of synaptic transmission and short-term plasticity, *Neuron*, **62**(4), 539–554 (2009).
62. S. M. Young Jr, and E. Neher, Synaptotagmin has an essential function in synaptic vesicle positioning for synchronous release in addition to its role as a calcium sensor, *Neuron*, **63**(4), 482–496 (2009).
63. D. Beutner, T. Voets, E. Neher and T. Moser, Calcium dependence of exocytosis and endocytosis at the cochlear inner hair cell afferent synapse, *Neuron*, **29**(3), 681–690 (2001).
64. G. Duncan, K. Rabl, I. Gemp, R. Heidelberger and W. B. Thoreson, Quantitative analysis of synaptic release at the photoreceptor synapse, *Biophys. J.*, **98**(10), 2102–2110 (2010).
65. C. Andreini, L. Banci, I. Bertini and A. Rosato, Counting the zinc-proteins encoded in the human genome, *J. Proteome Res.*, **5**(1), 196–201 (2006).
66. A. Klug, The discovery of zinc fingers and their development for practical applications in gene regulation and genome manipulation, *Q. Rev. Biophys.*, **43**(1), 1–21 (2010).
67. C. Andreini, L. Banci, I. Bertini and A. Rosato, Occurrence of copper proteins through the three domains of life: a bioinformatic approach, *J. Proteome Res.*, **7**(01), 209–216 (2007).
68. C. Andreini, L. Banci, I. Bertini, S. Elmi and A. Rosato, Non-heme iron through the three domains of life, *Proteins: Struct., Funct., Bioinf.*, **67**(2), 317–324 (2007).
69. T. Fukada and T. Kambe, Molecular and genetic features of zinc transporters in physiology and pathogenesis, *Metallomics*, **3**(7), 662–674 (2011).
70. D. E. K. Sutherland and M. J. Stillman, The "magic numbers" of metallothionein, *Metallomics*, **3**(5), 444–463 (2011).
71. S. S. Krishna, I. Majumdar and N. V. Grishin, Structural classification of zinc fingers SURVEY AND SUMMARY, *Nucleic Acids Res.*, **31**(2), 532–550 (2003).
72. T. Kambe, An Overview of a Wide Range of Functions of ZnT and zip zinc transporters in the secretory pathway, *Biosci. Biotech. Biochem.*, **75**(6), 1036–1043 (2011).
73. R. E. Dempski, The cation selectivity of the zip Transporters, *Metal Transporters*, **69**, 221 (2012).
74. L. Huang and S. Tepaamorndech, The SLC30 family of zinc transporters – A review of current understanding of their biological and pathophysiological roles, *Mol. Aspects Med.*, **34**(2), 548–560 (2013).
75. S. Ripa and R. Ripa, Zinc cellular traffic: physiopathological considerations, *Minerva Med.*, **86**(1-2), 37–43 (1995).
76. Y. Li and W. Maret, Human metallothionein metallomics, *J. Anal. At. Spectrom.*, **23**(8), 1055–1062 (2008).
77. A. Kręzel and W. Maret, Dual nanomolar and picomolar Zn (II) binding properties of metallothionein, *J. Am. Chem. Soc.*, **129**(35), 10911–10921 (2007).

78. W. Maret, Molecular aspects of human cellular zinc homeostasis: redox control of zinc potentials and zinc signals, *Biometals*, **22**(1), 149–157 (2009).

79. L. L. Pearce, K. Wasserloos, C. M. S. Croix, R. Gandley, E. S. Levitan and B. R. Pitt, Metallothionein, nitric oxide and zinc homeostasis in vascular endothelial cells, *J. Nutr.*, **130**(5), 1467S–1470S (2000).

80. E. Bossy-Wetzel, M. V. Talantova, W. D. Lee, M. N. Schölzke, A. Harrop, E. Mathews, T. Götz, J. Han, M. H. Ellisman, G. A. Perkins and S. A. Lipton, Crosstalk between nitric oxide and zinc pathways to neuronal cell death involving mitochondrial dysfunction and p38-activated K^+ channels, *Neuron*, **41**, 351–365 (2004).

81. A. Krżel, Q. Hao and W. Maret, The zinc/thiolate redox biochemistry of metallothionein and the control of zinc ion fluctuations in cell signaling, *Arch. Biochem. Biophys.*, **463**(2), 188–200 (2007).

82. R. J. Williams, Zinc: what is its role in biology? *Endeavour*, **8**(2), 65–70 (1984).

83. B. Roschitzki and M. Vašák, Redox labile site in a Zn4 cluster of Cu4, Zn4-metallothionein-3, *Biochemistry*, **42**(32), 9822–9828 (2003).

84. I. Korichneva, B. Hoyos, R. Chua, E. Levi and U. Hammerling, Zinc release from protein kinase C as the common event during activation by lipid second messenger or reactive oxygen, *Sci. Signal.*, **277**(46), 44327–44331 (2002).

85. S. Yamasaki, K. Sakata-Sogawa, A. Hasegawa, T. Suzuki, K. Kabu, E. Sato, T. Kurosaki, S. Yamashita, M. Tokunaga and K. Nishida, Zinc is a novel intracellular second messenger, *J. Cell Biol.*, **177**(4), 637–645 (2007).

86. S. Antala and R. E. Dempski, The human ZIP4 transporter has two distinct binding affinities and mediates transport of multiple transition metals, *Biochemistry*, **51**(5), 963–973 (2012).

87. D. Atar, P. H. Backx, M. M. Appel, W. D. Gao and E. Marban, Excitation-transcription coupling mediated by zinc influx through voltage-dependent calcium channels, *J. Biol. Chem.*, **270**(6), 2473–2477 (1995).

88. H. Haase, J. L. Ober-Blöbaum, G. Engelhardt, S. Hebel, A. Heit, H. Heine and L. Rink, Zinc signals are essential for lipopolysaccharide-induced signal transduction in monocytes, *J. Immunol.*, **181**(9), 6491–6502 (2008).

89. E. Tuncay, A. Bilginoglu, N. N. Sozmen, E. N. Zeydanli, M. Ugur, G. Vassort and B. Turan, Intracellular free zinc during cardiac excitation–contraction cycle: calcium and redox dependencies, *Cardiovasc. Res.*, **89**(3), 634–642 (2011).

90. Y. Li and W. Maret, Transient fluctuations of intracellular zinc ions in cell proliferation, *Exp. Cell Res.*, **315**(14), 2463–2470 (2009).

91. L. Petrie, J. Chesters and M. Franklin, Inhibition of myoblast differentiation by lack of zinc, *Biochem. J.*, **276**(Pt 1), 109–111 (1991).

92. H. Haase and W. Maret, Intracellular zinc fluctuations modulate protein tyrosine phosphatase activity in insulin/insulin-like growth factor-1 signaling, *Exp. Cell Res.*, **291**(2)doi, 289–298 (2003).

93. J. M. May and C. Contoreggi, The mechanism of the insulin-like effects of ionic zinc, *J. Biol. Chem.*, **257**(8), 4362–4368 (1982).

94. C. Hogstrand, P. Kille, R. Nicholson and K. Taylor, Zinc transporters and cancer: a potential role for ZIP7 as a hub for tyrosine kinase activation, *Trends Mol. Med.*, **15**(3), 101–111 (2009).

95. S.-J. Lee, K.-S. Cho, H. N. Kim, H.-J. Kim and J.-Y. Koh, Role of zinc metallothionein-3 (ZnMt3) in epidermal growth factor (EGF)-induced c-Abl protein activation and actin polymerization in cultured astrocytes, *J. Biol. Chem.*, **286**(47), 40847–40856 (2011).

96. J. M. Samet, L. M. Graves, J. Quay, L. A. Dailey, R. B. Devlin, A. J. Ghio, W. Wu, P. A. Bromberg and W. Reed, Activation of MAPKs in human bronchial epithelial cells exposed to metals, *Am. J. Physiol. Lung Cell. Mol. Physiol.*, **275**(3), L551–L558 (1998).

97. A. Baba, S. Etoh and H. Iwata, Inhibition of NMDA-induced protein kinase C translocation by a Zn^{2+} chelator: Implication of intracellular Zn^{2+}, *Brain Res.*, **557**(1), 103–108 (1991).

98. R. E. Carter, I. Aiba, R. M. Dietz, C. T. Sheline and C. W. Shuttleworth, Spreading depression and related events are significant sources of neuronal Zn^{2+} release and accumulation, *J. Cereb. Blood Flow Metab.*, **31**(4), 1073–1084 (2010).

99. G. Wellenreuther, M. Cianci, R. Tucoulou, W. Meyer-Klaucke and H. Haase, The ligand environment of zinc stored in vesicles, *Biochem. Biophys. Res. Commun.*, **380**(1), 198–203 (2009).

100. H. Haase, S. Hebel, G. Engelhardt and L. Rink, Flow cytometric measurement of labile zinc in peripheral blood mononuclear cells, *Anal. Biochem.*, **352**(2), 222–230 (2006).

101. S. L. Sensi, L. M. Canzoniero, S. P. Yu, H. S. Ying, J.-Y. Koh, G. A. Kerchner and D. W. Choi, Measurement of intracellular free zinc in living cortical neurons: routes of entry, *J. Neurosci.*, **17**(24), 9554–9564 (1997).

102. R. D. Palmiter, T. B. Cole and S. D. Findley, ZnT-2, a mammalian protein that confers resistance to zinc by facilitating vesicular sequestration, *EMBO J.*, **15**(8), 1784 (1996).

103. G. Danscher and M. Stoltenberg, Zinc-specific autometallographic in vivo selenium methods: tracing of zinc-enriched (ZEN) terminals, ZEN pathways, and pools of zinc ions in a multitude of other ZEN cells, *J. Histochem. Cytochem.*, **53**(2), 141–153 (2005).

104. H. Bandara, T. P. Walsh and S. C. Burdette, A second-generation photocage for Zn^{2+} inspired by TPEN: Characterization and insight into the uncaging quantum yields of ZinCleav chelators, *Chem. Eur. J.*, **17**(14), 3932–3941 (2011).

105. H. D. Bandara, D. P. Kennedy, E. Akin, C. D. Incarvito and S. C. Burdette, Photoinduced release of Zn^{2+} with ZinCleav-1: a nitrobenzyl-based caged complex, *Inorg. Chem.*, **48**(17), 8445–8455 (2009).

106. C. E. Outten and T. V. O'Halloran, Femtomolar sensitivity of metalloregulatory proteins controlling zinc homeostasis, *Science*, **292**(5526), 2488–2492 (2001).

107. C. Gwizdala, P. N. Basa, J. C. MacDonald and S. C. Burdette, Increasing the dynamic range of metal ion affinity changes in Zn^{2+} photocages using multiple nitrobenzyl groups, *Inorg. Chem.*, **52**, 8483–8494 (2013), DOI: 10.1021/ic400465g.

108. C. Gwizdala, D. P. Kennedy and S. C. Burdette, ZinCast-1: a photochemically active chelator for Zn^{2+}, *Chem. Commun.*, (**45**), 6967–6969 (2009).

109. C. Gwizdala, C. V. Singh, T. R. Friss, J. C. MacDonald and S. C. Burdette, Quantifying factors that influence metal ion release in photocaged complexes using ZinCast derivatives, *Dalton Trans.*, **41**, 8162–8174 (2012).

110. S. A. Siletsky and A. A. Konstantinov, Cytochrome c oxidase: Charge translocation coupled to single-electron partial steps of the catalytic cycle, *Biochem. Biophys. Acta, Bioenergetics*, **1817**(4), 476–488 (2012).

111. J. M. Leitch, P. J. Yick and V. C. Culotta, The Right to Choose: Multiple Pathways for Activating Copper, Zinc Superoxide Dismutase, *J. Biol. Chem.*, **284**(37), 24679–24683 (2009).

112. P. Bielli and L. Calabrese, Structure to function relationships in ceruloplasmin: a 'moonlighting' protein, *Cell. Mol. Life Sci.*, **59**(9), 1413–1427 (2002).

113. J. P. Klinman, The copper-enzyme family of dopamine beta-monooxygenase and peptidylglycine alpha-hydroxylating monooxygenase: Resolving the chemical pathway for substrate hydroxylation, *J. Biol. Chem.*, **281**(6), 3013–3016 (2006).

114. B.-E. Kim, T. Nevitt and D. J. Thiele, Mechanisms for copper acquisition, distribution and regulation, *Nat. Chem. Biol.*, **4**(3), 176–185 (2008).

115. Z. Xiao and A. G. Wedd, Metallo-oxidase Enzymes: Design of their active sites, *Aust. J. Chem.*, **64**(3), 231–238 (2011).

116. K. Jomova and M. Valko, Advances in metal-induced oxidative stress and human disease, *Toxicology*, **283**(2-3), 65–87 (2011).

117. C. T. Dameron and M. D. Harrison, Mechanisms for protection against copper toxicity, *Am. J. Clin. Nutrit.*, **67**(5), 1091S–1097S (1998).

118. C. J. Maynard, R. Cappai, I. Volitakis, R. A. Cherny, A. R. White, K. Beyreuther, C. L. Masters, A. I. Bush and Q.-X. Li, Overexpression of Alzheimer's disease amyloid-β opposes the age-dependent elevations of brain copper and iron, *J. Biol. Chem.*, **277**(47), 44670–44676 (2002).

119. I. L. Crawford, Zinc and the hippocampus: Histology, neurochemistry, pharmacology and putative functional relevance. In *Neurobiology of the Trace Elements: Trace Element Neurobiology and Deficiencies*, Dreosti, J., Smith, R., Eds. Humana Press: New Jersey, 1983; Vol. 1, pp. 163–212.

120. E. Bonilla, E. Salazar, J. J. Villasmil, R. Villalobos, M. Gonzalez and J. O. Davila, Copper distribution in the normal human brain, *Neurochem. Res.*, **9**(11), 1543–1548 (1984).

121. S. Lutsenko, N. L. Barnes, M. Y. Bartee and O. Y. Dmitriev, Function and regulation of human copper-transporting ATPases, *Physiol. Rev.*, **87**(3), 1011–1046 (2007).

122. B. Zhou and J. Gitschier, hCTR1: a human gene for copper uptake identified by complementation in yeast, *Proc. Nat. Acad. Sci. USA*, **94**(14), 7481–7486 (1997).

123. D. R. Brown and H. Kozlowski, Biological inorganic and bioinorganic chemistry of neurodegeneration based on prion and Alzheimer diseases, *Dalton Trans.*, (**13**), 1907–1917 (2004).

124. K. J. Barnham, W. J. McKinstry, G. Multhaup, D. Galatis, C. J. Morton, C. C. Curtain, N. A. Williamson, A. R. White, M. G. Hinds and R. S. Norton, Structure of the Alzheimer's disease amyloid precursor protein copper binding domain A regulator of neuronal copper homeostasis, *J. Biol. Chem.*, **278**(19), 17401–17407 (2003).

125. M. J. Petris, K. Smith, J. Lee and D. J. Thiele, Copper-stimulated endocytosis and degradation of the human copper transporter, hCtr1, *J. Biol. Chem.*, **278**(11), 9639–9646 (2003).

126. S.-J. Lin, R. A. Pufahl, A. Dancis, T. V. O'Halloran and V. C. Culotta, A role for the Saccharomyces cerevisiae ATX1 gene in copper trafficking and iron transport, *J. Biol. Chem.*, **272**(14), 9215–9220 (1997).

127. V. C. Culotta, L. W. Klomp, J. Strain, R. L. B. Casareno, B. Krems and J. D. Gitlin, The copper chaperone for superoxide dismutase, *J. Biol. Chem.*, **272**(38), 23469–23472 (1997).

128. D. M. Glerum, A. Shtanko and A. Tzagoloff, Characterization of COX17, a yeast gene involved in copper metabolism and assembly of cytochrome oxidase, *J. Biol. Chem.*, **271**(24), 14504–14509 (1996).

129. H. Zhou and D. J. Thiele, Identification of a novel high affinity copper transport complex in the fission yeast Schizosaccharomyces pombe, *J. Biol. Chem.*, **276**(23), 20529–20535 (2001).

130. D. J. Thiele and J. D. Gitlin, Assembling the pieces, *Nat. Chem. Biol.*, **4**(3), 145–147 (2008).

131. B. Sarkar, The malfunctioning of copper transport in Wilson and Menkes diseases. In *Neurodegenerative Diseases and Metal Ions: Metal Ions in Life Sciences*, Sigel, A., Sigel, H., Sigel, R. K. O., Eds. John Wiley & Sons, Ltd: Chichester, 2006; pp. 207–226.

132. K. L. Ciesienski, K. L. Haas, M. G. Dickens, Y. T. Tesema and K. J. Franz, A photo-labile ligand for light-activated release of caged copper, *J. Am. Chem. Soc.*, **130**(37), 12246–12247 (2008).

133. K. L. Ciesienski, K. L. Haas and K. J. Franz, Development of next-generation photo-labile copper cages with improved copper binding properties, *Dalton Trans.*, **39**(40), 9538–9546 (2010).

134. A. A. Kumbhar, A. T. Franks, R. J. Butcher and K. J. Franz, Light uncages a copper complex to induce nonapoptotic cell death, *Chem. Commun.*, **49**, 2460–2462 (2013).

135. S. C. Dodani, D. W. Domaille, C. I. Nam, E. W. Miller, L. A. Finney, S. Vogt and C. J. Chang, Calcium-dependent copper redistributions in neuronal cells revealed by a fluorescent copper sensor and X-ray fluorescence microscopy, *Proc. Nat. Acad. Sci. USA*, **108**(15), 5980–5985 (2011).

136. M. L. Schlief, A. M. Craig and J. D. Gitlin, NMDA receptor activation mediates copper homeostasis in hippocampal neurons, *J. Neurosci.*, **25**(1), 239–246 (2005).

137. H. W. Mbatia, H. D. Bandara and S. C. Burdette, CuproCleav-1, a first generation photocage for Cu^+, *Chem. Commun.*, **48**(43), 5331–5333 (2012).

138. K. H. Nealson and D. Saffarini, Iron and manganese in anaerobic respiration: environmental significance, physiology, and regulation, *Ann. Rev. Microbiol.*, **48**(1), 311–343 (1994).

139. J. Gao and D. R. Richardson, The potential of iron chelators of the pyridoxal isonicotinoyl hydrazone class as effective antiproliferative agents, IV: the mechanisms involved in inhibiting cell-cycle progression, *Blood*, **98**(3), 842–850 (2001).

140. D. R. Richardson, D. J. Lane, E. M. Becker, M. L.-H. Huang, M. Whitnall, Y. S. Rahmanto, A. D. Sheftel and P. Ponka, Mitochondrial iron trafficking and the integration of iron metabolism between the mitochondrion and cytosol, *Proc. Nat. Acad. Sci. USA*, **107**(24), 10775–10782 (2010).

141. M. Graziano, M. a. V. Beligni and L. Lamattina, Nitric oxide improves internal iron availability in plants, *Plant Physiol.*, **130**(4), 1852–1859 (2002).

142. H. W. Paerl, L. E. Prufert-Bebout and C. Guo, Iron-stimulated N_2 fixation and growth in natural and cultured populations of the planktonic marine cyanobacteria *Trichodesmium spp*, *App. Environ. Microbiol.*, **60**(3), 1044–1047 (1994).

143. B. Halliwell and J. Gutteridge, Oxygen free radicals and iron in relation to biology and medicine: some problems and concepts, *Arch. Biochem. Biophys.*, **246**(2), 501–514 (1986).

144. A. Terman and T. Kurz, Lysosomal iron, iron chelation, and cell death, *Antioxid. Redox Signaling*, **18**(8), 888–898 (2013).

145. K. Gkouvatsos, G. Papanikolaou and K. Pantopoulos, Regulation of iron transport and the role of transferrin, *Biochem. Biophys. Acta, Gen. Subjects*, **1820**(3), 188–202 (2012).

146. M. E. Conrad and J. N. Umbreit, Pathways of iron absorption, *Blood Cells Mol. Dis.*, **29**(3), 336–355 (2002).

147. P. Aisen and I. Listowsky, Iron transport and storage proteins, *Ann. Rev. Biochem.*, **49**(1), 357–393 (1980).

148. E. Morgan, Transferrin, biochemistry, physiology and clinical significance, *Mol. Aspects Med.*, **4**(1), 1–123 (1981).

149. J.-N. Octave, Y.-J. Schneider, A. Trouet and R. R. Crichton, Iron uptake and utilization by mammalian cells. I: Cellular uptake of transferrin and iron, *Trends Biochem. Sci.*, **8**(6), 217–220 (1983).

150. E. C. Theil, Iron, ferritin, and nutrition, *Ann. Rev. Nutr.*, **24**, 327–343 (2004).

151. M. L. Wallander, E. A. Leibold and R. S. Eisenstein, Molecular control of vertebrate iron homeostasis by iron regulatory proteins, *Biochem. Biophys. Acta, Mol. Cell Res.*, **1763**(7), 668–689 (2006).

152. M. Arredondo and M. T. Núñez, Iron and copper metabolism, *Mol. Aspects Med.*, **26**(4), 313–327 (2005).

153. P. T. Lieu, M. Heiskala, P. A. Peterson and Y. Yang, The roles of iron in health and disease, *Mol. Aspects Med.*, **22**(1), 1–87 (2001).

154. A.-S. Zhang, A. D. Sheftel and P. Ponka, Intracellular kinetics of iron in reticulocytes: evidence for endosome involvement in iron targeting to mitochondria, *Blood*, **105**(1), 368–375 (2005).

155. E. Nemeth and T. Ganz, Regulation of iron metabolism by hepcidin, *Ann. Rev. Nutr.*, **26**, 323–342 (2006).

156. N. C. Andrews, Disorders of iron metabolism, *N. Engl. J. Med.*, **341**(26), 1986–1995 (1999).

157. O. Kakhlon and Z. I. Cabantchik, The labile iron pool: characterization, measurement, and participation in cellular processes, *Free Radical Biol. Med.*, **33**(8), 1037–1046 (2002).

158. S. Epsztejn, O. Kakhlon, H. Glickstein, W. Breuer and Z. I. Cabantchik, Fluorescence analysis of the labile iron pool of mammalian cells, *Anal. Biochem.*, **248**(1), 31–40 (1997).

159. D. P. Kennedy, C. D. Incarvito and S. C. Burdette, FerriCast: a macrocyclic photocage for Fe^{3+}, *Inorg. Chem.*, **49**(3), 916–923 (2010).

160. C. Correnti and R. K. Strong, Mammalian siderophores, siderophore-binding lipocalins, and the labile iron pool, *J. Biol. Chem.*, **287**(17), 13524–13531 (2012).

161. K. Barbeau, E. L. Rue, C. G. Trick, K. W. Bruland and A. Butler, Photochemical reactivity of siderophores produced by marine heterotrophic bacteria and cyanobacteria based on characteristic Fe(III) binding groups, *Limnol. Oceanogr.*, **48**(3), 1069–1078 (2003).

162. K. Barbeau, E. L. Rue, K. W. Bruland and A. Butler, Photochemical cycling of iron in the surface ocean mediated by microbial iron(III)-binding ligands, *Nature*, **413**(6854), 409–413 (2001).

163. K. Barbeau, G. P. Zhang, D. H. Live and A. Butler, Petrobactin, a photoreactive siderophore produced by the oil-degrading marine bacterium *Marinobacter hydrocarbonoclasticus*, *J. Am. Chem. Soc.*, **124**(3), 378–379 (2002).

164. F. C. Kupper, C. J. Carrano, J. U. Kuhn and A. Butler, Photoreactivity of iron(III) - Aerobactin: Photoproduct structure and iron(III) coordination, *Inorg. Chem.*, **45**(15), 6028–6033 (2006).

165. H. Sayre, K. Milos, M. J. Goldcamp, C. A. Schroll, J. A. Krause and M. J. Baldwin, Mixed-donor, α-hydroxy acid-containing chelates for binding and light-triggered release of iron, *Inorg. Chem.*, **49**(10), 4433–4439 (2010).

166. J. L. Bricks, A. Kovalchuk, C. Trieflinger, M. Nofz, M. Büschel, A. I. Tolmachev, J. Daub and K. Rurack, On the development of sensor molecules that display FeIII-amplified fluorescence, *J. Am. Chem. Soc.*, **127**(39), 13522–13529 (2005).

167. D. P. Kennedy, C. Gwizdala and S. C. Burdette, Methods for preparing metal ion photocages: application to the synthesis of crowncast, *Org. Lett.*, **11**(12), 2587–2590 (2009).

168. H. W. Mbatia, D. P. Kennedy, C. E. Camire, C. D. Incarvito and S. C. Burdette, Buffering heavy metal ions with photoactive crowncast cages, *Eur. J. Inorg. Chem.*, **2010**(32), 5069–5078 (2010).

169. H. W. Mbatia, D. P. Kennedy and S. C. Burdette, Understanding the relationship between photolysis efficiency and metal binding using ArgenCast photocages, *Photochem. Photobiol.*, **88**, 844–850 (2012).

10

Release of Bioactive Molecules Using Metal Complexes

Peter V. Simpson and Ulrich Schatzschneider

Institut für Anorganische Chemie, Julius-Maximilians-Universität Würzburg, Germany

10.1   Introduction

The biological activity of molecules can be modulated by coordination to a transition metal centre. Ideally, this results in a completely inactive system, which acts as a prodrug, from which the bioactive component can be released by a specific trigger mechanism in a spatially and temporally controlled way (Fig. 10.1). Thus, the metal centre can be viewed as a protective group to prevent interaction of the bioactive molecule with its intended biological target. The trigger mechanism can be based on intrinsic properties of the biological system of interest, such as differences in pH or oxygen concentration between cells and tissues, or an external stimulus, such as application of a magnetic field or electromagnetic radiation able to penetrate the body. The release of the bioactive component from the metal coordination sphere inevitably generates a metal–coligand fragment with a free coordination site, which will rapidly become occupied by solvent or functional groups of bio(macro)molecules such as proteins or nucleic acids. Thus, metal–coligand fragments might have a biological activity of their own and should be tested in a separate control experiment.

The bioactive compounds that can be delivered by release from a metal centre range from small diatomic molecules such as nitric oxide (NO) and carbon monoxide (CO) to organic neurotransmitters and established pharmaceuticals. In contrast to the systems discussed in Chapter 9 (Photo-release of Metal Ions in Living Cells), where the metal ions liberated exert the biological activity, we will restrict ourselves here to compounds in which the activity is due to the component released from the metal coordination sphere, and assume that the metal–coligand fragment also generated only acts as an innocent spectator (although this should always be verified in a control experiment). Owing to the very

Inorganic Chemical Biology: Principles, Techniques and Applications, First Edition. Edited by Gilles Gasser.
© 2014 John Wiley & Sons, Ltd. Published 2014 by John Wiley & Sons, Ltd.

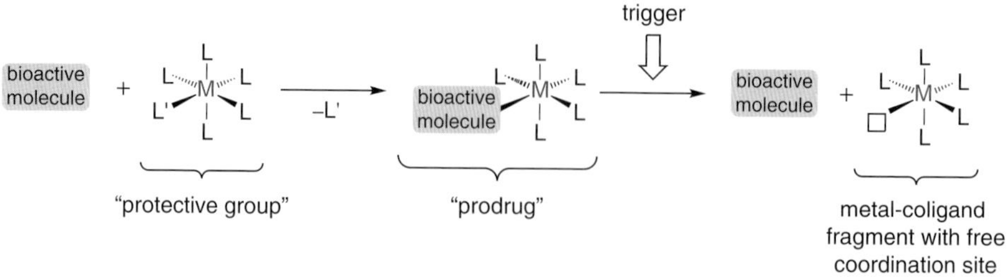

Figure 10.1 *General concept of a bioactive molecule coordinated to a transition metal centre to give an inactive "prodrug" and release of the active component by a specific trigger mechanism. Note the inevitable formation of a metal-coligand fragment with a free coordination site in the latter process*

fast progress of this field of research, no comprehensive treatment can be given. Instead, we will focus on three different applications, to highlight general concepts as well as current challenges in the area: (1) the release of small-molecule messengers such as NO and CO from the metal coordination sphere (2) the "photouncaging" of organic neurotransmitters from metal complexes, and (3) the hypoxia-activated release of anticancer-active compounds from metal-based prodrugs.

10.2   Small-molecule Messengers

10.2.1   Biological Generation and Delivery of CO, NO, and H_2S

Although generally known to the public only as highly toxic air pollutants, mostly resulting from the incomplete combustion of organic fuels, it is now well established that the small diatomic molecules nitric oxide (NO) and carbon monoxide (CO) are endogenously generated by tightly controlled enzymatic processes in higher organisms, including humans, and serve an important function in signaling processes and response to oxidative stress [1, 2]. Together with hydrogen sulfide (H_2S) [3], these small molecules have been somewhat unfortunately referred to as "gasotransmitters", although in biological systems they are actually not present in the form of gas bubbles or the like, but rather dissolved in body fluids and cytosol. The tissue-specific delivery of CO, NO, and H_2S faces particular challenges due to their unique reactivity [4]. Depending on pH, hydrogen sulfide can be present as a mixture of H_2S, HS^-, and S^{2-}, while nitric oxide as a radical species rapidly interacts with other constituents of biological systems. Although carbon monoxide is the most inert of the three, it is known for its facile binding to transition metal centres in metalloproteins, with the tissue accumulation of CO responsible for its general toxicity [5]. Furthermore, since these molecules by themselves do not have any handle for chemical modification, it is impossible to specifically direct them to particular tissues or sites of disease.

Thus, to utilize their many promising biological properties in the development of novel therapeutics [6, 7], they have to be bound to carrier molecules in the form of prodrugs. Owing to the complicated pH-dependent speciation and coordination chemistry of hydrogen sulfide, H_2S delivery is currently dominated by organic sulfur compounds and thus

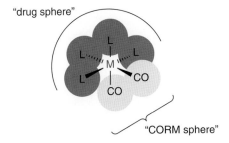

Figure 10.2   *"CORM sphere" (light grey) and "drug sphere" (dark grey) of a transition metal carbonyl complex*

will not be further discussed here [8, 9]. However, metal-based hydrogen sulfide delivery systems are certainly a promising target for future research. Carbon monoxide and nitric oxide, on the other hand, are well known to coordinate to a wide range of transition metals in different oxidation states and with various coligands, thus allowing a facile tuning of the properties of such systems. While the strength of the M−NO and M−CO bonds will mostly determine the release kinetics of these small-molecule messengers, functionalization of the "outer rim" of the coligands can be used to direct these compounds to specific biological target sites. In the case of metal−carbonyl complexes used as *CO-releasing molecules* (CORMs), these have been termed the "CORM sphere" and the "drug sphere", respectively, by Romao *et al.* (Fig. 10.2) [10], but the same principle also applies to nitric oxide carriers.

10.2.2   Metal−Nitrosyl Complexes for the Cellular Delivery of Nitric Oxide

Unaware of the underlying principle of NO-mediated bioactivity, nitric oxide donors were used long before NO was actually identified as the endothelium-derived relaxing factor (EDRF), a discovery subsequently honored with the Nobel Prize in Physiology to Furchgott, Ignarro, and Murat in 1998 [11−13]. Curiously, it has been reported that Alfred Nobel was prescribed nitroglycerine (glyercyl trinitrate, GTN) for the treatment of his angina pectoris and since then, a plethora of other mostly organic NO donors have been evaluated, many of which are now in clinical use [14−18]. A notable early example of a transition metal complex is sodium nitroprusside (SNP, $Na_2[Fe(CN)_5(NO)]$, **1**), which is used in intensive care medicine for the rapid reduction of arterial hypertension since it spontaneously releases nitric oxide at physiological pH values. However, for a more precise spatial and temporal control of NO release and biological action, photoactivatable metal−nitrosyl complexes have emerged as the focus of considerable research in recent years [19].

Owing to their rich coordination chemistry and tunable release kinetics, the focus has mostly been on iron and ruthenium nitrosyl compounds (Fig. 10.3), with either one or several NO ligands per molecular unit. In addition to SNP **1**, which only incorporates small, diatomic ligands, more complex NO delivery agents comprising dinitrosyl iron complexes (DNICs, **2**), iron−sulfur−cluster nitrosyls **3−5**, and porphyrin-based systems with axial NO ligands have been introduced. Many of the mononuclear iron and ruthenium nitrosyl compounds (e.g., **6** and **7**) can also generally be viewed as porphyrin analogues.

While the crystalline dihydrate of SNP **1** has very good long-term stability when stored dry and away from light, solutions thereof show fast photoinduced NO release, further

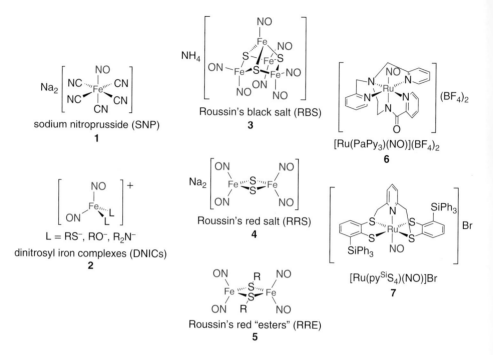

Figure 10.3 *Major classes of NO-releasing molecules*

accelerated by the presence of oxygen. In the body, however, it is the dark one-electron reduction that is the relevant trigger mechanism, with biogenic reducing agents such as thiols and ascorbate, but possibly also metalloproteins, inducing the NO release. These processes also lead to the liberation of cyanide and formation of peroxynitrite, which fortunately does not constitute a serious toxicity problem due to the very low doses of SNP used [14].

Dinitrosyl iron complexes (DNICs, **2**) are formed in biological systems, in particular under stress conditions, with the $Fe(NO)_2$ unit coordinated by a variety of anionic ligands, in particular *S*-donor groups from cysteine and gluthathione (GSH) [20]. Both paramagnetic and diamagnetic DNICs with core structures of $\{Fe(NO)_2\}^9$ and $\{Fe(NO)_2\}^{10}$, respectively, according to the Enemark–Feltham notation [21], are found, together with dimeric forms of the former one. Synthetic DNICs are currently considered as an interesting class of novel NO-donating drugs.

In addition to naturally occurring [Fe–S] clusters, which have very important functions in cells in both an electron transfer as well as a catalytic role, reaction with nitric oxide gives rise to iron–sulfur nitrosyl clusters. Such clusters can be grouped into the three classes of Roussin's black salts (RBS, **3**), Roussin's red salts (RRS, **4**), and Roussin's red esters (RRE, **5**), depending on the cluster core (Fig. 10.3). Known for almost 150 years, they have recently gained renewed interest as phototriggerable nitric oxide delivery systems [22]. Only the bridging thiolate groups in the RREs, however, provide a handle for further functionalization. They can easily be generated by alkylation of the μ_2-sulfido ligand in RRS or reaction of thiolates with $[Fe(CO)_2(NO)_2]$ with displacement of the two carbonyl ligands. Current work is directed at improving the water solubility and stability of such

RREs [23, 24] as well as their photophysical properties, for example by attachment of antennae to achieve high two-photon absorption (TPA) cross-sections for long-wavelength photoactivation [25].

A key problem, in particular with porphyrin-derived ligand systems, is the fast NO recombination with the metal–coligand fragment. Therefore, other polydentate ligands have been designed for facile photoinduced nitric oxide delivery (e.g., **6** and **7**) [19, 26, 27]. Because of their low inherent toxicity, and well-established and diverse coordination chemistry, manganese, iron, and ruthenium are the metal centres of choice and it is generally desirable to have a polydentate ligand with $(N - 1)$ donor groups if N is the preferred coordination number of the metal, thus tightly occupying all other coordination sites except one for the NO ligand. Following this strategy, the group of Mascharak, for example, studied complexes of the formula $[M(PaPy_3)(NO)]^{+/2+}$ (Fig. 10.3) with M = Mn, Fe, and Ru, in which the deprotonated carboxamido nitrogen group is always in trans position to the nitrosyl ligand. NO release from the iron compound could be triggered at 500 nm with a quantum efficiency of $\phi = 0.19$, which compares very favorably with SNP and RREs, which require higher energy irradiation (436 nm for SNP) but have much lower quantum yields [19]. Besides this success, these systems were hampered by ligand exchange reactions taking place in the dark with the solvent, and the oxidation of NO to NO_2, thus rendering them less useful as NO prodrugs. While the Ru analogue also requires low-wavelength excitation at around 350 nm, the Mn complex could be triggered at 532 nm with even higher quantum yield ($\phi = 0.55$), and showed greater stability in aqueous solution.

A vital requirement for photoinduced NO release is a low-spin $\{Mn(II)–NO^{\cdot}\}$ centre, with the excitation leading to a transition from an M–N(ligand) HOMO to an antibonding LUMO orbital of M–NO π^* character, thus weakening the metal–nitrosyl bonding. However, further shifting of the excitation wavelength to the red is highly desirable to reach the "phototherapeutic window" of the cellular system in the 600–800 nm range, in which light has a large tissue penetration depth and inflicts little damage to healthy cells. The main strategy is to attach organic chromophores with matching absorption as light-harvesting groups to the nitrosyl metal complexes for photosensitized NO release. Interesting heavy atom effects could be observed when an axial resorufin (7-hydroxy-3*H*-phenoxazin-3-one), *O*-coordinated trans to nitrosyl on a planar RuN_4 ligand core, was exchanged for the corresponding S- or Se-chromophores (thionol and selenophore, respectively), which lead to a shift in the main band from 590 to 610–612 nm [27].

In addition to red-light photoactivation, another common problem in NO delivery systems is the inevitable formation, in addition to nitric oxide, of a metal–coligand fragment, which might have a biological activity of its own. One strategy to ensure that only NO is released as the active species is the incorporation of the nitrosyl–metal–coligand system in a polymeric matrix [28]. This can be achieved in two ways. Either, a polymerizable functional group is introduced to the periphery of the coligand, for example using 4-vinylpyridine (4-vpy, as in **8**) [29], or the metal nitrosyl complex is non-covalently embedded in a macromolecular matrix. Another way of immobilizing NO delivery systems is the anchoring to surfaces, for example of gold nanoparticles, via peripheral thiol groups (see compound **9**) [30]. In all three cases, after proper triggering, the NO will diffuse away to its biological target site, leaving behind an inert metal–coligand fragment that remains bound to the solid support (Fig. 10.4).

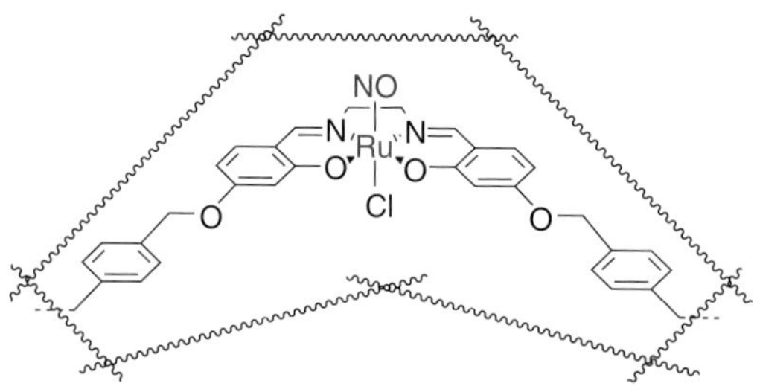

Figure 10.4 *Immobilization of metal nitrosyl complexes on a solid support, either by introduction of polymerizable groups (left, **8**), or thiolates with high affinity for gold surfaces (right, **9**)*

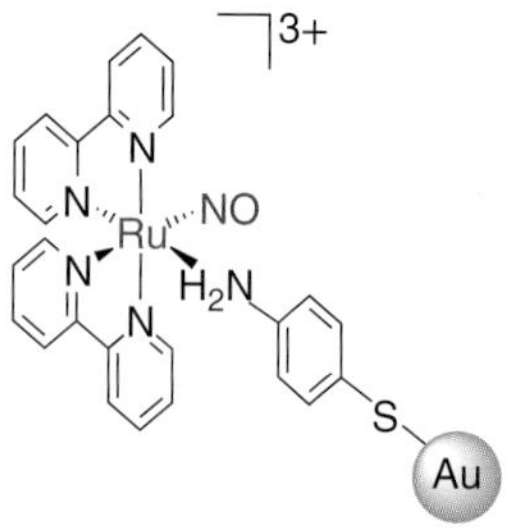

10.2.3 CO-releasing Molecules (CORMs)

In contrast to nitric oxide, no physiologically relevant systems are known for the cellular delivery of carbon monoxide that are not based on transition metal carbonyl complexes, with the exception of sodium boranocarbonate ($Na_2[H_3BCOO]$, CORM-A1, **10**) and the ester and amide derivatives, $Na[H_3BCOOR]$ and $Na[H_3BCONHR]$, reported by the group of Alberto [31–34]. The currently available *CO-releasing molecules* (CORMs), as these systems have been termed by Motterlini, span the whole range from mononuclear carbonyl complexes [$LM(CO)_n$], with n generally in the range of from 1 to 5, to dendrimers and metal carbonyl loaded polymers and nanomaterials. The metals employed usually belong to Groups 6 to 9, with a special focus on ruthenium, manganese, iron, and molybdenum, although chromium, cobalt, and tungsten complexes as well as, more recently, rhenium and iridium carbonyl compounds have also been explored, with all of them in low oxidation states of 0 to +II. Four different mechanisms to trigger CO release from the metal coordination sphere can be distinguished: ligand-exchange reactions with solvent, enzyme-mediated cleavage of remote bonds leading to increased susceptibility to oxidation (*enzyme-triggered CO-releasing molecules*, ET-CORMs), photoactivation (PhotoCORMs), and most recently, heating through an alternating magnetic field.

The first two metal carbonyl complexes studied for their CO-release behavior to biological systems, $Mn_2(CO)_{10}$ (CORM-1) and $Fe(CO)_5$, did require activation by light and thus qualify as PhotoCORMs (see below), but were quickly abandoned for other compounds due to their very poor biocompatibility [35]. After some initial attempts using CORM-2 **11**, which were however hampered by its poor water solubility and release of only one carbon monoxide per molecule [36], most biological studies since then have used CORM-3 **12**, which is soluble in water and induces a concentration-dependent relaxation of rat aortic ring preparations [37, 38]. Besides its broad use, however, this compound also illustrates some of the problems with recent generations of CO-releasing molecules. First of all, it has a very complicated and pH-dependent solution chemistry [39], and even more importantly, the decomposition products remaining after liberation of the carbon monoxide have not been sufficiently characterized to date. Usually, "aged" solutions that are no longer releasing CO are used as a negative control, which are termed *inactivated CORMs* (iCORMs). However, although ruthenium(II), chloride, and glycine are generally assumed to be benign, these fragments might interact with the constituents of biological systems and thus result in a CO-independent background reactivity. Upon incubation of CORM-3 with lysozyme, for example, a $[Ru(CO)_2(H_2O)_3]$ fragment bound to the His15 residue of the polypeptide chain could be identified as the major adduct by X-ray crystallography, in addition to two other sites of lower occupancy [40].

Therefore, it should become mandatory to independently prepare and comprehensively characterize these iCORMs resulting from the release of the carbon monoxide to unequivocally show that any biological effect observed is indeed only due to CO and not to the metal–coligand fragment inevitably generated in this process. A second problem arising with these first- and second-generation CORMs is due to the fact that they usually do not show any tissue specific activity and do not possess the properties required for drug molecules. For many additional compounds studied in this context, the reader is referred to the excellent reviews by Alberto, Mann, Motterlini, Romao, and Zobi [10, 36, 41–43]. Here, only three recent examples will be discussed further, which are distinctive in their special bioactivity and targeting properties (Fig. 10.5).

CORM-401 **13** is a neutral manganese(I) tetracarbonyl complex that is relatively stable in water and PBS buffer, with only about 0.3 CO equivalents released from the compound over the course of 4 h. Addition of myoglobin, however, leads to a substantial decrease in the half-life and three out of four carbon monoxides are readily delivered to the protein iron centre in less than 2 min [44]. A similar effect has been observed for CORM-3. It is important to ensure that the CO is readily delivered to its biological target with as little as possible freely diffusing around or prematurely released. The precise reason for this different activity in the presence of myoglobin, however, requires further study.

The second compound to be discussed here is ALF-794 **14**, a molybdenum(0) tricarbonyl complex with three isocyanide coligands, whose structure can be widely varied to tune the solubility and target specificity. This CORM was developed as part of a large family of related complexes by Romao and coworkers at the spin-off company ALFAMA with a specific medical condition in mind, namely the treatment of acetaminophen

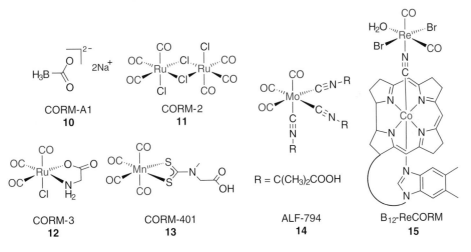

Figure 10.5 *CO-releasing molecules (CORMs) triggered by ligand exchange with solvent*

(APAP)-induced acute liver failure [45], resulting from the metabolites of an overdose of paracetamol, a common prescription-free analgesic. By careful variation of the isocyanide substituents, CORM uptake in different tissues could be optimized for high liver accumulation, which largely was found to depend on the number of methyl groups in the $N{\equiv}C-C(CH_3)_nH_{2-n}-COOH$ ligand. Liver CO concentrations increased by a factor of 0.3 ($n = 0$) to 1 ($n = 1$) and 5.3 ($n = 2$) compared with the blood carbon monoxide level, respectively. Furthermore, for $n = 2$, an increased liver over kidney specificity was also observed. Isocyanide ligands with two anionic carboxylates, on the other hand, preferentially accumulated in the kidney. Further differential activity came from activation of the CO-release from the metal coordination sphere by rat liver microsomes. In a mouse model, ALF-794 affected up to a 60% reduction in ALT (alanine aminotransferase) levels at $30\,mg\,kg^{-1}$ body weight compared with an untreated control, which is usually taken as a measure of hepatocellular death, with other compounds showing even higher levels of ALT reduction [45]. These results show that by careful tuning of the "drug sphere" of a CORM (see Fig. 10.2), a remarkable tissue specificity and biological activity can be achieved, with structure-activity relationships (SAR) delineated as in organic drug development.

Another interesting approach has recently been followed by the group of Zobi. Instead of tuning the drug sphere of a CORM, a novel rhenium(II) complex based on the $[ReBr_2(CO)_2]$ fragment was attached to cyanocobalamin (vitamin B_{12}) via a bridging cyanide ligand to form B_{12}–ReCORM (**15**, Fig. 10.5). This essential molecule was used as a biocompatible scaffold for the delivery of the CO-releasing moiety [43]. One of the two CO ligands in **15** was fully released after about 2 h. Additionally, **15** showed interesting activity on neonatal rat cardiomycetes and an ischemia-reperfusion injury model, both relevant models in the context of stroke and myocardial infarction ("heart attack"). When $30\,\mu M$ of B_{12}–ReCORM was administered at the beginning of the reperfusion conditions, cell death was reduced by 80% compared with the control.

These results show that by going beyond simple CO-release experiments and with careful tuning of the drug sphere of molecules or attachment to biocompatible or bio-derived carrier systems, CO-releasing molecules (CORMs) can be tuned to achieve high tissue specificity and biological activity. Enzyme expression profiles can vary significantly between different tissues and thus be exploited for a site-selective activation of prodrugs. The group of Schmalz has recently introduced this concept to the field of CO-releasing molecules with iron carbonyl complexes they have termed *enzyme-triggered CORMs* (ET-CORMs) [46–48]. The general idea is to have an $Fe(CO)_3$ moiety η^4-coordinated to a 1,3-butadiene coligand, which is functionalized with a hydroxy group in either the 1- or 2-position. When the hydroxy group is "trapped" as a carboxylic acid ester or phosphate, these piano-stool metal-carbonyl compounds are stable, in particular when the 1,3-butadiene group is incorporated in ring systems such as 1,4-cyclohexadiene (Fig. 10.6). Enzymatic cleavage of the ester or phosphate will then release the dienol, which tautomerizes to the 1,2-unsaturated ketone. This leads to a hapticity shift of the iron tricarbonyl moiety from η^4 to η^2, which is associated with a significant increase in susceptibility towards oxidative degradation, leading to the ultimate release of three equivalents of carbon monoxide in addition to ferric iron and the organic ligand.

In initial work, it was demonstrated that the ET-CORMs tested had moderate to low cytotoxicity on macrophages, released CO to myoglobin in the presence of suitable esterases,

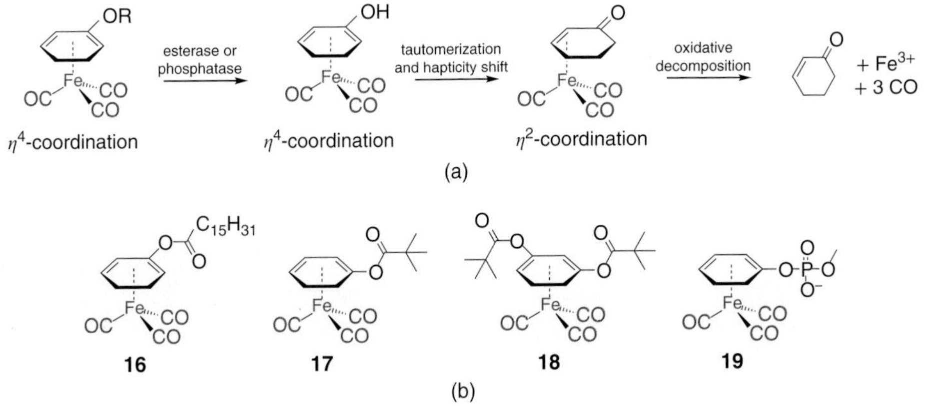

Figure 10.6　(A) General concept and (B) some examples of enzyme-triggered CORMs (ET-CORMs)

with a *Candida rugosa* lipase (LCR) being most active, and showed inhibition of NO formation due to CO-binding to inducible nitric oxide synthease (iNOS) [46]. In further studies, structure–activity relationships were determined, with a particular focus on the 1- versus 2-position of the hydroxy group and the type of carboxylic acid attached (**16** versus **17**) as well as substitution adjacent to the 1,3-butadiene moiety by a 5,5-dimethyl functionalization, and introduction of two hydroxy groups to both the 1- and 3-position **18** [47]. Again, most compounds tested had only mid-micromolar IC_{50} values but could inhibit enzymatic NO production by up to 50–60% at 10 μM for the most active compounds. The best balance between high inhibition of NO production and low cytotoxicity was found for the $Fe(CO)_3$ complex of 1-hydroxy-5,5-dimethyl-1,3-cyclohexadiene. Geminal dimethylation tended to reduce the cytotoxicity, but also somewhat reduced activity, while the diesters were more toxic but also had greater NO-inhibition power. The corresponding phosphates, such as **19**, could also be activated enzymatically and showed CO release as well as NO formation inhibition, but interestingly plant phosphatases were found to be most efficient to trigger the release, suggesting that the spectrum of enzyme activation needs to be further investigated.

In addition to the CORMs described in the preceding section, where CO-release is triggered intrinsically, by interaction of the metal carbonyl complex with solvent or the constituents of cellular systems such as proteins and enzymes, researchers have also started to explore externally applied triggers. A particular focus has been on photoactivation, since this allows a precise spatial and temporal control of the biological action (Fig. 10.7). The two metal carbonyl complexes $Mn_2(CO)_{10}$ **20** and $Fe(CO)_5$ **21** initially explored by Motterlini *et al.* belong to this category [35]. Owing to very poor bioavailability and the use of a "cold light source", these compounds were quickly abandoned for CORM-2 and CORM-3 (see Fig. 10.5).

It was only half-a-decade after the initial concept of metal carbonyl complexes as CO-releasing molecules (CORMs) was introduced that, inspired by organic "caged" compounds (see also Section 10.3.1), the group of Schatzschneider reported the photoactivated release of carbon monoxide from a manganese(I) tricarbonyl complex,

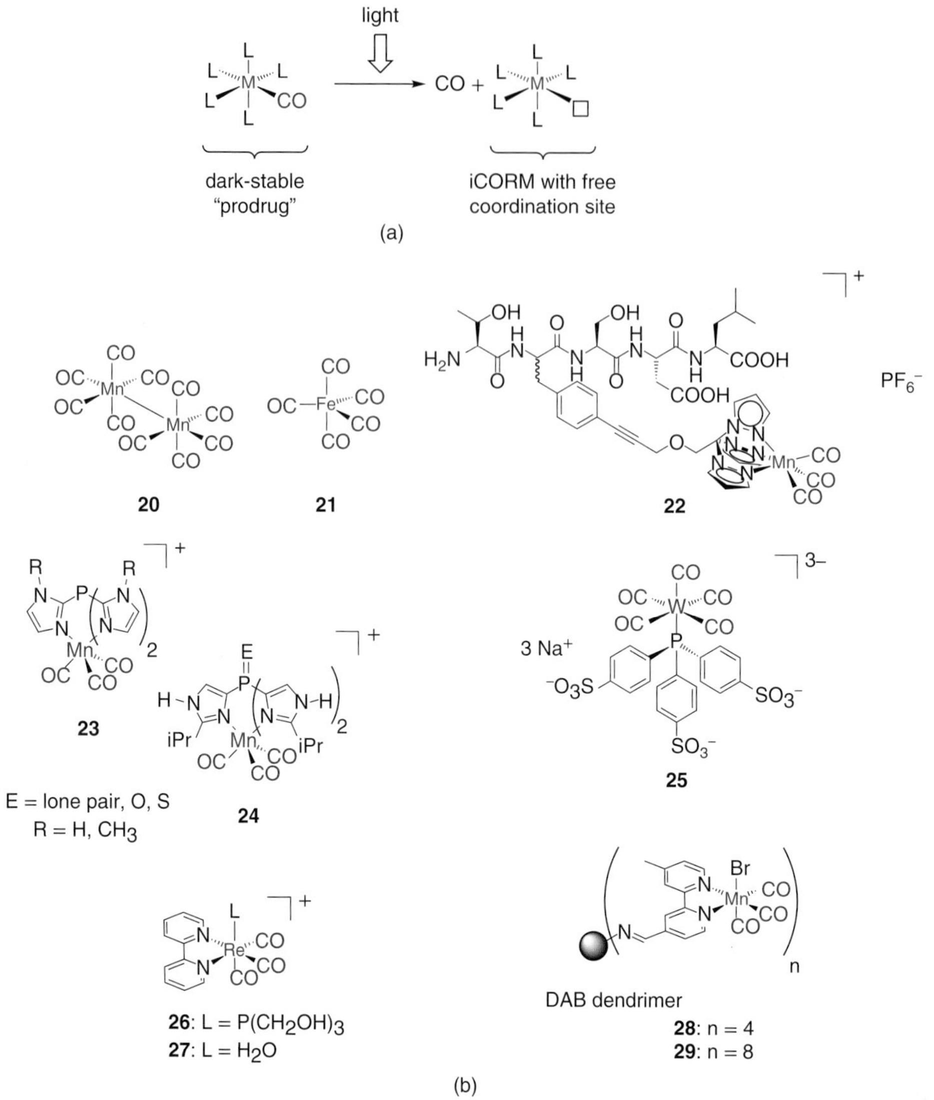

Figure 10.7 *(a) General concept and (b) some examples of photoactivatable CO-releasing molecules (PhotoCORMs)*

as in **22**, with a facial tridentate tris(pyrazolyl)methane (tpm) ligand. Using 365 nm light conveniently generated by a UV hand lamp normally used to visualize TLC plates, they could show that two out of three CO equivalents could be released from the metal coordination sphere [49]. Furthermore, efficient passive uptake by HT-29 human cancer cells could be demonstrated using atomic absorption spectroscopy (AAS) with the Mn metal serving as an intrinsic marker. A differential bioactivity was observed in cell culture experiments. While the complex was inactive on HT-29 cell at a concentration of 100 μM with a 48 h incubation time, a 10 min irradiation at 365 nm in the middle of a 48 h incubation lead to a significant reduction in cell biomass as evidenced by the crystal violet assay and comparable to that of organic standard drug 5-fluorouracil.

While it is normally the cytoprotective activity of CO that is in the focus, these experiments show that highly localized toxic carbon monoxide concentrations can also be beneficial if an aberrant cell population is to be eradicated from the body. Very recent UV-pump/IR-probe experiments at the femtosecond timescale showed that initially, however, only one of the three CO ligands is lost from the metal coordination sphere upon photoexcitation and that some molecules may undergo geminate recombination [50]. The final iCORM products are mononuclear manganese(II) tpm complexes formed by oxidative loss of the remaining carbon monoxide ligands, as evidenced from EPR studies, which finally dimerize on further oxidation to oxo-bridged manganese(III) dimers [51].

In addition to the tpm compounds, analogous $Mn^I(CO)_3$ complexes with tris(imidazolyl) ligands such as **23** and **24** have also been studied in detail by the group of Kunz, to investigate the effect of substituents on the ligand [52, 53], and were also attached to polymer carrier systems by use of a polymerizable ligand [54]. Along the same lines, the tpm parent compound was attached to peptides, silica, and carbon nanomaterial carriers for improved and targeted cellular delivery (Fig. 10.7) [55–57].

As the first non-manganese compound studied in this context, in 2010 Ford and coworkers reported on the tris-anionic complex $[W(CO)_5(tppts)]^{3-}$ **25**, in which the tris(sulfonatophenyl)phosphine (tppts) ligand confers the required water solubility to the otherwise neutral tungsten(0) pentacarbonyl fragment. The compound was stable in aerated aqueous buffer for several hours when kept in the dark, while 313 nm irradiation lead to the release of one of the five carbonyl ligands with subsequent formation of the aquo product, which undergoes further irreversible oxidation in the presence of dioxygen [58]. Although some photoactivated CO-release carrier compounds had been reported previously, the first use of the term PhotoCORMs for these systems has to be credited to this report. Since then, additional PhotoCORMs have appeared in the literature, especially those based on iron and ruthenium [59–64]. In particular the Mascharak group has been very successful in the study of CORM congeners of the manganese nitrosyl complexes already discussed in Section 10.2.2 (such as **6** in Fig. 10.3). By careful ligand design, CO release could be achieved using low-power visible light in the 400–550 nm range, which is essential for more advanced biological applications.

A major problem with CORMs is the lack of sensitive analytical methods to study the cellular uptake and intracellular distribution of the metal carbonyl complexes as well as the fate of the carbon monoxide ligand. In general, M–(CO) vibrations appear in a spectroscopic window between about 1850 and 2300 cm^{-1}, where the constituents of bio(macro)molecules show no signals and cells are thus transparent for IR or Raman spectroscopy. Therefore, the $[Mn(CO)_3(tpm)]^+$ complex could be visualized in living, unfixed human colon cancer cells using confocal Raman microscopy [65], but the concentrations required for good signal-to-noise are in the low millimolar to high micromolar range and thus about 2–3 orders of magnitude higher than normally applied with physiological CORM studies. This is due to the inherently low sensitivity of Raman spectroscopy and has to be alleviated by the use of near-field techniques such as surface-enhanced Raman scattering (SERS). There has been some current activity in that direction, but experiments thus far unfortunately used only fixed and dried cells [66, 67], which might lead to considerable intracellular redistribution of the compound and therefore clearly does not represent a native state. At this stage, we refer interested readers to Chapter 5, which describes in detail the opportunity to image living cells with metal carbonyl complexes.

In 2012, Pierri and Ford and coworkers presented a new and very innovative Photo-CORM [Re(bpy)(CO)$_3$(P(OCH$_2$OH)$_3$)]$^+$ **26**, which undergoes photoinduced release of the axial carbon monoxide ligand, trans to the phosphine, stereospecifically forming the corresponding aquo compound [Re(bpy)(CO)$_2$(P(OCH$_2$OH)$_3$)(H$_2$O)]$^+$ **27** [68, 69]. Quantum yields of 0.21 and 0.11 were determined for 365 and 405 nm excitation, respectively. Even more interestingly, both compounds are highly luminescent at different wavelengths. Upon excitation at 355 nm, the rhenium(I) tricarbonyl complex **26** has an emission maximum of 515 nm, the aquo product **27** shows a red-shifted emission at 585 nm, which was also retained in an aerated aqueous medium. In a first study of its kind, confocal fluorescence microscopy was used to follow both the PhotoCORM and its iCORM product present in living PPC-1 cells in real time. Importantly, both compounds also showed low cytotoxicity at concentrations up to 100 μM. In addition to this first inherently fluorescent CORM, very recently two luminescent probes to study intracellular CO have been reported by the Chang and He groups, respectively [70–72]. Whether these probes will serve as universally applicable tools to study carbon monoxide delivery to biological systems at low micromolar concentrations, in the presence of a complex biological matrix, remains to be seen.

In an extension into the field of macromolecular chemistry, the groups of Smith and Schatzschneider recently reported on dendrimer conjugates of the [MnBr(bpy^{R,R})(CO)$_3$] parent compound (Fig. 10.7) [73]. Introduction of an aldehyde group to the functionalized 2,2'-bipyridine ligand allowed attachment to the terminal amino groups of poly(propylene)imine dendrimers with a 1,4-diaminobutane (DAB) core. The generation G1 and G2 compounds **28** and **29**, with four and eight manganese tricarbonyl moieties, respectively, could be prepared and characterized, incorporating a total of 12 and 24 carbonyl ligands. Using an LED light source at 410 nm, two out of three CO ligands per Mn(CO)$_3$ unit could be photolytically released. Interestingly, the absence of any scaling effects showed that all manganese moieties behaved as independent PhotoCORMs without any intramolecular communication.

A very innovative strategy to trigger CO release from superparamagnetic iron oxide nanoparticles (SPIONs) functionalized with metal carbonyl complexes by localized heating with an alternating magnetic field was recently reported by Janiak and coworkers [74]. An amino-acid derived ligand, 3,4-dihydroxyphenylalanine, was anchored to maghemite SPIONs via the chelating ortho-catecholate group, leaving the nanoparticles presenting a bidentate N,O ligand set, which resembles that of CORM-3. Reaction with [RuCl(μ-Cl)(CO)$_3$]$_2$ then led to the RuCl(CO)$_3$-functionalized nanoparticles **30** (Fig. 10.8). In addition, a phenylalaninato model compound was also prepared for comparison.

The loading was determined to be about 400 RuCl(CO)$_3$ moieties per nanoparticle, which had an average diameter of about 10 nm, translating to functionalization of 60% of the 3,4-dihydroxyphenylalanine moieties with the RuCl(CO)$_3$ units. Even upon this coating, the particles retained their superparamagnetic behavior. As determined with the standard myoglobin assay, the CORM@SPION nanoparticles showed some CO release even under field-free conditions, yet the determined half-life was longer than observed for the molecule model compound as well as CORM-3. The application of an external alternating magnetic field (about 250 kHz and 40 mT) led to a decrease in half-life by a factor of two for the

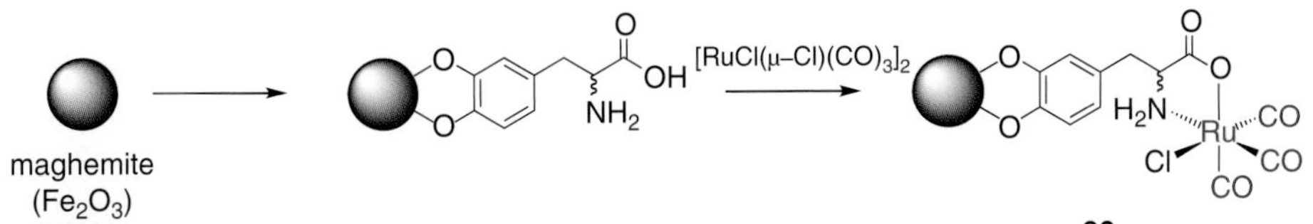

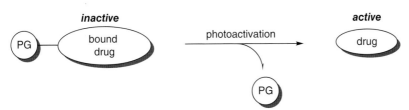

Figure 10.8 *Synthesis of RuCl(CO)$_3$-functionalized superparamagnetic iron oxide nanoparticles for CO-release triggered by heating in an alternating magnetic field*

functionalized nanoparticles while $t_{1/2}$ for the model compound remained constant within the experimental error with and without the field.

10.3 "Photouncaging" of Neurotransmitters from Metal Complexes

10.3.1 "Caged" Compounds

A "caged" compound consists of a bioactive molecule that has been inactivated by covalent bonding to a photolabile protective group (see also Chapter 9) [75–79]. This choice of name is rather unfortunate, as essentially none of these compounds are actually container molecules in which the bioactive component is released from a closed inner cavity. In any case, the name has stuck since the first "caged" compounds were described more than 30 years ago [80]. The "uncaging" is usually triggered by photoexcitation leading to dissociation of the protective group, thus allowing precise spatial and temporal control of the biological activity of the moiety released (Fig. 10.9). This concept was first exploited in a biological context by Hoffman and coworkers in 1978 when they reported the photolytic release of adenosine-5'-triphosphate (ATP) from "caged-ATP" containing 2-nitrobenzyl phosphate or 1-(2-nitro)phenylethyl protective groups attached to the γ-phosphate of ATP by an ester bond using 340 nm irradiation [81]. ATP liberated from the "caged" molecules could activate Na,K-ATPase in red blood cells depleted of internal energy stores, whereas the "caged" molecule had no effect. Since then, a plethora of other photolabile protective groups have been studied including nitroaryl, arylmethyl, coumarin-4-ylmethyl, and arylcarbonylmethyl groups [82, 83], but the use of metal complexes in this role has been largely neglected so far.

Figure 10.9 *Mechanism of photoactivated "uncaging" of a drug molecule from a photolabile protective group PG*

While many organic photolabile protective groups can only be removed at wavelengths below 400 nm, the absorption spectra of metal complexes often possess comparatively red-shifted low energy bands, thus allowing lower energy light to be used for the photorelease. As the tissue penetration depth of light is directly proportional to the wavelength employed [84–86], the use of metal complexes as photolabile protective groups holds much promise for more efficient drug release and the lower excitation energy is also expected to reduce photodamage to healthy tissue. This section will focus on the release and activity of important bioactive organic molecules from metal centres by photoactivation. For a discussion of the release of small molecule messengers such as CO and NO from metals, the reader is directed to the preceding section.

10.3.2   "Uncaging" of Bioactive Molecules

The use of metal complexes as photolabile protective groups thus far has almost exclusively revolved around ruthenium(II) compounds containing polypyridyl ligands, with one or more coordination sites occupied by the bioactive agent to be "uncaged". In fact, photochemically induced ligand dissociation from polypyridyl complexes of ruthenium has been widely studied, and involves the transition between the MLCT state to a lower-energy d–d state, which promotes ligand release [87, 88]. Only recently, however, has this property been explored in a biological context. During the last 10 years, the group of Etchenique in particular has expertly exploited the rich photochemistry of ruthenium polypyridyl systems to prepare inorganic "caged" compounds that release neuroactive molecules upon photoexcitation. The rapid and localized release of these neuroactive molecules has allowed the study of receptor distribution, ion channel kinetics, and other processes. This was first achieved by the synthesis and study of $[Ru(bpy)_2(4AP)_2]^{2+}$ **31** [89], containing 4-aminopyridine (4AP), a neuroactive molecule that blocks certain K^+ channels [90], promoting depolarization and increasing neuron activity. Upon treatment of the central ganglion of the medicinal leech *Hirudo medicinalis* with the dark stable complex **31**, characteristic action potentials were observed, as measured by the single-cell transmembrane potential, following visible light pulses from a Xe flash lamp. These workers showed that one molecule of 4AP was released from the metal centre upon photoexcitation, whereas the complex $[Ru(bpy)_3]^{2+}$ showed no activity under identical conditions. It was later reported that two-photon excitation (TPE) from a high power laser source at 720 nm could also photodissociate one molecule of 4AP from **31** under physiological conditions, while two-photon fluorescence was observed upon excitation at 720, 800, and 950 nm [91].

Similar complexes of the type $[Ru(bpy)_2L_2]^{2+}$ containing other bioactive amines such as tryptamine, tyramine, serotonin, γ-aminobutyric acid (GABA), and butylamine were also found to release one molecule of amine, albeit with a relatively low quantum yield of dissociation (Fig. 10.10) [92]. For a detailed computational study of the photodissociation of 4AP, butylamine, and γ-aminobutyric acid from complexes of the type $[Ru(bpy)(L_2)_2]^{2+}$, the reader is directed to the report by Salassa *et al.* [93]. By modifying the ruthenium polypyridyl ligand sphere to allow only one coordination site for binding of a bioactive amine, Etchenique and coworkers reported on $[Ru(bpy)_2(PPh_3)(GABA)]^{2+}$ **32**, which released GABA upon excitation at 450 nm [94]. The quantum yield of GABA release from **32** of $\phi = 0.21$ is significantly higher than for **31** ($\phi = 0.036$) or commercially

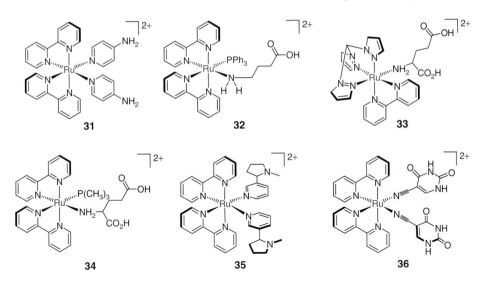

Figure 10.10 *Some examples of bioactive molecules "caged" by ruthenium protective groups*

available γ-aminobutyric acid α-carboxy-2-nitrobenzyl ester (*O*-(CNB-caged) GABA, ϕ = 0.16) [95], thus combining efficient photorelease with a trigger wavelength in the visible region. Upon photoexcitation, complex **32** induced membrane ionic current changes in frog oocytes expressing the GABA$_c$ receptor in a similar fashion to that of free GABA.

Another neurotransmitter that is important for learning and memory function is glutamate (GlutH$_2$), and the ability to release this compound from a metal based "caging" group with low energy visible light represents a major step forward. Glutamate, however, is able to chelate ruthenium by binding of both the amino and α-carboxylate groups, thus forming a robust bidentate complex that has a very low quantum yield of dissociation upon photo-excitation [96]. To block the chelation of GlutH$_2$ to ruthenium, one of the bipyridine ligands in the Ru(bpy)$_2$ system was replaced by a tridentate tris(pyrazolyl)methane (tpm) ligand, thus allowing metal coordination of GlutH$_2$ exclusively through the amino group. [Ru(bpy)(tpm)(GlutH$_2$)]$^{2+}$ **33** releases glutamate with a quantum yield of 0.035 upon 450 nm irradiation, 17 times greater than the 2-(dimethylamino)-5-nitrophenyl (DANP) protecting group at the same wavelength [97]. It is very important that the release of the bioactive compound from the metal coordination sphere is rapid, particularly for neurotransmitters, as it sets the timeframe for the processes to be studied or controlled. Complex [Ru(bpy)$_2$(PMe$_3$)(GlutH$_2$)]$^{2+}$ **34** releases glutamate in less than 50 ns after excitation at 532 nm, and also via two-photon absorption at 800 nm, with a functional cross-section of 0.14 GM [98, 99]. To demonstrate the effectiveness of this system, a laser pulse was multiplexed into five closely spaced beamlets and directed onto cortical pyramidal neurons treated with **34**, causing the generation of action potentials consistent with the activation of glutamate receptors.

An area of pharmacology that is not entirely understood is the mechanism of nicotine addiction [100, 101], an avoidable condition associated with tobacco smoking, the principal cause of lung cancer. Nicotine acts as an agonist for nicotinic acetylcholine receptors (nAChR), ligand-gated ion channels present in muscle, ganglia, and the brain, while also interacting with the dopaminergic system. To understand the molecular mechanisms underlying nicotine addiction and potentially aid in the development of new therapeutic strategies, it is essential that rapid release and precise spatial distribution of nicotine can be effectively controlled in living tissue. Early studies have shown that irradiation of $[Ru(bpy)_2(nic)_2]^{2+}$ **35** using a 532 nm Nd–YAG pulsed laser affected the rapid release (approximately 17 ns) of one molecule of nicotine from **35**. Subsequent experiments using 473 nm blue laser pulses elicited action potential firing in the Retzius neurons of a leech ganglion, consistent with nicotine activation [102].

Recently, a ruthenium polypyridyl complex has been used to photolytically release 5-cyanouracil (5CNU) [103], previously shown to inhibit pyrimidine catabolism *in vivo*, and an analogue of the anti-cancer agent 5-fluorouracil. Turro and coworkers [103] prepared $[Ru(bpy)_2(5CNU)_2]^{2+}$ **36** which, unlike the systems studied by Etchenique, released both equivalents of the bioactive molecule upon irradiation with visible light ($\lambda_{irr} \geq 395$ nm). It was further shown that the diaqua product $[Ru(bpy)_2(H_2O)_2]^{2+}$, present following photodissociation of 5CNU, could bind linearized pUC18 plasmids whereas samples incubated in the dark had no effect. This complex represents a new design principle, constituting a dual-action approach where biological activity could be affected through both release of bioactive 5CNU as well as binding of DNA by $[Ru(bpy)_2(H_2O)_2]^{2+}$ following photoexcitation.

Although the concept of using metal complexes as photolabile protective groups for "caged" bioactive molecules is now 10 years old, there remains much scope for further investigation, in particular the use of metals other than ruthenium. Metal complexes offer many advantages over traditional organic photolabile protective groups, particularly due to photorelease triggered by longer wavelength light more suitable for *in vivo* use due to deeper tissue penetration and less damage to the constituents of normal cells. The utility of ruthenium polypyridyl systems in this role has been elegantly explored, and even offers the possibility of multiple action photochemotherapy by the generation of bioactive molecules released from the metal centre and a resulting metal fragment capable of binding to DNA from one initially inert species. In the future, significant focus should be placed on the generation of metal photolabile protective groups possessing high two-photon absorption (TPA) cross-sections, to facilitate "uncaging" with low energy visible or NIR light. Although this has been extensively studied for organic "caged" molecules [104, 105], the intrinsic donor properties of some metal complexes may lead to more suitable candidates.

10.4   Hypoxia Activated Cobalt Complexes

10.4.1   Bioreductive Activation of Cobalt Complexes

In recent years, a significant expansion in the area of cobalt-based drugs has focused on the exploitation of the favorable redox chemistry of this metal. The different electronic properties between Co(II) and Co(III) species can lead to a drastic difference in the lability of associated ligands. By taking advantage of the specific physiological conditions in tumor tissue to activate cobalt complexes, one could potentially design a drug that possesses a high activity at the site of interest whilst remaining relatively inert in other areas of the body, thus reducing unwanted side-effects of chemotherapy associated with non-selective toxicity.

The selective activation of a prodrug in tumors occurs ideally when a molecule has been designed to form the active component only in the specific chemical microenvironment found in the tumor itself, which is otherwise not found in normal tissue. A variety of strategies are based on the targeting of tumor-specific enzymes or antigens (and thus antibodies), pH differences, therapeutic radiation, and tumor hypoxia [106, 107]. The first two of these strategies present difficulties due to the variable expression levels of enzymes in tumors and low levels of the targeted antigens, and will therefore not be discussed here. There are several cases that detail the activation of a metal-based prodrug by exploiting the pH differences within solid tumors, or by using radiation therapy to produce reducing species through the radiolysis of water, and these will be discussed later. It is known that imperfect neovascularization and high interstitial pressures lead to regions of hypoxia (low oxygen concentrations) in solid tumors, difficult to reach areas where tumor cells are not rapidly dividing. This property seems to occur quite universally and uniquely in solid tumors, rendering them an excellent site for selective prodrug activation. Many organic compounds have been developed that selectively form active drugs under these hypoxic conditions including quinones [108, 109], and aromatic [110–113] and aliphatic N-oxides [114, 115], some of which have reached clinical trials [116].

The activation of a prodrug in a hypoxic environment occurs by single-electron reduction, and is depicted in Fig. 10.11. In the body, this reduction can occur via a number of reductase enzymes (xanthine oxidase and cytochrome P450 reductase) that are present in all tissues [117]. In the normally oxygenated environment of healthy tissues, the reduced, activated species are rapidly re-oxidized by oxygen to re-form the inactive prodrug. It has been postulated that the activation of a drug in a hypoxic environment is dependent primarily on the rate of re-oxidation instead of the rate of initial reduction [107]. In areas of reduced O_2 levels, however, such as the hypoxic regions of solid tumors, the rate of re-oxidation by oxygen is greatly reduced, thus allowing the activated drug to affect its action [118]. Although this general mechanism has been questioned for at least one cobalt complex [119], it certainly appears that single electron redox cycling is the primary mechanism through which

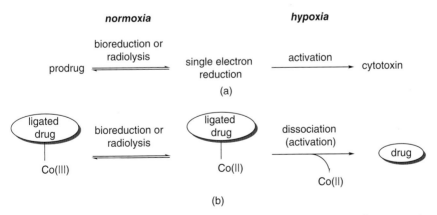

Figure 10.11 *Proposed mechanism of release of bioactive species via hypoxia-activated selectivity*

oxygen inhibits metabolic reduction. By using a green fluorescent protein (GFP) redox probe, Østergaard *et al.* have revealed that the cytosolic environment of a eukaryotic cell is highly reducing (−298 mV versus NHE), meaning that any hypoxia-activated prodrug would need to be designed to be reducible under these conditions [120].

Although there are report of ruthenium, copper, and rhenium complexes being investigated as potential hypoxia-activated prodrugs [121, 122], the majority of cases involve cobalt [106, 107, 118, 123, 124]. Cobalt complexes have shown the most potential in this context, and therefore they will be the only ones discussed in this chapter. The electronic properties of cobalt are extremely favorable for use in this context. In simple terms, the higher oxidation state of Co(III) is substitutionally inert (low spin $3d^6$ metal centre), while the lower oxidation state Co(II) easily exchanges ligands due to the $3d^7$ electronic configuration. Deactivation of a bioactive molecule by coordination to the relatively inert Co(III) centre allows the complex to act as a prodrug, which can then release the active species upon bioreduction to the labile Co(II) state in hypoxic environments, as depicted in Fig. 10.11 [118]. Facile substitution of ligands by water would lead to the essentially irreversible formation of $[Co^{II}(H_2O)_6]^{2+}$ due to the very high oxidation potential of this species ($E^0 =$ 1800 mV) [125].

To ensure that cobalt complexes designed as hypoxia-activated prodrugs can perform successfully in this function, the ancillary ligands around the cobalt centre can be finely tuned to alter the Co(II)/Co(III) redox potential, solubility (hydro- or lipophilicity), and overall charge [118]. Cobalt can coordinate a variety of high- and moderate-field ligands, thus allowing a wide range of complexes to be prepared with significantly different physiological properties.

10.4.2   Hypoxia-activated Cobalt Prodrugs of DNA Alkylators

For a more in-depth summary of the use of Co(III) complexes to deliver cytotoxic agents to tumors up to 2007, the reader is directed to the eloquent review by Hambley and coworkers [118]. Early attempts to design Co(III) prodrugs for increased antitumor activity focused on incorporation of alkylating moieties in the coordination sphere of the metal. Teicher *et al.*, who previously found that Co(III) complexes bearing nitro ligands were effective radiosensitizers [126], went on to describe the synthesis and antitumor activity of the Co(III) complex, *trans*-[Co(acac)₂(NO₂)(bca)] **37** (acac = acetylacetonato), possessing the nitrogen mustard, bis(2-chloroethyl)amine (bca) ligand (Fig. 10.12) [127]. Nitrogen mustards have high cytotoxicity due to the formation of interstrand cross-links between the strands of the

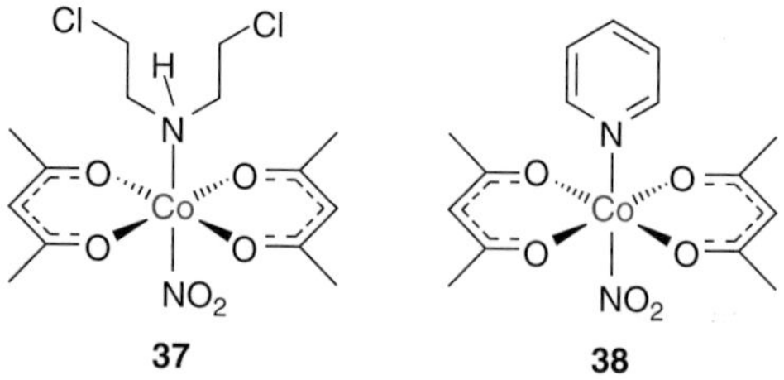

Figure 10.12   *Hypoxia-activated cobalt prodrug **37** and control complex **38***

DNA double helix, and have therefore been used as chemotherapeutic agents against cancer for a long time. It was envisaged that the bca ligand would be deactivated upon coordination to Co(III), as the nitrogen lone pair is no longer available, but upon bioreduction of Co(III) to Co(II), the cytotoxic mustard would be released. These workers also prepared the Co(III) complex *trans*-[Co(acac)$_2$(NO)$_2$(py)] **38** (py = pyridine) as a control, possessing an inactivated pyridine, as a means to elucidate the individual contribution of the mustard and the cobalt–coligand fragment to the overall biological activity.

They also found that cobalt complex **37** showed a higher activity than complex **38** against EMT6 mouse mammary tumor cells, although both complexes were more active under oxygenated than hypoxic conditions. This contrasted with the activity of the bca ligand alone, which showed the same cytotoxicity in oxygenated and hypoxic cells. Complex **37** was, however, an effective radiosensitizer, exhibiting an increased activity in hypoxic cells upon irradiation *in vitro*, with a dose-modifying factor of 2.4 compared with the absence of **37**. Furthermore, complex **37** showed good activity against the SCC-25/HN2 nitrogen mustard-resistant subline of the SCC-25 human squamous carcinoma, indicating that the method of nitrogen mustard deactivation in resistant cell lines is bypassed when a cobalt prodrug complex is employed.

The majority of other reports on Co(III) complexes bearing nitrogen mustard ligands comes from the group of Ware and coworkers. Although initially investigating Co(III) complexes of deactivated aziridine ligands [128], the researchers thereafter began to focus on bidentate nitrogen mustards. It was anticipated that chelating mustard ligands would provide greater kinetic stability to the reduced Co(II) species to allow re-oxidation in normal tissue by oxygen, a prerequisite for hypoxia-activated prodrugs. A series of complexes of this type, possessing *N,N*-bis(2-chloroethyl)ethylenediamine (dce) and *N,N'*-bis(2-chloroethyl)ethylenediamine (bce) ligands, respectively, as well as different substituents on the acac ligands, were synthesized (**39** and **40**, Fig. 10.13) [128].

The complexes were tested in a growth-inhibition assay against the AA8 cell line [129] and its mutant subline UV4, the latter being especially sensitive to DNA alkylating agents [130]. Complexes of type **39** and **40** had similar activities compared with the respective free dce and bce ligands, but the unsymmetrical **39** was more cytotoxic than the symmetrical complexes of type **40**, possibly due to more positive reduction potentials of the dce complexes facilitating more rapid release. Both dce and bce complexes were more active against the UV4 cell line, confirming that release of the active mustard ligands was responsible for the cytotoxicity. It should be noted that the reduction potentials of the type **39**

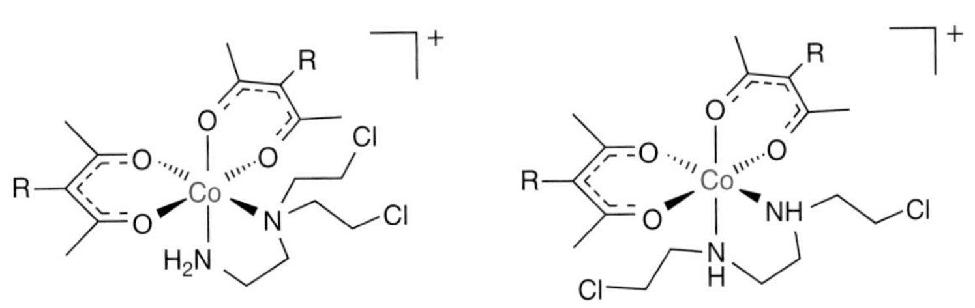

39 **40**

Figure 10.13 Hypoxia-activated cobalt prodrugs containing bidentate nitrogen mustards

complexes are around −300 mV versus NHE, while reduction potentials for type **40** compounds are generally higher, indicating a narrow range of acceptable reduction potentials that should not exceed the reducing conditions present in the cell. Complex **39** possessing methyl groups on the acac ligands displayed a 20-fold increased potency under hypoxic conditions compared to oxygenated cells, whereas very little selectivity was observed for the bce type ligands. This complex also showed high hypoxic-activated potency against intact EMT6 spheroids (three-dimensional aggregates of tumor cells that display many of the characteristics of solid tumors), with further evidence that the complex released diffusible cytotoxic mustards inside the hypoxic regions of the solid tumor core [131].

Ware *et al.* then went on to report the synthesis and biological evaluation of Co(III) complexes bearing tridentate mustard ligands, which would increase the stability of the reduced species under aerobic conditions [132]. Unfortunately, these complexes were found to be less active than complex **39**, with only a fivefold hypoxic selectivity observed in one case. Similar complexes to **39** with the acac ligand replaced by carbonato [133], oxalato [133], and tropolonato [134] ligands all showed significantly reduced hypoxic selectivity.

The same group also reported the use of ionizing radiation to generate reductive radicals by the radiolysis of water that could reduce the cobalt complexes and thus cause release of a cytotoxic agent. The method brings with it certain advantages in that very specific areas could be targeted for cytotoxic drug release. In this context, Co(III) complexes bearing 8-hydroxyquinoline (8-HQ) and various macrocyclic ancillary ligands were prepared (Fig. 10.14) [135]. 8-HQ, itself a cytotoxic compound, served as a model compound for azachloromethylbenzindoline (azaCBI) cytotoxin, a synthetic analogue of the potent DNA minor groove alkylator duocarmycin [136]. Therapeutically relevant doses of radiation were found to efficiently reduce the complexes and release 8-HQ in a variety of media, but they possessed low cytotoxicity compared with free 8-HQ. Furthermore, the complex incorporating a macrocyclic cyclen ligand **41** was stable under metabolic conditions of both high and low oxygen concentrations.

These workers followed up by preparing the cobalt complex **42** with azaCBI replacing the 8-HQ ligand [137]. The complex displayed significant radiolytic reduction of **42** and concurrent release of azaCBI in hypoxic formate buffer as well as human plasma, and was almost as potent as free azaCBI in a proliferation assay against UV4 cells. Against human colon carcinoma cell line HT29, however, complex **42** was significantly less potent than

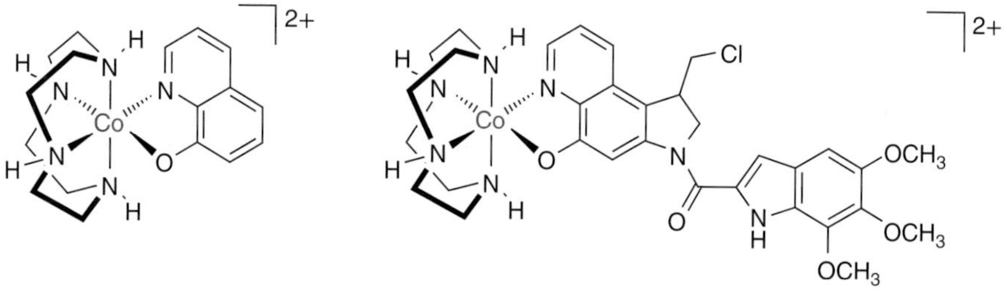

41 **42**

Figure 10.14 Hypoxia-activated cobalt prodrugs containing cytotoxic molecules

free azaCBI, but demonstrated a 20-fold hypoxic selectivity and could also be activated in hypoxic media without the presence of ionizing radiation. In an effort to understand the intracellular mechanism of reduction, complex **42** was incubated in hypoxic suspensions of wild type A549 cells, and hypoxic A549 cells with overexpressed cytochrome P450 reductase (thought to be the major one electron reductase in mammalian cells). In both cases, the rate of reductive metabolism of complex **42** was not significantly different, prompting the authors to speculate that other reductants such as ascorbate, thiols or NAD(P)H might play a significant role in the reduction of Co(III) hypoxia-activated prodrugs.

A recent report by Denny and coworkers detailed the synthesis and biological evaluation of a series of analogues of **42** where various substituents had been added to the axial nitrogen atoms of the macrocyclic ligand [138]. They varied the overall charge and solubility of the complexes by adding methyl, sulfonate, phosphate, or carboxylate groups, and compared their activities with **42** and free azaCBI against SKOV3 and HT29 human ovarian and colon carcinoma cell lines, respectively. Interestingly, all compounds were more active than the parent complex **42**, with IC_{50} values under hypoxic conditions comparable to free azaCBI. Only one compound, however, with *N*-methyl substituents in the axial positions, showed a higher hypoxic selectivity than **42**, perhaps due to the complex having the same overall charge and a greater stability under physiological conditions.

10.4.3 Hypoxia-activated Cobalt Prodrugs of MMP Inhibitors

In an alternative strategy to the targeting of DNA by cytotoxic DNA-alkylators, more recent research has focused on exploiting cellular processes that are unique to, or exacerbated in, cancer cells. These strategies may include the inhibition of specific enzymes responsible for the accelerated growth of tumors or the proliferation of tumors throughout the body by metastasis. Metastasis involves the spread of cancer cells from one part of the body to another and is the major cause of death from malignant diseases. The growing cancer cells first break through the epithelial basement membrane and initiate angiogenesis, allowing expansion of the primary tumor. The cells may then enter the bloodstream and eventually migrate to other organs or tissue and grow into secondary tumors, or metastases [139]. A number of new antimetastatic drugs have been designed to inhibit this process, including matrix metalloproteinase (MMP) inhibitors.

Matrix metalloproteinases are a family of metallo-enzymes that are able to cause degradation of connective tissue and extracellular matrix proteins, and are important in processes such as tissue remodeling and wound healing. However, MMPs are also thought to be involved in metastasis, facilitated by tumor cells sequestering MMPs on the surface of their leading edge as they invade neighboring tissue and degrade collagen that forms the major structural component of basement membranes [139]. The active site of MMPs is composed of a catalytically important Zn(II) ion that is bound by three histidine residues, while possessing a further two free sites for substrate binding to occur. From the point of view of designing a drug candidate for the selective inhibition of metastasis, the targeting of MMPs is promising due the substantial overexpression and/or activation of MMPs in invasive tumor cells [140–144]. In fact, overexpression of MMPs is associated with an adverse prognosis in patients with cancer [145]. Many MMP inhibitors contain a hydroxamate group that is able to strongly chelate the Zn ion in MMPs, thus blocking the essential catalytic domain of the proteinases and preventing tumor cells from escaping into the bloodstream [146]. One such compound that attracted substantial attention as an antimetastatic drug was marimastat **43** (Fig. 10.15), which reached phase III clinical trials before eventually being dropped due to poor patient response.

One problem associated with the use of marimastat is its non-selectivity for MMPs in cancer cells, with inflammatory polyarthritis (clinically manifested as arthralgia, erythema, and joint swelling with severe limitation of function) common dose-dependent side effects [147]. In an effort to improve the selectivity of marimastat, Failes and Hambley have investigated the feasibility of using Co(III)–marimastat conjugates as a means to selectively release marimastat in hypoxic environments within solid tumors. Initial studies of a number of Co(III) complexes containing simple hydroxamate and hydroximate ligands as models for marimastat revealed irreversible reduction processes, indicating ligand dissociation from the metal centre, at potentials suitable for cellular bioreductants but not too low as to be unstable before reaching the site of action [148]. The same authors subsequently reported the preparation of cobalt complex **44** bearing tris(methylpyridyl)-amine (tpa) and marimastat ligands, and its selective release of marimastat under hypoxic conditions [149].

Complex **44** was obtained as the doubly deprotonated hydroximate form as verified by X-ray crystallography. In solution, two isomers with the deprotonated "hydroxyl" group either cis or trans to the tertiary tpa nitrogen were observed in an approximate 1:1 ratio. The isomers could not be separated, but as this was not expected to alter the biological activity of complex, a mixture was used in subsequent testing. Complex **44** possessed an irreversible reduction wave, as seen by cyclic voltammetry, at −863 mV (versus Ag/AgCl), compared with −166 mV for the parent complex $[CoCl(H_2O)(tpa)](ClO_4)_2$, indicating a large degree of stability conferred by the hydroximate ligand. Whether this potential is optimal for hypoxia-selective bioreductive release is unclear since this prodrug is expected to act extracellularly where the local reduction potential is not precisely known [149].

These workers performed an *in vitro* inhibition study of MMP-9 in a non-reducing environment using complex **44** and free marimastat **43**. The IC_{50} value of marimastat was found to be 7 nM, comparable to the reported value of 3 nM, whereas complex **44** had an IC_{50} of 900 nM. The large difference in activity is presumably due to the coordination of the

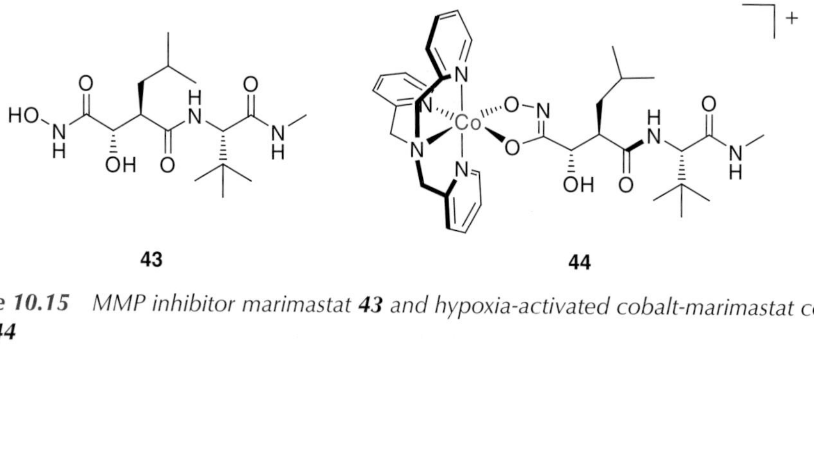

Figure 10.15 *MMP inhibitor marimastat **43** and hypoxia-activated cobalt-marimastat conjugate **44***

marimastat to the cobalt centre, which prevents it from binding the catalytic zinc ion in MMP-9. It was speculated that the observed activity of complex **44** was due to marimastat that has been released from the complex, as opposed to any intrinsic activity of the complex itself. A related Fe(III) complex containing a marimastat ligand was found to have an IC_{50} value of 190 nM, consistent with the increased lability of Fe(III) relative to Co(III) [150]. *In vivo* testing of complex **44** in mice with 4T1.2 tumor implants showed that **44** inhibited tumor growth relative to the control and free marimastat, suggestive of increased delivery of marimastat to the active site of the MMPs. Rather surprisingly, however, upon harvesting of the spine, lungs, and femurs at the end of the experiment, increased levels of metastasis (measured as tumor burden using real time polymerase chain reaction) were observed for marimastat and **44**. Although it is unclear why complex **44** would potentiate metastasis in this case, the authors have previously observed that high concentrations of Co(II) can increase the proteolytic activity of MMP-9, consistent with the replacement of Zn(II) ion in MMP-9 with the similar sized Co(II) ion [151].

To better understand the process of uptake, distribution, and activation of cobalt hypoxia-activated complexes within isolated cells and solid tumors, Hambley and coworkers have reported complexes **45** to **50** as elegant fluorescent probes and prodrugs for cytotoxic agents (Fig. 10.16) [152, 153]. In the free form, the functionalized anthraquinone and Coumarin-343 ligands display intense fluorescence that is quenched when bound to the cobalt centre, thus allowing simple visualization of areas in the cell or tumor where the ligand has dissociated from the metal. By driving the expression of photo-convertible green–red fluorescent protein EosFP with subsequent transfection in DLD-1 human colon carcinoma cells, a polyclonal HRE-Eos cell line was generated that fluoresces red in areas of local hypoxia. Treatment of spheroids of this cell line with **46** and free Coumarin-343 revealed different distribution profiles. While fluorescence due to Coumarin-343 was

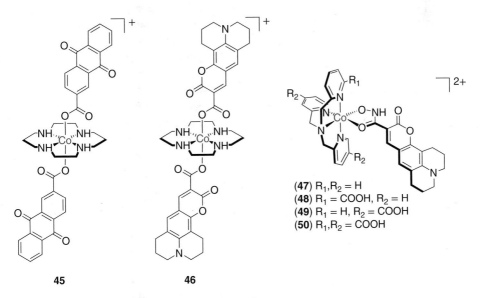

(47) R$_1$,R$_2$ = H
(48) R$_1$ = COOH, R$_2$ = H
(49) R$_1$ = H, R$_2$ = COOH
(50) R$_1$,R$_2$ = COOH

45   **46**

Figure 10.16 *Hypoxia-activated fluorescent probes and cytotoxic agents*

observed within the spheroids, it was not present in the central 150 μm diameter where hypoxia induced fluorescence due to EosFP was seen. Fluorescence due to complex **46**, however, was seen in the centre of the spheroids in regions shared with EosFP fluorescence, indicating enhanced penetration into the spheroid followed by bioreduction and release of Coumarin-343 in a hypoxia-selective manner [152].

The delivery of cytotoxic agents using cobalt prodrugs was further refined with **47** to **50**, each containing a tpa ligand functionalized with varying numbers of nicotinic acid groups. The authors postulated that using a weakly acid prodrug with pK_a around 7 would allow for selective uptake and accumulation in the acidic tumor microenvironment [153]. By manipulating the overall charge of the complexes at physiological pH through incorporation of nicotinic acid groups (nicotinic acid pK_a about 4.6) [154], it could be shown that complex **49** was selectively accumulated in the hypoxic regions of the tumor spheroids (Fig. 10.17). At neutral pH, the hydroxamic acid amine of **13** is essentially deprotonated and the charged complex penetrates through the structure. Upon reaching the acidic core of the spheroids, however, the hydroximate amine is then protonated to form an essentially neutral species, which facilitates cellular uptake and subsequent release of the free fluorophore either by bioreduction or ligand exchange. This is an excellent example of the rational design of a metal complex prodrug candidate that exhibits targeted accumulation in a difficult to reach region within a tumor, and then selectively releases a bioactive molecule through a change in external conditions. The ability to deliver a variety of anticancer cytotoxins in this way is an exciting prospect and one that no doubt will be of intense interest in the future.

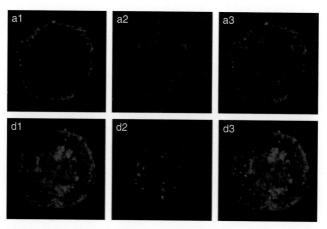

Figure 10.17 *Uptake of cobalt complexes in spheroids is dependent on the charge of the compound. Confocal microscopy images of HRE-Eos DLD-1 spheroids treated with (a) free fluorophore and (d) complex **13**. Column 1 shows the fluorescence due to the ligand in the blue channel, column 2 shows the cellular expression EosFP in response to hypoxia in the red channel, and column 3 shows the overlay. Adapted with permission from [153] © 2013 American Chemical Society*

10.5   Summary

Transition metal complexes can serve as protective groups for bioactive molecules, with the resulting systems serving as prodrugs for a targeted cellular delivery. The release from the metal coordination sphere can be achieved by a number of different trigger mechanisms, either intrinsically due to changes in the cellular microenvironment, or externally applied, in particular in the form of light. The latter requires long wavelength irradiation to ensure deep tissue penetration and minimization of damage to normal tissue. In all cases, in addition to the bioactive molecule released, a metal–coligand fragment is also generated that can have an activity of its own and thus needs to be independently prepared and tested in a control experiment.

Acknowledgments

P. S. thanks the Alexander-von-Humboldt Foundation for a postdoctoral fellowship.

References

1. A. K. Mustafa, M. M. Gadalla, S. H. Snyder, *Sci. Signal.* **2009**, *2*, re2.
2. M. Kajimura, R. Fukuda, R. M. Bateman, T. Yamamoto, M. Suematsu, *Antioxid. Redox Signal.* **2010**, *13*, 157–192.
3. L. Li, P. Rose, P. K. Moore, *Ann. Rev. Pharmacol. Toxicol.* **2011**, *51*, 169–187.
4. J. M. Fukuto, S. J. Carrington, D. J. Tantillo, J. G. Harrison, L. J. Ignarro, B. A. Freeman, A. Chen, D. A. Wink, *Chem. Res. Toxicol.* **2012**, *25*, 769–793.
5. R. Foresti, R. Motterlini, *Curr. Drug Targets* **2010**, *11*, 1595–1604.
6. R. Motterlini, L. E. Otterbein, *Nature Rev. Drug Discovery* **2010**, *9*, 728–743.
7. B. Wegiel, D. W. Hanto, L. E. Otterbein, *Trends Mol. Med.* **2013**, *19*, 3–11.
8. M. Whiteman, S. Le Trionnair, M. Chopra, B. Fox, J. Whatmore, *Clinical Sci.* **2011**, *121*, 459–488.
9. M. Whiteman, P. G. Winyard, *Exp. Rev. Clin. Pharmacol.* **2011**, *4*, 13–32.
10. C. C. Romao, W. A. Blättler, J. D. Seixas, G. J. L. Bernardes, *Chem. Soc. Rev.* **2012**, *41*, 3571–3583.
11. R. F. Furchgott, *Angew. Chem. Int. Ed.* **1999**, *38*, 1870–1880.
12. L. J. Ignarro, *Angew. Chem. Int. Ed.* **1999**, *38*, 1882–1892.
13. F. Murad, *Angew. Chem. Int. Ed.* **1999**, *38*, 1856–1868.
14. P. G. Wang, M. Xian, X. Tang, X. Xu, Z. Wen, T. Cai, A. J. Janczuk, *Chem. Rev.* **2002**, *102*, 1091–1134.
15. C. Napoli, L. J. Ignarro, *Ann. Rev. Pharmacol. Toxicol.* **2003**, *43*, 97–123.
16. D. A. Riccio, M. H. Schoenfisch, *Chem. Soc. Rev.* **2012**, *41*, 3731–3741.
17. A. W. Carpenter, M. H. Schoenfisch, *Chem. Soc. Rev.* **2012**, *41*, 3742–3752.
18. P. N. Cnoeski, M. H. Schoenfisch, *Chem. Soc. Rev.* **2012**, *41*, 3753–3758.
19. M. J. Rose, P. K. Mascharak, *Curr. Opin. Chem. Biol.* **2008**, *12*, 238–244.

20. H. Lewandowska, M. Kalinowska, K. Brzoska, K. Wojciuk, G. Wojciuk, M. Kruszewski, *Dalton Trans.* **2011**, *40*, 8273–8289.
21. J. H. Enemark, R. D. Feltham, *Coord. Chem. Rev.* **1974**, *13*, 339–406.
22. P. C. Ford, J. Bourassa, K. Miranda, B. Lee, I. Lorkovic, S. Boggs, S. Kudo, L. Laverman, *Coord. Chem. Rev.* **1998**, *171*, 185–202.
23. Y.-J. Chen, W.-C. Ku, L.-T. Feng, M.-L. Tsai, C.-H. Hsieh, W.-H. Hsu, W.-F. Liaw, C.-H. Hung, Y.-J. Chen, *J. Am. Chem. Soc.* **2008**, *130*, 10929–10938.
24. H. H. Chang, H. J. Huang, Y. L. Ho, Y. D. Wen, S. N. Huang, S. J. Chiou, *Dalton Trans.* **2009**, 6396–6402.
25. S. R. Wecksler, A. Mikhailovsky, D. Korystov, F. Buller, R. Kannan, L. S. Tan, P. C. Ford, *Inorg. Chem.* **2007**, *46*, 395–402.
26. M. J. Rose, P. K. Mascharak, *Coord. Chem. Rev.* **2008**, *252*, 2093–2114.
27. N. L. Fry, P. K. Mascharak, *Acc. Chem. Res.* **2011**, *44*, 289–298.
28. D. Crespy, K. Landfester, U. S. Schubert, A. Schiller, *Chem. Commun.* **2010**, *46*, 6651–6662.
29. J. T. Mitchell-Koch, T. M. Reed, A. S. Borovik, *Angew. Chem. Int. Ed.* **2004**, *43*, 2806–2809.
30. A. Diaz-Garcia, M. Fernandez-Oliva, M. Ortiz, R. Cao, *Dalton Trans.* **2009**, 7870–7872.
31. R. Alberto, K. Ortner, N. Wheatley, R. Schibli, A. P. Schubiger, *J. Am. Chem. Soc.* **2001**, *123*, 3135–3136.
32. R. Motterlini, P. Sawle, S. Bains, J. Hammad, R. Alberto, R. Foresti, C. J. Green, *FASEB J.* **2004**, *18*, 284–286.
33. T. S. Pitchumony, B. Spingler, R. Motterlini, R. Alberto, *Chimia* **2008**, *62*, 277–279.
34. T. S. Pitchumony, B. Spingler, R. Motterlini, R. Alberto, *Org. Biomol. Chem.* **2010**, *8*, 4849–4954.
35. R. Motterlini, J. E. Clark, R. Foresti, P. Sarathchandra, B. E. Mann, C. J. Green, *Circ. Res.* **2002**, *90*, e17–e24.
36. B. E. Mann, *Organometallics* **2012**, *31*, 5728–5735.
37. R. Foresti, J. Hammad, J. E. Clark, T. R. Johnson, B. E. Mann, A. Friebe, C. J. Green, R. Motterlini, *Br. J. Pharmacol.* **2004**, *142*, 453–460.
38. J. E. Clark, P. Naughton, S. Shurey, C. J. Green, T. R. Johnson, B. E. Mann, R. Foresti, R. Motterlini, *Circ. Res.* **2003**, *93*, e2–e8.
39. T. R. Johnson, B. E. Mann, I. P. Teasdale, H. Adams, R. Foresti, C. J. Green, R. Motterlini, *Dalton Trans.* **2007**, 1500–1508.
40. T. Santos-Silva, A. Mukhopadhyay, J. D. Seixas, G. J. L. Bernardes, C. C. Romao, M. J. Romao, *J. Am. Chem. Soc.* **2011**, *133*, 1192–1195.
41. R. Alberto, R. Motterlini, *Dalton Trans.* **2007**, 1651–1660.
42. B. E. Mann, in *Top. Organomet. Chem. Vol. 32*, eds. N. Metzler-Nolte and G. Jaouen, Springer, Berlin, **2010**, pp. 247–285.
43. F. Zobi, *Future Med. Chem.* **2013**, *5*, 175–188.
44. S. H. Crook, B. E. Mann, A. J. H. M. Meijer, H. Adams, P. Sawle, D. Scapens, R. Motterlini, *Dalton Trans.* **2011**, *40*, 4230–4235.

45. A. R. Marques, L. Kromer, D. J. Gallo, N. Penacho, S. S. Rodrigues, J. D. Seixas, G. J. L. Bernardes, P. M. Reis, S. L. Otterbein, R. A. Ruggieri, A. S. G. Goncalves, A. M. L. Goncalves, M. N. De Matos, I. Bento, L. E. Otterbein, W. A. Blättler, C. C. Romao, *Organometallics* **2012**, *31*, 5810–5822.

46. S. Romanski, B. Kraus, U. Schatzschneider, J. Neudörfl, S. Amslinger, H.-G. Schmalz, *Angew. Chem. Int. Ed.* **2011**, *50*, 2392–2396.

47. S. Romanski, B. Kraus, M. Guttentag, W. Schlundt, H. Rücker, A. Adler, J.-M. Neudörfl, R. Alberto, S. Amslinger, H.-G. Schmalz, *Dalton Trans.* **2012**, *41*, 13862–13875.

48. S. Romanski, H. Rücker, E. Stamellou, M. Guttentag, J. Neudörfl, R. Alberto, S. Amslinger, B. Yard, H.-G. Schmalz, *Organometallics* **2012**, *31*, 5800–5809.

49. J. Niesel, A. Pinto, H. W. Peindy N'Dongo, K. Merz, I. Ott, R. Gust, U. Schatzschneider, *Chem. Commun.* **2008**, 1798–1800.

50. P. Rudolf, F. Kanal, J. Knorr, C. Nagel, J. Niesel, T. Brixner, U. Schatzschneider, P. Nürnberger, *J. Phys. Chem. Lett.* **2013**, *4*, 596–602.

51. H.-M. Berends, P. Kurz, *Inorg. Chim. Acta* **2012**, *380*, 141–147.

52. P. C. Kunz, W. Huber, A. Rojas, U. Schatzschneider, B. Spingler, *Eur. J. Inorg. Chem.* **2009**, 5358–5366.

53. W. Huber, R. Linder, J. Niesel, U. Schatzschneider, B. Spingler, P. C. Kunz, *Eur. J. Inorg. Chem.* **2012**, 3140–3146.

54. N. E. Brückmann, M. Wahl, G. J. Reiß, M. Kohns, W. Wätjen, P. C. Kunz, *Eur. J. Inorg. Chem.* **2011**, 4571–4577.

55. H. Pfeiffer, A. Rojas, J. Niesel, U. Schatzschneider, *Dalton Trans.* **2009**, 4292–4298.

56. G. Dördelmann, H. Pfeiffer, A. Birkner, U. Schatzschneider, *Inorg. Chem.* **2011**, *50*, 4362–4367.

57. G. Dördelmann, T. Meinhardt, T. Sowik, A. Krüger, U. Schatzschneider, *Chem. Commun.* **2012**, *48*, 11528–11530.

58. R. D. Rimmer, H. Richter, P. C. Ford, *Inorg. Chem.* **2010**, *49*, 1180–1185.

59. R. Kretschmer, G. Gessner, H. Görls, S. H. Heinemann, M. Westerhausen, *J. Inorg. Biochem.* **2011**, *105*, 6–9.

60. V. P. Lorett-Velasquez, T. M. A. Jazzazi, A. Malassa, H. Görls, G. Gessner, S. H. Heinemann, M. Westerhausen, *Eur. J. Inorg. Chem.* **2012**, 1072–1078.

61. C. S. Jackson, S. Schmitt, Q. P. Dou, J. J. Kodanko, *Inorg. Chem.* **2011**, *50*, 5336–5338.

62. M. A. Gonzalez, N. L. Fry, R. Burt, R. Davda, A. Hobbs, P. K. Mascharak, *Inorg. Chem.* **2011**, *50*, 3127–3134.

63. M. A. Gonzalez, S. J. Carrington, N. L. Fry, J. L. Martinez, P. K. Mascharak, *Inorg. Chem.* **2012**, *51*, 11930–11940.

64. M. A. Gonzalez, M. A. Yim, S. Cheng, A. Moyes, A. J. Hobbs, P. K. Mascharak, *Inorg. Chem.* **2012**, *51*, 601–608.

65. K. Meister, J. Niesel, U. Schatzschneider, N. Metzler-Nolte, D. A. Schmidt, M. Havenith, *Angew. Chem. Int. Ed.* **2010**, *49*, 3310–3312.

66. C. Policar, J. B. Waern, M.-A. Plamont, S. Clede, C. Mayer, R. Prazeres, J.-M. Ortega, A. Vessiéres, A. Dazzi, *Angew. Chem. Int. Ed.* **2011**, *123*, 890–894.

67. S. Clede, F. Lambert, C. Sandt, Z. Gueroui, M. Refregiers, M.-A. Plamont, P. Dumas, A. Vessiéres, C. Policar, *Chem. Commun.* **2012**, *48*, 7729–7731.

68. A. E. Pierri, A. Pallaoro, G. Wu, P. C. Ford, *J. Am. Chem. Soc.* **2012**, *134*, 18197–18200.

69. R. D. Rimmer, A. E. Pierri, P. C. Ford, *Coord. Chem. Rev.* **2012**, *256*, 1509–1519.

70. B. W. Michel, A. R. Lippert, C. J. Chang, *J. Am. Chem. Soc.* **2012**, *134*, 15668–15671.

71. J. Wang, J. Karpus, B. S. Zhao, Z. Luo, P. R. Chen, C. He, *Angew. Chem. Int. Ed.* **2012**, *51*, 9652–9656.

72. L. Yuan, W. Lin, L. Tan, K. Zheng, W. Huang, *Angew. Chem. Int. Ed.* **2013**, *52*, 1628–1630.

73. P. Govender, S. Pai, U. Schatzschneider, G. Smith, *Inorg. Chem.* **2013**, *52*, 5470–5478.

74. P. C. Kunz, H. Meyer, J. Barthel, S. Sollazzo, A. M. Schmidt, C. Janiak, *Chem. Commun.* **2013**, *49*, 4896–4898.

75. S. R. Adams, R. Y. Tsien, *Ann. Rev. Physiol.* **1993**, *55*, 755–784.

76. G. C. R. Ellis-Davies, *Nat. Methods* **2007**, *4*, 619–628.

77. H. Yu, J. Li, D. Wu, Z. Qiu, Y. Zhang, *Chem. Soc. Rev.* **2010**, *39*, 464–473.

78. C. Brieke, F. Rohrbach, A. Gottschalk, G. Meyer, A. Heckel, *Angew. Chem. Int. Ed.* **2012**, *51*, 8446–8476.

79. P. Klan, T. Solomek, C. G. Bochet, A. Blanc, R. Givens, M. Rubina, V. Popik, A. Kostikov, J. Wirz, *Chem. Rev.* **2013**, *113*, 119–191.

80. H. A. Lester, J. M. Lerbonne, *Ann. Rev. Biophys. Bioeng.* **1982**, *11*, 151–175.

81. J. H. Kaplan, B. Forbush, J. F. Hoffman, *Biochemistry* **1978**, *17*, 1929–1935.

82. N. Hoffmann, *Chem. Rev.* **2008**, *108*, 1052–1103.

83. P. Klán, T. Šolomek, C. G. Bochet, A. Blanc, R. Givens, M. Rubina, V. Popik, A. Kostikov, J. Wirz, *Chem. Rev.* **2013**, *113*, 119–191.

84. R. Weissleder, V. Ntziachristos, *Nat. Med.* **2003**, *9*, 123–128.

85. K. Szacilowski, W. Macyk, A. Drzewiecka-Matuszek, M. Brindell, G. Stochel, *Chem. Rev.* **2005**, *105*, 2647–2694.

86. P. Agostinis, K. Berg, K. A. Cengel, T. H. Foster, A. W. Girotti, S. O. Gollnick, S. M. Hahn, M. R. Hamblin, A. Juzeniene, D. Kessel, M. Korbelik, J. Moan, P. Mroz, D. Nowis, J. Piette, B. C. Wilson, J. Golab, *Cancer J. Clini.* **2011**, *61*, 250–281.

87. V. Balzani, V. Carassitti, *Photochemistry of Coordination Compounds*, Academic Press, New York, **1970**.

88. D. V. Pinnick, B. Durham, *Inorg. Chem.* **1984**, *23*, 1440–1445.

89. L. Zayat, C. Calero, P. Alborés, L. Baraldo, R. Etchenique, *J. Am. Chem. Soc.* **2003**, *125*, 882–883.

90. M. Müller, P. W. Dierkes, W.-R. Schlue, *Brain Res.* **1999**, *826*, 63–73.

91. V. Nikolenko, R. Yuste, L. Zayat, L. M. Baraldo, R. Etchenique, *Chem. Commun.* **2005**, 1752–1754.

92. L. Zayat, M. Salierno, R. Etchenique, *Inorg. Chem.* **2006**, *45*, 1728–1731.

93. L. Salassa, C. Garino, G. Salassa, R. Gobetto, C. Nervi, *J. Am. Chem. Soc.* **2008**, *130*, 9590–9597.

94. L. Zayat, M. G. Noval, J. Campi, C. I. Calero, D. J. Calvo, R. Etchenique, *ChemBioChem* **2007**, *8*, 2035–2038.

95. K. R. Gee, R. Wieboldt, G. P. Hess, *J. Am. Chem. Soc.* **1994**, *116*, 8366–8367.

96. M. Salierno, C. Fameli, R. Etchenique, *Eur. J. Inorg. Chem.* **2008**, *2008*, 1125–1128.

97. A. Banerjee, C. Grewer, L. Ramakrishnan, J. Jäger, A. Gameiro, H.-G. A. Breitinger, K. R. Gee, B. K. Carpenter, G. P. Hess, *J. Org. Chem.* **2003**, *68*, 8361–8367.

98. E. Fino, R. Araya, D. S. Peterka, M. Salierno, R. Etchenique, R. Yuste, *Front. Neural Circuits* **2009**, *3*, 1–9.

99. M. Salierno, E. Marceca, D. S. Peterka, R. Yuste, R. Etchenique, *J. Inorg. Biochem.* **2010**, *104*, 418–422.

100. J. A. Dani, S. Heinemann, *Neuron* **1996**, *16*, 905–908.

101. J.-P. Changeux, *Nat. Rev. Neurosci.* **2010**, *11*, 389–401.

102. O. Filevich, M. Salierno, R. Etchenique, *J. Inorg. Biochem.* **2010**, *104*, 1248–1251.

103. R.N. Garber, J.C. Gallucci, K.R. Dunbar, C. Turro, *Inorg. Chem.* **2011**, *50*, 9213–9215.

104. D. Warther, S. Gug, A. Specht, F. Bolze, J. F. Nicoud, A. Mourot, M. Goeldner, *Bioorg. Med. Chem.* **2010**, *18*, 7753–7758.

105. F. Bolze, J. F. Nicoud, C. Bourgogne, S. Gug, X. H. Sun, M. Goeldner, A. Specht, L. Donato, D. Warther, G. F. Turi, A. Losonczy, *Optical Mat.* **2012**, *34*, 1664–1669.

106. W. A. Denny, *Eur. J. Med. Chem.* **2001**, *36*, 577–595.

107. W. A. Denny, *Cancer Invest.* **2004**, *22*, 604–619.

108. I. Antonini, T. S. Lin, L. A. Cosby, Y. R. Dai, A. C. Sartorelli, *J. Med. Chem.* **1982**, *25*, 730–735.

109. E. Hatzigrigoriou, M. V. Papadopoulou, D. Shields, W. D. Bloomer, *Oncol. Res.* **1993**, *5*, 29–36.

110. A. Monge, F. J. Martinez-Crespo, A. Lopez de Cerain, J. A. Palop, S. Narro, V. Senador, A. Marin, Y. Sainz, M. Gonzalez, *J. Med. Chem.* **1995**, *38*, 4488–4494.

111. J. S. Daniels, K. S. Gates, *J. Am. Chem. Soc.* **1996**, *118*, 3380–3385.

112. J. M. Brown, *Cancer Res.* **1999**, *59*, 5863–5870.

113. J.-T. Hwang, M. M. Greenberg, T. Fuchs, K. S. Gates, *Biochemistry* **1999**, *38*, 14248–14255.

114. W. R. Wilson, P. Van Zijl, W. A. Denny, *Int. J. Radiat. Oncol. Biol. Phys.* **1992**, *22*, 693–696.

115. P. J. Smith, N. J. Blunt, R. Desnoyers, Y. Giles, L. H. Patterson, *Cancer Chemother. Pharmacol.* **1997**, *39*, 455–461.

116. Q.-T. Le, A. Taira, S. Budenz, M. Jo Dorie, D. R. Goffinet, W. E. Fee, R. Goode, D. Bloch, A. Koong, J. Martin Brown, H. A. Pinto, *Cancer* **2006**, *106*, 1940–1949.

117. A. V. Patterson, M. P. Saunders, E. C. Chinje, L. H. Patterson, I. J. Stratford, *Anti-Cancer Drug Des.* **1998**, *13*, 541–573.

118. M. D. Hall, T. W. Failes, N. Yamamoto, T. W. Hambley, *Dalton Trans.* **2007**, 3983–3990.

119. R. F. Anderson, W. A. Denny, D. C. Ware, W. R. Wilson, *Br. J. Cancer* **1996**, *74*, S48–S51.

120. H. Østergaard, C. Tachibana, J. R. Winther, *J. Cell. Biol.* **2004**, *166*, 337–345.

121. P. J. Blower, J. R. Dilworth, R. I. Maurer, G. D. Mullen, C. A. Reynolds, Y. Zheng, *J. Inorg. Biochem.* **2001**, *85*, 15–22.

122. L. L. Parker, S. M. Lacy, L. J. Farrugia, C. Evans, D. J. Robins, C. C. O'Hare, J. A. Hartley, M. Jaffar, I. J. Stratford, *J. Med. Chem.* **2004**, *47*, 5683–5689.

123. T. W. Hambley, *Dalton Trans.* **2007**, 4929–4937.

124. N. Graf, S. J. Lippard, *Adv. Drug Delivery Rev.* **2012**, *64*, 993–1004.

125. D. C. Ware, B. D. Palmer, W. R. Wilson, W. A. Denny, *J. Med. Chem.* **1993**, *36*, 1839–1846.

126. B. A. Teicher, J. L. Jacobs, K. N. S. Cathcart, M. J. Abrams, J. F. Volano, D. H. Picker, *Radiat. Res.* **1987**, *109*, 36–46.

127. B. A. Teicher, M. J. Abrams, K. W. Rosbe, T. S. Herman, *Cancer Res.* **1990**, *50*, 6971–6975.

128. D. C. Ware, B. G. Siim, K. G. Robinson, W. A. Denny, P. J. Brothers, G. R. Clark, *Inorg. Chem.* **1991**, *30*, 3750–3757.

129. L. Thompson, J. Rubin, J. Cleaver, G. Whitmore, K. Brookman, *Somat. Cell Mol. Genet.* **1980**, *6*, 391–405.

130. C. A. Hoy, L. H. Thompson, C. L. Mooney, E. P. Salazar, *Cancer Res.* **1985**, *45*, 1737–1743.

131. W. R. Wilson, J. W. Moselen, S. Cliffe, W. A. Denny, D. C. Ware, *Int. J. Radiat. Oncol. Biol. Phys.* **1994**, *29*, 323–327.

132. D. C. Ware, P. J. Brothers, G. R. Clark, W. A. Denny, B. D. Palmer, W. R. Wilson, *J. Chem. Soc., Dalton Trans.* **2000**, 925–932.

133. P. R. Craig, P. J. Brothers, G. R. Clark, W. R. Wilson, W. A. Denny, D. C. Ware, *Dalton Trans.* **2004**, *0*, 611–618.

134. D. C. Ware, H. R. Palmer, P. J. Brothers, C. E. F. Rickard, W. R. Wilson, W. A. Denny, *J. Inorg. Biochem.* **1997**, *68*, 215–224.

135. G. O. Ahn, D. C. Ware, W. A. Denny, W. R. Wilson, *Radiat. Res.* **2004**, *162*, 315–325.

136. D. L. Boger, D. S. Johnson, *Angew. Chem. Int. Ed.* **1996**, *35*, 1438–1474.

137. G. O. Ahn, K. J. Botting, A. V. Patterson, D. C. Ware, M. Tercel, W. R. Wilson, *Biochem. Pharmacol.* **2006**, *71*, 1683–1694.

138. G.-L. Lu, R. J. Stevenson, J. Y.-C. Chang, P. J. Brothers, D. C. Ware, W. R. Wilson, W. A. Denny, M. Tercel, *Bioorg. Med. Chem.* **2011**, *19*, 4861–4867.

139. B. R. Zetter, *Ann. Rev. Med.* **1998**, *49*, 407–424.

140. S. R. Bramhall, *Int. J. Pancreatol.* **1997**, *21*, 1–12.

141. J. Trédaniel, P. Boffetta, E. Buiatti, R. Saracci, A. Hirsch, *Int. J. Cancer* **1997**, *72*, 565–573.

142. B. Davidson, I. Goldberg, P. Liokumovich, J. Kopolovic, W. H. Gotlieb, L. Lerner-Geva, I. Reder, G. Ben-Baruch, R. Reich, *Int. J. Gynecol. Pathol.* **1998**, *17*, 295–301.

143. A. Lochter, M. J. Bissell, *APMIS* **1999**, *107*, 128–136.

144. L. J. van 't Veer, H. Dai, M. J. van de Vijver, Y. D. He, A. A. M. Hart, M. Mao, H. L. Peterse, K. van der Kooy, M. J. Marton, A. T. Witteveen, G. J. Schreiber, R. M. Kerkhoven, C. Roberts, P. S. Linsley, R. Bernards, S. H. Friend, *Nature* **2002**, *415*, 530–536.

145. M. Hidalgo, S. G. Eckhardt, J. Natl. *Cancer Inst.* **2001**, *93*, 178–193.

146. M. Whittaker, C. D. Floyd, P. Brown, A. J. H. Gearing, *Chem. Rev.* **1999**, *99*, 2735–2776.

147. E. Rosenbaum, M. Zahurak, V. Sinibaldi, M. A. Carducci, R. Pili, M. Laufer, T. L. DeWeese, M. A. Eisenberger, *Clin. Cancer Res.* **2005**, *11*, 4437–4443.
148. T. W. Failes, T. W. Hambley, *Dalton Trans.* **2006**, 1895–1901.
149. T. W. Failes, C. Cullinane, C. I. Diakos, N. Yamamoto, J. G. Lyons, T. W. Hambley, *Chem. Eur. J.* **2007**, *13*, 2974–2982.
150. T. W. Failes, T. W. Hambley, *J. Inorg. Biochem.* **2007**, *101*, 396–403.
151. C. K. Underwood, D. Min, J. G. Lyons, T. W. Hambley, *J. Inorg. Biochem.* **2003**, *95*, 165–170.
152. B. J. Kim, T. W. Hambley, N. S. Bryce, *Chem. Sci.* **2011**, *2*, 2135–2142.
153. N. Yamamoto, A. K. Renfrew, B. J. Kim, N. S. Bryce, T. W. Hambley, *J. Med. Chem.* **2012**, *55*, 11013–11021.
154. R. W. Green, H. K. Tong, *J. Am. Chem. Soc.* **1956**, *78*, 4896–4900.

11

Metal Complexes as Enzyme Inhibitors and Catalysts in Living Cells

Julien Furrer,[a] Gregory S. Smith[b] and Bruno Therrien[c]
[a]Department of Chemistry and Biochemistry, University of Berne, Switzerland
[b]Department of Chemistry, University of Cape Town, South Africa
[c]Institute of Chemistry, University of Neuchatel, Switzerland

11.1 Introduction

The catalytic potential of metal complexes has been exploited for decades, and nowadays, metal-based catalysts are used by synthetic chemists on an everyday basis [1]. Nobel prizes in 2001 awarded to Knowles–Noyori–Sharpless for asymmetric catalysis, in 2005 to Chauvin–Grubbs–Schrock for the metathesis methodology and in 2010 to Heck–Negishi–Suzuki for palladium-catalysed cross coupling reactions are renowned examples on how metal-based catalytic processes have changed the daily life of synthetic chemists. These examples among others illustrate how metal-based catalysts are now playing a fundamental role in chemistry.

When added to biological media, metal complexes can also act as catalysts for various reactions, such as the reduction of nicotinamide adenine dinucleotide (NAD^+) to form 1,4-NADH or the oxidation of glutathione (GSH) to form oxidised glutathione (GSSG), to name just a few [2]. These intracellular catalytic reactions can disturb natural functions and consequently can become interesting biological targets for the development of new metal-based drugs. In this chapter, an overview of the catalytic reactions that can occur

Inorganic Chemical Biology: Principles, Techniques and Applications, First Edition. Edited by Gilles Gasser.
© 2014 John Wiley & Sons, Ltd. Published 2014 by John Wiley & Sons, Ltd.

within cells and organisms is given and examples in which metal complexes have been found to act as catalysts are discussed.

In cells and organisms, metal complexes can also interact with biomolecules and accordingly disrupt important biological processes. These interactions between metal complexes and biomolecules offer another approach for targeting specific biological functions and diseases [3, 4]. For example, the interaction of cisplatin and DNA accounts for its efficacy against cancers by blocking cell replication [5]. While several metal complexes can in principle interact with DNA, often these metal complexes also interact with additional biomolecules such as proteins.

Enzymes are an important class of proteins; they are the natural biocatalysts of the organisms [6]. The active site of enzymes is in general a cavity in which the substrate binds weakly before being transformed and released. These specific binding sites can be blocked by replacing the natural substrate by an inhibitor. A perfect knowledge of the three-dimensional structure of these active sites allows the design of highly efficient inhibitors, thus offering new challenges for chemists.

Several detailed reviews have dealt with metal complexes as enzymes inhibitors [3, 7–9] and therefore, instead of trying to prepare a similar review focussing on recent publications that were not covered in these excellent reviews, we have chosen to emphasise to the readers, the strategies that have been developed by researchers to target enzyme inhibition with metal complexes.

Overall, this chapter is aimed at undergraduate students or new-comers in the field, focussing on giving them mainly the concepts and strategies that can be used to design new metal complexes as enzyme inhibitors or as catalysts in cells and organisms. The selected examples from the literature are used to illustrate the different concepts and we acknowledge that they may not provide an exhaustive coverage of the field; they are merely the personal choices of the authors.

11.2   Metal-based Inhibitors: From Serendipity to Rational Design

Coordination chemistry offers great diversity and versatility to chemists for preparing highly potent enzyme inhibitors. The introduction of selected ligands with structural or biological functions is relatively easy through coordination chemistry. Therefore, some of the strategies that have been developed so far to obtain metal complexes as enzyme inhibitors are discussed in this section.

11.2.1   Mimicking the Structure of Known Enzyme Binders

Several known organic inhibitors of enzymes can be found in the literature and some of them are even used in the clinic to treat biological disorders. The three-dimensional structure of these inhibitors can be used as a model to design metal-based analogues. This strategy has been elegantly exploited by Meggers [3], using staurosporine as a model and transition metals such as ruthenium, osmium or iridium as a metal centre to build structurally analogous complexes to staurosporine (Fig. 11.1). Staurosporine prevents ATP from binding to the active site of protein kinases, an important class of enzymes involved in the phosphorylation of various proteins.

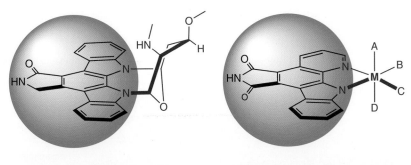

staurosporine (kinase inhibitor)   metallo-staurosporine-analogue

Figure 11.1   *Metal complexes mimicking the staurosporine structure. Adapted with permission from [3] © 2009 Royal Society of Chemistry*

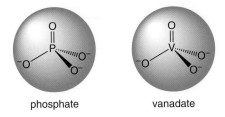

phosphate   vanadate

Figure 11.2   *Structural analogy between phosphate and vanadate ions [11]*

Following this strategy, several highly potent organometallic inhibitors for kinases were prepared by the group of Meggers [3]. Crystal structures of metal complexes within kinases suggest that the metal centre plays only a structural role, ensuring the three-dimensional arrangement of the selected ligands and adding rigidity to the systems.

The strong similarities between phosphate (PO_4^{3-}) and vanadate (VO_4^{3-}) have been exploited to prepare phosphatase inhibitors (Fig. 11.2) [10]. During the hydrolysis of phosphate within the cavity of phosphoryl transfer enzymes, a pentacoordinated phosphorus is formed, and therefore, a stable pentacoordinated vanadate complex can interact strongly and inhibit the enzyme. The inhibition effect and the selectivity can even be increased by adding an appropriate ligand on the vanadium atom. Indeed, introduction of two 3-hydroxypicolinate ligands, one water molecule and an oxo ligand around the vanadium centre generates a highly selective inhibitor ($IC_{50} = 35\,nM$) for the phosphinositide 3-phosphatase enzyme [11].

These affinities of vanadate complexes for phosphatases, and its ability to oxidise the catalytic cysteine moiety within the active site of protein tyrosine phosphatase, appear to be the key elements for the insulin-mimetic effect of vanadium complexes [12].

11.2.2   Coordinating Known Enzymatic Inhibitors to Metal Complexes

Glutathione transferase (GST) is an important class of enzymes involved in the detoxification processes of cells. These sulfur-containing enzymes are known to interact with metal complexes, forming strong M−S bonds. This is particularly relevant for cisplatin, which

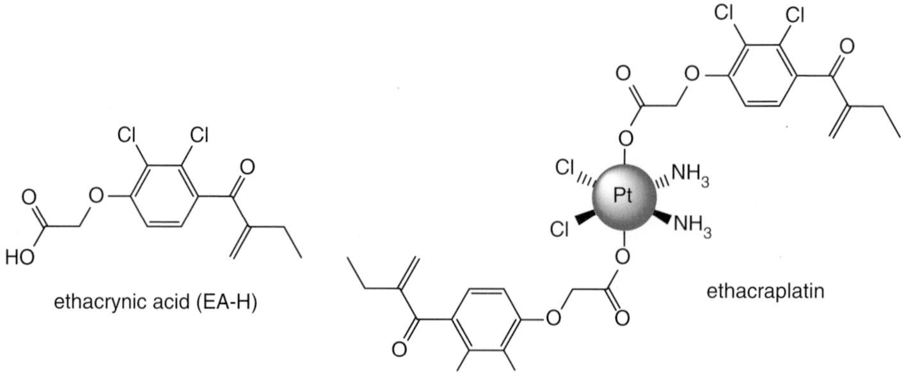

Figure 11.3 *Ethacrynic acid (EA-H) and ethacraplatin complex. Adapted with permission from [13] © 2005 American Chemical Society*

tends to be intercepted by GST, thus reducing the efficacy of cisplatin against cancer cells. Therefore, to overcome this limitation, ethacrynic acid (EA-H), a known inhibitor of GST, was recently coordinated to a platinum(IV) centre [13]. The $Pt(NH_3)_2Cl_2(EA)_2$ complex (Fig. 11.3), named ethacraplatin by the authors, shows a high inhibition activity of GST, and after decomplexation of the EA units and reduction of the Pt(IV) centre to Pt(II), a cisplatin analogous complex is released.

11.2.3   Exchanging Ligands to Inhibit Enzymes

Phosphole-containing gold complexes have been used to inhibit human disulfide reductase enzymes (Fig. 11.4) [14]. The weakly coordinated phosphole and chloro ligands are released stepwise in biological media and after decomplexation of the initial ligands the gold atom can sit between two cys–thiolate units in the active site of the enzyme, thus forming an almost linear Cys-S-Au-S-Cys complex, as demonstrated by X-ray structure analysis. The pyridyl derivative was extremely potent on human glutathione reductase and thioredoxin reductase.

11.2.4   Controlling Conformation by Metal Coordination

Controlling the orientation of peptidic groups to mimic naturally occurring peptidic β-turn functions similar to those found in proteins is of great interest to Marshall and his

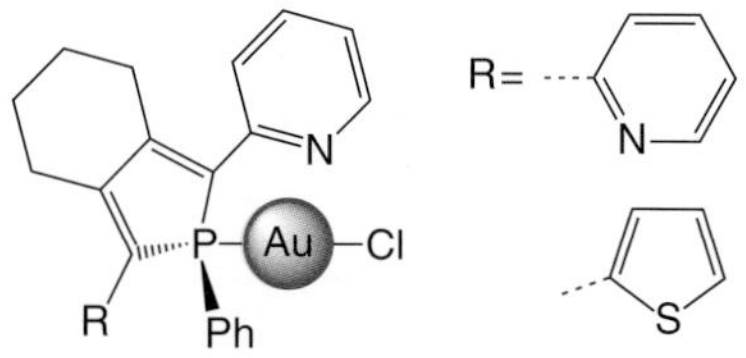

Figure 11.4 *Phosphole-containing gold(I) complexes. Adapted with permission from [14] © 2006 Wiley-VCH Verlag GmbH & Co. KGaA, Weinheim*

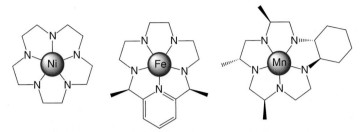

Figure 11.5 *Selected penta-azacrown metal complexes. Adapted with permission from [15]* © 2007, John Wiley and Sons

coworkers [15]. These β-turn motifs are often involved in protein–protein interactions, and therefore, artificial β-turn motifs can block recognition processes. To achieve their goals, cyclic pentapeptides were prepared, and after reduction of the amide bonds, a metal was added in the core of the azacrown motif (Fig.11.5). The metal (Mn, Fe, Co, Ni, Cu or Zn) introduces rigidity to the system, and allows the peptide groups to be properly oriented to form β-turn analogues. It was suggested that these systems can serve as inhibitors of amylase.

11.2.5 Competing with Known Metallo-Enzymatic Processes

Several families of enzymes possess a metal-based active site, and, therefore, coordination chemistry can be a powerful tool to disrupt these processes. Three strategies can be envisaged: (i) replacement of the active metal ion by a different one; (ii) insertion of a strong ligand to block the metal-based active site of the enzyme; (iii) coordination of a metal complex to the active site via cysteine or selenocysteine units.

Among metallo-enzymes, zinc finger proteins are interesting targets for biological applications. For example, a gold(III) phenanthroline (phen) complex, [AuCl$_2$(phen)]Cl, was evaluated as a potential inhibitor of poly(adenosine diphosphate (ADP)-ribose) polymerases (PARPs) (Fig. 11.6) [16]. The studies suggest that the zinc ion within the

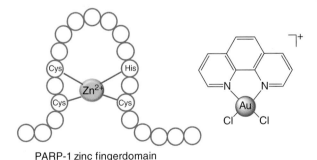

PARP-1 zinc fingerdomain

Figure 11.6 *Gold phenanthroline complex of zinc finger proteins. Adapted with permission from [16]* © 2011 American Chemical Society

active site of the PARPs is replaced by a gold ion. The formation of a gold adduct reduces significantly the activity of the enzyme.

11.3   The Next Generation: Polynuclear Metal Complexes as Enzyme Inhibitors

Structural studies have revealed that many metallo-enzymes utilise more than one metal ion in their active sites, often in different oxidation states [17–22]. In fact, as we begin to understand the mechanistic aspect of how these enzymes operate, it is becoming clear that the presence of several metals is essential to the enzyme's function. Based on this notion, another strategy to inhibit enzymatic processes is the application of polynuclear complexes as enzyme inhibitors. The premise behind this type of inhibitory effect is to prepare complexes containing metals that are or can easily be converted into the oxidation state of the endogenous metals that specific enzymes need to carry out their function. Depending on the nature of the ligand systems used for their preparation, these complexes can then competitively bind to the enzyme's active site thus blocking its function. This type of specific targeting of enzymes is extremely useful for the development of effective drug therapies.

The aforementioned sections in this chapter have highlighted the roles that metal complexes play, particularly mononuclear complexes, as enzyme inhibitors and in the latter sections are highlighted catalysts in living cells and organisms. Recent approaches have focussed on combining two or more metal centres via bridging ligands or as part of clusters to generate novel polynuclear complexes. These new polynuclear complexes offer new approaches for chemotherapy with novel modes of action that are often not accessible with mononuclear metal species.

Generally, metallo-enzymes contain several metallic active sites, which are often made of the same metal ions. However, there are examples for which the active enzymatic sites contain different types of metal ions. The presence of two or more metal centres in enzymes, such as urease, allows the multinuclear complexes to activate electrophilic and nucleophilic substrates [23]. The close proximity and cooperative interaction between the metal centres allow for superior activity and selectivity, and also for the conversion of challenging substrates in a number of metallo-catalysed processes [19, 20, 22].

Multinuclear metal complexes that are reported to inhibit enzymes often utilise protein targets such as histone deacetylase, telomerases, topoisomerases and protein kinases. Several comprehensive literature reviews have already adequately summarised the application of polynuclear complexes as pharmacological agents [24–26]. It is important to stress here that the purpose of this section is to emphasise the potential of polynuclear complexes as enzyme inhibitors because the application of polynuclear complexes as biological agents is still a relatively new area and research into their specific targets is still ongoing. The intent of the subsequent discussion is to give the reader insight into the potential of polynuclear enzyme inhibitors. Thus, in this section, we will highlight either a general class of complexes with enzymatic inhibitory effects or a specific enzyme that is being targeted by polynuclear complexes in general.

11.3.1   Polyoxometalates: Broad Spectrum Enzymatic Inhibitory Effects

11.3.1.1   Structural aspect

Arguably one of the largest classes of versatile multimetallic enzyme inhibitors is the poly-oxometalates (POMs). These polynuclear compounds are transition metal clusters that are bridged by oxygen atoms [27–30]. Their structures predominantly comprise octahedral geometry and typical structures involved in enzyme inhibition are shown as ball and stick molecules in Fig. 11.7 [31]. POMs are characterised as anionic complexes with relatively high stability even in aqueous media at biological pH values. Structurally, they can be divided into two groups, isopolyanions or -oxometalates and heteropolyanions or -oxometalates. The isopolyanions are comprised of only one type of d^0 metal cation, typically Mo^{6+}, W^{6+} or V^{5+}, which are bridged by oxide anions. The heteropolyanion complexes have one or more additional p-, d- or f-block elements as well as a large number of elements giving them much higher structural diversity compared with isopolyanions. The class of heteropolyanion POMs have been studied as enzyme inhibitors because it is easier to modify their structure.

11.3.1.2   Enzyme inhibitory activity

A variety of POMs have displayed inhibitory activities in the micro- to nanomolar concentration range against enzymes such as phosphatases, kinases, nucleases and proteases, to name just a few. Their enzyme inhibitory action is also thought to contribute to their pharmacological activities as anticancer, antibacterial, antiprotazoal, antiviral and antidiabetic activities. These complexes have received considerable attention in the past 40 years and a recent review published by Stephan *et al.* has highlighted the great potential of these compounds and their possible enzymatic targets [31]. POMs have been proposed to interact with a variety of biomolecules. Certain phosphatases reside extracellularly and thus POMs can interact with them even if they are unable to enter cells [32, 33]. Some POMs have been investigated as *ecto*-nucleotidase inhibitors and most notably were found to impede ATP breakdown in rat cerebellar and hippocampal slice preparations [34]. They also effectively

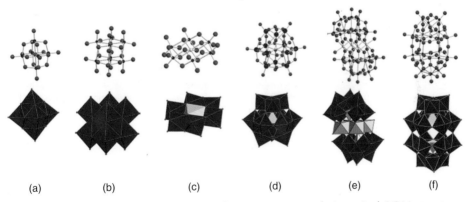

(a)	(b)	(c)	(d)	(e)	(f)

Figure 11.7   *Ball-and-stick and polyhedral representations of six typical POM structures. Reproduced with permission from [31] © 2013 Wiley-VCH Verlag GmbH & Co. KGaA, Weinheim*

inhibited nucleoside triphosphate diphosphohydrolase (NTPDase) found in the pathogenic bacteria *Legionella pneumophila* [35].

Protein kinases are upregulated in cancer cells and have become an important target for drug design. POMs such as $[P_2Mo_{18}O_{62}]^{6-}$ have shown inhibition of the protein kinase CK2 in a non-competitive fashion [36–39]. Inhibitory interactions of POMs with sulfotransferases and sialyltranferases have also been identified [40]. These particular enzymes modify sugar chains on cell surfaces, which is especially important as many microorganisms utilize sialylated/sulfated glycans on cell surfaces for the preliminary step of infection [40].

The interactions of POMs with intracellular enzymes have been proposed, but at this stage, it is still uncertain whether the activities observed can be extended to studies *in vivo*. Histone deacetylases (HDAC) are one such example. In a high-throughput screening against HDAC, many strong POM inhibitors were identified but it is yet to be determined if this inhibition will be effective in cellular systems or in animals [41]. Much interest has also been expressed in their ability to inhibit nucleases, DNA and RNA polymerases and proteases *in vitro* [28, 42–48]. However, these enzymes are intracellular targets and at present there is no evidence to support POMs being able to access them. The design and study of POMs as enzyme inhibitors is still in the relatively early stages. As with most other polynuclear complexes currently garnering interest for their potential enzymatic interaction, the preparation of POMs that can access intracellular targets *in vivo* is yet to be accomplished but in our opinion is not far off.

Traditional polyoxometalates have demonstrated an affinity for a variety of enzymes and in so doing have led to an interest in the use of other metals for POM preparation [49, 50]. Wong *et al.* synthesised and studied the enzymatic activity of the tetranuclear polyanionic ruthenium-oxo oxalate cluster $Na_7[Ru_4(\mu_3\text{-}O)_4(C_2O_4)_6]$ (Fig. 11.8) [50]. This cluster displayed anti-HIV activity in a dose dependent manner. The cluster was evaluated for its antiviral activity toward R5-tropic HIV-1(BaL) infection/replication and was found to

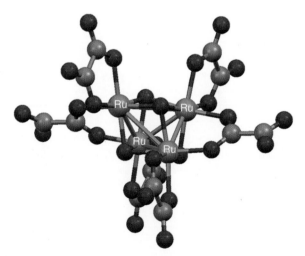

Figure 11.8   *Ball and stick representation of the anion $[Ru_4(\mu_3\text{-}O)_4(C_2O_4)_6]^{7-}$. Adapted with permission from [50] © 2006 American Chemical Society*

inhibit over 98% of viral replication. Further investigation of its effect on HIV-1 reverse transcriptase revealed it to reduce HIV-1 RT activity more effectively than the organic inhibitor, 3'-azido-3'-deoxythymidine 5'-triphosphate (AZT-TP) [50].

The exact mode of action of POMs remains unclear [31]. The cellular uptake of POMs has been occasionally observed, however, the electrostatic repulsion between negatively charged cell membranes and anionic POMs should not favour cellular uptake. Therefore, it was suggested that the enzymatic inhibition by POMs is more likely to happen with extra-cellular enzymes. In addition, the stability of POMs is also an important factor to consider, as fragments can be on several occasions the active inhibitors. Nevertheless, although still in its infancy, this approach represents a new area of research that is receiving prolific attention and further investigations will be needed to better understand the interaction of POMs with enzymes.

11.3.2   Polynuclear G-quadruplex DNA Stabilizers: Potential Inhibitors of Telomerase

Telomerase is a ribonucleo-protein responsible for the addition of DNA sequence repeats to the 3' end of DNA strands in the telomere regions. It has also been evidenced that this enzyme plays a role in the upregulation of 70 genes that could be implicit in cancerous growth and metastases as well as the activation of glycolysis, which allows cancer cells to quickly consume sugars to facilitate their programmed growth rate [51]. Telomeric stabilisation assists the preservation of unconstrained cell proliferation potential. Structurally, telomerase is a multimeric enzyme consisting of different components: the RNA subunit human telomerase RNA component (hTERC, also known as hTR or hTER), the catalytic subunit human telomerase reverse transcriptase (hTERT) and associated proteins [52]. It has been observed that most human tumours display a strong telomerase upregulation (they are overexpressed in 85–90 % of cancer cells) yet somatic cells do not express this enzyme. Telomerase inhibition is therefore a potential target for anticancer therapies [53].

G-quadruplexes are made from guanine-rich DNA sequences and are abundant in the human telomere and in gene promoters (see also Chapter 6 which deals with the probing of DNA using metal complexes) [54]. Guanine (G) bases have the ability to self-assemble in tetrads and these tetrads can stack on top of each other to form quadruplex DNA structures. These quadruplex structures are further stabilised by cations (K^+) centred between neighbouring tetrads. These G-quadruplexes have a variety of biological functions including the inhibition of telomerase [55, 56]. Recently, interest in polynuclear complexes as potential telomerase inhibitors has begun to emerge. Water-soluble triosmium organometallic clusters containing quinolone, 3-aminoquinoline, quinoxoline or phenanthroline ligands (Fig. 11.9) were assayed for telomerase inhibition [57, 58]. The complexes containing sulfonated phosphine ligands were found to be the best inhibitors of telomerase function when tested on enzymes in a cell free medium. However, they were found to have decreased activity when tested for telomerase inhibition in the MCF-7 breast cancer cell line.

Four tetranuclear Pt(II) metalla-squares (Fig. 11.10) were found to selectively bind to *htelo* G-quadruplexes over promoter G-quadruplexes (*bcl2*) and duplex DNA in cell free assays [59]. Further *in vitro* assessments of these compounds using telomeric repeat amplification protocol (TRAP) and the 3-(4,5-dimethylthiazol-2-yl)-2,5-diphenyltetrazolium bromide (MTT) assays revealed high telomerase inhibition and anticancer activity [59].

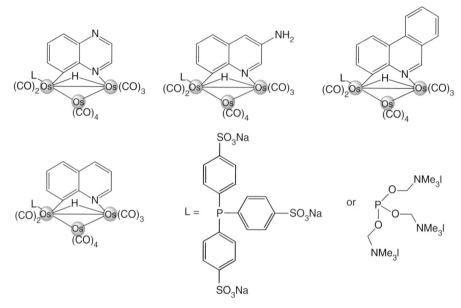

Figure 11.9 Osmium clusters tested for anti-telomerase activity in a cell free medium. Adapted with permission from [57] © 2005 Elsevier and adapted with permission from [58] © 2011 Royal Society of Chemistry

Di- and polynuclear ruthenium complexes have been proposed to be better G-quadruplex DNA binders since they exhibit low sensitivity to changes in salt concentration [60]. Ionic strength around duplex DNA is much lower compared with G-quadruplex DNA, thus complexes that can effectively bind to G-quadruplex DNA must be resilient to high salt concentrations. The dinuclear complex (structure **1**, Fig. 11.11) was able to stimulate the formation of an antiparallel G-quadruplex structure both in the absence and presence of Na^+ and K^+ buffers [61].

The dinuclear complexes **2** and **3** (Fig. 11.11) were able to bind to both G-quadruplex and duplex DNA [62] and the complex **4** (Fig. 11.12), which contains a flexible non-cyclic crown ether chain, displayed moderate G-quadruplex stabilisation [63]. Fluorescence studies of **4** revealed that the complex was selective for G-quadruplex structures over DNA duplex structures; there was a large distinction in fluorescence between the DNA duplex structures and the quadruplex structures. Trinuclear complexes, **5** and **6** (Fig. 11.12), have been reported to bring about significant conformational changes in human telomeric DNA and stabilise the G-quadruplex $AG_3(T_2AG_3)_3$ motif [64].

A series of guanidinium-modified Zn-phthalocyanine derivatives (structure **7**, Fig. 11.13) has been prepared by the group of Luedtke [65]. Interestingly, these Zn-phthalocyanine derivatives show selective fluorescence response to quadruplex DNA. A 5000-fold higher affinity for *c-Myc* (oncogenic promoter) quadruplex DNA than for duplex calf thymus DNA was observed. The intracellular localisation of the Zn–phthalocyanine complex was different from duplex DNA probes, being consistent with quadruplex-mediated promoter deactivation. However, further studies will be needed to confirm the exact mode of action.

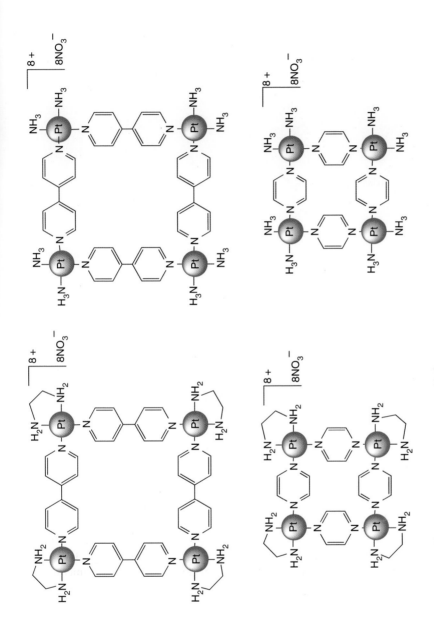

Figure 11.10 *Metalla-squares showing promising telomerase inhibition in vitro. Adapted with permission from [59] © 2012 Royal Society of Chemistry*

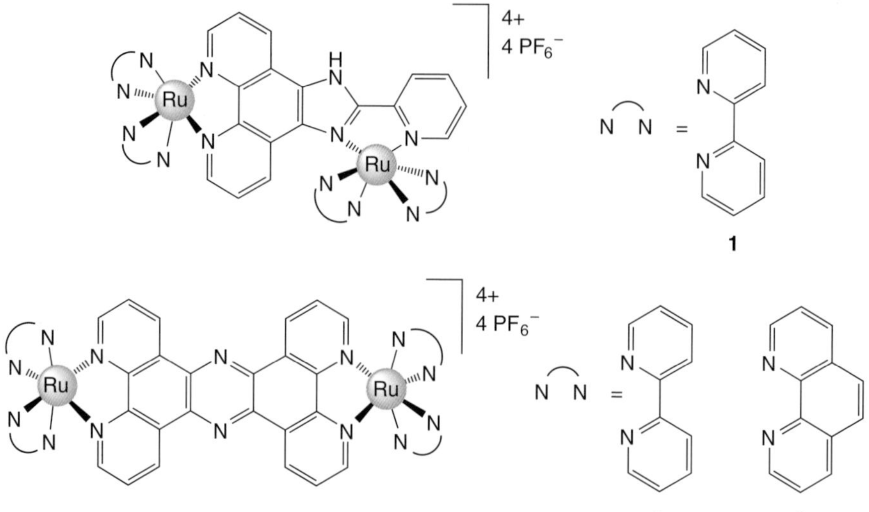

Figure 11.11   *Dinuclear ruthenium complexes as G-quadruplex stabilisers. Adapted with permission from [61] © 2008 American Chemical Society*

A similar dual quadruplex DNA binder and optical switch has recently been described by Vilar and coworkers [66]. The cyclometallated platinum complex (structure **8**, Figure 11.13) shows a 1000-fold selectivity for *c-Myc* over a 26-mer duplex DNA sequence. Despite this high affinity for quadruplex DNA in DMSO, this poorly water-soluble complex was not taken up by osteosarcoma U2OS cells. To overcome this limitation, the platinum complex was encapsulated inside a water-soluble metalla-cage [67], thus ensuring delivery of the complex to cancer cells. After internalisation and release of the platinum complex from the hexaruthenium cage compound; confocal fluorescence microscopy has shown localisation of the complex in the nucleoli, where it can potentially interact with quadruplexes.

11.3.3   Polynuclear Polypyridyl Ruthenium Complexes: DNA Topoisomerase II Inhibitors

Complexes containing ruthenium have gained considerable attention as pharmacological agents. Ruthenium can access a range of oxidation states (Ru(II), Ru(III) and Ru(IV)) under physiological conditions and its complexes have also shown low toxicity [68–72]. The application of polynuclear Ru(II) complexes in biology is a burgeoning area of research [26]. As investigations into these complexes' biological function expand, it is becoming clear that polynuclear ruthenium complexes have a diverse set of targets. Type II topoisomerases (Topo II) provide an integral function in DNA synthesis. These enzymes cut both strands of DNA concurrently in one double helix followed by linkage to Topo II via a tyrosyl bond. A second DNA duplex is then passed through the break. Topo II thus is able to control DNA tangles and supercoils. This is achieved by hydrolysis of ATP and during this process Topo II is able to increase or decrease the linking number of a DNA loop by

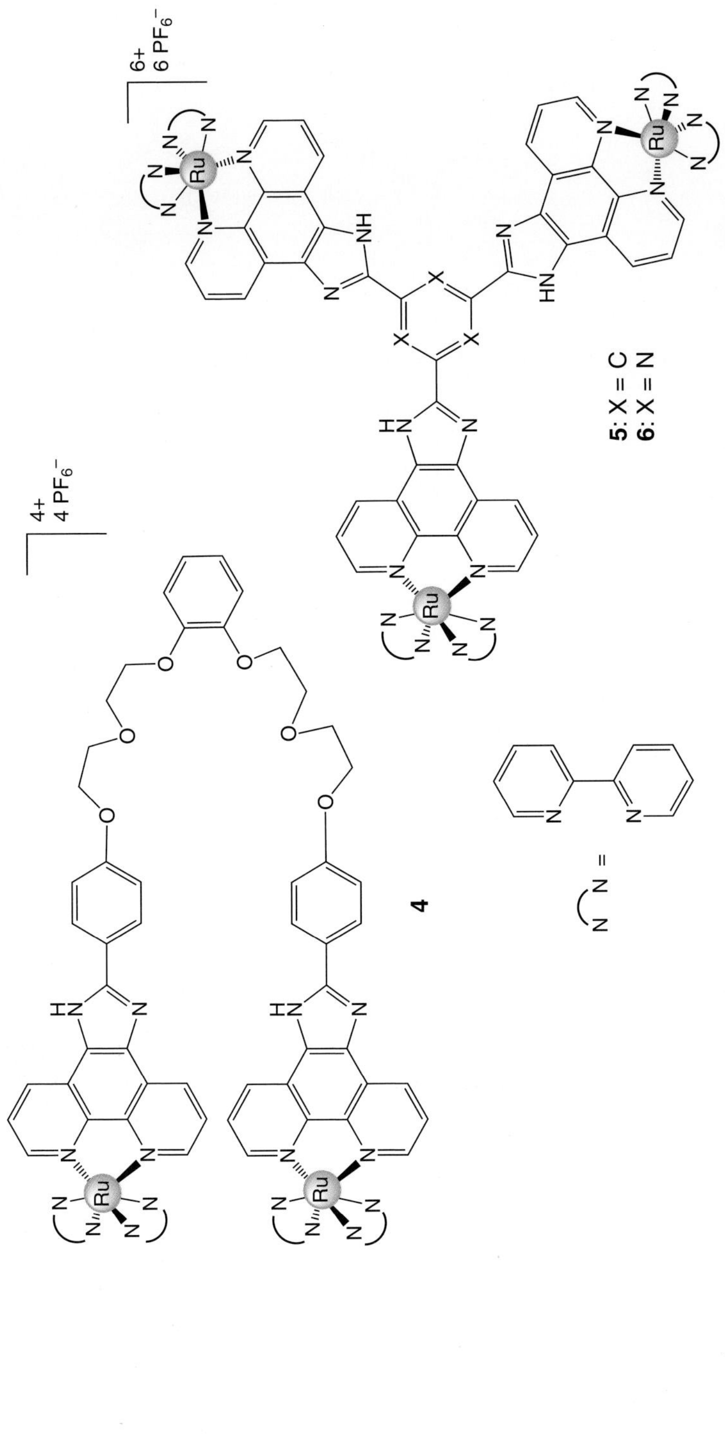

Figure 11.12 Complexes **4–6** used for G-quadruplex stabilisation. Adapted with permission from [63] © 2009 Royal Society of Chemistry and adapted with permission from [64] © 2010, Elsevier

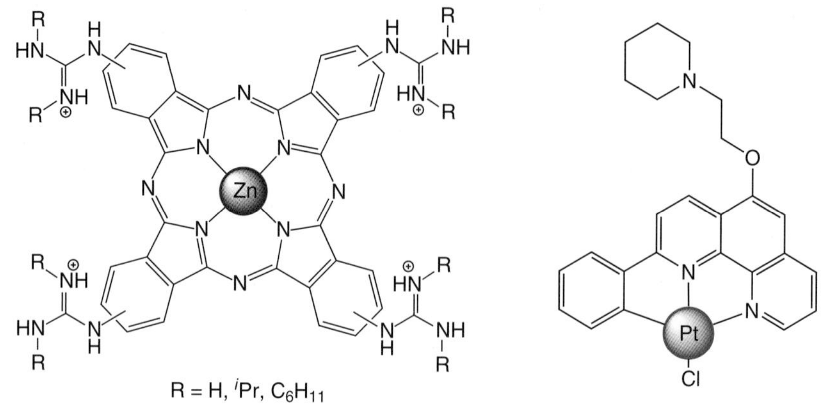

R = H, ^{i}Pr, C$_6$H$_{11}$

7

8

Figure 11.13 *Complexes **7** and **8** used as G-quadruplex stabilisers and as optical switches. Adapted with permission from [65] © 2009 Wiley-VCH Verlag GmbH & Co. KGaA, Weinheim and adapted with permission from [66] © 2012, KGaA, Weinheim Wiley-VCH Verlag GmbH & Co.*

two units. In addition, Topo II favours chromosome disentanglement. The entire process is completed by re-ligation, which is ATP dependent [63, 74].

Inhibitors of Topo II target ATPase activity by non-competitive binding to ATP. Some polypyridyl binuclear complexes have displayed this ability. The dinuclear complex (structure **9**, Fig. 11.14) was evaluated *in vitro* for DNA Topo II inhibition of the filarial parasite

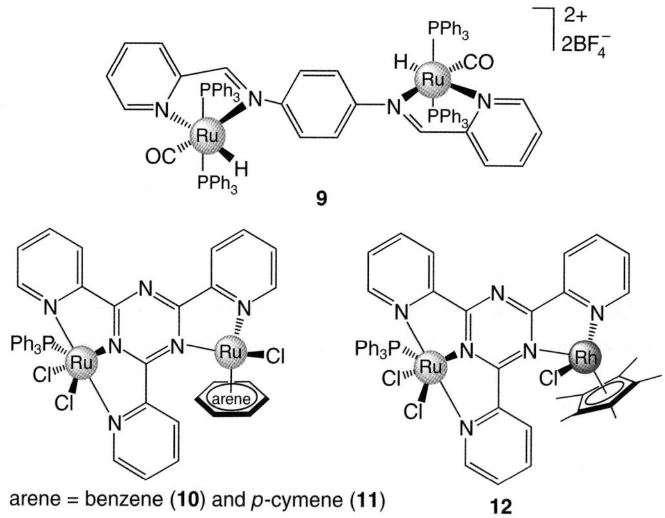

9

arene = benzene (**10**) and *p*-cymene (**11**)   **12**

Figure 11.14 *Dinuclear complexes **9–12** exhibiting an affinity for Topo II enzyme. Adapted with permission from [75] © 2004, Elsevier and adapted with permission from [76] © 2008 American Chemical Society*

S. cervei. After a brief incubation of the complex with the enzyme, pBR322 DNA was added. It was found that **9** was able to inhibit Topo II activity by 75 % [75]. Using an enzyme-mediated supercoiled pBR322 relaxation assay; the influence of complexes **10–12** (Fig. 11.14) on Topo-II DNA activity on the filarial parasite *S. cervei* was evaluated by Sharma *et al.* [76]. Both the hetero- and homo-dinuclear complexes exhibited a strong affinity for complex formation with the Topo II-DNA complex even at concentrations less than $2\,\mu g\,ml^{-1}$. The authors proposed that the complexes may influence Topo II-DNA activity by one of three mechanisms: binding to either (i) the enzyme, (ii) the Topo II-DNA complex or (iii) DNA itself [76].

11.4   Metal Complexes as Catalysts in Living Cells

In a very interesting perspective review, Alessio and coworkers have suggested a categorisation of metal anticancer compounds into five classes based on their mode of action, with the hope that this would help researchers for future rational design of anticancer drugs [77]. The proposed categories are: (i) functional compounds, such as cisplatin, NAMI-A or KP1019, in which the metal binds to the biological target; (ii) structural compounds, in which the metal determines the global shape of the compound; (iii) carrier compounds, in which the metal acts as a carrier for active ligands that are expected to be delivered *in vivo*; (iv) catalytic compounds, in which the metal compound acts as a catalyst; and (v) photoactive compounds, in which the metal compound is photoactive and behaves as a photo-sensitizer. While categories (i) functional compounds and (ii) structural compounds encompass numerous metal complexes and have been extensively discussed in this and other reviews [78–86], the category (iv), catalytic compounds, has remained a relatively unexplored domain. However, a recent feature article has presented and discussed synthetic metal complexes that can catalyse chemical transformations in living organisms [2], showing that significant progress has been made towards the application of non-biological reactions in living systems. Metal complex catalysis currently includes only a few compounds, and it is supposed that the metal behaves as a catalyst *in vivo*, for example, through the production of reactive oxygen species (ROS) that cause cell damage, or oxidation/reduction of important biomolecules, such as for instance nicotinamide adenine dinucleotide (NAD^+) and glutathione (GSH).

11.4.1   Catalysis of NAD^+/NADH

Nicotinamide adenine dinucleotide (NAD^+), known as a coenzyme NAD^+/NADH, is involved in many cellular redox reactions (Fig. 11.15), but it can also act as a substrate for bacterial DNA ligases that use it to remove acetyl groups from proteins [87, 88]. Because of the importance of these functions, the enzymes involved in NAD^+ metabolism have become interesting and promising targets for drug discovery. Moreover, the coenzyme 1,4-NADH is required for many enzymatic reduction reactions that are useful for the stereo-selective synthesis of organic compounds [89, 90]. In recent years, interest in the possibility that organometallic compounds could interfere with NAD^+/NADH hydride transfer reactions in cells as a novel mechanism of action has received much attention.

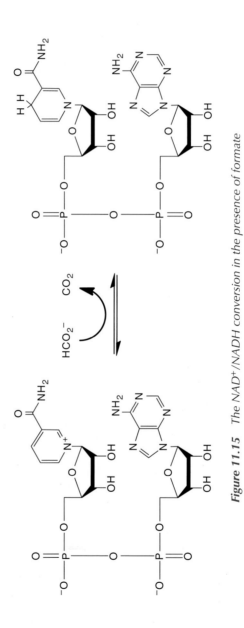

Figure 11.15   *The NAD⁺/NADH conversion in the presence of formate*

The ruthenium(II) arene anticancer complexes of Sadler's group [(η^6-arene)Ru(en)Cl] PF$_6$ (arene = hexamethylbenzene, *p*-cymene, indane; en = ethylenediamine) [91–94] were shown to efficiently catalyse regioselective reduction of NAD$^+$ by formate in water to form 1,4-NADH under physiological conditions [95]. The reaction rates obtained were, however, up to 50 times slower than those catalysed by rhodium(III) pentamethylcyclopentadienyl complexes [96]. Interestingly, the A549 lung cancer cells used for testing the compounds were remarkably tolerant to formate even at millimolar concentrations. This study has therefore opened the possibility of using arene ruthenium complexes coadministered with formate as catalytic drugs.

In an attempt to improve the efficiency of [(η^6-arene)Ru(en)Cl]$^+$ complexes as catalysts for hydride transfer from formate to NAD$^+$, Sadler and coworkers have observed the reverse reaction, namely the transfer of hydride from 1,4-NADH to organometallic complexes [97] (Fig. 11.15). Thus, the developed half-sandwich Ru(II) arene and Ir(III) cyclopentadienyl complexes can use NADH as a hydride source for the reduction of ketones. It clearly appears from this study that these new complexes may be efficient for modulating the redox characteristics of cells, which in turn lead to apoptosis.

Recently, the same group has synthesized a series of neutral ruthenium(II) half-sandwich complexes of the type [(η^6-arene)Ru(N∩N′)Cl] but with modified ethylenediamine chelating ligands (Fig. 11.16). They have added a sulfonamide group that resembles Noyori's ligand with the goal of improving the regioselective catalysis of the transfer hydrogenation of NAD$^+$ to give 1,4-NADH in the presence of formate. The series of compounds showed improved catalytic activity that is about an order of magnitude higher than that of the ethylenediamine analogues [98].

11.4.2   Oxidation of the Thiols Cysteine and Glutathione

Redox mechanisms can be involved in the activation of ruthenium–arene thiolato complexes. This exciting feature of ruthenium complexes was mainly demonstrated by Sadler and coworkers. For example, reactions of glutathione with [(η^6-bip)Ru(en)Cl]$^+$ was shown

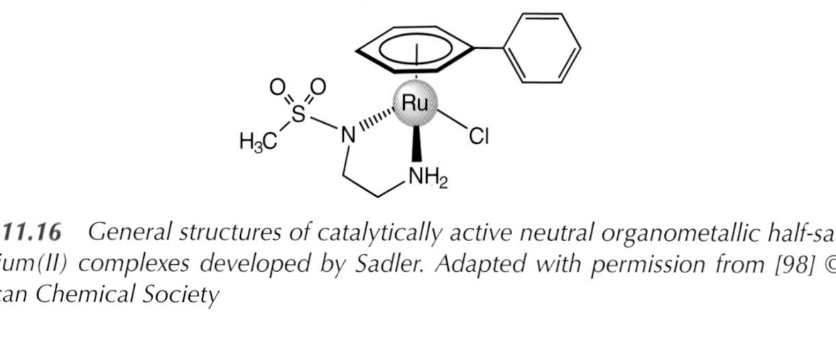

Figure 11.16   *General structures of catalytically active neutral organometallic half-sandwich ruthenium(II) complexes developed by Sadler. Adapted with permission from [98] © 2012, American Chemical Society*

to yield the thiolato complex $[(\eta^6\text{-bip})\text{Ru(en)(GS)}]^+$ [99]. This complex subsequently underwent oxidation to give a very unusual sulfenato complex $[(\eta^6\text{-bip})\text{Ru(en)}\{\text{GS(O)}\}]$, usually unstable, but in that particular case stabilised by H-bonding as well as by coordination to the ruthenium(II) centre [100, 101]. The sulfenate–Ru adduct can subsequently react with the standard target of guanine, the nitrogen N7. Interestingly, these authors showed that the cGMP adduct $[(\eta^6\text{-bip})\text{Ru(en)(cGMP-N7)}]^+$ remained the major product even upon the addition of a large molar excess of GSH. Therefore, the facile displacement of *S*-bound glutathione by N7 of guanine via the sulfenato intermediate may provide a potential route for RNA and DNA ruthenation *in vivo*. Recently, the same group has performed competition experiments between glutathione and DNA oligonucleotides for two complexes $[(\eta^6\text{-bip})\text{Ru(en)Cl}]^+$ and $[(\eta^6\text{-bip})\text{Ru(tha)Cl}]^+$ (tha = tetrahydroanthracene). The results indicate that the reaction of both complexes with single-stranded oligonucleotides always gave rise to mono-ruthenated oligonucleotides, irrespective of the GSH concentration. Their findings reveal therefore a potentially contrasting role for GSH in the mechanism of action of these ruthenium anticancer complexes [102].

Glutathione was also shown to promote oxidation of the thiolato complex $[(\eta^6\text{-hmb})\text{Ru(en)SR}]^+$ (R = ^{i}Pr) to give the sulfenato complex $[(\eta^6\text{-hmb})\text{Ru(en)}\{\text{S(O)R}\}]^+$, oxidation was apparently mediated by the O_2–GSH couple under physiological conditions. Ruthenium arene complexes can also induce oxygenation of cysteine residues in proteins, for example Cys34 of human serum albumin (HsA). The complex $[(\eta^6\text{-}p\text{-cymene})\text{Ru(en)Cl}]^+$ is able to induce oxidation of the only free thiol group in albumin, Cys34, to the sulfinate. On the other hand, no oxidation of Cys34 is observed with $[(\eta^6\text{-bip})\text{Ru(en)Cl}]^+$. The reason advanced by the authors, conceivably, is the hindered access to Cys34 because of the bigger arene [103].

The possibility of developing purely catalytic drugs was also pioneered by Sadler and coworkers [104]. They have developed a family of organometallic half-sandwich ruthenium(II) compounds of the type $[(\eta^6\text{-arene})\text{Ru(azpy)I}]^+$ (azpy = *N,N*-dimethylphenylazopyridine, arene = *p*-cymene or biphenyl) (Fig. 11.17).

The most remarkable feature of these complexes is that they are highly cytotoxic against human ovarian A2780 and human lung A549 cancer cell lines (IC_{50} in the μM range), despite being substitutionally very inert [104]. They are for instance unable to undergo activation by aquation. Remarkably, Sadler and coworkers have demonstrated that the cytotoxicity of the complexes are derived directly from an increase in reactive oxygen species (ROS) generated through a redox cycle in which the complex acts as a catalyst for the

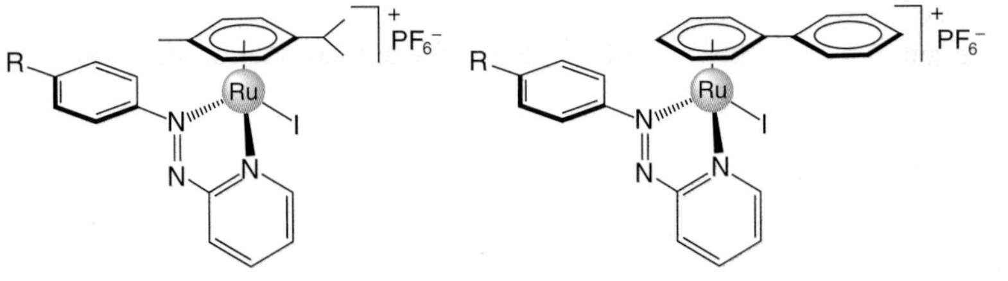

Figure 11.17 *General structures of catalytically active cationic organometallic half-sandwich ruthenium(II) compounds developed by Sadler and coworkers, R = NMe$_2$, OH or H. Adapted with permission from [104] © 2008, National Academy of Sciences, U.S.A.*

$$R\text{-}SH + R'\text{-}SH \xrightarrow{\text{Ru-complex}} RS\text{-}SR' + H_2$$

Scheme 11.1   *General scheme for the oxidation reaction of thiols catalysed by ruthenium complexes*

oxidation of glutathione (GSH) to oxidised glutathione, GSSG. Indeed, ^{1}H-NMR spectra showed that incubation of 10 mM GSH with 100 μM of complexes led to a steady oxidation of about 4 mM of GSH to GSSG, as demonstrated by the gradual disappearance of the original resonances of Cys and the appearance of new Cys resonances corresponding to the α-CH and β-CH$_2$ of Cys in GSSG [104]. This oxidation reaction produces H$_2$ (Scheme 11.1), which could however not be directly characterised. Nevertheless, the authors noticed bubble formation in the reaction tubes during the course of the reaction, thus strongly suggesting that H$_2$ was produced.

Even without generation of ROS, the catalytic conversion of GSH to GSSG may be directly related to the anticancer activity. Indeed, cancer cells are known to have a higher glutathione pool than healthy cells. In all living cells more than 90% of the total glutathione pool is in the reduced form (GSH) and less than 10% exists in the disulfide form (GSSG) [105]. An increased GSSG-to-GSH ratio is considered indicative of oxidative stress, that damages all components of the cell, including proteins, lipids and DNA, and which may lead to apoptosis.

Recently, the groups of Süss-Fink and Furrer have also observed that water-soluble and air-stable arene ruthenium complexes of the type $[(\eta^6\text{-}p\text{-cymene})_2\text{Ru}_2(\mu_2\text{-SR})_3]^+$ and $[(\eta^6\text{-}p\text{-cymene})_2\text{Ru}_2(\mu_2\text{-SR}')_2(\mu_2\text{-SR}'')]^+$ (Fig. 11.18) were surprisingly highly cytotoxic against human ovarian A2780 cancer cells despite their complete inertness to ligand substitution [106–108]. It is worth pointing out that the IC$_{50}$ values observed for some of these compounds (as low as 30 nM) place them among the most cytotoxic arene ruthenium compounds reported so far. Such a strong activity despite a complete inertness to ligand substitution may appear surprising at first glance. However, a recent study performed by Hartinger and coworkers suggests an inverse correlation between metallodrug–protein interaction and cytotoxicity against tumour cells [109], a finding that seems to apply also for other types of complexes.

Similarly to the complexes of Sadler, these trithiolato complexes are also highly efficient catalysts for the oxidation of GSH in aqueous solution. Overall, the results presented suggest that part of the high *in vitro* anticancer activity of these complexes may be due to their catalytic potential for the oxidation of GSH, but it also seems that some other properties/mechanisms such as Hammett's constants (which contains substituent constants such as electronegativity and mesomeric effects) and the lipophilicity parameters are involved since the authors could not provide evidence for any correlation between the IC$_{50}$ and TOF$_{50}$ (turn over frequency at 50% conversion) values. More precisely, a linear regression between IC$_{50}$ values, the Hammett constants σ_p and the lipophilicity parameters log P for more than 20 complexes shows a good non-linear determination, and the optimal region encompassing the most favourable values for both the Hammett constants σ_p ($-0.2 < \sigma_p < 0$) and log P values (log $P > 3.0$) leading to the lowest IC$_{50}$ values was apparent. With those results in hand, the authors could assume that the ruthenium compounds probably alter the behaviour of certain enzymes in the cells after the formation of reactive oxygen species, with uptake by the cancer cells influenced by the lipophilicity properties [106–108].

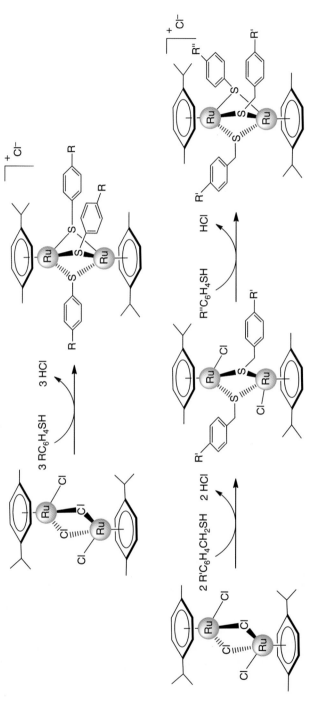

Figure 11.18 *Synthesis of complexes* $[(\eta^6\text{-}p\text{-}cymene)_2Ru_2(\mu_2\text{-}SC_6H_4R)_3]^+$ *(top) and* $[(\eta^6\text{-}p\text{-}cymene)_2Ru_2(\mu_2\text{-}SCH_2C_6H_4R')_2(\mu_2\text{-}SC_6H_4R'')]^+$ *(bottom) as the chloride salts, adapted from reference [108]*

A potential drawback claimed for these metal-centred catalysts is that they may have the propensity to be inactivated (poisoned) in the complex biological media. For one of their complexes, the groups of Süss-Fink and Furrer have shown that it can be recovered unchanged as the chloride salt. In addition, the stability of the complex was remarkable, since the TOF$_{50}$ drops by only 15% after the fifth run [106]. Therefore, these complexes might, on the contrary, have the advantage of being very stable and robust and therefore might have greater potential for biological activity.

11.4.3   Cytotoxicity Controlled by Oxidation

In a very interesting paper, Sadler and coworkers have reported the preparation and characterisation of ruthenium(II) arene complexes containing *o*-phenylenediamine (*o*-pda), *o*-benzoquinonediimine (*o*-bqdi), or 4,5-dimethyl-*o*-phenylenediamine (dmpda) as chelating ligands (Fig. 11.19), and have studied the effect of the variation of the arene and halides on the oxidation of coordinated *o*-pda, as well as their effect on cytotoxicity against A2780 human ovarian and A549 human lung cancer cell lines [110]. They have demonstrated that ruthenium arene complexes containing the diamine *o*-pda chelating ligand can lose their cytotoxic activity towards cancer cells upon oxidation to give the reduced *o*-bqdi diimine complexes. The main and most exciting result of this report was that oxidation can be controlled through changes in the electronic properties of the other ligands (arene and monodentate ligands) in the complex.

11.5   Catalytic Conversion and Removal of Functional Groups

The recent development of compounds possessing a protecting group that can be removed catalytically within living biological systems represents an attractive goal, since it

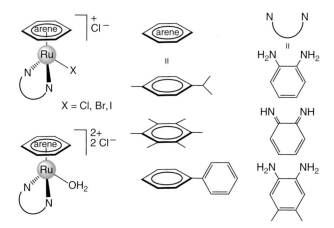

Figure 11.19   *Ruthenium complexes containing o-pda, o-benzoquinonediimine (o-bqdi), or 4,5-dimethyl-o-phenylenediamine (dmpda) as chelating ligands, isolated as their PF$_6^-$, Cl$^-$ or I$^-$ salts. Adapted with permission from [111] © 2006 Wiley-VCH Verlag GmbH & Co. KGaA, Weinheim*

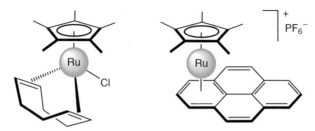

Figure 11.20 *Complexes [Cp*Ru(COD)Cl] and [Cp*Ru(η^6-pyrene)]PF$_6$. Adapted with permission from [111] © 2006 Wiley-VCH Verlag GmbH & Co. KGaA, Weinheim*

would allow the design of catalytically activable prodrugs. The group of Meggers have successively developed ruthenium half-sandwich complexes of the general formula [Cp*Ru(COD)Cl] and [Cp*Ru(η^6-pyrene)]PF$_6$ with Cp* = pentamethylcyclopentadienyl and COD = 1,5-cyclooctadiene (Fig. 11.20). These complexes can cleave allylcarbamates to their respective primary amines in the presence of water, air and even thiols. Interestingly, [Cp*Ru(η^6-pyrene)]PF$_6$ was successfully used as a catalyst in HeLa cells [111, 112].

The same group has also investigated fascinating iron meso-tetraarylporphines compounds that act as efficient catalysts for the reduction of aromatic azides to their amines. Interestingly, the authors found that the reduction can be performed under physiological conditions and even in living mammalian cells. The reaction uses thiols as reducing agents and tolerates, amongst others, water and air. However, the authors also noticed an *in vivo* metabolic reduction of aromatic azides, thus limiting this method. Applications of this new catalytic reaction to cellular imaging and catalytic drug release are currently investigated in the same laboratory [113].

11.6   Catalytically Controlled Carbon–Carbon Bond Formation

The possibility to form C–C bonds within a living biological system represents an exciting research area for the field of medicinal chemistry and organometallic chemistry, for instance to assemble a drug from two or more components at its final cellular destination. In this domain, palladium catalysis, probably inspired by the range of work done on palladium-catalysed cross-coupling reactions, represents the large majority of the metal catalysis in living biological systems reported so far. More than 20 years ago, palladium catalysis was first reported for the hydrogenation of membrane lipids of cyanobacteria [114]. Recently, palladium nanoparticles trapped within polystyrene microspheres were reported, with the ability to enter mammalian cells and catalyse an intracellular Suzuki–Miyaura cross-coupling reaction (Scheme 11.2) [115, 116]. Davis and coworkers have reported palladium-catalysed labelling of the cell surface of *Escherichia coli*, and Lin and coworkers recently reported a copper-free Sonogashira cross-coupling reaction in bacterial cells mediated by palladium catalysts [117, 118]. Interestingly, the palladium-catalyst was shown to be effective at low concentrations, for which no toxicity was observed [117]. Other metals, such as gold, also seem to be promising. For instance, Kim and coworkers have recently developed and applied a fluorescent gold sensor to the

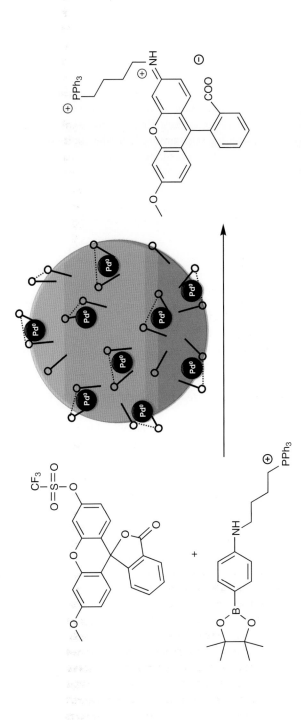

Scheme 11.2 *Intracellular Suzuki–Miyaura cross-coupling reaction using palladium nanoparticles trapped in polystyrene microspheres. Adapted from reference [115]*

fluorescence detection of gold ions in human keratinocyte cells. The formation of the final sensor is primarily based on a catalytic intramolecular hydroarylation [119].

11.7   Conclusion

For decades, DNA was believed to be the main target for designing biologically active metal-based drugs. However, as highlighted in this book chapter, it is evident that other targets can be envisaged for metal complexes. Enzymes are one of several potential targets and nowadays several research groups are studying the ability of metal complexes to inhibit enzymatic processes, using different strategies to generate metal-based inhibitors. Initiating catalytic reactions within cells is another approach for obtaining new biologically relevant metal complexes. All prospects are that adding selectivity and specificity to metal-based compounds are certainly the keys to the future development of bioinorganic chemistry.

References

1. C. Bolm, Cross-coupling reactions, *J. Org. Chem.*, **77**, 5221−5223 (2012).
2. P. K. Sasmal, C. N. Streu and E. Meggers, Metal complex catalysis in living biological systems, *Chem. Commun.*, **49**, 1581−1587 (2013).
3. E. Meggers, Targeting proteins with metal complexes, *Chem. Commun.*, 1001−1010 (2009).
4. J. F. Norman and T. W. Hambley, Targeting strategies for metal-based therapeutics, in E. Alessio (Ed.), *Bioinorganic Medicinal Chemistry*, Wiley-VCH, Weinheim, Germany, pp. 49−78 (2011).
5. J. Reedijk, Why does cisplatin reach guanine-N7 with competing S-donor ligands available in the cell? *Chem. Rev.*, **99**, 2499−2510 (1999).
6. S. J. Benkovic and S. Hammes-Schiffer, A perspective on enzyme catalysis, *Science*, **301**, 1196−1202 (2003).
7. K. J. Kilpin and P. J. Dyson, Enzyme inhibition by metal complexes: concepts, strategies and applications, *Chem. Sci.*, **4**, 1410−1419 (2013).
8. G. Gasser and N. Metzler-Nolte, Metal compounds as enzyme inhibitors, in E. Alessio (Ed.), *Bioinorganic Medicinal Chemistry*, Wiley-VCH, Weinheim, Germany, pp. 351−382 (2011).
9. A. Y. Louie and T. J. Meade, Metal complexes as enzyme inhibitors, *Chem. Rev.*, **99**, 2711−2734 (1999).
10. D. Rehder, The trigonal-bipyramidal NO_4 ligand set in biologically relevant vanadium compounds and their inorganic models, *J. Inorg. Biochem.*, **102**, 1152−1158 (2008).
11. M. Nakai, M. Obata, F. Sekiguchi, M. Kato, M. Shiro, A. Ichimura, I. Kinoshita, M. Mikuriya, T. Inohara, K. Kawabe, H. Sakurai, C. Orvig and S. Yano, Synthesis and insulinomimetic activities of novel mono- and tetranuclear oxovanadium(IV) complexes with 3-hydroxypyridine-2-carboxylic acid, *J. Inorg. Biochem.*, **98**, 105−112 (2004).

12. K. H. Thompson, J. Lichter, C. LeBel, M. C. Scaife, J. H. McNeil and C. Orvig, Vanadium treatment of type 2 diabetes: A view to the future, *J. Inorg. Biochem.*, **103**, 554–558 (2009).

13. W. H. Ang, I. Khalaila, C. S. Allardyce, L. Juillerat-Jeanneret and P. J. Dyson, Rational design of platinum(IV) compounds to overcome glutathione-S-transferase mediated drug resistance, *J. Am. Chem. Soc.*, **127**, 1382–1383 (2005).

14. S. Urig, K. Fritz-Wolf, R. Réau, C. Herold-Mende, K. Tóth, E. Davioud-Charvet and K. Becker, Undressing of phosphine gold(I) complexes as irreversible inhibitors of human disulfide reductases, *Angew. Chem. Int. Ed.*, **45**, 1881–1886 (2006).

15. Y. Che, B. R. Brooks, D. P. Riley, A. J. H. Reaka and G. R. Marshall, Engineering metal complexes of chiral pentaazacrowns as privileged reverse-turn scaffolds, *Chem. Biol. Drug Des.*, **69**, 99–110 (2007).

16. F. Mendes, M. Groessl, A. A. Nazarov, Y. O. Tsybin, G. Sava, I. Santos, P. J. Dyson and A. Casini, Metal-based inhibition of poly(ADP-ribose) polymerase – The guardian angel of DNA, *J. Med. Chem.*, **54**, 2196–2206 (2011).

17. R. H. Holm, P. Kennepohl and E. I. Solomon, Structural and Functional Aspects of Metal Sites in Biology, *Chem. Rev.*, **96**, 2239–2314 (1996).

18. P. M. Vignais and B. Billoud, Occurrence, classification, and biological function of hydrogenases: An overview, *Chem. Rev.*, **107**, 4206–4272 (2007).

19. N. Mitić, S. J. Smith, A. Neves, L. W. Guddat, L. R. Gahan and G. Schenk, The catalytic mechanisms of binuclear metallohydrolases, *Chem. Rev.*, **106**, 3338–3363 (2006).

20. L. R. Gahan, S. J. Smith, A. Neves and G. Schenk, Phosphate ester hydrolysis: Metal complexes as purple acid phosphatase and phosphotriesterase analogues, *Eur. J. Inorg. Chem.*, 2745–2758 (2009).

21. D. E. Wilcox, Binuclear metallohydrolases, *Chem. Rev.*, **96**, 2435–2458 (1996).

22. C. Belle and J.-L. Pierre, Asymmetry in bridged binuclear metalloenzymes: Lessons for the chemist, *Eur. J. Inorg. Chem.*, 4137–4146 (2003).

23. E. Jabri, M. B. Carr, R. P. Hausinger and P. A. Karplus, The crystal structure of urease from Klebsiella aerogenes, *Science*, **268**, 998 (1995).

24. C. G. Hartinger, A. D. Phillips and A. A. Nazarov, Polynuclear ruthenium, osmium and gold complexes. The quest for innovative anticancer chemotherapeutics, *Curr. Topics Med. Chem*, **11**, 2688–2702 (2011).

25. P. Govender, B Therrien and G. S. Smith, Bio-metallodendrimers - Emerging strategies in metal-based drug design, *Eur. J. Inorg. Chem.*, 2853–2862 (2012).

26. G. S. Smith and B. Therrien, Targeted and multifunctional arene ruthenium chemotherapeutics, *Dalton Trans.*, **40**, 10793–10800 (2011).

27. C. E. Müller, J. Iqbal, Y. Baqi, H. Zimmermann, A. Röllich and H. Stephan, Polyoxometalates−a new class of potent ecto-nucleoside triphosphate diphosphohydrolase (NTPDase) inhibitors, *Bioorg. Med. Chem. Lett.*, **16**, 5943–5947 (2006).

28. D. A. Judd, J. H. Nettles, N. Nevins, J. P. Snyder, D. C. Liotta, J. Tang, J. Ermolieff, R. F. Schinazi and C. L. Hill, Polyoxometalate HIV-1 protease inhibitors. A new mode of protease inhibition, *J. Am. Chem. Soc.*, **123**, 886–897 (2001).

29. M. T. Pope and A. Müller, Chemistry of polyoxometallates. Actual variation on an old theme with interdisciplinary references, *Angew. Chem. Int. Ed.*, **30**, 34–48 (1991).

30. M. T. Pope; *Heteropoly and Isopoly Oxometalates*, Springer, New York, 1983.
31. H. Stephan, M. Kubeil, F. Emmerling and C. E. Müller, Polyoxometalates as versatile enzyme inhibitors, *Eur. J. Inorg. Chem.*, 1585−1594 (2013) and references therein.
32. T. L. Turner, V. H. Nguyen, C. C. McLauchlan, Z. Dymon, B. M. Dorsey, J. D. Hooker and M. A. Jones, Inhibitory effects of decavanadate on several enzymes and Leishmania tarentolae In Vitro, *J. Inorg. Biochem.*, **108**, 96−104 (2012).
33. J. D. Foster, S. E. Young, T. D. Brandt and R. C. Nordlie, Tungstate: a potent inhibitor of multifunctional glucose-6-phosphatase, *Arch. Biochem. Biophys.*, **354**, 125−132 (1998).
34. M. J. Wall, G. Wigmore, J. Lopatar, B. G. Frenguelli and N. Dale, The novel NTP-Dase inhibitor sodium polyoxotungstate (POM-1) inhibits ATP breakdown but also blocks central synaptic transmission, an action independent of NTPDase inhibition, *Neuropharmacology*, **55**, 1251−1258 (2008).
35. F. M. Sansom, P. Riedmaier, H. J. Newton, M. A. Dunstone, C. E. Müller, H. Stephan, E. Byres, T. Beddoe, J. Rossjohn, P. J. Cowan, A. J. F. d'Apice, S. C. Robson and E. L. Hartland, Enzymatic properties of an ecto-nucleoside triphosphate diphosphohydrolase from Legionella pneumophila, *J. Biol. Chem.*, **283**, 12909−12918 (2008).
36. R. Prudent, V. Moucadel, B. Laudet, C. Barette, L. Lafanechère, B. Hasenknopf, J. Li, S. Bareyt, E. Lacôte, S. Thorimbert, M. Malacria, P. Gouzerh and C. Cochet, Identification of polyoxometalates as nanomolar noncompetitive inhibitors of protein kinase CK2, *Chem. Biol.*, **15**, 683−692 (2008).
37. D. W. Boyd, K. Kustin and M. Niwa, Do vanadate polyanions inhibit phosphotransferase enzymes? *Biochim. Biophys. Acta*, **827**, 472−475 (1985).
38. G. Choate and T. E. Mansour, Subunit structure of functional porin oligomers that form permeability channels in the outer membrane of Escherichia coli, *J. Biol. Chem.*, **254**, 1457−1462 (1979).
39. R. Prudent, C. F. Sautel and C. Cochet, Structure-based discovery of small molecules targeting different surfaces of protein-kinase CK2, *Biochim. Biophys. Acta*, **1804**, 493−498 (2010).
40. A. Seko, T. Yamase and K. Yamashita, Polyoxometalates as effective inhibitors for sialyl- and sulfotransferases, *J. Inorg. Biochem.*, **103**, 1061−1066 (2009).
41. Z. X. Dong, R. K. Tan, J. Cao, Y. Yang, C. F. Kong, J. Du, S. Zhu, Y. Zhang, J. Lu, B. Q. Huang and S. X. Liu, Discovery of polyoxometalate-based HDAC inhibitors with profound anticancer activity in vitro and in vivo, *Eur. J. Med. Chem.*, **46**, 2477−2484 (2011).
42. Y. Inouye, Y. Tokutake, J. Kunihara, T. Yoshida, T. Yamase, A. Nakata and S. Nakamura, Suppressive effect of polyoxometalates on the cytopathogenicity of human immunodeficiency virus type 1 (HIV-1) in vitro and their inhibitory activity against HIV-1 reverse transcriptase, *Chem. Pharm. Bull.*, **40**, 805−807 (1992).
43. S. G. Sarafianos, U. Kortz, M. T. Pope and M. J. Modak, Mechanism of polyoxometalate-mediated inactivation of DNA polymerases: an analysis with HIV-1 reverse transcriptase indicates specificity for the DNA-binding cleft, *Biochem. J.*, **319**, 619−626 (1996).
44. C. Schoeberl, R. Boehner, B. Krebs, C. Mueller and A. Barnekow, A new polyoxometalate complex inhibits retrovirus encoded reverse transcriptase activity in vitro and in vivo, *Int. J. Oncol.*, **12**, 153−160 (1998).

45. J. M. Messmore and R. T. Raines, Decavanadate inhibits catalysis by ribonuclease A, *Arch. Biochem. Biophys.*, **381**, 25–30 (2000).
46. A. Bartholomeusz, E. Tomlinson, P. J. Wright, C. Birch, S. Locarnini, H. Weigold, S. Marcuccio and G. Holan, Use of a Flavivirus RNA-dependent RNA polymerase assay to investigate the antiviral activity of selected compounds, *Antiviral Res.*, **24**, 341–350 (1994).
47. D. Hu, C. Shao, W. Guan, Z. M. Su and J. Z. Sun, Studies on the interactions of Ti-containing polyoxometalates (POMs) with SARS-CoV 3CLpro by molecular modeling, *J. Inorg. Biochem.*, **101**, 89–94 (2007).
48. A. Flutsch, T. Schroeder, M. G. Grutter and G. R. Patzke, HIV-1 protease inhibition potential of functionalized polyoxometalates, *Bioorg. Med. Chem. Lett.*, **21**, 1162–1166 (2011).
49. J. A. F. Gamelas, H. M. Carapuça, M. S. Balula, D. V. Evtuguin, W. Schlindwein, F. G. Figueiras, V. S. Amaral and A. M. V. Cavaleiro, Synthesis and characterisation of novel ruthenium multi-substituted polyoxometalates: α,β-[SiW$_9$O$_{37}$Ru$_4$(H$_2$O)$_3$Cl$_3$]$^{7-}$, *Polyhedron*, **29**, 3066–3073 (2010).
50. E. L. -M. Wong, R. W. -Y. Sun, N. P. -Y. Chung, C. -L. S. Lin, N. Zhu and C. -M. Che, A mixed-valent ruthenium–oxo oxalato cluster Na$_7$[Ru$_4$(μ^3-O)$_4$(C$_2$O$_4$)$_6$] with potent anti-HIV activities, *J. Am. Chem. Soc.*, **128**, 4938–4939 (2006).
51. E. H. Blackburn, Telomeres and telomerase: their mechanisms of action and the effects of altering their functions, *FEBS Lett.*, **579**, 859–862 (2005).
52. T. R. Cech, Life at the end of the chromosome: telomeres and telomerase, *Angew. Chem., Int. Ed.*, **39**, 34–43 (2000).
53. E. K. Parkinson, R. F. Newbold and W. N. Keith, The genetic basis of human keratinocyte immortalisation in squamous cell carcinoma development: the role of telomerase reactivation, *Eur. J. Cancer*, **33**, 727–734 (1997).
54. J. L. Huppert and S. Balasubramanian, Prevalence of quadruplexes in the human genome, *Nucleic Acids Res.*, **33**, 2908–2916 (2005).
55. W. H. Zhou, N. J. Brand and L. M. Ying, G-quadruplexes-novel mediators of gene function, *J. Cardiovasc. Transl. Res.*, **4**, 256–270 (2011).
56. A. M. Zahler, J. R. Williamson, T. R. Cech and D. M. Prescott, Inhibition of telomerase by G-quartet DNA structures, *Nature*, **350**, 718–720s (1991).
57. D. Colangelo, A. Ghiglia, A. Ghezzi, M. Ravera, E. Rosenberg, F. Spada and D. Osella, Water-soluble benzoheterocycle triosmium clusters as potential inhibitors of telomerase enzyme, *J. Inorg. Biochem.*, **99**, 505–512 (2005).
58. E. Rosenberg and R. Kumar, New methods for functionalizing biologically important molecules using triosmium metal clusters, *Dalton Trans.*, **41**, 714–722 (2012).
59. X.-H. Zheng, Y.-F. Zhong, C.-P. Tan, L.-N. Ji and Z.-W. Mao, Pt(II) squares as selective and effective human telomeric G-quadruplex binders and potential cancer therapeutics, *Dalton Trans.*, **41**, 11807–11812 (2012).
60. J. Zhang, F. Zhang, H. Li, C. Liu, J. Xia, L. Ma, W. Chu, Z. Zhang, C. Chen, S. Li and S. Wang, Recent progress and future potential for metal complexes as anticancer drugs targeting G-quadruplex DNA, *Curr. Med. Chem.*, **19**, 2957–2975 (2012).
61. S. Shi, J. Liu, T. Yao, X. Geng, L. Jiang, Q. Yang, L. Cheng and L. N. Ji, Promoting the formation and stabilization of G-quadruplex by dinuclear RuII complex Ru$_2$(obip)L$_4$, *Inorg. Chem.*, **47**, 2910–2912 (2008).

62. T. Wilson, M. P. Williamson and J. A. Thomas, Differentiating quadruplexes: binding preferences of a luminescent dinuclear ruthenium(II) complex with four-stranded DNA structures, *Org. Biomol. Chem.*, **8**, 2617–2621 (2010).

63. L. Xu, D. Zhang, J. Huang, M. Deng, M. G. Zhang and X. Zhou, High fluorescence selectivity and visual detection of G-quadruplex structures by a novel dinuclear ruthenium complex, *Chem. Commun.*, **46**, 743–745 (2010).

64. L. Xu, G. L. Liao, X. Chen, C. -Y. Zhao, H. Chao and L. N. Ji, Trinuclear Ru(II) polypyridyl complexes as human telomeric quadruplex DNA stabilizers, *Inorg. Chem. Commun.*, **13**, 1050–1053 (2010).

65. J. Alzeer, B. R. Vummidi, P. J. C. Roth and N. W. Luedtke, Guadinium-modified phthalocyanines as high-affinity G-quadruplex fluorescent probes and transcriptional regulators, *Angew. Chem. Int. Ed.*, **48**, 9362–9365 (2009).

66. K. Suntharalingam, A. Łęczkowska, M. A. Furrer, Y. Wu, M. K. Kuimova, B. Therrien, A. J. P. White and R. Vilar, A cyclometallated platinum complex as a selective optical switch for quadruplex DNA, *Chem. Eur. J.*, **18**, 16277–16282 (1012).

67. B. Therrien, Drug delivery by water-soluble organometallic cages, *Top. Curr. Chem.*, **319**, 35–56 (2012).

68. W. H. Ang and P. J. Dyson, Classical and non-classical ruthenium-based anticancer drugs: towards targeted chemotherapy, *Eur. J. Inorg. Chem.*, 4003–4018 (2006).

69. C. S. Allardyce and P. J. Dyson, Ruthenium in medicine: Current clinical uses and future prospects, *Platinum Met. Rev.*, **45**, 62–69 (2001).

70. I. Kostova, Ruthenium complexes as anticancer agents, *Curr. Med. Chem.*, **13**, 1085–1107 (2006).

71. C. S. Allardyce, A. Dorcier, C. Scolaro and P. J. Dyson, Development of organometallic (organo-transition metal) pharmaceuticals, *Appl. Organomet. Chem.*, **19**, 1–10 (2005).

72. M. Galanski, V. B. Arion, M. A. Jakupec and B. K. Keppler, Recent developments in the field of tumor-inhibiting metal complexes, *Curr. Pharm. Des.*, **9**, 2078–2089 (2003).

73. H. M. R. Robinson, S. Bratlie-Thoresen, R. Brown and D. A. F. Gillespie, Chk1 is required for G2/M checkpoint response induced by the catalytic topoisomerase II inhibitor ICRF-193, *Cell Cycle*, **6**, 1265–1267 (2007).

74. C. L. Baird, M. S. Gordon, D. M. Andrenyak, J. F. Mareceki and J. E. Lindsley, The ATPase reaction cycle of yeast DNA topoisomerase II, *J. Biol. Chem.*, **276**, 27893–27898 (2001).

75. M. Chandra, A. N. Sahay, D. S. Pandey, R. P. Tripathi, J. K. Saxena, V. J. M. Reddy, M. C. Peurta and P. Valegra, Potential inhibitors of DNA topoisomerase II: ruthenium(II) poly-pyridyl and pyridyl-azine complexes, *J. Organomet. Chem.*, **689**, 2256–2267 (2004).

76. S. Sharma, S. K. Singh and D. S. Pandey, Ruthenium(II) polypyridyl complexes: Potential precursors, metalloligands, and Topo II inhibitors, *Inorg. Chem.*, **47**, 1179–1189 (2008).

77. T. Gianferrara, I. Bratsos and E. Alessio, A categorization of metal anticancer compounds based on their mode of action, *Dalton Trans.*, **37**, 7588–7598 (2009).

78. E. S. Antonarakis and A. Emadi, Ruthenium-based chemotherapeutics: are they ready for prime time? *Cancer Chemother. Pharmacol.*, **66**, 1–9 (2010).

79. G. Gasser, I. Ott and N. Metzler-Nolte, Organometallic anticancer compounds, *J. Med. Chem.*, **54**, 3–25 (2011).

80. A. Bergamo and G. Sava, Ruthenium anticancer compounds: myths and realities of the emerging metal-based drugs, *Dalton Trans.*, **40**, 7817–7823 (2011).

81. E. A. Hillard and G. Jaouen, Bioorganometallics: Future trends in drug discovery, analytical chemistry, and catalysis, *Organometallics*, **30**, 20–27 (2011).

82. G. Sava, A. Bergamo and P. J. Dyson, Metal-based antitumour drugs in the post-genomic era: what comes next? *Dalton Trans.*, **40**, 9069–9075 (2011).

83. C. G. Hartinger, N. Metzler-Nolte and P. J. Dyson, Challenges and opportunities in the development of organometallic anticancer drugs, *Organometallics*, **31**, 5677–5685 (2012).

84. G. Sava, G. Jaouen, E. A. Hillard and A. Bergamo, Targeted therapy *vs.* DNA-adduct formation-guided design: thoughts about the future of metal-based anticancer drugs, *Dalton Trans.*, **41**, 8226–8234 (2012).

85. A. L. Noffke, A. Habtemariam, A. M. Pizarro and P. J. Sadler, Designing organometallic compounds for catalysis and therapy, *Chem. Commun.*, **48**, 5219–5246 (2012).

86. A. Bergamo, C. Gaiddon, J. H. M. Schellens, J. H. Beijnen and G. Sava, Approaching tumour therapy beyond platinum drugs: Status of the art and perspectives of ruthenium drug candidates, *J. Inorg. Biochem.*, **106**, 90–99 (2012).

87. D. Westerhausen, S. Herrmann, W. Hummel and E. Steckhan, Formate-driven, non-enzymatic NAD(P)H regeneration for the alcohol dehydrogenase catalyzed stereoselective reduction of 4-phenyl-2-butanone, *Angew. Chem. Int. Ed.*, **31**, 1529–1531 (1992).

88. C. Wong, D. G. Drueckhammer and H. M. Sweers, Enzymatic vs. fermentative synthesis: thermostable glucose dehydrogenase catalyzed regeneration of NAD(P)H for use in enzymatic synthesis, *J. Am. Chem. Soc.*, **107**, 4028–4031 (1985).

89. U. Kragl, D. Vasic-Racki and C. Wandrey, Continuous production of L-*tert*-leucine in series of two enzyme membrane reactors, *Bioprocess Eng.*, **14**, 291–297 (1996).

90. P. S. Wagenknecht and E. J. Sambriski, *Recent Res. Dev. Inorg. Chem.*, **3**, 35–50 (2003).

91. R. E. Morris, R. E. Aird, P. del S. Murdoch, H. Chen, J. Cummings, N. D. Hughes, S. Parsons, A. Parkin, G. Boyd, D. I. Jodrell and P. J. Sadler, Inhibition of cancer cell growth by ruthenium(II) arene complexes, *J. Med. Chem.*, **44**, 3616–3621 (2001).

92. R. Fernández, M. Melchart, A. Habtemariam, S. Parsons and P. J. Sadler, Use of chelating ligands to tune the reactive site of half-sandwich ruthenium(II)–arene anticancer complexes, *Chem. Eur. J.*, **10**, 5173–5179 (2004).

93. Y. K. Yan, M. Melchart, A. Habtemariam and P. J. Sadler, Organometallic chemistry, biology and medicine: ruthenium arene anticancer complexes, *Chem. Commun.*, **38**, 4764–4776 (2005).

94. F. Wang, A. Habtemariam, E. P. L. van der Geer, R. Fernández, M. Melchart, R. J. Deeth, R. Aird, S. Guichard, F. P. A. Fabbiani, P. Lozano-Casal, I. D. H. Oswald, D. I. Jodrell, S. Parsons and P. J. Sadler, Controlling ligand substitution reactions of organometallic complexes: tuning cancer cell cytotoxicity, *Proc. Natl. Acad. Sci. U.S.A.*, **102**, 18269–18274 (2005).

95. Y. K. Yan, M. Melchart, A. Habtemariam, A. F. A. Peacock and P. J. Sadler, Catalysis of regioselective reduction of NAD⁺ by ruthenium(II) arene complexes under biologically relevant conditions, *J. Biol. Inorg. Chem.*, **11**, 483–488 (2006).

96. H. C. Lo, C. Leiva, O. Buriez, J. B. Kerr, M. M. Olmstead and R. H. Fish, Bioorganometallic chemistry. 13. Regioselective reduction of NAD⁺ models, 1-benzylnicotinamde triflate and β-nicotinamide ribose-5′-methyl phosphate, with in situ generated [Cp*Rh(bpy)H]⁺: Structure–activity relationships, kinetics, and mechanistic aspects in the formation of the 1,4-NADH derivatives, *Inorg. Chem.*, **40**, 6705–6716 (2001).

97. S. Betanzos-Lara, Z. Liu, A. Habtemariam, A. M. Pizarro, B. Qamar and P. J. Sadler, Organometallic ruthenium and iridium transfer-hydrogenation catalysts using coenzyme NADH as a cofactor, *Angew. Chem. Int. Ed.*, **51**, 3897–3900 (2012).

98. J. J. Soldevila-Barreda, P. C. A. Bruijnincx, A. Habtemariam, G. J. Clarkson, R. J. Deeth and P. J. Sadler, Improved catalytic activity of ruthenium–arene complexes in the reduction of NAD⁺, *Organometallics*, **31**, 5958–5967 (2012).

99. F. Wang, J. Xu, A. Habtemariam, J. Bella and P. J. Sadler, Competition between glutathione and guanine for a ruthenium(II) arene anticancer complex: detection of a sulfenato intermediate, *J. Am. Chem. Soc.*, **127**, 17734–17743 (2005).

100. H. Petzold and P. J. Sadler, Oxidation induced by the antioxidant glutathione (GSH), *Chem. Commun.*, **37**, 4413–4415 (2008).

101. H. Petzold, J. Xu and P. J. Sadler, Metal and ligand control of sulfenate reactivity: arene ruthenium thiolato-mono-*s*-oxides, *Angew. Chem., Int. Ed.*, **47**, 3008–3011 (2008).

102. F. Wang, J. Xu, K. Wu, S. K. Weidt, C. L. Mackay, P. R. R. Langridge-Smith and P. J. Sadler, Competition between glutathione and DNA oligonucleotides for ruthenium(II) arene anticancer complexes, *Dalton Trans.*, **42**, 3188–3195 (2013).

103. W. Hu, Q. Luo, X. Ma, K. Wu, J. Liu, Y. Chen, S. Xiong, J. Wang, P. J. Sadler and F. Wang, Arene control over thiolate to sulfinate oxidation in albumin by organometallic ruthenium anticancer complexes, *Chem. Eur. J.*, **15**, 6586–6594 (2009).

104. S. J. Dougan, A. Habtemariam, S. E. McHale, S. Parsons and P. J. Sadler, Catalytic organometallic anticancer complexes, *Proc. Natl. Acad. Sci. USA*, **105**, 11628–11633 (2008).

105. N. Satoh, N. Watanabe, A. Kanda, M. Sugaya-Fukasawa and H. Hisatomi, Expression of glutathione reductase splice variants in human tissues, *Biochem. Genet.*, **48**, 816–821 (2010).

106. F. Giannini, G. Süss-Fink and J. Furrer, Efficient oxidation of cysteine and glutathione catalyzed by a dinuclear areneruthenium trithiolato anticancer complex, *Inorg. Chem.*, **50**, 10552–10554 (2011).

107. F. Giannini, J. Furrer, A.-F. Ibao, G. Süss-Fink, B. Therrien, O. Zava, M. Baquie, P. J. Dyson and P. Šepnička, Highly cytotoxic trithiophenolatodiruthenium complexes of the type [(η⁶-*p*-MeC₆H₄Pr^i)₂Ru₂(SC₆H₄-*p*-X)₃]⁺: synthesis, molecular structure, electrochemistry, cytotoxicity, and glutathione oxidation potential, *J. Biol. Inorg. Chem.*, **17**, 951–960 (2012).

108. F. Giannini, L. E. H. Paul and J. Furrer, Insights into the mechanism of action and cellular targets of ruthenium complexes from NMR spectroscopy, *Chimia*, **66**, 775–780 (2012).

109. S. M. Meier, M. Hanif, W. Kandioller, B. K. Keppler and C. G. Hartinger, Biomolecule binding *vs.* anticancer activity: reactions of Ru(arene)[(thio)pyr-(id)one] compounds with amino acids and proteins, *J. Inorg. Biochem.*, **108**, 91–95 (2012).

110. T. Bugarcic, A. Habtemariam, R. J. Deeth, F. P. A. Fabbiani, S. Parsons and P. J. Sadler, Ruthenium(II) arene anticancer complexes with redox-active diamine ligands, *Inorg. Chem.*, **48**, 9444–9453 (2009).

111. C. N. Streu and E. Meggers, Ruthenium-induced allylcarbamate cleavage in living cells, *Angew. Chem., Int. Ed.*, **45**, 5645–5648 (2006).

112. P. K. Sasmal, S. Carregal-Romero, W. J. Parak and E. Meggers, Light-triggered ruthenium-catalyzed allylcarbamate cleavage in biological environments, *Organometallics*, **31**, 5968–5970 (2012).

113. P. K. Sasmal, S. Carregal-Romero, A. A. Han, C. N. Streu, Z. Lin, K. Namikawa, S. L. Elliott, R. W. Köster, W. J. Parak and E. Meggers, Catalytic azide reduction in biological environments, *ChemBioChem*, **13**, 1116–1120 (2012).

114. L. Vigh, F. Joó and Á. Cséplő, Modulation of membrane fluidity in living protoplasts of *Nicotiana plumbaginifolia* by catalytic hydrogenation, *Eur. J. Biochem.*, **146**, 241–244 (1985).

115. R. M. Yusop, A. Unciti-Broceta, E. M. V. Johansson, R. M. Sánchez-Martín and M. Bradley, Palladium-mediated intracellular chemistry, *Nat. Chem.*, **3**, 239–243 (2011).

116. A. Unciti-Broceta, E. M. V. Johansson, R. M. Yusop, R. M. Sánchez-Martín and M. Bradley, Synthesis of polystyrene microspheres and functionalization with Pd(0) nanoparticles to perform bioorthogonal organometallic chemistry in living cells, *Nat. Protocols*, **7**, 1207–1218 (2012).

117. C. D. Spicer, T. Triemer and B. G. Davis, Palladium-mediated cell-surface labeling, *J. Am. Chem. Soc.*, **134**, 800–803 (2012).

118. N. Li, R. K. V. Lim, S. Edwardraja and Q. Lin, Copper-free Sonogashira cross-coupling for functionalization of alkyne-encoded proteins in aqueous medium and in bacterial cells, *J. Am. Chem. Soc.*, **133**, 15316–15319 (2011).

119. J. H. Do, H. N. Kim, J. Yoon, J. S. Kim and H.-J. Kim, A rationally designed fluorescence turn-on probe for the gold(III) ion, *Org. Lett.*, **12**, 932–934 (2010).

12

Other Applications of Metal Complexes in Chemical Biology

Tanmaya Joshi, Malay Patra and Gilles Gasser
Department of Chemistry, University of Zurich, Switzerland

12.1 Introduction

In addition to the applications presented in the first 11 chapters of this book, metal complexes also find other important uses in chemical biology. However, these applications are either still in their infancy or have been recently reviewed in detail elsewhere. Therefore, in this relatively short chapter, our aim is to present these other applications of metal complexes in chemical biology as well as highlighting their potential. As has been the case throughout this book, we will emphasize the specificity that metal complexes can provide to this field of research.

12.2 Surface Immobilization of Proteins and Enzymes

Proteins and enzymes have been and continue to be immobilized on surfaces for applications in different areas of research, such as bionanotechnology, biomedicine, biosensing, biocatalysis, and biofuel cells [1–4]. To this aim, several immobilization techniques have been employed [1–7]. However, covalent attachment of the enzyme/protein to the solid support using cross-linking agents (e.g., glutaraldehyde, carbodiimide derivatives, etc.) or physisorption methods [5] (e.g., adsorption into the material pores) are the most popular methods. The covalent immobilization of the protein or enzyme prevents its leaching, the technique thus presenting an advantage over the physisorption method. Nevertheless, alongside this is a drawback that the synthetic conditions employed for

Inorganic Chemical Biology: Principles, Techniques and Applications, First Edition. Edited by Gilles Gasser.
© 2014 John Wiley & Sons, Ltd. Published 2014 by John Wiley & Sons, Ltd.

covalent immobilization can potentially induce conformational changes in the surface bound enzyme and thus decreasing the enzyme activity compared with the in-solution activity [1]. In order to avoid these particular problems and to therefore obtain an "ideal" enzyme immobilization (e.g., no perturbation of the structure of the enzyme as well as no hindered diffusion of the substrate and product to and from the active site), alternative approaches were investigated. Interestingly, a method using metal complexes was found to allow immobilization of enzymes and proteins on different surfaces including, among others, mesoporous silicates (MPS) [8–10], agarose beads [11], a polyvinylpyrrolidone-based matrix [12] or gold [13–17]. At this stage, we invite the interested readers to also look at Chapter 1 of this book, which presents, in detail, the immobilized metal ion chromatography (IMAC) technique, an extremely elegant method to purify enzymes and proteins, in particular, using immobilized metal complexes.

To immobilize a protein or an enzyme, generally speaking, the surface is first synthetically modified from its original form to introduce sites where the metal complex can be attached. After metal complex attachment this modified surface thus acts as a template for binding with specific protein sites. To illustrate the concept of using metal complexes for enzyme/protein immobilization, we will present, as an example, the deposition of a small protease inhibitor (Spi) from the human pathogen *Streptococcus pyogenes* on MPS [8]. The mesoporous silicate surface, generated from a non-ionic surfactant template (SBA-15), was first functionalized with 3-iodo-trimethoxypropylsilane (Fig. 12.1). The iodo groups on the silicate surface were then reacted with 1,4,8,11-tetraazacyclotetradecane (cyclam) to give the cyclam-modified SBA-15. The latter could be then complexed with Ni^{2+} to give SBA-15-Ni–cyclam [8]. Interestingly, the entire preparation of SBA-15-Ni-cyclam can be conveniently followed by observation of the colour changes of the (modified) MPS, as described in Fig. 12.1 [8]. The authors could then immobilize the enzyme, His-tagged Spi in this case, on SBA-15-Ni-cyclam. As in the IMAC technique, the protein to be deposited needs to contain a His-tag to allow metal coordination. Moreover, several parameters such as pH, ionic strength, and protein concentration must be adjusted to avoid non-specific interactions of the proteins on the surface (e.g., hydrophobic interactions). These workers noticed that the addition of PEG_{400} (PEG = polyethylene glycol) during the incubation was

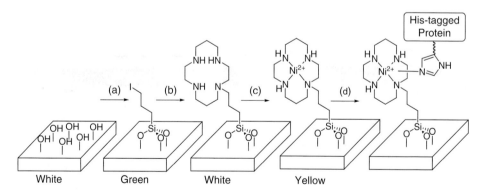

Figure 12.1 *Schematic diagram of the immobilization of a protein on metal-modified mesoporous silicates. The color of the (modified) silicates is also indicated. (a) $I(CH_2)_3Si(OCH_3)_3$, Toluol (b) cyclam, K_2CO_3, CH_3CN (c) $NiCl_2$, H_2O (d) His_6-Spi, buffer, PEG400. Reprinted with permission from [1] and [8] © 2013 and 2010 Royal Society of Chemistry*

essential to abolish non-specific binding of His-tagged Spi to unfunctionalized SBA-15. Using a similar approach, the same group could also successfully immobilize His-tagged alanine racemase [18] and His-tagged *Candida antartica* lipase B [18] on MPS.

In addition to the approach just described, which takes advantage of very specific metal coordination affinity of amino acid side chains to deposit proteins and enzyme onto the surface, the immobilization of proteins and peptides onto gold surfaces functionalized with supramolecular host molecules was recently achieved by Kim, Brunsveld, Jonkheijm and their coworkers through *supramolecular host–guest interactions* with organometallic compounds [13–17]. As recently discussed in two different reviews [4, 19], the advantage of this strategy, as in the metal coordination binding approach, is that it allows a reversible immobilization since the proteins/peptides are non-covalently bound to the surface.

More specifically, these groups employed the ferrocenyl derivative–cucurbit[7]uril host–guest system developed by Kaifer and coworkers to achieve the immobilization [20–22]. This host–guest system is based on the ability of the doughnut-shaped (also sometimes described as pumpkin-shaped!) symmetric host molecule cucurbit[7]uril (CB[7], Fig. 12.2) to recognize hydrophobic moieties such as ferrocenyl (Fc) derivatives in water with high affinity (in the nano- to picomolar range!) [4]. Compared with the other better known and more widely used "biotin-avidin–streptavidin" host–guest system, the advantage of the Fc/CB[7] recognition motif is that it is small, bioorthogonal, not denaturated by organic solvents or at elevated temperatures, and that the financial costs involved are lower [15].

With this in mind, the group working with Kim was the first to demonstrate that glucose oxidase unspecifically labeled with Fc (an average of 19 Fc moieties per protein) could indeed be anchored onto a gold surface, which was prefunctionalized with sulfur-containing CB[7] derivatives [15]. Interestingly, they demonstrated the use of this

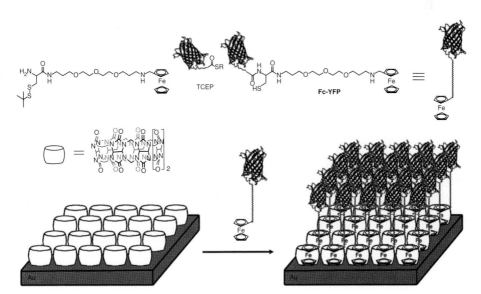

Figure 12.2 *Ligation of a ferrocene–cysteine derivative with YFP and immobilization of the resulting Fc–YFP onto a CB[7] monolayer. Reproduced with permission from [14] © 2010 Wiley-VCH Verlag GmbH & Co. KGaA, Weinheim*

system as a glucose sensor [15]. Brunsveld, Jonkheijm, and coworkers went a step further and utilized the approach reported by Zhang and coworkers to selectively immobilize peptides or proteins on CB[7] monolayers efficiently self-assembled on gold surfaces, without any modification or special treatment of the CB[7] molecules [13, 14, 16, 23]. Contrary to Kim's work where the number of ferrocenyl moieties was not well defined, in this case peptide and protein bioconjugates containing either one [13, 14] or two [16] Fc moieties were used. As shown in Fig. 12.2 as an example, the yellow fluorescent protein (YFP) could also be labeled with an Fc derivative using native chemical ligation to form the **Fc–YFP** bioconjugate. The successful immobilization of **Fc–YFP** onto the CB[7] monolayer was confirmed by X-ray photoelectron (XPS) spectroscopy, infrared reflection absorption spectroscopy (IR-RAS), cyclic voltammetry (CV) and water contact angle (WCA) measurements [14]. Interestingly, these workers could pattern the immobilization of **Fc–YFP** (Fig. 12.3). Furthermore, as expected, the immobilization was found to be reversible. Indeed, upon addition of an excess of an Fc derivative ($FcCH_2NMe_3I$), **Fc–YFP** could be removed (Fig. 12.3) [14]. Significantly, immobilization could similarly be performed with a thio-ether functionalized β-cyclodextrin (βCD) host system instead of CB[7] [16].

In a follow up to their initial work on Fc-YFP [14], the same researchers demonstrated that this concept could be extended even further to control the orientation of the immobilized proteins [17]. Contrary to CB[7], complexation of Fc with βCD is strongly reduced upon oxidation of Fc to ferrocenium (Fc^+) [20, 21]. These workers employed this property to control the orientation of the immobilization of ferrocene-tagged proteins [17]. More specifically, as described in more detail in Fig. 12.4A, they generated five different Fc–YFP constructs, which have the possibility to either dimerize (**Fc–dYFPs**) or not (**Fc–eYFPs**). The successful immobilization and micro-patterning of the YFPs on the gold surface was confirmed by surface plasmon resonance (SPR) and fluorescence microscopy in combination with electrochemistry [17]. Last but not least, a more stable and reversible protein immobilization could be obtained by application of a dynamic covalent disulfide lock between two YFP proteins. In this way, a switch from monovalent to divalent ferrocene interactions with the βCD surface could be formed (Fig. 12.4B) [17].

Very recently, the same research groups further capitalized on these previous results to control cell adhesion on gold surfaces [13]. As shown in Fig. 12.5, after immobilization of

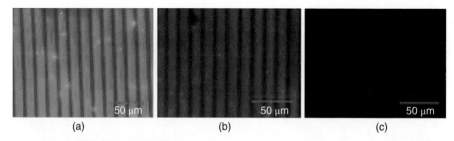

(a)	(b)	(c)

Figure 12.3   *FM images of **Fc-YFP** on a CB[7] SAM (A) directly after printing, (B) after 24 h of washing in phosphate buffer solution, and (C) after 24 h of washing in $FcCH_2NMe_3I$ (5 mM in phosphate buffer) solution. Reproduced with permission from [14] © 2010 Wiley-VCH Verlag GmbH & Co. KGaA, Weinheim*

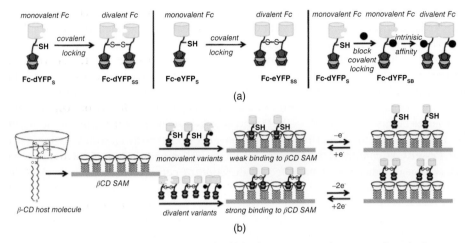

Figure 12.4 *(a) Covalent locking (via disulfide formation) and non-covalent locking (via intrinsic affinity) yield divalent Fc−YFP variants starting from their monovalent counterparts; two YFP variants are used, one prone to dimerizing (dYFP) and one suppressed from dimerization (enhanced YFP (eYFP)). (b) Thioether-functionalized βCD monolayers are self-assembled on gold slides, and variants of Fc-YFP are complexed to the βCD SAMs with different binding affinities; following oxidation and reduction of the Fc groups, the proteins can be desorbed and (re-)adsorbed, respectively. Abbreviations: SS, disulfide formation; S, free cysteine; SB, cysteine reacted with N-methylmaleimide). Reprinted with permission from [17] © 2012 American Chemical Society*

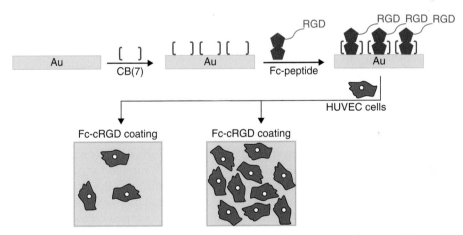

Figure 12.5 *Coating of gold surfaces with a CB[7] monolayer, pre-incubation with Fc−cRGD and reference compound Fc−cRAD, and subsequent endothelial cell adhesion to the supramolecularly functionalized surfaces [13]. Reprinted with permission from [13] © 2013 Royal Society of Chemistry*

ferrocene-containing integrin binding cyclic RGD (**Fc–cRGD**) peptide bioconjugates on CB[7] coated gold surfaces, adhesion of human umbilical vein endothelial cells (hUVECs) could be achieved [13]. As expected, the adhesion of hUVECS on CB[7] coated surfaces incubated with truncated cRGD peptide ferrocenyl bioconjugate **Fc–cRAD** was found to be less effective. The surfaces showed less spreading of cells and threefold less adhered cells compared with **Fc–cRGD** [13]. Of great interest is that it could be demonstrated, by a live–dead assay, that the hUVECs remain viable on the self-assembled CB[7] monolayer complexed with **Fc–cRGD** in comparison with control tissue culture plates (TCP) and fibronectin coated TCP [13]. Moreover, a wound assay on surfaces with CB[7] and **Fc–cRGD** confirmed a full recovery of the cell monolayer within 8 h [13]. This observation is indicative of the effective direction of cell growth by this immobilized supramolecular system.

12.3   Metal Complexes as Artificial Nucleases

The estimated half-life for uncatalyzed cleavage of the phosphate diester bond in deoxyribonucleic acid (DNA), under neutral conditions, is tens to hundreds of billions of years [24]. In ribonucleic acid (RNA), despite the presence of an internal nucleophile in the form of 2′-OH, the rate for this cleavage is still in the order of 10^{-10} s^{-1}, accounting for a half-life of over 100 years [24]. While it is essential for these naturally occurring phosphate ester bonds to maintain inertness towards their spontaneous degradation, it is equally necessary for resources to be available for these bonds to be cleaved in an efficient manner, as and when required. Hydrolysis of phosphate ester bonds is indeed vital for several cellular functions such as DNA repair, transcription, degradation of mutated DNA, signal transduction, and metabolism [25, 26]. Nature can competently perform the task of cleaving the phosphate esters with the help of enzymes known as *nucleases*. Most of these nucleases, specifically referred to as *metallonucleases*, contain metal ions (mainly Mg^{2+}, Ca^{2+}, Fe^{2+}, Mn^{2+} or Zn^{2+}) as cofactors [25, 26]. In metallonucleases, metal ion participation is believed to play a crucial role in promoting the enzymatically catalyzed phosphoester hydrolysis. However, beyond this, the exact role of metal ions in the catalytic cycle of hydrolysis has not yet been unraveled for all naturally occurring metallonucleases. Whilst obtaining a precise mechanistic understanding on the role of metal complexes is not a straightforward task, biomimicking the hydrolysis activity of natural metallonucleases is equally challenging.

  To derive a less complex solution to this two-dimensional puzzle, researchers have studied a number of low molecular weight metal complexes as synthetic nucleases, designed as simple structural and functional models for complex natural metallo-enzymes. Although still uncertain, the information obtained so far from these systematically conducted studies has helped enormously in the evolution of our understanding of the role of metal ions in the catalytic activity of nucleases. Research in this direction has, over past decades, produced a number of elegant synthetic metallonuclease systems [26, 28–34]. Development of metal complexes as phosphoesterases is also relevant for conformational probes, or therapeutic and footprinting agents, in view of biotechnological and molecular biology applications [25, 34]. A variety of metal complexes, employing metal ions from the transition metal series as well as a few selected lanthanide and actinide ions, have been investigated as synthetic nucleases. A cluster of reviews covering the topics of structural and functional relationships in these synthetic metallonucleases, the hurdles being faced in mimicking the catalytic activity of naturally occurring metallonucleases, and strategic attempts for overcoming these, have been published in recent years [25–31, 35–52].

A full discussion of biomimetic metallonucleases is outside the scope of this chapter; in this section our purpose is to outline current progress using selected examples. Impressive rates for phosphate ester hydrolysis have been obtained mainly from copper(II), zinc(II), and lanthanide complexes, and for this reason, we have restricted our discussion to these complexes. In principle, the rich coordination chemistry of these, and all metal ions in general, can be utilized for: (i) Lewis acid/electrostatic activation of the phosphodiester bond; (ii) facilitating nucleophilic attack, and stabilization of the transition state and the leaving group; and/or (iii) providing a powerful nucleophile at neutral pH to promote/assist hydrolysis (Fig. 12.6) [24, 35]. Thus, we direct our readers to the cited reviews for an extensive insight into the design and development of artificial metallonucleases. Even though the progress made so far towards the preparation of efficient artificial metallonucleases is impressive, there is still a remarkable difference of almost 4–5 orders of magnitude in activity between the synthetic and natural metallonucleases [25, 35]. This difference leaves ample scope for tuning of the coordination sphere and ligand environment of artificial metallonucleases in order to enhance their effectiveness. The naturally occurring metallonucleases cleave the phosphate ester bonds hydrolytically; therefore, to put things

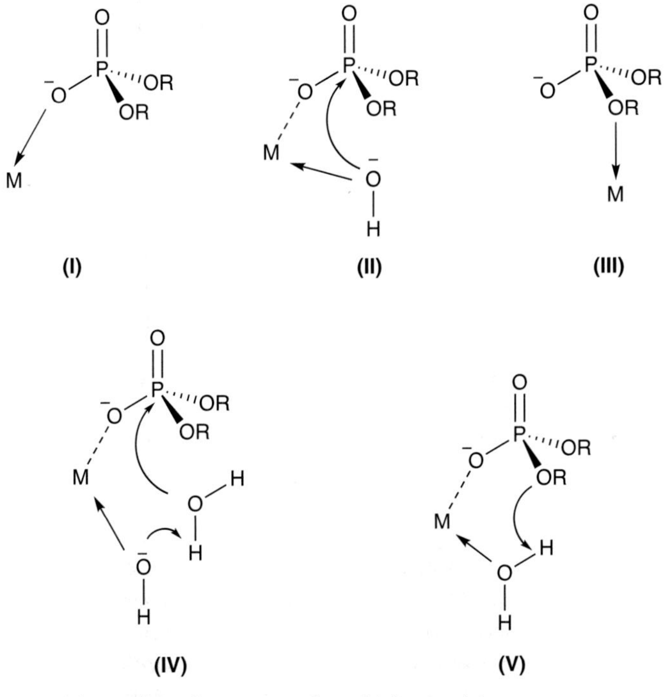

Figure 12.6 *Possible modes of activation by which metal ions can promote phosphoester hydrolysis: (I) Lewis acid activation; (II) nucleophile activation; (III) leaving group activation; (IV) coordination with hydroxides to form activated base; and (V) coordination with water molecules to form activated acids [24]*

into perspective, we have limited our discussion to model systems mediating hydrolytic cleavage. However, at this point, readers should note that an alternative mechanism for degradation of nucleic acids can also be oxidative or photo-induced, this being the subject of discussion in many excellent reviews [40, 53–56]. For a brief but pertinent discussion on photo-induced DNA damage, the readers are also referred to Chapter 6 of this book.

In the literature, a number of non-natural phosphate esters have been used as model substrates to probe the kinetics of phosphate ester bond cleavage by metal complexes [25]. These substrates have proven to be extremely useful models for a preliminary assessment on the cleavage activity of the metal complexes. However, it is beyond the scope of this book chapter to provide an in-depth analysis of all the studies done so far. Instead, to provide a concise overview of the mechanistic information gathered so far on the metallonuclease mediated phosphate ester bond cleavage, selective examples where either the oligonucleotide fragments of naturally occurring DNA or RNA sequences or the double-stranded plasmid DNA have been employed as substrates will be presented.

12.3.1 Mono- and Multinuclear Cu(II) and Zn(II) Complexes

Mechanistic revelations from the pioneering studies conducted by Burstyn and coworkers on phosphate ester cleavage activity of the Cu(II) complex of 1,4,7-triazacyclononane (tacn, **1** in Fig. 12.7) [28] have driven research towards the development of other macrocyclic complexes as cleavage agents. They established that, at near-physiological conditions, the Cu(II)–tacn complex is capable of cleaving both single-stranded and double-stranded DNA. In addition, it was also shown that the introduction of steric bulk onto the macrocyclic amine can result in an alteration of the DNA cleavage activity [28, 57]. Complementary studies conducted on a range of *N*-substituted Cu(II)–tacn complexes (Fig. 12.7) using model phosphate esters as substrates showed that the rate of hydrolysis is enhanced with an increase in steric bulk, in correlation with a decrease in the rate of formation of inactive hydroxo-bridged dimers in solution [25, 28].

Cu(II) and Zn(II) complexes of polypyridyl ligands have also been assessed for their phosphate ester cleavage ability. Cationic Cu(II) complexes of 2,2′-bipyridine (bpy, **12** in Fig. 12.8) derivatives bearing electropositive pendants (structures **13–19** in Fig. 12.8) also cleave DNA [58, 59]; the activity is dependent on the relative bulk of the incorporated pendant arms and their ability to enhance substrate binding through electrostatic and/or charge-assisted hydrogen bonding interactions.

The Cu(II)–bpy complexes Cu(II)–**18**, Cu(II)–**13**, and Cu(II)–**15**, bearing guanidinium, ammonium, and tetraalkylammonium functionalities, respectively, showed almost tenfold higher activities compared with the parent unmodified complex (at pH 7.2 and 37 °C, pseudo-Michaelis–Menten parameter $k_{cat} = 1.23 \times 10^{-3}$, 1.17×10^{-3}, and $1.15 \times 10^{-3}\,s^{-1}$, respectively). Similar studies on Zn(II) complexes Zn(II)–**18** and Zn(II)–**19** showed that they also exhibit nuclease activity, although their activity was lower than that of the Cu(II) counterparts, indicating metal ion influence [60]. A copper(II) complex of 2,2′,6′,2″-terpyridine (terpy, **20** in Fig. 12.8) was also found to be capable of hydrolytically cleaving the RNA substrate poly(A)$_{12-18}$ at neutral pH and at 37 °C with a pseudo-first order rate constant of $1.69 \times 10^{-5}\,s^{-1}$ [61]. Of particular relevance here is another study by Burstyn and coworkers [62] on RNA cleavage by Cu(II)–tacn complex **1**, which

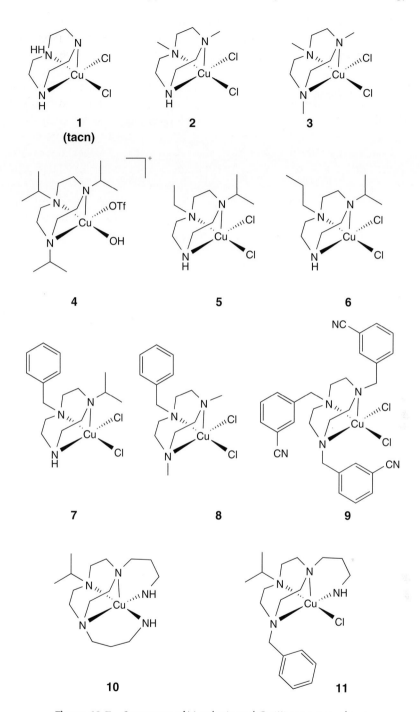

Figure 12.7 *Structures of N-substituted Cu(II)-tacn complexes*

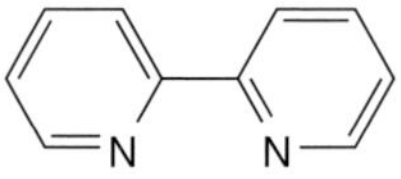

12
(bpy)

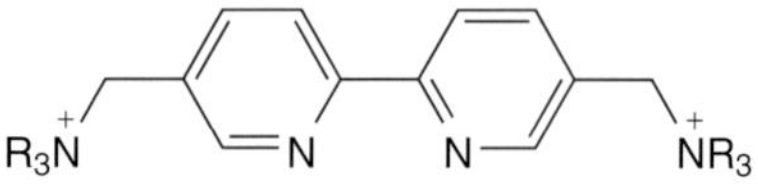

13 R = H
14 R = Me
15 R = Et
16 R = Bu

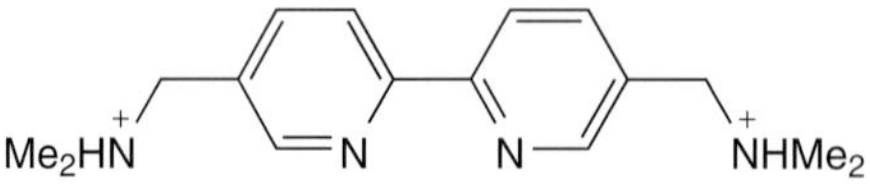

17

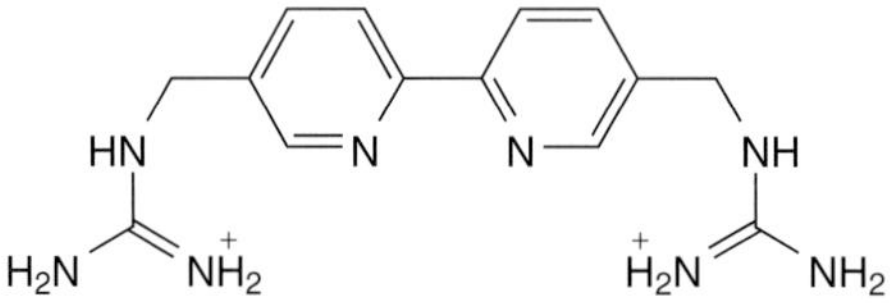

18

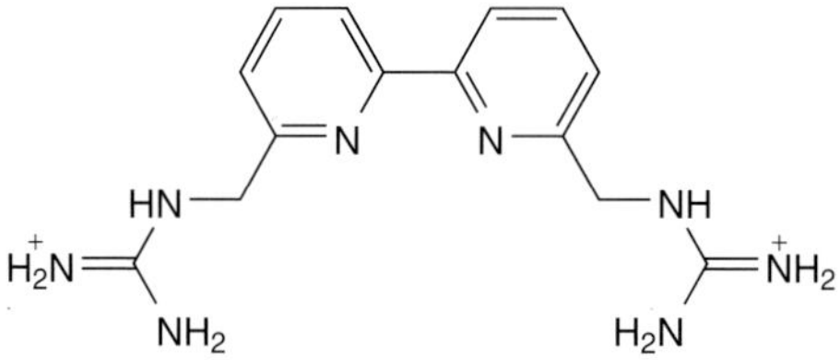

19

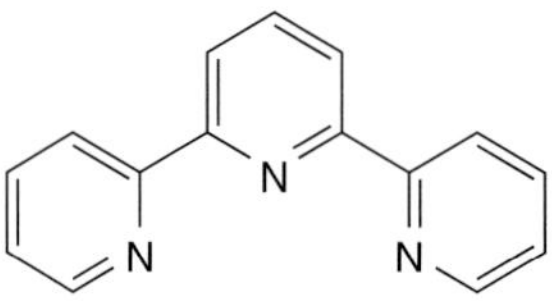

20
(terpy)

Figure 12.8 *Series of polypyridyl ligands explored as metallonuclease scaffolds*

was found to actively cleave both the single- and double-stranded RNA oligonucleotide, although the cleavage was less efficient compared with the Cu(II)–terpy complex [63].

In efforts to improve the cleavage activity of Cu(II) complexes of macrocyclic ligands, bifunctional cooperative catalysis has also been adapted. Staphylococcal nuclease (SNase) is an elegant example of the naturally occurring metallonucleases where DNA hydrolysis follows this approach. The Cu(II) complex of aza-oxo-crown **21** (Fig. 12.9), bearing two

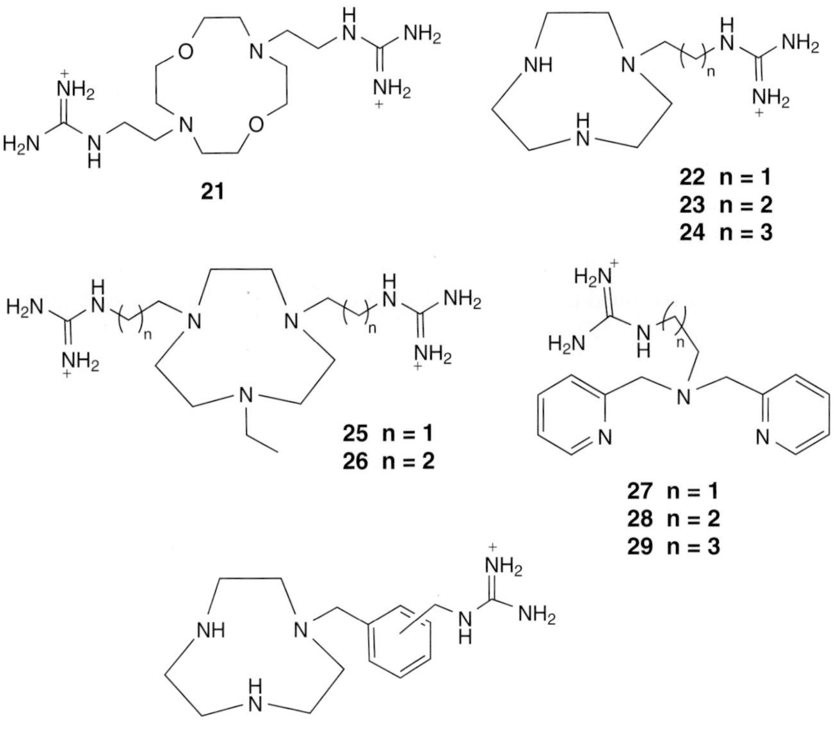

21

22 n = 1
23 n = 2
24 n = 3

25 n = 1
26 n = 2

27 n = 1
28 n = 2
29 n = 3

30 -o-xylyl
31 -m-xylyl
32 -p-xylyl

Figure 12.9 *Examples of ligands with guanidine pendants*

ethyl guanidinium pendant arms, was found to cleave DNA with a 10^8-fold enhancement in rate (at pH 7.5 and 37 °C, $k_{cat} = 1.65 \times 10^{-4}\ s^{-1}$) over the uncatalyzed supercoiled DNA cleavage [64]. Through separately conducted studies on a series of Cu(II) complexes of macrocyclic/non-macrocyclic ligands bearing one (or more) guanidinium pendant (**22–32**, Fig. 12.9), Spiccia, Graham, and coworkers have shown that the length and position of the guanidinium pendant can have significant influence on the cleavage activity of the complex [25].

Using alkyl spacers of varied lengths for appending the guanidinium moiety to bi*s*(2-pyridylmethyl)amine (dpa) or tacn backbone, Spiccia, Graham, and coworkers could observe a detrimental effect in DNA cleaving activity of the corresponding Cu(II) complexes as the possibility for guanidinium group to coordinate to the Cu(II) center increased. The coordination of guanidinium pendant to the Cu(II) center in Cu(II)–**22** and Cu(II)–**25** could also be confirmed by X-ray crystallography. Regardless of such Cu(II)–guanidinium coordination, the rate of DNA cleavage was found to be comparable to or faster than the parent complex, which suggests a degree of cooperation between the Cu(II) center and the guanidinium moiety in DNA phosphate ester hydrolysis [25].

In another case study, the Cu(II)–tacn complex Cu(II)–**30** with an *ortho*-xylyl linked guanidinium group was found to cleave pBR322 plasmid DNA approximately 22-times faster than the nonguanidinylated Cu(II) complex ($k_{obs} = 2.7 \times 10^{-4}$ and 1.2×10^{-5} s^{-1}, respectively at pH = 7.0 and 37 °C) [65]. Similarities between the distance of the guanidinium group from the copper(II) center in complex Cu(II)–**30** and that of adjacent phosphate ester groups in polyanionic DNA was also believed to partly contribute towards the activity enhancement. It was postulated that with these matching distances, the guanidinium pendant can be involved in a charge-assisted hydrogen bonding interaction with a phosphodiester group with favorable complex geometry, allowing hydroxide on Cu(II) center to be well positioned to carry out nucleophilic attack on the phosphate atom of the adjacent phosphodiester [65].

Studies have shown that Cu(II) and Zn(II) complexes of the 1,3,5-triaminocyclohexane (tach) ligand (see Fig. 12.10) can also promote hydrolysis of phosphate ester bonds in DNA [66, 67]. The rate constant for hydrolysis of phage DNA by Cu(II)–tach was 1.2×10^{-3} s^{-1} (pH 8.1 and 35 °C) while the Zn(II)–tach complex showed significantly less hydrolytic cleavage activity with a rate constant of 2.0×10^{-6} s^{-1} at pH 7.0 and 35 °C. Interestingly, the Cu(II) complex of the *trans,cis*-1,3,5-triaminocyclohexane (*t*-tach) ligand (structure **34**, Fig. 12.10), where one of the NH$_2$ groups is trans-positioned, can effectively cleave plasmid DNA at pH 7.0; the measured rate constant k_{obs} is 5.5×10^{-5} s^{-1} at 35 °C [68]. The copper(II) complex of a closely related all-*cis*-2,4,6-triamino-1,3,5-trihydroxycyclohexane (taci, structure **36** in Fig. 12.10) ligand mediates plasmid DNA cleavage at the rate of 2.3×10^{-3} s^{-1} [69], its efficient hydrolytic activity being attributed to the increased affinity of the complex towards the DNA substrate due to the presence of hydroxyl groups. With anthraquinone (a known DNA intercalator) tethered to triaminocyclohexane via a flexible spacer in **37–39** (see Fig.12.10), the corresponding Zn(II) complex Zn(II)–**39** showed a 15-fold increase in cleavage activity versus the unmodified Zn(II) complex, under physiological conditions [67].

Examples of Cu(II)–bpy complexes with appended nucleobases (structures **42–44**, Fig. 12.11), which exhibit impressive rates of pBR322 plasmid DNA cleavage, are also

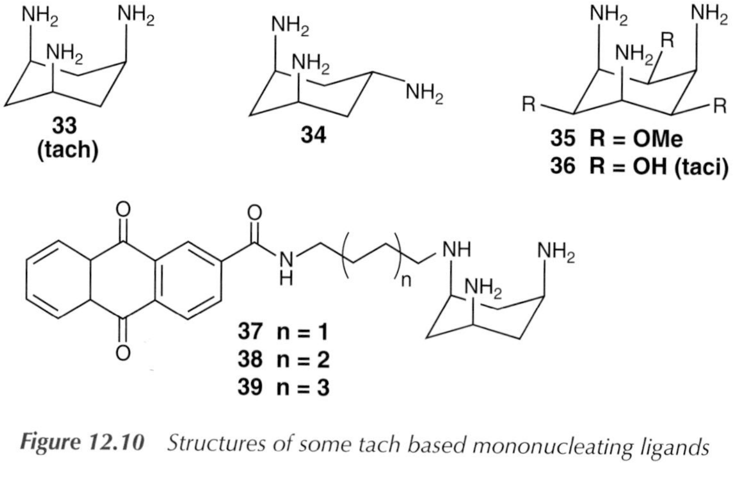

Figure 12.10   *Structures of some tach based mononucleating ligands*

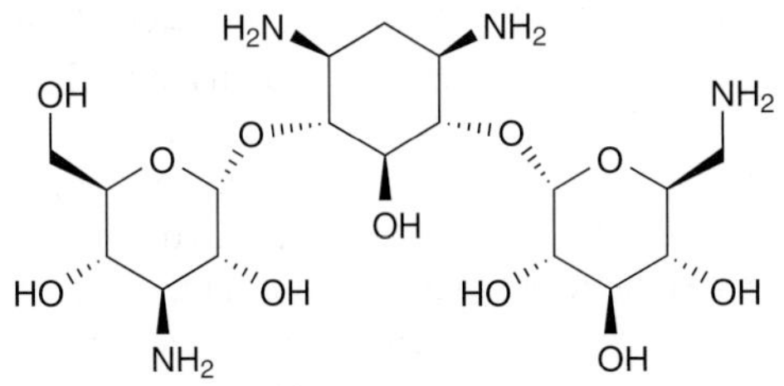

40
(Kanamycin)

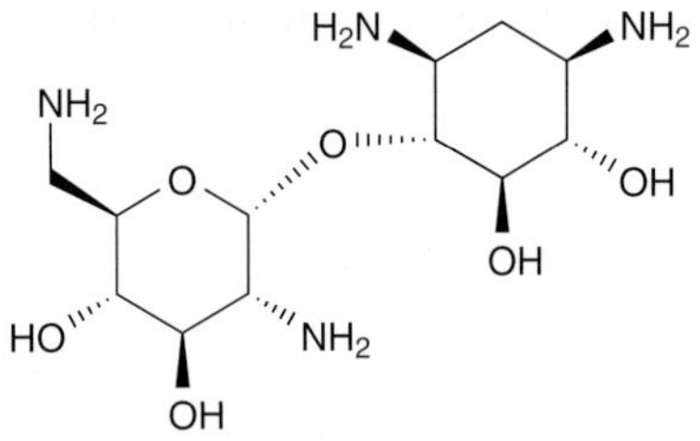

41
(Neamine)

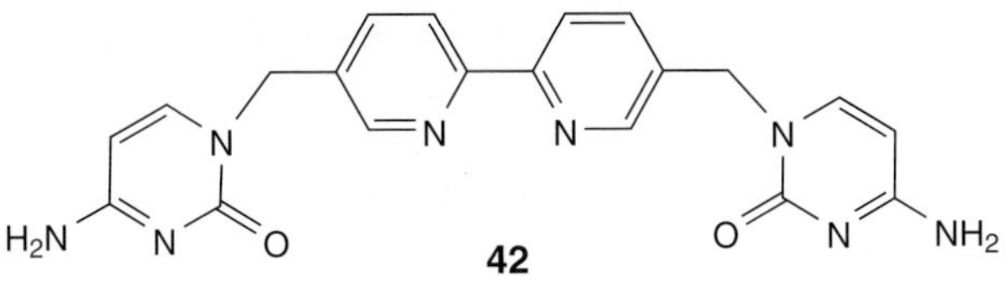

42

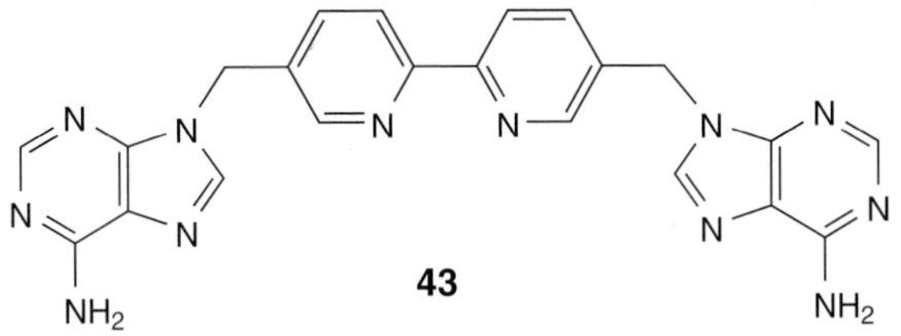

43

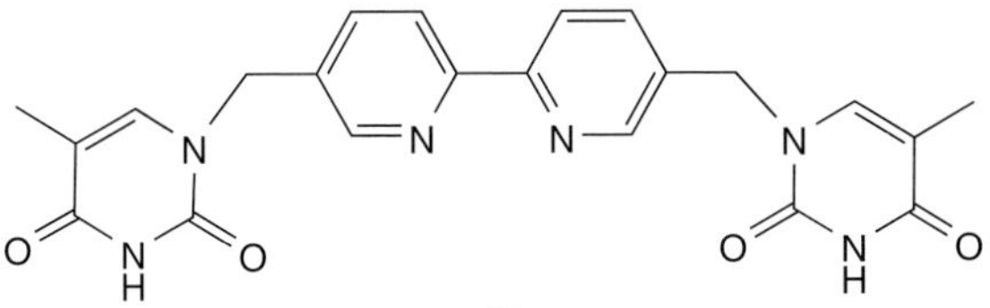

44

Figure 12.11 *Mononucleating ligands with nucleic acid affinity (**40–41**) or featuring nucleobase pendants (**42–44**)*

available [25]. Cu(II) complexes of aminoglycosides kanamycin and neamine (Fig. 12.11), naturally occurring RNA-binding antibiotics, have also shown the ability to perform catalyzed cleavage of DNA and RNA substrates [70].

In an attempt to improve the nuclease activity of metal complexes, multimetallic cooperativity has also been explored [24, 26, 30, 33–35, 37, 39, 42]. This, however, should not be a surprising choice if we consider that natural metallonucleases also employ more than one substrate–enzyme interaction to perform the cleavage. Multinuclear complexes based on tacn and other macrocycles have been routinely tested for their DNA cleaving ability (Fig. 12.12) [71–73]. Dinuclear Cu(II) and Zn(II) complexes of the bis(tacn) ligand **45** (Fig. 12.12) [72], for example, show the ability to hydrolyze the RNA construct GpppG with up to a 100-fold acceleration in rate for Cu(II)$_2$–**45** in comparison with

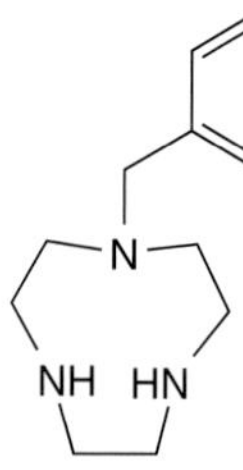

Figure 12.12 *Representative examples of macrocyclic bi- and tri-nucleating ligands*

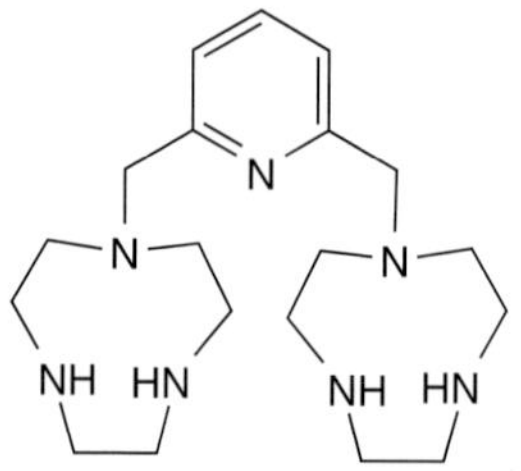

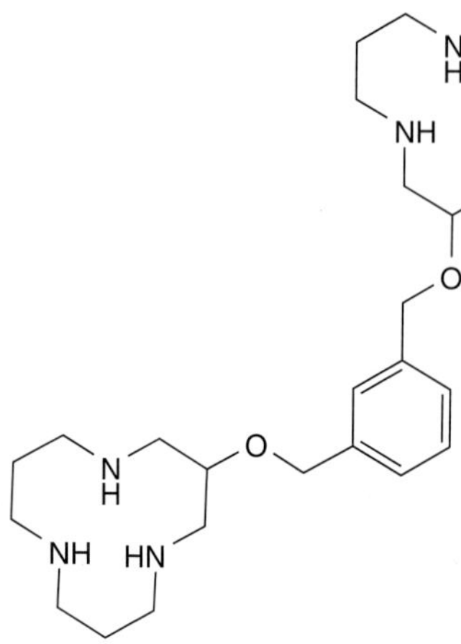

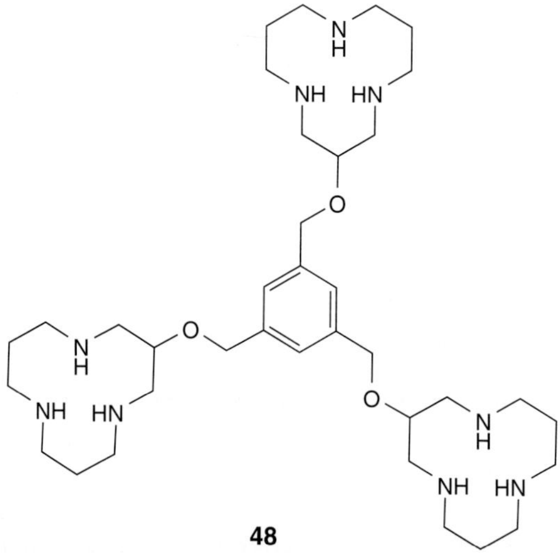

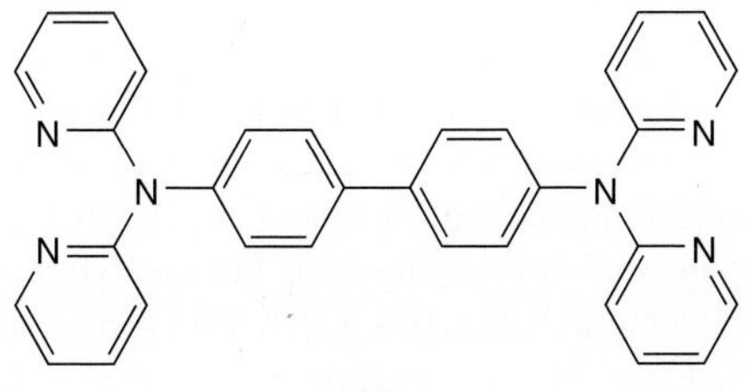

Figure 12.13 *Structures of pyridyl- or dpa- (bis(2-pyridylmethyl)amine) based bi- and tri-nu-cleating ligands*

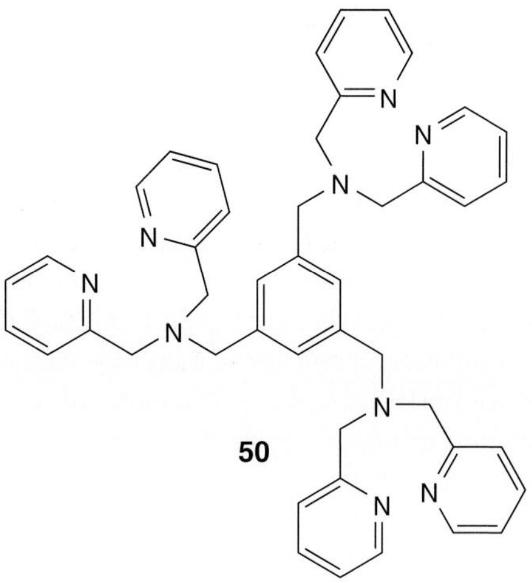

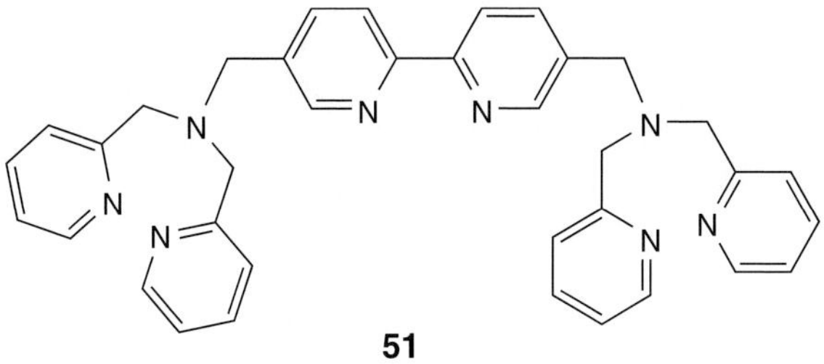

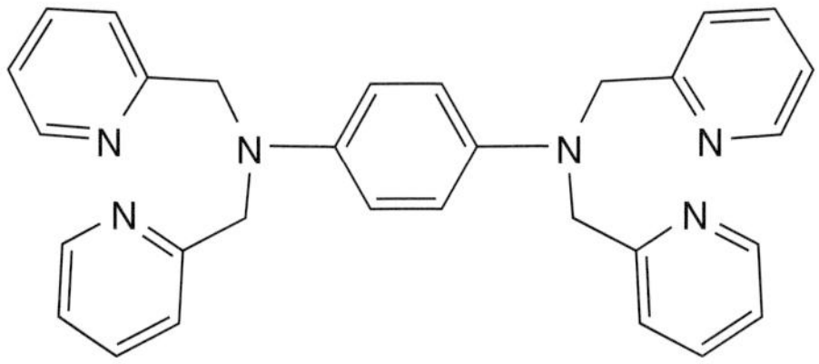

the unmodified mononuclear Cu(II)–tacn complex [72]. Experimental results validate a certain degree of bimetallic cooperativity in such systems [71]. It is important to note that the activity of such bimetallic systems is not always superior to their mononuclear analogues [73].

Examples of phosphoester cleaving binuclear and trinuclear Cu(II)/Zn(II) complexes built on the chelating dpa ligand (Fig. 12.13) are also available in the literature [74–76]. The multinuclear copper(II) complexes Cu(II)$_3$–**50**, Cu(II)$_3$–**51**, and Cu(II)$_2$–**52** have been observed to cleave plasmid DNA with a greater than 100-fold increment versus the mononuclear Cu(II)–dpa complex [74, 75].

12.3.2   Lanthanide Complexes

Hard Lewis acidity, high coordination number, fast ligand exchange kinetics, and absence of any inherent redox chemistry, generally speaking, make the lanthanide ions promising candidates for use as cofactors in phosphate ester hydrolysis [32, 35, 43, 45, 47]. Use of free lanthanide ions, however, is not ideal due to their limited (or poor) solubility and high toxicity. On the other hand, fast ligand exchange kinetics can be an obstacle in designing Ln(III) based highly active systems where the overall thermodynamic and kinetic stability of the complex is maintained [30, 32, 43]. Good fundamental knowledge on lanthanide coordination chemistry is therefore vital to design substitutionally inert but catalytically active Ln(III) complexes. Examples of trivalent lanthanides complexes as hydrolytic agents that are available in the literature generally include polyaminocarboxylates, macrocylic Schiff bases, crown ethers, or azacrown macrocycles as ligands (see Fig. 12.14).

A dicerium(IV) complex of a polyaminocarboxylate ligand, 5-methyl-2-hydroxy-1,3-xylene-α,α-diamine-N,N,N',N'-tetraacetic acid (HXTA, **54** in Fig. 12.14) was found to hydrolyze Litmus 29 plasmid DNA at $1.4 \times 10^{-4}\,s^{-1}$ at pH 8 and 37 °C, with double-strand scission being preferred [77]. Similarly, the experiments also showed that, at 55 °C, di(lanthanide) complexes $Ce_2(HPTA)$ and $La_2(HPTA)$ are also capable of performing double-strand DNA hydrolysis (HPTA = 1,3-diamino-2-hydroxypropane-N,N,N',N'-tetraacetate, structure **53** in Fig. 12.14) [78].

Mono- and bi-nuclear Ln(III) complexes of Schiff base macrocycles have been shown to be effective metallo(ribo)nucleases. Morrow *et al.* [47] showed that the Eu(III), Tb(III), Lu(III), and Gd(III) complexes of the hexadentate ligand **56** (Fig. 12.14), while being inert to metal release, showed promising catalytic activity for hydrolysis of the RNA oligomer $A_{12}-A_{18}$ (pseudo-first order rate constant $(k_{obs}) = 4.17 \times 10^{-4}\,s^{-1}$ for $[Eu(\mathbf{56})]^{3+}$). Similarly, binuclear Ho_2 and Er_2 complexes of the hydroxyl group containing the Schiff base macrocycle **57** (Fig. 12.14) have also been demonstrated to be active regarding the cleavage of double-stranded plasmid DNA into linear products [79].

Cleavage capacity of azacrown complexes of Eu^{3+} and Pr^{3+} ions have also been investigated by Schneider and coworkers [80, 81]. Whilst the inclusion of carboxylate arms on the ligand backbone in **58–61** (see Fig. 12.14) had no favorable effects on the catalytic activity of the respective mononuclear Eu^{3+}–azacrown ether complexes [80], structural modifications in the ligand framework in order to introduce a bimetallic cooperativity for cleavage in **62** (Fig. 12.15) [81] resulted in an approximately fourfold increase of activity. Working along the same lines, Janda and coworkers studied a series of lanthanide(III) complexes of azacrown ether based ligands (**63–66**, Fig. 12.14) for their ability to hydrolyze the model phosphate esters and plasmid DNA [82]. In general, Eu, Gd, or Tb based systems showed better cleavage activity among the screened lanthanide ions. Overall, the highest cleavage rate was obtained for the β-naphthyl containing complex Gd(III)–**65**. The k_{cat} calculated following a Michaelian kinetic model for double stranded DNA cleavage was $7.5 \times 10^{-3}\,s^{-1}$.

Ln(III) complexes of derivatives of DOTA (DOTA = 1,4,7,10-tetraazacyclododecane-1,4,7,10-tetraacetic acid), an azamacrocyclic ligand, show promising kinetic inertness towards in-solution ligand exchange [43]. Investigations on one such complex $[La(TCMC)]^{3+}$ (TCMC = 1,4,7,10-tetrakis(2-carbamoylmethyl)-1,4,7,10-tetraazacyclododecane, **67** in Fig. 12.14) showed it to cleave the RNA oligomer $A_{12}-A_{18}$ at the rate of $1.58 \times$

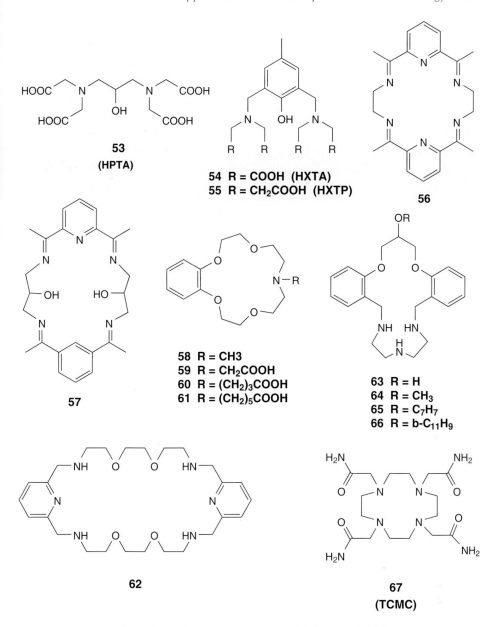

Figure 12.14   *Series of ligands used in lanthanide based artificial nucleases*

$10^{-4}\,s^{-1}$ [83]. Surprisingly, the La(III) ion when replaced with either Eu(III) or Dy(III) in $[Eu(TCMC)]^{3+}$ and $[Dy(TCMC)]^{3+}$, respectively, rendered the corresponding complexes inactive.

To conclude, from a detailed survey of the research in field of metallonucleases, it emerges that high reactivity and selective affinity towards the substrate are key to high efficiency synthetic metallonucleases. The issues of reactivity and affinity have, in general, been addressed through appropriate choice of the metal ions, cooperativity in multimetallic assemblies, and well-positioned inclusion of organic functional groups in the ligand design (for cooperative second sphere interactions with the metal center). To accommodate the issues of selectivity, conjugation of active metal complexes with biopolymers, such as complimentary oligonucleotides and peptide nucleic acid (PNA) oligomers (anti-sense approach) or sequence selective peptide/protein motifs, has been attempted [25, 32, 35, 41, 43, 84–87]. However, as a wide gap between the natural and artificial systems continues to exist, an important concern in recent times has been the rather critical but under-investigated contributory impact of the localized microenvironment of the active enzyme site [25, 35]. It is another crucial parameter for future consideration when designing metallonuclease constructs capable of establishing new benchmarks of catalytic efficiencies.

12.4   Cellular Uptake Enhancement Using Metal Complexes

Over recent decades, substantial progress has been made in the development of new strategies for efficient intracellular delivery of therapeutic molecules. It has been shown that a variety of medicinally/biologically important peptides, proteins [88, 89], oligonucleotides [90, 91], and nanoparticles [92, 93] can be transported successfully inside the cell using suitable carriers such as polymers [94, 95], lysosomes [96, 97] or cell penetrating peptides (CPPs) [98–101]. The strategy has also been successfully extended to the delivery of medicinally/biologically important metal complexes. It has been proven, with various examples, that not only can an enhanced cellular uptake be achieved but also an intracellular distribution of a metal complex can be formulated, if required, by attaching them to a suitable vector [102, 103]. In contrast, there are only a handful of examples testing the reverse scenario, that is, metal complexes being used to enhance the cellular uptake and tune the intracellular distribution of a biomolecule of interest. Although still in its infancy, the "proof of principle" has been demonstrated with a few metal peptide/peptide nucleic acid (PNA) conjugates.

For PNAs, their poor cellular uptake and endosomal entrapment are the main limitations that restrain their fast progress towards therapeutic applications as antisense and antigene agents. All around the world a tremendous amount of research has been devoted to developing new strategies for efficient intra-cellular delivery of PNAs [104]. In this context, PNAs conjugated with cationic peptides, fatty acids, and nanoparticles have shown improvement in the cellular uptake [104]. A newly introduced but rather promising approach for PNA delivery has been direct/indirect metallation of these synthetic oligonucleotides. The strategy relies on the conjugation or complexation of chemically modified PNAs with metal complexes or ions, respectively [105, 106].

Kramer and coworkers for the first time reported that the cellular uptake of tpy (tpy = 2,2′:6′2′-terpyridine) tethered PNA oligomers is significantly improved upon complexation with Zn^{2+} ions in a cell culture medium [105]. They prepared a series of *N*- and the *C*-terminal modified PNA oligomers (**PNA 1a–3b**, Fig. 12.15a). For **PNA 1a–3a**, a tpy chelator and a fluorophore were attached to the *N*- and the *C*-terminus, respectively (Fig. 12.15a). The cellular uptake of the modified PNA oligomers in HeLa cells was monitored using flow cytometry technique. As shown in Fig. 12.15b, it was found that **PNA**

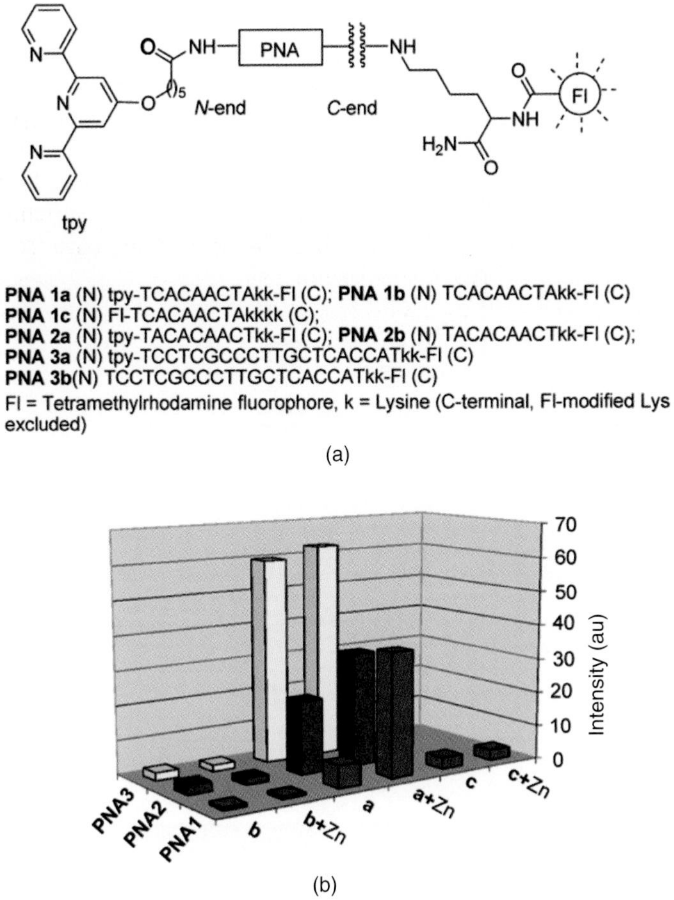

N
N
N
tpy
O
NH
PNA
NH
5
N-end
C-end
O
Fl
NH
H2N
O
PNA 1a (N) tpy-TCACAACTAkk-Fl (C); PNA 1b (N) TCACAACTAkk-Fl (C)
PNA 1c (N) Fl-TCACAACTAkkkk (C);
PNA 2a (N) tpy-TACACAACTkk-Fl (C); PNA 2b (N) TACACAACTkk-Fl (C);
PNA 3a (N) tpy-TCCTCGCCCTTGCTCACCATkk-Fl (C)
PNA 3b(N) TCCTCGCCCTTGCTCACCATkk-Fl (C)
Fl = Tetramethylrhodamine fluorophore, k = Lysine (C-terminal, Fl-modified Lys excluded)
(a)
70
60
50
40
30
20
10
0
Intensity (au)
PNA3
PNA2
PNA1
b
b+Zn
a
a+Zn
c
c+Zn
(b)

Figure 12.15 *a) Modified PNA oligomers [105]. b) Mean cellular fluorescence in flow cytometry analysis of HeLa cells incubated with 2.5 μM probes for 1 h. Reprinted with permission from [105] © 2006 American Chemical Society*

1a–3a were taken up more efficiently than **PNA 1b–3b** lacking the tpy unit. Importantly, the uptakes of **PNA 1a–2a** were significantly enhanced when incubated with 1 equiv. of Zn^{2+}. The authors suggested that the increase in uptake is due to the Zn^{2+} complex formation with the attached tpy moiety. As expected, in the absence of a PNA conjugated tpy moiety, Zn^{2+} does not alter the uptake of **PNA 1b–3b** or **PNA 1c**. In order to establish the optimum concentration of Zn^{2+} required to reach the highest cellular uptake, HeLa cells were incubated with **PNA 1a** and varying concentrations of Zn^{2+} for 1 h. The highest cellular uptake was achieved using 0.5 equiv. of Zn^{2+} with respect to **PNA 1a**. This is an indication that [Zn(**PNA 1a**)$_2$]$^{2+}$ is the species preferentially taken up by the HeLa cells.

Similarly, in another case study on cellular uptake of metal–peptide conjugates, Metzler-Nolte and coworkers showed that a cobaltocenium conjugate of an NLS (nuclear localization signal) peptide (Fig. 12.16a, NLS-2) was taken up readily by cells via an active

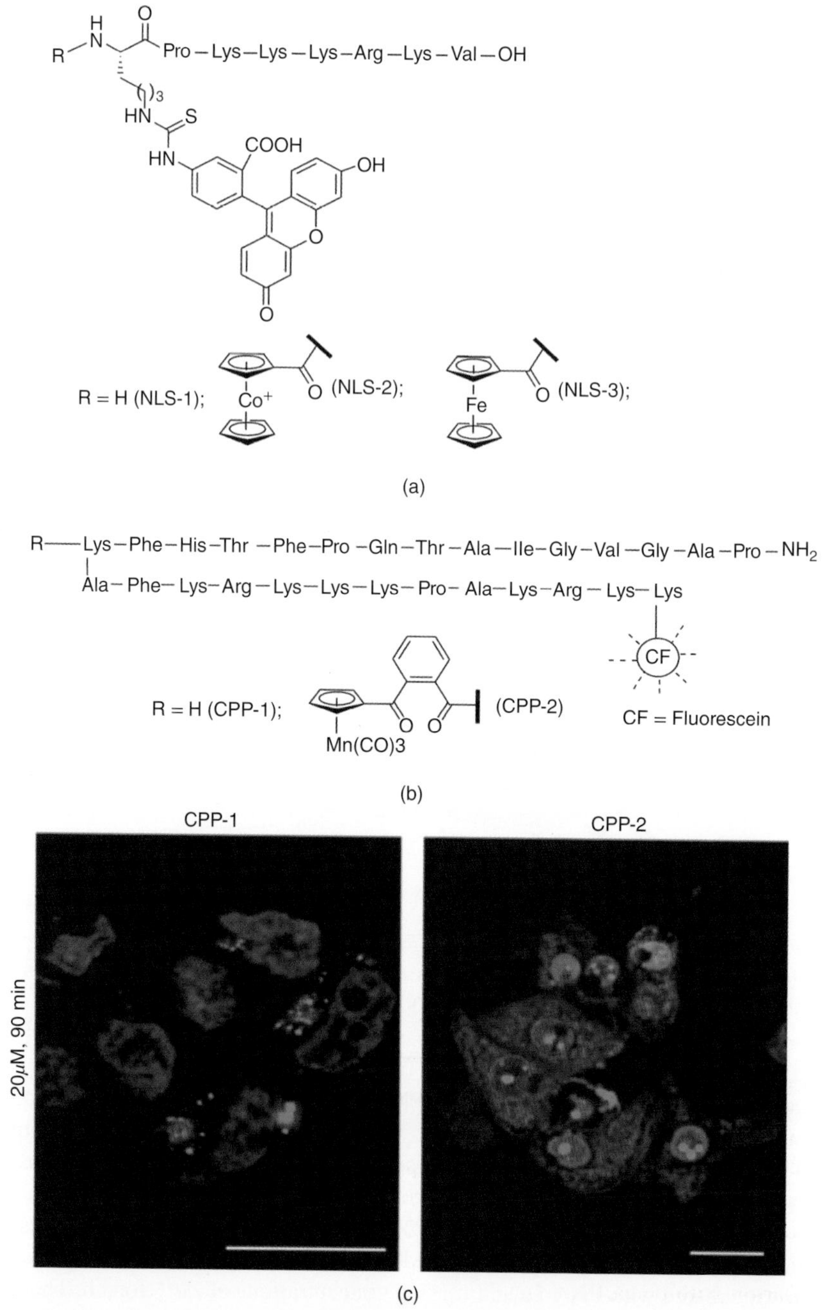

(a)

(b)

(c)

Figure 12.16   a) *Structures of NLS conjugates (NLS-1, 2, and 3) prepared by Metzler-Nolte and coworkers [106, 107]. b) Structures of hCT (18–32)-k7 conjugates (CPP-1 and CPP-2) prepared by Schatzschneider and coworkers [110]. c) Fluorescence microscopy images of living unfixed MCF-7 cells after incubation with CPP-1 (left) and CPP-2 (right) at a 20 μM concentration for 90 min. Reprinted with permission from [110] © 2008 Royal Society of Chemistry*

mechanism, namely endocytosis [107]. The presence of the fluorophore (fluorescein) in the cobaltocenium-NLS conjugate allowed monitoring of the cellular uptake in HepG2 cells using fluorescence microscopy. Furthermore, this group confirmed that NLS-2 mainly accumulates in the nucleus. This suggests that NLS-2 is capable of escaping from the endosomes to the cytosol. The presence of the metallocene moiety in NLS-2 increased the endosomal-membrane permeability, because the reference compound NLS-1, which lacks the organometallic moiety, was also found to be taken up by the cells but was trapped in the endosomes. Encouraged by these preliminary findings, this same group studied the cellular uptake of an analogous NLS conjugate containing a ferrocenyl moiety (NLS-3, Fig. 12.16a). Despite significant differences between the two metallocenes in terms of their net charge and redox potential, similar behavior for the ferrocene-containing NLS-3 to that seen previously for the NLS-2 was recorded [108].

However, when the cell penetrating peptide (CPP) TAT (48–60), which is derived from the HIV virus, was used as the sequence of interest, the ferrocene and its isostructural cobaltocenium conjugates showed contrasting cellular uptake patterns [109]. Whilst the ferrocene–TAT derivative gained entry into the cell and is localized in the cytoplasm, the cobaltocenium–TAT derivative did not show any cellular uptake. At this stage, it is important to mention that the coupling of the metallocene to the peptides did not result in cellular toxicity. The cells were found to be healthy up to a highest test concentration of 1 mM of metallo–NLS peptide conjugates [108].

Worthy of note is the uptake and intracellular distribution of a CpMn(CO)$_3$ conjugate of the CPP hCT(18-32)-k7 in human breast adenocarcinoma cells MCF-7 which has also been investigated, using fluorescence microscopy, by Schatzschneider and coworkers (Fig. 12.16b) [110]. As presented in Fig. 12.16c, a relatively weak and dot-like distribution pattern with vesicles close to the nucleus was observed for the fluorescein (CF)-labeled reference peptide CPP-1, suggesting a vesicular uptake by endocytosis, as reported earlier for similar peptides [111, 112]. However, the organometallic derivatization of CPP-1 to give CPP-2 changed the scenario. The manganese tricarbonyl containing peptide CPP-2 was not trapped inside the endosomes, but was distributed throughout cytosol. In addition, a significant amount was found to accumulate in the nucleus. Based on these results, it was concluded that the organometallic derivatization helps the CPP, in this particular case, to cross the endosomal as well as the nuclear membranes, resulting in an efficient release of the peptide from the endosomes followed by its nuclear localization. However, in contrast to the previously mentioned metallo–NLS conjugates [107], the hCT(18-32)-k7-CpMn(CO)$_3$ conjugate was found to be moderately cytotoxic with an IC$_{50}$ value of 36 μM against MCF-7 cells.

All metal–bioconjugates discussed in the previous paragraphs contained a fluorescent dye. However, its influence on the uptake or intracellular localization was not discussed in the works cited. It is important to note that the presence of an organic fluorophore can modify the intracellular distribution pattern of a peptide, as recently observed by Puckett and Barton with octaarginine–ruthenium peptide conjugates [113]. Therefore, direct analytical methods such as atomic absorption spectroscopy (AAS) or inductively coupled plasma mass spectrometry (ICP-MS), to name a few, where the presence of an external tag (fluorophore) is not a pre-requisite are more reliable techniques to study the uptake of metal-containing biomolecules in cells (the readers are invited to refer to Chapter 3 for a detailed presentation of these analytical techniques).

12.5 Conclusions

As the few examples presented in this chapter showcase, metal complexes have a pivotal role to play in the future in a variety of areas of research other than those exclusively presented in the first 11 chapters of this book. We have succinctly highlighted these other applications of metal complexes in chemical biology, which, at this stage, have either not yet sufficiently evolved to have their "own" chapter, or have recently been reviewed elsewhere quite extensively. In more general terms, inorganic chemists will undoubtedly have to continue exploring the spectacularly unique physicochemical properties that metal complexes can present in a biological environment. Furthermore, there is a great need for aggressive promotion in order to expand the frontiers of Inorganic Chemical Biology beyond those already existing.

Acknowledgments

This work was supported by the Swiss National Science Foundation (SNSF Professorship PP00P2_133568 to G.G.), the Stiftung für wissenschaftliche Forschung of the University of Zurich and the University of Zurich.

References

1. E. Magner, Immobilisation of enzymes on mesoporous silicate materials, *Chem. Soc. Rev.*, **42**, 6213–6222 (2013).
2. P. Jonkheijm, D. Weinrich, H. Schröder, C. M. Niemeyer, H. Waldmann, Chemical strategies for generating protein biochips, *Angew. Chem. Int. Ed.*, **47**(50), 9618–9647 (2008).
3. M. J. W. Ludden, D. N. Reinhoudt, J. Huskens, Molecular printboards: versatile platforms for the creation and positioning of supramolecular assemblies and materials, *Chem. Soc. Rev.*, **35**(11), 1122–1134 (2006).
4. D. A. Uhlenheuer, K. Petkau, L. Brunsveld, Combining supramolecular chemistry with biology, *Chem. Soc. Rev.*, **39**(8), 2817–2826 (2010).
5. L. Cao. Carrier-bound Immobilized Enzymes: Principles, Application and Design. Weinheim: Wiley-VCH; 2005.
6. U. Hanefeld, L. Gardossi, E. Magner, Understanding enzyme immobilisation, *Chem. Soc. Rev.*, **38**(2), 453-468 (2009).
7. D. Gaffney, J. Cooney, E. Magner, Modification of mesoporous silicates for immobilization of enzymes, *Top. Catal.*, **55**, 1101–1106 (2012).
8. D. A. Gaffney, S. O'Neill, M. C. O'Loughlin, U. Hanefeld, J. C. Cooney, E. Magner, Tailored adsorption of His6-tagged protein onto nickel(ii)-cyclam grafted mesoporous silica, *Chem. Commun.*, **46**(7), 1124–1126 (2010).
9. K. E. Cassimjee, M. Trummer, C. Branneby, P. Berglund, Silica-immobilized His6-tagged enzyme: Alanine racemase in hydrophobic solvent, *Biotechnol. Bioeng.*, **99**(3), 712–716 (2008).

10. E. Kang, J.-W. Park, S. J. McClellan, J.-M. Kim, D. P. Holland, G. U. Lee, E. I. Franses, K. Park, D. H. Thompson, Specific adsorption of histidine-tagged proteins on silica surfaces modified with Ni^{2+}/NTA-derivatized poly(ethylene glycol), *Langmuir*, **23**(11), 6281–6288 (2007).

11. J. Nahalka, Z. Liu, X. Chen, P. G. Wang, Superbeads: Immobilization in "sweet" chemistry, *Chem. Eur. J.*, **9**(2), 372–377 (2003).

12. G. Drager, C. Kiss, U. Kunz, A. Kirschning, Enzyme-purification and catalytic transformations in a microstructured PASSflow reactor using a new tyrosine-based Ni-NTA linker system attached to a polyvinylpyrrolidinone-based matrix, *Org. Biomol. Chem.*, **5**(22), 3657–3664 (2007).

13. P. Neirynck, J. Brinkmann, Q. An, D. van der Schaft, P. Jonkheijm, L. G. Milroy, L. Brunsveld, Supramolecular control over cell adhesion via ferrocene-cucurbit[7] uril host-guest binding on gold surfaces, *Chem. Commun.*, **49**, 3679–3681 (2013).

14. J. F. Young, H. D. Nguyen, L. Yang, J. Huskens, P. Jonkheijm, L. Brunsveld, Strong and Reversible Monovalent Supramolecular Protein Immobilization, *ChemBioChem*, **11**, 180–183 (2010).

15. I. Hwang, K. Baek, M. Jung, Y. Kim, K. M. Park, D.-W. Lee, N. Selva-palam, K. Kim, Noncovalent immobilization of proteins on a solid surface by cucurbit[7]uril-ferrocenemethylammonium pair, a potential replacement of biotin-avidin pair, *J. Am. Chem. Soc.*, **129**, 4170–4171 (2007).

16. D. Wasserberg, D. Uhlenheuer, P. Neirynck, J. Cabanas-Danés, J. Schenkel, B. Ravoo, Q. An, J. Huskens, L.-G. Milroy, L. Brunsveld, P. Jonkheijm, Immobilization of ferrocene-modified SNAP–fusion proteins, *Int. J. Mol. Sci.*, **14**(2), 4066–4080 (2013).

17. L. Yang, A. Gomez-Casado, J. F. Young, H. D. Nguyen, J. Cabanas-Danés, J. Huskens, L. Brunsveld, P. Jonkheijm, Reversible and oriented immobilization of ferrocene-modified proteins, *J. Am. Chem. Soc.*, **134**, 19199–19206 (2012).

18. D. A. Gaffney, J. C. Cooney, F. R. Laffir, K. E. Cassimjee, P. Berglund, U. Hanefeld, E. Magner, *Microporous Mesoporous Mater.*, submitted (2013).

19. M. Patra, G. Gasser, Organometallic compounds, an opportunity for chemical biology, *ChemBioChem*, **13**, 1232–1252 (2012).

20. W. Ong, A. E. Kaifer, Unusual electrochemical properties of the inclusion complexes of ferrocenium and cobaltocenium with cucurbit[7]uril, *Organometallics*, **22**(21), 4181–4183 (2003).

21. W. S. Jeon, K. Moon, S. H. Park, H. Chun, Y. H. Ko, J. Y. Lee, E. S. Lee, S. Samal, N. Selvapalam, M. V. Rekharsky, V. Sindelar, D. Sobransingh, Y. Inoue, A. E. Kaifer, K. Kim, Complexation of ferrocene derivatives by the cucurbit[7]uril host: A comparative study of the cucurbituril and cyclodextrin host families, *J. Am. Chem. Soc.*, **127**(37), 12984–12989 (2005).

22. D. Sobransingh, A. E. Kaifer, Binding interactions between the host cucurbit[7]uril and dendrimer guests containing a single ferrocenyl residue, *Chem. Commun.*, (**40**), 5071–5073 (2005).

23. Q. An, G. Li, C. Tao, Y. Li, Y. Wu, W. Zhang, A general and efficient method to form self-assembled cucurbit[n]uril monolayers on gold surfaces, *Chem. Commun.*, (**17**), 1989–1991 (2008).

24. N. H. Williams, B. Takasaki, M. Wall, J. Chin, Structure and nuclease activity of simple dinuclear metal complexes: Quantitative dissection of the role of metal ions, *Acc. Chem. Res.*, **32**(6), 485–493 (1999).

25. D. Desbouis, I. P. Troitsky, M. J. Belousoff, L. Spiccia, B. Graham, Copper(II), zinc(II) and nickel(II) complexes as nuclease mimetics, *Coord. Chem. Rev.*, **256**(11-12), 897–937 (2012).

26. C. Liu, L. Wang, DNA hydrolytic cleavage catalyzed by synthetic multinuclear metallonucleases, *Dalton Trans.*, 227–239 (2009).

27. J. A. Cowan, Metal activation of enzymes in nucleic acid biochemistry, *Chem. Rev.*, **98**(3), 1067–1088 (1998).

28. E. L. Hegg, J. N. Burstyn, Toward the development of metal-based synthetic nucleases and peptidases: a rationale and progress report in applying the principles of coordination chemistry, *Coord. Chem. Rev.*, **173**(1), 133–165 (1998).

29. F. Mancin, P. Scrimin, P. Tecilla, U. Tonellato, Artificial metallonucleases, *Chem. Commun.*, 2540–2548 (2005).

30. C. Liu, M. Wang, T. Zhang, H. Sun, DNA hydrolysis promoted by di- and multi-nuclear metal complexes, *Coord. Chem. Rev.*, **248**(1-2), 147–168 (2004).

31. A. Dallas, Principles of nucleic acid cleavage by metal ions, *Nucleic Acids Mol. Biol.*, **13**, 61 (2004).

32. J. A. Cowan, Chemical nucleases, *Curr. Opin. Chem. Biol.*, **5**(6), 634–642 (2001).

33. F. Mancin, P. Tecilla, Zinc(II) complexes as hydrolytic catalysts of phosphate diester cleavage: from model substrates to nucleic acids, *New J. Chem.*, **31**(6), 800 (2007).

34. T. Shell, D. L. Mohler, Hydrolytic DNA cleavage by non-lanthanide metal complexes, *Curr. Org. Chem.*, **11**(17), 1525 (2007).

35. F. Mancin, P. Scrimin, P. Tecilla, Progress in artificial metallonucleases, *Chem. Commun.*, **48**(45), 5545–5559 (2012).

36. W. Yang, Nucleases: diversity of structure, function and mechanism, *Q. Rev. Biophys.*, **44**(1), 1–93 (2011).

37. C. M. Dupureur, One is enough: insights into the two-metal ion nuclease mechanism from global analysis and computational studies, *Metallomics*, **2**(9), 609–620 (2010).

38. J. R. Morrow, Speed limits for artificial ribonucleases, *Comments Inorg. Chem.*, **29**(5–6), 169–188 (2008).

39. C. M. Dupureur, Roles of metal ions in nucleases, *Curr. Opin. Chem. Biol.*, **12**(2), 250–255 (2008).

40. Q. Jiang, N. Xiao, P. Shi, Y. Zhu, Z. Guo, Design of artificial metallonucleases with oxidative mechanism, *Coord. Chem. Rev.*, **251**(15-16), 1951–1972 (2007).

41. J. R. Morrow, O. Iranzo, Synthetic metallonucleases for RNA cleavage, *Curr. Opin. Chem. Biol.*, **8**(2), 192–200 (2004).

42. J. Suh, Synthetic artificial peptidases and nucleases using macromolecular catalytic systems, *Acc. Chem. Res.*, **36**(7), 562 (2003).

43. S. J. Franklin, Lanthanide-mediated DNA hydrolysis, *Curr. Opin. Chem. Biol.*, **5**(2), 201–208 (2001).

44. C.-H. Chen, Artificial nucleases, *ChemBioChem*, **2**(10), 735 (2001).

45. R. Haener, D. Huesken, J. Hall, Development of artificial ribonucleases using macrocyclic lanthanide complexes, *CHIMIA Int. J. Chem.*, **54**(10), 569–573 (2000).

46. M. Komiyama, N. Takeda, T. Shiiba, Y. Takahashi, Y. Matsumoto, M. Yashiro, Rare earth metal ions for DNA hydrolyses and their use to artificial nuclease, *Nucleo.s Nucleot.*, **13**(6-7), 1297–1309 (1994).

47. J. R. Morrow, L. A. Buttrey, V. M. Shelton, K. A. Berback, Efficient catalytic cleavage of RNA by lanthanide(III) macrocyclic complexes: toward synthetic nucleases for in vivo applications, *J. Am. Chem. Soc.*, **114**(5), 1903–1905 (1992).

48. B. Lippert, From cisplatin to artificial nucleases. The role of metal ion-nucleic acid interactions in biology, *BioMetals*, **5**(4), 195 (1992).

49. L. Basile, Metallonucleases: real and artificial, *Met. Ions Biol. Syst.*, **25**(Interrelat. Met. Ions, Enzymes, Gene Expression), 31 (1989).

50. M. Costas, M. P. Mehn, M. P. Jensen, L. Que, Dioxygen activation at mononuclear nonheme iron active sites: Enzymes, models, and intermediates, *Chem. Rev.*, **104**(2), 939–986 (2004).

51. G. Parkin, Synthetic analogues relevant to the structure and function of zinc enzymes, *Chem. Rev.*, **104**(2), 699–768 (2004).

52. L. M. Mirica, X. Ottenwaelder, T. D. P. Stack, Structure and spectroscopy of copper–dioxygen complexes, *Chem. Rev.*, **104**(2), 1013–1046 (2004).

53. B. Armitage, Photocleavage of nucleic acids, *Chem. Rev.*, **98**(3), 1171–1200 (1998).

54. D. R. McMillin, K. M. McNett, Photoprocesses of copper complexes that bind to DNA, *Chem. Rev.*, **98**(3), 1201–1220 (1998).

55. W. K. Pogozelski, T. D. Tullius, Oxidative strand scission of nucleic acids: Routes initiated by hydrogen abstraction from the sugar moiety, *Chem. Rev.*, **98**(3), 1089–1108 (1998).

56. L. J. K. Boerner, J. M. Zaleski, Metal complex–DNA interactions: from transcription inhibition to photoactivated cleavage, *Curr. Opin. Chem. Biol.*, **9**(2), 135–144 (2005).

57. E. L. Hegg, S. H. Mortimore, C. L. Cheung, J. E. Huyett, D. R. Powell, J. N. Burstyn, Structure–reactivity studies in copper(II)-catalyzed phosphodiester hydrolysis, *Inorg. Chem.*, **38**(12), 2961–2968 (1999).

58. J. He, P. Hu, Y.-J. Wang, M.-L. Tong, H. Sun, Z.-W. Mao, L.-N. Ji, Double-strand DNA cleavage by copper complexes of 2,2′-dipyridyl with guanidinium/ammonium pendants, *Dalton Trans.*, (24), 3207–3214 (2008).

59. Y. An, M.-L. Tong, L.-N. Ji, Z.-W. Mao, Double-strand DNA cleavage by copper complexes of 2,2′-dipyridyl with electropositive pendants, *Dalton Trans.*, (17), 2066–2071 (2006).

60. J. He, J. Sun, Z.-W. Mao, L.-N. Ji, H. Sun, Phosphodiester hydrolysis and specific DNA binding and cleavage promoted by guanidinium-functionalized zinc complexes, *J. Inorg. Biochem.*, **103**(5), 851–858 (2009).

61. M. K. Stern, J. K. Bashkin, E. D. Sall, Hydrolysis of RNA by transition metal complexes, *J. Am. Chem. Soc.*, **112**(13), 5357–5359 (1990).

62. E. L. Hegg, K. A. Deal, L. L. Kiessling, J. N. Burstyn, Hydrolysis of Double-stranded and single-stranded RNA in hairpin structures by the copper(II) macrocycle Cu([9]aneN$_3$)Cl$_2$, *Inorg. Chem.*, **36**(8), 1715–1718 (1997).

63. L. A. Jenkins, J. K. Bashkin, M. E. Autry, The embedded ribonucleotide Assay: A chimeric substrate for studying cleavage of RNA by transesterification, *J. Am. Chem. Soc.*, **118**(29), 6822–6825 (1996).

64. X. Sheng, X.-M. Lu, Y.-T. Chen, G.-Y. Lu, J.-J. Zhang, Y. Shao, F. Liu, Q. Xu, Synthesis, DNA-binding, cleavage, and cytotoxic activity of new 1,7-dioxa-4, 10-diazacyclododecane artificial receptors containing bisguanidinoethyl or diaminoethyl double side arms, *Chem. Eur. J.*, **13**(34), 9703–9712 (2007).

65. L. Tjioe, A. Meininger, T. Joshi, L. Spiccia, B. Graham, Efficient plasmid DNA cleavage by copper(II) complexes of 1,4,7-triazacyclononane ligands featuring xylyl-linked guanidinium groups, *Inorg. Chem.*, **50**(10), 4327–4339 (2011).

66. T. Itoh, H. Hisada, T. Sumiya, M. Hosono, Y. Usui, Y. Fujii, Hydrolytic cleavage of DNA by a novel copper(II) complex with *cis,cis*-1,3,5-triaminocyclohexane, *Chem. Commun.*, 677–678 (1997).

67. E. Boseggia, M. Gatos, L. Lucatello, F. Mancin, S. Moro, M. Palumbo, C. Sissi, P. Tecilla, U. Tonellato, G. Zagotto, Toward efficient Zn(II)-based artificial nucleases, *J. Am. Chem. Soc.*, **126**(14), 4543–4549 (2004).

68. T. Kobayashi, S. Tobita, M. Kobayashi, T. Imajyo, M. Chikira, M. Yashiro, Y. Fujii, Effects of N-alkyl and ammonium groups on the hydrolytic cleavage of DNA with a Cu(II)TACH (1,3,5-triaminocyclohexane) complex. Speciation, kinetic, and DNA-binding studies for reaction mechanism, *J. Inorg. Biochem.*, **101**(2), 348–361 (2007).

69. C. Sissi, F. Mancin, M. Gatos, M. Palumbo, P. Tecilla, U. Tonellato, Efficient plasmid DNA cleavage by a mononuclear copper(II) complex, *Inorg. Chem.*, **44**(7), 2310–2317 (2005).

70. A. Sreedhara, J. A. Cowan, Catalytic hydrolysis of DNA by metal ions and complexes, *J. Biol. Inorg. Chem.*, **6**(4), 337 (2001).

71. X. Sheng, X. Guo, X.-M. Lu, G.-Y. Lu, Y. Shao, F. Liu, Q. Xu, DNA binding, cleavage, and cytotoxic activity of the preorganized dinuclear zinc(II) complex of triazacyclononane derivatives, *Bioconjug. Chem.*, **19**(2), 490–498 (2008).

72. K. P. McCue, J. R. Morrow, Hydrolysis of a model for the 5′-Cap of mRNA by dinuclear copper(II) and zinc(II) complexes. Rapid hydrolysis by four copper(II) ions, *Inorg. Chem.*, **38**(26), 6136–6142 (1999).

73. M. Laine, K. Ketomaki, P. Poijarvi-Virta, H. Lonnberg, Base moiety selectivity in cleavage of short oligoribonucleotides by di- and tri-nuclear Zn(II) complexes of azacrown-derived ligands, *Org. Biomol. Chem.*, **7**(13), 2780–2787 (2009).

74. Y. An, S.-D. Liu, S.-Y. Deng, L.-N. Ji, Z.-W. Mao, Cleavage of double-strand DNA by linear and triangular trinuclear copper complexes, *J. Inorg. Biochem.*, **100**(10), 1586–1593 (2006).

75. D. Li, J. Tian, Y. Kou, F. Huang, G. Chen, W. Gu, X. Liu, D. Liao, P. Cheng, S. Yan, Synthesis, X-ray crystal structures, magnetism, and DNA cleavage properties of copper(II) complexes with 1,4-tpbd ligand, *Dalton Trans.*, 3574–3583 (2009).

76. V. Uma, M. Kanthimathi, J. Subramanian, B. Unni Nair, A new dinuclear biphenylene bridged copper(II) complex: DNA cleavage under hydrolytic conditions, *Biochim. Biophys. Acta (BBA) - General Subjects*, **1760**(5), 814–819 (2006).

77. M. E. Branum, A. K. Tipton, S. Zhu, L. Que, Double-strand hydrolysis of plasmid DNA by dicerium complexes at 37 °C, *J. Am. Chem. Soc.*, **123**(9), 1898–1904 (2001).

78. M. E. Branum, L. Que Jr,, Double-strand DNA hydrolysis by dilanthanide complexes, *JBIC J. Biol. Inorg. Chem.*, **4**(5), 593–600 (1999).

79. B. Zhu, D.-Q. Zhao, J.-Z. Ni, Q.-H. Zeng, B.-Q. Huang, Z.-L. Wang, Binuclear lan-thanide complexes as catalysts for the hydrolysis of double-stranded DNA, *Inorg. Chem. Commun.*, **2**(8), 351–353 (1999).

80. A. Roigk, O. V. Yescheulova, Y. V. Fedorov, O. A. Fedorova, S. P. Gromov, H.-J. Schneider, Carboxylic groups as cofactors in the lanthanide-catalyzed hydrolysis of phosphate esters. Stabilities of europium(III) complexes with aza-benzo-15-crown-5 ether derivatives and their catalytic activity vs bis(p-nitrophenyl)phosphate and DNA, *Org. Lett.*, **1**(6), 833–835 (1999).

81. K. G. Ragunathan, H.-J. Schneider, Binuclear lanthanide complexes as catalysts for the hydrolysis of bis(*p*-nitrophenyl)-phosphate and double-stranded DNA, *Angew. Chem. Int. Ed.*, **35**(11), 1219–1221 (1996).

82. T. Berg, A. Simeonov, K. D. Janda, A combined parallel synthesis and screening of macrocyclic lanthanide complexes for the cleavage of phospho di- and triesters and double-stranded DNA, *J. Com. Chem.*, **1**(1), 96–100 (1998).

83. S. Amin, J. R. Morrow, C. H. Lake, M. R. Churchill, Lanthanide(III) tetraamide macrocyclic complexes as synthetic ribonucleases: Structure and catalytic properties of [La(tcmc)(CF$_3$SO$_3$)(EtOH)](CF$_3$SO$_3$)$_2$, *Angew. Chem. Int. Ed.*, **33**(7), 773–775 (1994).

84. T. Niittymaki, H. Lonnberg, Artificial ribonucleases, *Org. Biomol. Chem.*, **4**(1), 15–25 (2006).

85. A. Whitney, G. Gavory, S. Balasubramanian, Site-specific cleavage of human telomerase RNA using PNA-neocuproine·Zn(II) derivatives, *Chem. Commun.*, 36–37 (2003).

86. M. Murtola, M. Wenska, R. Strömberg, PNAzymes that are artificial RNA restriction enzymes, *J. Am. Chem. Soc.*, **132**(26), 8984–8990 (2010).

87. M. Murtola, R. Stromberg, PNA based artificial nucleases displaying catalysis with turnover in the cleavage of a leukemia related RNA model, *Org. Biomol. Chem.*, **6**(20), 3837–3842 (2008).

88. V. P. Torchilin, Intracellular delivery of protein and peptide therapeutics, *Drug Discov. Today: Tech.*, **5**(2-3), e95–e103 (2008).

89. C. Y. Looi, M. Imanishi, S. Takaki, M. Sato, N. Chiba, Y. Sasahara, S. Futaki, S. Tsuchiya, S. Kumaki, Octa-arginine mediated delivery of wild-type Lnk protein inhibits TPO-induced M-MOK megakaryoblastic leukemic cell growth by promoting apoptosis, *PLoS ONE*, **6**(8), e23640 (2011).

90. S. EL Andaloussi, T. Lehto, I. Mäger, K. Rosenthal-Aizman, I. I. Oprea, O. E. Simon-son, H. Sork, K. Ezzat, D. M. Copolovici, K. Kurrikoff, J. R. Viola, E. M. Zaghloul, R. Sillard, H. J. Johansson, F. Said Hassane, P. Guterstam, J. Suhorutšenko, P. M. D. Moreno, N. Oskolkov, J. Hälldin, U. Tedebark, A. Metspalu, B. Lebleu, J. Lehtiö, C. I. E. Smith, Ü. Langel, Design of a peptide-based vector, PepFect6, for efficient delivery of siRNA in cell culture and systemically in vivo, *Nucleic Acids Res.*, **39**(9), 3972–3987 (2011).

91. S. Trabulo, S. Resina, S. Simões, B. Lebleu, M. C. Pedroso de Lima, A non-covalent strategy combining cationic lipids and CPPs to enhance the delivery of splice cor-recting oligonucleotides, *J. Control. Release*, **145**(2), 149–158 (2010).

92. H. Yukawa, Y. Kagami, M. Watanabe, K. Oishi, Y. Miyamoto, Y. Okamoto, M. Tokeshi, N. Kaji, H. Noguchi, K. Ono, M. Sawada, Y. Baba, N. Hamajima,

S. Hayashi, Quantum dots labeling using octa-arginine peptides for imaging of adipose tissue-derived stem cells, *Biomaterials*, **31**(14), 4094–4103 (2010).

93. K. T. Yong, Y. Wang, I. Roy, H. Rui, M. T. Swihart, W. C. Law, S. K. Kwak, L. Ye, J. Liu, S. D. Mahajan, J. L. Reynolds, Preparation of quantum dot/drug nanoparticle formulations for traceable targeted delivery and therapy, *Theranostics*, **2**, 681–694 (2012).

94. W. B. Liechty, D. R. Kryscio, B. V. Slaughter, N. A. Peppas, Polymers for drug delivery systems, *Annu. Rev. Chem. Biomol. Eng.*, **1**(1), 149–173 (2010).

95. Q. Zhu, F. Qiu, B. Zhu, X. Zhu, Hyperbranched polymers for bioimaging, *RSC Adv.*, **3**(7), 2071–2083 and references therein (2013).

96. V. P. Torchilin, Recent approches to intracellular delivery of drugs and DNA and organelle targeting, *Ann. Rev. Biomed. Eng.*, **8**(1), 343–375 (2006).

97. J. Connor, L. Huang, pH-sensitive Immunoliposomes as an efficient and target-specific carrier for antitumor drugs, *Cancer Res.*, **46**(7), 3431–3435 (1986).

98. E. Koren, V. P. Torchilin, Cell-penetrating peptides: breaking through to the other side, *Trends Mol. Med.*, **18**(7), 385–393 (2012).

99. R. Johnson, S. Harrison, D. Maclean. Therapeutic applications of cell-penetrating peptides. In: Langel Ü. (ed.) Cell-Penetrating Peptides: Humana Press; 2011. pp. 535–551.

100. W. L. Munyendo, H. Lv, H. Benza-Ingoula, L. D. Baraza, J. Zhou, Cell penetrating peptides in the delivery of biopharmaceuticals, *Biomolecules*, **2**(2), 187–202 (2012).

101. M. C. Morris, S. Deshayes, F. Heitz, G. Divita, Cell-penetrating peptides: from molecular mechanisms to therapeutics, *Bio. Cell*, **100**(4), 201–217 (2008).

102. D. E. Reichert, J. S. Lewis, C. J. Anderson, Metal complexes as diagnostic tools, *Coord. Chem. Rev.*, **184**(1), 3–66 (1999).

103. N. Metzler-Nolte, Medicinal applications of metal-peptide bioconjugates, *Chimia*, **61**, 736–741 (2007).

104. E. Rozners, Recent advances in chemical modification of peptide nucleic acids, *J. Nucleic Acids*, **2012**, 8 and references therein (2012).

105. A. Füssl, A. Schleifenbaum, M. Görtz, A. Riddell, C. Schultz, R. Krämer, Cellular uptake of PNA-terpyridine conjugates and its enhancement by Zn^{2+} Ions, *J. Am. Chem. Soc.*, **128**(18), 5986–5987 (2006).

106. G. Gasser, A. M. Sosniak, N. Metzler-Nolte, Metal-containing peptide nucleic acid conjugates, *Dalton Trans.*, **40**, 7061–7076 and references therein (2011).

107. F. Noor, A. Wüstholz, R. Kincherf, N. Metzler-Nolte, A cobaltocenium-peptide bioconjugate shows enhanced cellular uptake and directed nuclear delivery, *Angew. Chem. Int. Ed.*, **44**(16), 2429–2432 (2005).

108. F. Noor, R. Kincherf, G. A. Bonaterra, S. Walczak, S. Wölfl, N. Metzler-Nolte, Enhanced cellular uptake and cytotoxicity studies of organometallic bioconjugates of the NLS peptide in HepG2 cells, *ChemBioChem*, **10**, 493–502 (2009).

109. G. Jaouen, N. Metzler-Nolte. Medicinal Organometallic Chemistry. Topics in Organometallic Chemistry. 1st edn. Heidelberg: Springer; 2010.

110. I. Neundorf, J. Hoyer, K. Splith, R. Rennert, H. W. P. N'Dongo, U. Schatzschneider, Cymantrene conjugation modulates the intracellular distribution and induces high cytotoxicity of a cell-penetrating peptide, *Chem. Commun.*, 5604–5606 (2008).

111. R. Rennert, I. Neundorf, A. G. Beck-Sickinger, Calcitonin-derived peptide carriers: Mechanisms and application, *Adv. Drug Delivery Rev.*, **60**(4–5), 485–498 (2008).

112. C. Foerg, U. Ziegler, J. Fernandez-Carneado, E. Giralt, R. Rennert, A. G. Beck-Sickinger, H. P. Merkle, Decoding the entry of two novel cell-penetrating peptides in HeLa cells: Lipid raft-mediated endocytosis and endosomal escape, *Biochemistry*, **44**(1), 72–81 (2004).

113. C. A. Puckett, J. K. Barton, Fluorescein redirects a ruthenium-octaarginine conjugate to the nucleus, *J. Am. Chem. Soc.*, **131**(25), 8738–8739 (2009).

Index